AF903651

Emerging Optimization Techniques in Production Planning and Control

Emerging Optimization Techniques in Production Planning and Control

Godfrey C Onwubolu
Professor and Chair of Engineering
Department of Engineering
The University of the South Pacific

ICP
Imperial College Press

Published by

Imperial College Press
57 Shelton Street
Covent Garden
London WC2H 9HE

Distributed by

World Scientific Publishing Co. Pte. Ltd.
P O Box 128, Farrer Road, Singapore 912805
USA office: Suite 1B, 1060 Main Street, River Edge, NJ 07661
UK office: 57 Shelton Street, Covent Garden, London WC2H 9HE

British Library Cataloguing-in-Publication Data
A catalogue record for this book is available from the British Library.

EMERGING OPTIMIZATION TECHNIQUES IN PRODUCTION PLANNING AND CONTROL

Copyright © 2002 by Imperial College Press

All rights reserved. This book, or parts thereof, may not be reproduced in any form or by any means, electronic or mechanical, including photocopying, recording or any information storage and retrieval system now known or to be invented, without written permission from the Publisher.

For photocopying of material in this volume, please pay a copying fee through the Copyright Clearance Center, Inc., 222 Rosewood Drive, Danvers, MA 01923, USA. In this case permission to photocopy is not required from the publisher.

ISBN 1-86094-266-0

Printed by FuIsland Offset Printing (S) Pte Ltd, Singapore

Preface

The concept of adaptive memory programming (AMP) is proposed for grouping a number of generic optimization techniques for combinatorial problems. The same common features that are the use of memory and a local search procedure drive these *Emerging Optimization Techniques* which include artificial neural networks, genetic algorithms, tabu search, and ant systems. The primary motivation for adaptive memory programming, therefore, is to group and unify all these emerging optimization techniques for enhancement of the computational capabilities that they offer to combinatorial problems which are encountered in real life in the area of production planning and control.

One memoryless technique, simulated annealing, is also included in the book manuscript for completeness.

The text, *Emerging Optimization Techniques in Production Planning and Control* describes the theoretical aspects of the adaptive memory programming together with relevant production planning and control applications. These production planning and control applications are taken from journal articles that the author has already published or submitted for publication in international journals, as well as from lecture-notes that he has prepared for teaching undergraduate and post-graduate courses in operations/production planning and control. Algorithms for the computer programs written for each application are also given. This way, the text covers the techniques, applications and algorithms. It is therefore possible for readers to adapt the algorithms presented for other areas of application.

The main objectives of the text are to serve as an instructional material for upper division undergraduates, entry-level graduate student, and a resource material for production planners and controllers, research scientists, practicing engineers, operations researchers, management scientists, applied mathematicians, and computer scientists. To achieve the first objective, the eight chapters have been written in such a way that it can serve as an instructional text to students and those who are taking tuition on their own. The numerical examples given are first solved manually to enhance the readers understanding of the material, followed by the description of the algorithms and computer results. This way, the students can fully follow the materials presented. In addition, problems are given at the end of each chapter for students to practice with. Because all production planning and control applications in the text have been computerized, solutions for all problems are given in a separate Reference Manual. This way, the Lecturers can use the text as both instructional text and a Reference Manual for setting examinations and tutorials. To achieve the second objective, the algorithms described for each application are useful to both interested students and practitioners in grasping how to implement similar applications in computer code using the Emerging Optimization Techniques.

Those to purchase the book include upper division undergraduates or entry-level graduate students, professionals and academics of disciplines listed above, and libraries.

Our approach throughout the text is to present in a systematic manner the superiority of implementing the Emerging Optimization Techniques over other known heuristics available in the literature. To this end, additional intractable (complex) real life problems encountered in practice in the area of production planning and control are presented and solved using Emerging Optimization Techniques in each chapter.

The book has three parts: one, two and three. The text commences with part one, which contains an introductory chapter (Chapter 1) that gives an overview of Adaptive Memory Programming and Production Planning and Control. The second part of the book contains Chapters 2 and 3, respectively. The basic components of Production Planning System are presented in Chapter 2, while the basic components of Production Control System are presented in Chapter 3. These two chapters are quite adequate for a two-semester course on Production Planning and Control at the final year undergraduate level or first year of a graduate program. Part three of the book contains Chapters 4 to 8, dealing with Emerging Optimization Techniques and their applications to Production Planning and Control. Chapter 9 deals with programming techniques. Beginning with neural networks in Chapter 4, four basic types which include back propagation networks (BPN), self-organizing map (SOM), adaptive resonance theory (ART 1 and ART 2), and Hopfield networks are presented. Practical examples treated in this chapter are the application of Hopfield networks to machine tooling and production sequencing in manufacturing cell planning, as well as flow shop and job shop scheduling. Chapter 5 presents the basic theory of genetic algorithms (GAs) and we then apply GAs to the aggregate production planning-problem. In Chapter 6, tabu search (TS) theory is presented. In this chapter, TS is applied to resource planning and allocation based on constraint management principles. This is altogether, a new application. Then follows Chapter 7 on Ant Systems and its underlying theory. The chapter presents the application of Ants Systems (AS) to scheduling bi-criteria flow shop problems. Again, this is altogether a new application. Recent advances in Emerging Optimization Techniques include hybrid systems. As has been mentioned, one approach that does not have memory structures is included in the text for completeness. This is Simulated Annealing (SA) in Chapter 8, applied to cell formation problem. Chapter 9 discusses programming techniques, and presents source code of one of the memory-based optimization techniques presented in this book.

Godfrey C. Onwubolu

Acknowledgements

The journey from the time I muted the idea of writing this book to the time of its publication is both long and complex. This is because, I have tried to bring together a number of emerging optimization techniques and attempted to apply them to the area of production planning and control. Grasping the concepts of these varied areas takes time. It is therefore logical that the list of acknowledgements will be also long.

I want to first, thank all those who helped me to understand the emerging optimization techniques presented in this book. In this regard, I thank Professor Jacques A. Ferland of the University of Montreal, Quebec, Canada, who made the concept of *tabu search* clear to me. He also provided me with initial materials for *simulated annealing*. My initial contact with Professor Khalil S. Hindi of Brunnel University also helped me grasp simulated annealing technique. The textbooks by James A. Freeman and David M. Skapura on *neural networks* motivated me to explore the concepts of these brain-emulators. The simplicity and clarity of the book on *genetic algorithms* by Dr David E. Goldberg helped me to understand the concept genetic algorithms. I eventually, applied genetic-algorithms to a wide range of engineering problems. My contacts with Professor Marco Dorigo of the Universite Libre de Brussels, and Professor Eric Taillard of Ecole d'Ingenieurs du Canton de Baud, Switzerland, helped me a great deal to understand *ant systems* fundamentals and application to combinatorial optimization problems. Professor Dorigo is the originator of ant systems. That is the list on the side of emerging optimization techniques. Who are those to be acknowledged on the side of applications of these emerging optimization techniques?

Professor Asbjorn Rolstadas of the Norwegian University of Science and Technology, Trondheim introduced me to the *production planning and control* community. As the then Editor of the *International Journal of Planning and Control*, Professor Asbjorn Rolstadas appointed me to join the Editorial Board of the journal, without having known me previously, expect through my submission of articles to the journal. I thank staff at the Warwick Manufacturing Group of the University of Warwick, Coventry, where I was a Visiting Fellow. The integrated production planning and control project helped me much in my career.

Publishing this book was not going to be possible if not for Tony Moore, Book Editor (Physical Science and Engineering), Taylor and Francis Ltd. UK. He helped me contact Imperial College Press (ICP). I thank Dr. John Navas, then Manager/Commissioning Editor of ICP for his enthusiasm to publish the book before he left the company. I am particularly grateful to the three anonymous reviewers of my initial book proposal for their positive comments, which encouraged me to complete the book. They suggested changes to the title of the book. I also thank Ellen Haigh of ICP, Yeow Hwa Quek and David Loo of World Scientific Publishing Company and all other personnel of WSP who helped with the production of this book.

A difficult aspect of writing a book, is that of word processing. In this regard, I thank staff of Sight and Sound College (Educational Limited), Birmingham, UK, for giving me training on typing to achieve 35–40 w.p.m. copy-typing and taped-speed, when I was studying for my graduate program. They did a tremendous job.

I thank the National University of Science and Technology (NUST), Bulawayo, Zimbabwe, where I spent almost six years, teaching production planning and control and manufacturing systems. NUST provided most of the resources such as Research Grants for funding some of the research projects, and Research Assistants, that helped a great deal toward publishing this book. I thank my colleagues, former students, and research assistants in the Department of Industrial Engineering at NUST for their supports. Mr. Samson Mhlanga helped me with some literature when he was undergoing his graduate program at Brunel University. My association with Professor Akpofure Taigbenu helped me a great deal in forging ahead to bring this book to completion. Having gone through the tedium of publishing a textbook, he was always encouraging me to carry-on! He also helped with word-processing environment at some point. I thank him.

There is a proverb that "behind every successful man, is a woman". Ngozi, my beloved wife has supported me beyond measure in this regard and she deserves a special "thank you". I look forward to also helping her so that it can be said that, "behind every successful woman, is a man". Not only do I thank Ngozi, I also thank our children, Chioma, Chineye, Chukujindu, Chinwe and Chinedu for their tolerance of my spending much time away from them to concentrate on this book.

As I read materials over the years on the five different emerging optimization techniques presented in this book: neural networks, genetic algorithms, ant system, tabu search, (and simulated annealing for completeness), one thing became clear! It is, that the ideas that motivated the particular structures of each methodology of the emerging optimization techniques presented by the original inventors come from connections with creatures of God Almighty! One cannot but exclaim "Before the mountains were brought forth, or ever thou hadst formed the earth and the world, even from everlasting to everlasting, thou art God" (Psalm 90:2).

Godfrey C. Onwubolu
University of the South Pacific
Suva, FIJI

October 2001

Contents

PART 1

Introduction

Chapter 1 presents the conventional optimization techniques for solving combinatorial problems, and their limitations. Adaptive memory programming approaches that address these shortfalls are then presented. The fundamentals of adaptive memory programming are fully discussed. These are generally accepted as the yardstick for determining whether or not a method qualifies as being intelligent. Neural-networks, genetic algorithms, tabu search, and ant systems are presented as being adaptive memory programming. These are briefly discussed. Hybrid systems, which are extension of adaptive memory programming are presented. A full treatment of production planning and control is given so that readers can understand the underlying theory of the areas of application of the book. The material on production planning and control is sufficient for two semester modules at the undergraduate final year level. This chapter therefore, sets the scene for what follows in the remaining parts of the book.

Chapter 1

Introduction to Adaptive Memory Programming and Production Planning and Control

1.1. Production Planning Control within Integrated Manufacturing Framework

Computer Integrated Manufacturing (CIM) requires integration among the basic manufacturing functions (technical and operational tasks of an industrial

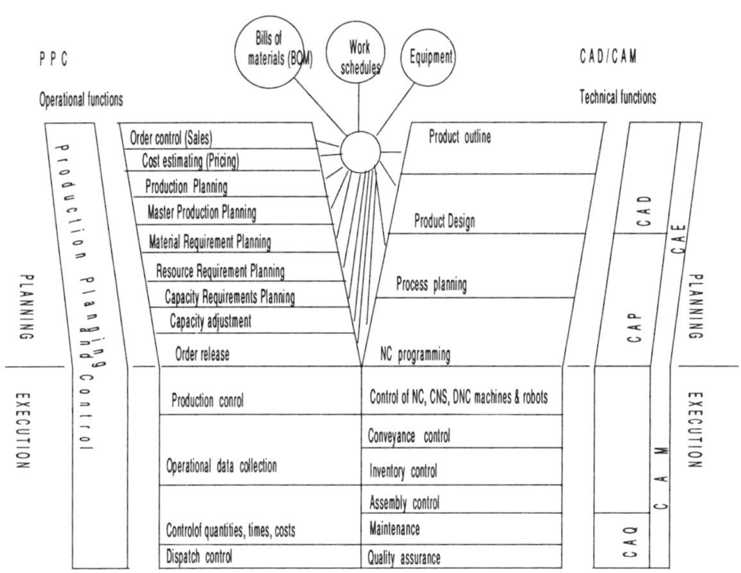

Fig. 1.1. Basic integrated manufacturing functions.

enterprise) shown in Fig. 1.1. Production planning and control system (PPC) refers to the operational tasks represented in the left part of the Y. The more technical activities are characterized by the various computer aided (CAX)-concepts in the right fork of the Y. The PPC system is normally determined by order handling, whereas the CAX-components support product description and production resources.

As can be seen from the integrated manufacturing framework in Fig. 1.1, production planning and control encompasses important aspects of an industrial enterprise. Production planning and control is the hub that ties together product design and manufacturing engineering. It is concerned with translating business plan into reality. The planning horizon involved is shorter, usually in terms of the next three to six months in the near future. The functions of production planning and control (PPC) system are shown in Fig. 1.2. In order to understand the whole process involved in production planning and

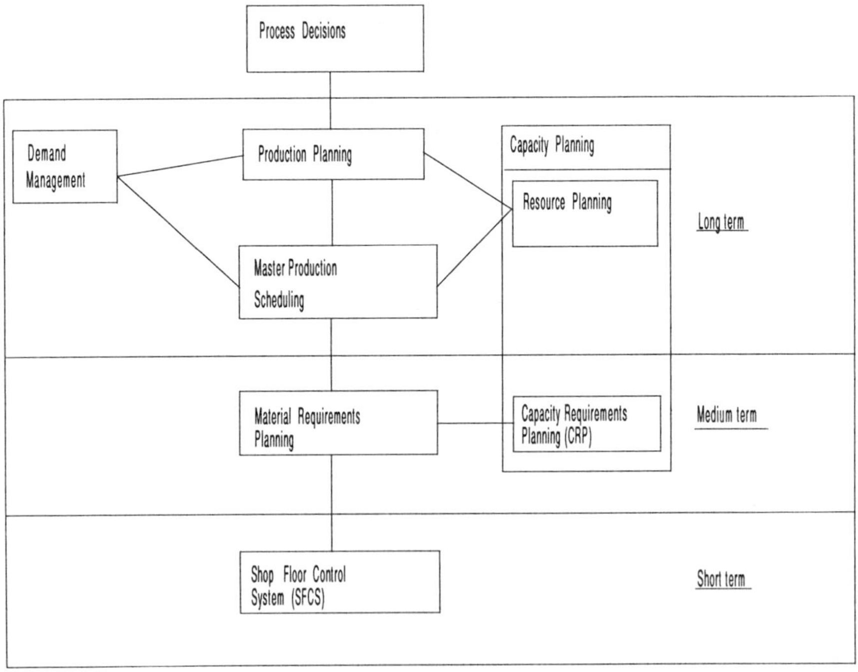

Fig. 1.2. Production planning and control framework.

control, the different functions are summarized in this chapter, but are given full treatment in Chapters 2 and 3.

1.1.1. *Demand management*

Forecasting is the prediction of the future by considering past data. It is an input to production planning, hence accurate methods of forecasting have to be adopted by management in order to achieve a good plan. Forecasting has impact on planning for capacity, schedule and procurement of raw materials. There are basically two groups of forecasting methods. The first, is the time-series in which history is considered to repeat itself, so that the future is considered to be some kind of continuation of the past. The second, is the explanatory type, in which the future is predicted by simply understanding the factors that give explanation for variation in some variables of interest. Simple moving average, single exponential smoothing, and seasonal exponential smoothing techniques fall under the first group, while simple regression falls under the second group. These forecasting techniques are discussed in Chapter 2 as important tools for business planning in the remaining part of this section.

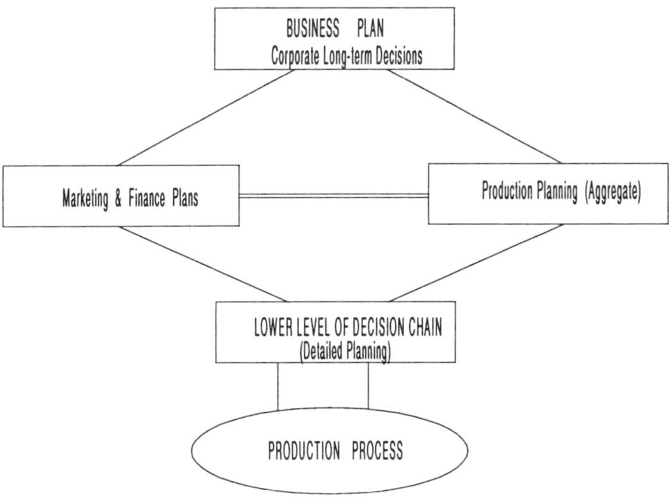

Fig. 1.3. Simplified decision-making process.

1.1.2. *Production planning*

Within the hierarchical decision process in the production planning and control framework, one of the first actions of planning is the elaboration of an aggregated production plan. Figure 1.3 summarizes some of the main modules that constitute a typical strategic level within a hierarchical framework (Filho, 1999). The production plan module interacts strongly with the upper and lower levels of the framework. The production plan aims at achieving the goals set by the business plan. This production plan is used as a target to be followed by the lower level. This target becomes the constraint to be satisfied during development of the detailed planning.

The production plan is conceived over a long-term horizon, between one to two years with monthly or quarterly discrete time buckets. Under this time horizon, enormous quantity of data is available to state the production plan, but a great part of it is not exactly known. Hence, the planning is done using both aggregate and forecasting raw data from marketing department.

The main objective of the aggregate production plan is to tune the resources available in the industrial enterprise so as to meet the fluctuating market demand requirements and minimize the expected total production costs. The advantages realized from aggregated forecast demand for production planning are as follows (Gelders and van Wassonhove, 1981):

- improvement of the plant manager's overall insights
- minimizes the plant manager's overall computational efforts
- reduced variance.

1.1.3. *Master production scheduling*

While the production plan was considered as being clustered, with regard to material, capacity requirements, and product groups, the master production schedule (MPS) is de-clustered. The MPS is a specific statement of the exact individual end items that are to be manufactured in each time period. These time periods are now much shorter than the production planning time horizon and are usually in terms of weeks. The summation of the de-clustered items must sum up to the same clustered end result items.

1.1.4. *Final assembly scheduling*

When customer requirements for finished products can be predicted with some degree of certainty, it will be more beneficial to plan the final assembly schedule (FAS) of subassemblies or groups of options. FAS has the same features as MPS except that the planning horizon is usually as short as a week ahead of time, with exact plan for the option combinations for each finished end item to be produced. In just-in-time (JIT) operations, the time horizon for FAS reduces to as short as hours within each working day of the week.

1.1.5. *Material requirement planning*

The MPS is input to material requirement planning (MRP). In order to execute the MPS, it is necessary to plan to have the material required, parts, and assemblies required for production of the finished products. Decisions have to be reached on which parts or assemblies are to be bought from outside the company and which ones are to be manufactured in-house. Orders have to be placed in good time for these parts, assemblies, or materials so that the MPS can be accomplished. Parts, materials, or assemblies that are to be supplied by outside vendors have to be ordered in good time by the purchasing department. Similar items that are to be manufactured in-house have to be scheduled into production and made on various machines in good time in order to fit into the planning horizon of the final assembly schedule. In MRP, the basic question to answer, is, when do we place order for materials, parts or assemblies so that they can be received for production within the lead-time in order to meet customer requirement?

1.1.6. *Resource requirement planning and allocation*

Resource requirement planning is a typical long-term activity, which provides the capacity required for manufacturing products and meeting market demand. Using a production plan as an input, resource requirement planning carries out an evaluation of aggregate resources (such as gross labour-hours, capital expansion, machines, warehouse space, etc.) in preparation for future production.

1.1.7. *Rough-cut-capacity planning*

Strictly related to resource requirement planning is rough-cut-capacity planning which, even though on a shorter term, provides information on how resource levels should be changed to ensure the execution of the master production schedule (Vollmann *et al.*, 1991). The concept of rough cut capacity planning (RCCP) used in production planning applies to operational planning. The essence of RCCP is to check that the MPS takes into consideration capacity requirements. However, since MPS planning horizon is in terms of weeks it may not be possible to accurately estimate the capacity of individual machines until a further detailed level of planning is done.

1.1.8. *Capacity requirement planning*

When material requirement planning is completed, it would be necessary to check if there are available capacity to meet both the MPS and MRP. Capacity requirement planning accomplishes this by estimating total capacity needed by the individual machine or work centre and equating it to the capacity required to balance the MPS and MRP. Unloaded machines and work centres are identified and it may be necessary at this stage of planning to move jobs from an over-loaded machine or work centre to under-loaded machine or work centre. The planning at this stage is very dynamic and usually puts much pressure on operations managers, because last minute decisions must be taken to meet production requirements. It may be necessary to schedule overtime or hire temporary workers. When the MPS is simply non-doable, it may be altogether revised in the interest of the company.

1.1.9. *Production control*

Close to the shop floor operations is the production scheduling which is the actual planning of the sequence of production of the various parts on machines or work centres. Production scheduling involves detailed determination of the sequence in which finished products will be produced, the employee work times and size, and the individual machine operation times. Mistake made at this time costs the company much money, and hence this aspect of planning is given much attention by management. This is where the strategic plan, which has cascaded through the business plan and production plan, down to the operational plan is executed.

For a more extensive coverage of production planning and control, interested readers may consult Chapters 2 and 3, Melnyk and Denzler (1996), Vonderembse and White (1988) and Wild (1992).

1.1.10. *New developments*

We have presented an overview of production planning and control as it takes place in the industrial enterprise that has the goal of delivering finished product or service to a satisfied customer. It is projected that in the near future, an engineer will perform a multitude of highly specialized tasks, with each of these having more than one disciplinary flavor. The search for building systems that can function automatically has attracted a lot of attention over the centuries and created continuous research activity. The current trend is to build autonomous systems that can adapt to changes in their environment. Although there is much to be done before we can reach this point, the desire to achieve fully automated manufacturing system is here to stay.

Computer Integrated Manufacturing (CIM) systems of the 21st century will demand more flexibility in product design, process planning, process planning and control, manufacturing operations and process control. This may be achieved through integrated software and hardware architectures that generate current decisions based on information collected from the manufacturing systems environment and then execute these decisions by converting them into signals transferred through a communication network.

With regard to CIM, the realization of this integration principle necessitates the creation of the relevant data links between the technical areas of product design, process planning and production and the corresponding operational processes, such as production planning and control. This means that information systems which are in themselves already partly integrated must now be unified with each other, due to the fact that technical and operational activities involve increasing interaction within the processing chain of a given customer order. Integrated manufacturing has not yet reached this state. However, the urge for achieving this strategic vision is transferred into the term "intelligent systems" that we are now using in the late 1990s and early 2000s.

Knowledge-based systems, our first effort in this endeavour, were not sufficient in the 1980s to generate the "intelligence" required, so our quest still continues. The approach to improve our quest for intelligent systems is to merge the old with the new technologies. Adaptive memory programming are other technologies that need to be integrated to improve our quest for intelligent systems.

Newly emerging-technologies, such as adaptive memory programming, will be some of the contributions in the quest of achieving fully adaptive or intelligent manufacturing systems in the 21st century. This book, *Emerging Optimization Techniques in Production Planning and Control*, presents in a unique way, *adaptive memory programming* approaches and their impact in the design of autonomous manufacturing units to realize our quest of building adaptive systems.

1.2. Conventional Combinatorial Optimization Techniques

The theory of optimization encompasses the quantitative study of optima and methods for finding them. Therefore, when we say we want to optimize a function or process, we are in effect seeking to improve performance toward some optimal point or points. Two issues are important in optimization: the optimal point which is the goal, and seeking improvement on how to get there. Over the past two centuries, theorists interested in optimization have emphasized convergence, that is, reaching the optimum. They paid less attention to *provisory improvements* that drive the process of optimization towards the destination or optimum.

Many real life optimization problems encountered in practical settings such as transportation, logistics, production and task scheduling, telecommunications and financial planning have large search spaces. For any search problem, the search space is the set of possible solutions, while the solution space is the subset of the search space that contains actual solutions. The underlying problem in real life optimization problems such as in the examples cited above, is that it is impossible to sample every point in the search space. If all the points in a search space are examined, it is possible to find an optimum solution. But as this is impossible for this class of problems, we must reject techniques that emphasis convergence. This leads to the problem of selecting points to search, which give a "good enough" performance in a reasonable time. In difficult optimization problems encountered in practical settings, convergence to the optimum is not an issue. What is important is getting a *good enough* solution, within a reasonable time and so improve the expectation of overall return in the long run.

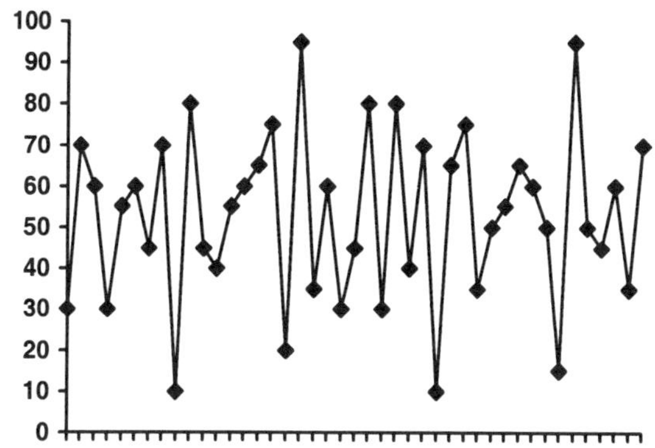

Fig. 1.4. Noisy and discontinuous functions that confound traditional methods.

We have only looked at one aspect of difficult real life optimization problem, which is the large search space that is impossible to search on a point-to-point basis for the optimum solution. The other aspect of this class of problems is that they abound with discontinuities and vast multi-modal, noisy search spaces as shown in Fig. 1.4. These search spaces are clearly described by less friendly calculus functions. Here, we are faced with functions of *discrete* variables. Combinatorial optimization is the search for optima of functions of *discrete* variables. From now onward, we take it for granted that the problems we are solving are: (1) combinatorial optimization problems (2) of infinite search spaces that cannot be exhaustively searched by a finite machine such as the computer. We conclude that we are faced with a dilemma of putting aside all traditional techniques of seeking for an optimum solution and seeking methods that will search only subset of the search space in order to reach a near-optimum that is *good enough* within a reasonable time.

The remaining parts of this chapter discuss conventional combinatorial optimization techniques, intelligent optimization fundamentals, adaptive memory programming, and combinations of adaptive memory programming approaches that result in hybrid systems.

We have concluded that we are left in a situation, where in practical settings many larger-scale combinatorial optimization problems cannot be solved to optimality, due to the constraints that prohibitive amounts of computation times are required to search for the optimum solution. For example, finding the best sequence for producing n jobs on a machine for a given performance measure (e.g. completion time) requires $n!$ combinations. Suppose there are 100 jobs, how many combinations are required? The answer is simple: there are 100! (factorial) combinations that have to be considered. We begin to see the difficulty faced by the computer power if an attempt is made to search each point in the search space.

There are two conventional combinatorial optimization techniques: optimization algorithms and approximation algorithms. These are now discussed in some details in order to have understanding in issues involved in solving this class of problems.

1.2.1. *Optimization algorithms*

Optimization algorithms are capable of solving a combinatorial optimization problem, yielding a globally optimal solution in a possible prohibitive amount of computation time. Two well-known optimization algorithms are *branch-and-bound* and *A^* algorithm*. Both of these schemes use tree structures made up of nodes connected by arcs. In the A^* algorithm, all the existing paths that have been found so far are evaluated and the shortest one is chosen and advanced. In this way some branches of the tree are never explored, hence the search time is reduced. Moreover since many alternative paths are explored and are never entirely dropped, optimal solution is guaranteed. The A^* algorithm is therefore, a very powerful breath-first tree-search technique when evaluation is available at each node of the tree structure. The *branch-and-bound* technique commences with a partial solution at the root node. As branches progress into more complete partial solutions, new terminal nodes replace old ones. Although these optimization algorithms guarantee optimum solutions, they are less attractive in solving many large-scale combinatorial optimization problems. This is because, this class of problems are solvable by an amount of computational effort that is bounded by a polynomial function of the size of the problem.

1.2.2. *Approximation algorithms or heuristics*

When designing a heuristic for a combinatorial optimisation problem one can either build a method that is specific to the problem under study or adapt some general existing schemes. In the latter case constructive, sequential and adaptive techniques are search strategies that have successfully been used for the solution of difficult problems.

Constructive methods build feasible solutions by starting from an "empty" solution and by inserting repeatedly a new component into the current partial solution. Sequential methods start from an initial feasible solution and move step by step from one solution to a neighbour solution by performing some elementary modifications.

Approximation algorithms can be divided into two categories: *constructive* and *neighbourhood* or *local search.* These two categories are now discussed.

Neighbourhood or *local search* techniques have an operation called a *move.* The move operation defines a neighbourhood N_i for each configuration i in one iteration. Commencing from a given configuration, a possible transition from the current configuration to a configuration chosen from the neighbourhood of the current configuration is examined. If the objective function evaluated for this neighboring configuration is less than the objective function evaluated for the current configuration, then this neighbour replaces the current configuration, otherwise another neighbour is selected and compared for its evaluated objective function. The algorithm continues until it finds a configuration whose evaluated objective function is better than its neighbours. Such a neighbourhood search that we have just described is the *steepest descent.* Although the steepest-descent approach has attractive features it is impracticable in certain practical settings because it is computationally too expensive, as where the neighbourhood contains many elements or each element is expensive to evaluate. Local search method also suffers from the disadvantage of terminating in a local optimum for which we may not know how far it is from the global optimum solution. Neighbourhood describes all moves that can be made from a current solution. Suppose in 6-job flow shop problem, the current solution (i.e. sequence) is {2, 3, 1, 6, 5, 4}. If an insert-move is adopted, then removing a job from the *ith* position and inserting it in the *jth* position describes a neighbour. For example, removing a job from the fifth position and inserting it in the second position results in a sequence {2, 5, 3, 1, 6, 4}. The *move* described illustrates the concept of neighbourhood as follows:

Neighbourhood 1	2	3	1	6	5	4
Neighbourhood 2	2	5	3	1	6	4

The first neighbourhood is made up of sequence {2, 3, 1, 6, 5, 4}. By inserting the fifth job in the second position, a new neighbourhood is obtained which is made up of sequence {2, 5, 3, 1, 6, 4}. We defer detailed discussions on neighbourhood or local search until we get to later chapters.

Constructive methods build feasible solutions from an initial partial solution by inserting repeatedly a new component into current partial solution. Most constructive methods can be conceived to be descent methods, where each constructive step toward a more complete solution therefore is an improving step in the neighbourhood being considered. The final constructive step becomes a local optimum since no better solution is obtained by going beyond a complete construction. The concept of constructive method is best understood using the curtailed-enumeration algorithm of NEH (Nawaz, Enscore, and Ham, 1983) for scheduling n jobs on m machines. We defer detailed discussions on curtailed-enumeration algorithm of NEH until we get to later chapters.

Both constructive and neighborhood or local search techniques give "good solutions" in short computation times, for combinatorial optimization problems. But we recall that the main goal of optimization is to seek for interim improvements that drive the process towards the destination, the optimum solution. This underlying philosophy is the motivation for research in seeking for adaptive memory programming in the past few decades. This is the very essence of this book: to present modern or emerging-intelligent techniques that are preferred to both constructive and neighborhood or local search techniques for solving practical real life combinatorial optimization problem. That is, we emphasis on improvements on provisory solutions with the goal of moving nearer to the optimum solution in a short time. Constructive and neighborhood or local search methods may obtain "reasonable" solution in very short time but such solution may be local optima for which we are not sure how far they are from the global optimum. We need some mechanisms that can facilitate the "jumping out" from local optimum with the hope that our solution moves toward the global optimum. Not all techniques that can provide this "jumping out" mechanisms qualify for adaptive memory programming. The remaining parts of this chapter explain the fundamentals and types of adaptive memory programming approaches currently in use.

1.3. Intelligent Optimization Fundamentals

Intelligence is the ability to respond to varying conditions, or of changing decisions over time as a function of multiple considerations. Adaptive memory programming (AMP) suggests that an *adaptive memory* and a *responsive exploration* need to be integral part of a search algorithm. Adaptive memory means ability to selectively remember key elements of previous information stored in the memory while responsive exploration means ability to make strategic choices concerning the direction to follow. This contrast between the adaptive memory programming orientation and the local optimality orientation is significant. Responsive exploration exploits the information desirable from a selected strategy. Such information may be more substantial, even if the selected strategy is in some sense a bad strategy, than the outcome from a good random choice. In an optimization technique that uses memory (such as adaptive memory programming), a bad choice based on strategy provides opportunity to intelligently modify the strategy. Memory-based systems provide some clues as to how one can modify the strategy. Additional features of adaptive memory programming include: the kind of neighbourhood exploration used, intensification and diversification mechanisms, the number of current solutions transported from one iteration to the next, and ability to improve on initial seed solution from an inferior heuristic.

Figure 1.5 shows the classification of the six fundamental features of adaptive memory programming under strategic, tactical, and operational levels. At strategic level, an adaptive memory programming scheme, is driven by *adaptive memory* and a *strategic guidance* (*exploration*) mechanism. At tactical level, an adaptive memory programming searches over a restricted search-space for

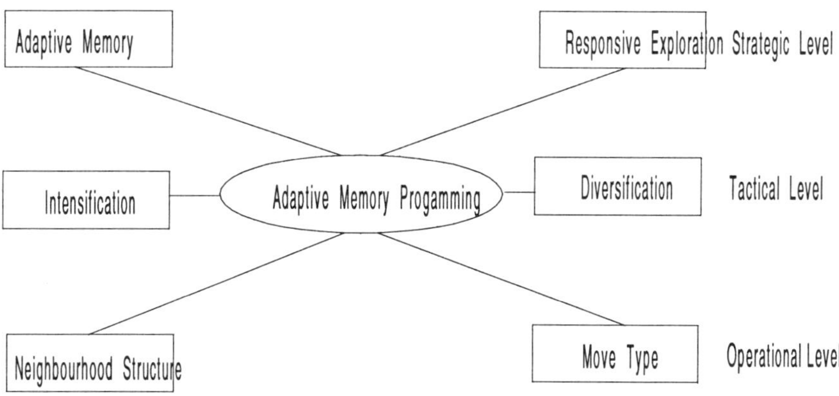

Fig. 1.5. Features and levels of Adaptive Memory Programming.

each iteration known as *neighbourhood.* It should be able to utilize intensification and diversification mechanisms to explore better solutions within the solution space. At operational level, the *move type* determines the number of current solution carried forward from one iteration to the next. It should also be able to consider initial solution from constructive moves. These fundamental features are now discussed in more details.

1.3.1. *Adaptive memory*

The contrast between the adaptive memory programming orientation and the local optimality orientation is significant. Iterative heuristics that give local optimum are often referred to as ascent methods, descent methods, steepest descent methods, or local search methods. The term "local search" is currently sometimes being used in a very loose way to include search that is not limited to being local. Therefore, care must be taken to understand the context in which the term "local search" is used in the literature.

Memory is no doubt an integral component of any search that qualifies to be called "intelligent". Adaptive memory depends intimately on four elements of recency, frequency, quality, and influence. Recency-based memory keeps track of solutions attributes that have changed during the recent past. Frequency-based memory provides useful information that facilitates choosing preferred moves. Quality-based memory differentiates the merit of solutions visited during the search. When operationalizing quality-based memory, inducements provide reinforcement actions that drive the algorithm towards good solutions and penalties are provided to discourage action that lead to poor solutions. Influence considers the impact of strategic exploration on both quality and structure.

Three types of memory are commonly used: associative, attributive and explicit. In associative memory structure, given a pair of vectors, an input of one of the vectors should result in the output of the other corresponding vector. Moreover, a noisy, or corrupted, version of one of the input vectors should also recall the corresponding output vector. Associative memory based structures store vectors so that later, if a noisy vector, or an incomplete vector is used as an input, the algorithm will detect the stored vector closest in some sense to the input vector. In attribute memory structure, information regarding solution attributes are recorded as the attributes change in transition from one solution to another. In explicit memory structure, the entire attractive solutions are memorized and used to expand the local search.

1.3.2. *Strategic exploration*

Strategic exploration or choice requires a strategic guidance mechanism. An adaptive memory programming may also modify the neighbourhood of moves in the search space by selectively pruning some neighbours and including others. This approach enlarges a standard neighbourhood that allows moves only among feasible solutions to include unrealizable solutions. The search is then strategically driven beyond the feasibility boundary into the unrealizable region. The search progresses to some depth in the unrealizable solution space and then retracts into the feasible solution space and progresses in the direction of increased feasibility. The process is repeated as the search progresses. The strategic guidance mechanism that drives the search in and out of unrealizable solution spaces requires that a change in the nature of evaluation be made.

1.3.3. *Intensification*

Intensification and diversification strategies constitute the two important components of adaptive memory programming. In the adaptive memory programming framework presented in Fig. 1.5, intensification and diversification strategies are positioned at the tactical level.

Intensification is one of the important components of adaptive memory programming. Intensification strategies are based on modifying choice rules in order to favour move combinations and elite solutions. Intensification strategies may also return to attractive or green regions to search them more thoroughly. This concept agrees with natural phenomenon. During intensification, the neighbours of elite solutions are examined more thoroughly. Two things happen when the neighbours of elite solutions are examined: (1) neighbours to elite solutions are considered using standard move mechanism, (2) neighbours are generated by using modified evaluation or choice strategies that favour the introduction of such components into a current solution.

1.3.4. *Diversification*

Diversification is another one of the important components of adaptive memory programming. In the framework of adaptive memory programming discussed in Fig. 1.5, intensification and diversification are positioned at the tactical level, thus, showing their importance. Diversification favours the search process to examine unvisited regions in the solution space and generates solutions that differ in various significant ways from those already obtained.

1.3.5. *Neighbourhood*

The neighbourhood N_i for each solution i, is made up of all solutions that can be reached from i in one transition. Adaptive memory programming explores the solution space by repeatedly making moves from one solution to another solution located in the neighbourhood. These moves are performed with an ultimate goal of reaching a good solution by the evaluation of some objective function to be optimized (maximized or minimized). However, contrary to other traditional iterative methods, in adaptive memory programming, the subsequent function evaluated need not be better in solution quality than the current function evaluated, in every iteration.

There are three commonly described neighbourhood structures: exhaustive, randomized, and dynamic. In exhaustive neighbourhood structure, the whole neighbourhood space is searched. Exhaustive searches are impracticable for large-scale combinatorial optimization problems. In randomized neighbourhood structure, not all the neighbours are examined. Here, neighbours are selected at random, and the objective function at that neighbour is evaluated. If the value is the smallest value that has been found so far it will be stored as the minimum value. The variable neighbourhood structure is the result of maintaining a selective history of the state encountered during the search. Operationally, the modified neighbourhood may be shrunk as a subset of the current neighbourhood or may be expanded to include solutions not ordinarily found in the current neighbourhood. The latter case is a description of a dynamic neighbourhood. It is dynamic because the neighbourhood can change according to the history of the search.

1.3.6. *Move type*

There are two types of moves that are commonly considered: non-solution space-based move and solution space-based move. In the former approach, the solution search moves from one current solution to the next after each iteration and uses the properties of the solution for searching intelligently in the solution space. In the latter approach, collective properties of a group of distinguishable solutions, called solution space, are used for searching intelligently in the solution space.

1.3.7. *Solution from constructive methods*

Constructive methods have already been discussed. As can be seen, their solutions are obtained in a static manner. Constructive methods cannot improve their solution. An adaptive memory programming approach should be able to receive a seed solution from a constructive method and then improve on this. The idea of merging constructive methods and adaptive memory programming in order to refine a seed solution is now popular amongst researchers. It is a good way of expanding the scope of hybrid systems.

1.3.8. *Generic scheme of an adaptive memory programming*

An adaptive memory programming works as follows: a new solution of the combinatorial optimization problem being solved is constructed by first creating an intermediate solution using information in the memory and then this intermediate solution is improved with a local search procedure. Finally, the improved solution is used to update the memory and the process is continued until a stopping criterion is reached. The generic scheme of an adaptive memory programming is given in Fig. 1.6.

Initialise the *memory* or pheromone trail.
Repeat
  Construct a *new provisory solution*, using the information contained in the *memory*
  Improve the *provisory solution* with a *local search.*
  Update the *memory.*
Until a stop criterion is satisfied:

Fig. 1.6. General scheme of an Adaptive Memory Programming.

Now that the fundamental features of adaptive memory programming have been discussed in some details, it becomes a simple matter to classify extant approaches used for solving combinatorial optimization problems.

1.4. Adaptive Memory Programming

There are many optimization techniques that are labeled "intelligent" whereas they are not. The notion of intelligence presupposes that memory exists. It is not enough to have memory. It is how the memory is used that determines whether a system is intelligent or not. We have already defined what intelligence means. It is not sufficient for human beings to remember everything. Over the years we come across a number of events, perhaps in their thousands. Irrelevant events should not be remembered because they do not add value to our lives. However, important events need to be remembered and properly utilized. Suppose that some years ago, you were driving your car and then you were stopped by the police who ask you to produce your driving license and your vehicle particulars. You neither carried your driving license nor your vehicle particulars, nor the photocopies. The police then gave you a ticket for not having the documents in your possession. Of course, this event should be remembered. When next you want to drive out for a leisure trip through a highway where police usually stop cars randomly for some checks, you want to be sure you have your vehicle particulars in your possession. Why? Because you remember that unpleasant incidence when you were served with a ticket when you did not have your vehicle particulars in your possession. Assuming some years back you went into the city centre for shopping and saw some people cleaning their floors with hovers. Or you went to the post office and saw some workers selling stamps and telephone cards across the counters. After a year or two do you need to remember that you saw some cleaners cleaning the floors of their shops or some post office clerks selling stamps and telephone cards over sales counters? The answer is obviously, no. Why? Because remembering these events do not add value in any way. This is essentially the point we are emphasizing in intelligent or emerging optimization techniques. One of the strategic features of adaptive memory programming is that of *adaptive memory.* Adaptive memory means remembering selective events that add value to the goal of the system. The system we are referring to here, is search system. This one key feature disqualifies some search techniques that are labeled "intelligent". Although it is not the only feature that qualifies an optimization technique as "intelligent", it is a highly important one and perhaps the most important. We recapitulate the features that qualify an optimization as being "intelligent". They are as follows: (1) adaptive memory (2) strategic exploration or choice (3) intensification mechanism, (4) diversification mechanism (5) neighbourhood structure (6) the number of current solutions transported from one iteration to the next (7) and, ability to improve on an initial seed solution from an inferior heuristic. The first two are at the strategic level, the third and fourth are at the tactical level, while the last two are at the operational level.

The past one-decade has witnessed tremendous research works and publications in the area termed "modern search techniques". There has been proliferation in the use of terminology to describe these search techniques. In the literature, you will come across the terms such as "modern heuristics", "randomized algorithms", "local search", etc., used to describe these recent search techniques that have the power to solve large-scale combinatorial optimization problems. They all seek to introduce mechanisms through which the search techniques escape local optimality. Commonly listed modern optimization techniques include simulated annealing, neural networks, genetic algorithms, tabu search, scatter search, and ant systems.

This book is not intended to be a monograph of optimization search techniques classifications. However, we have laid the foundation for determining to some extent what techniques pass the test of being referred to as adaptive memory programming. Of all the techniques listed only simulated annealing does not incorporate memory in the context we have discussed so far. It is therefore, excluded from the list of adaptive memory programming.

Adaptive memory programming approaches have connections with physical and natural phenomena. Neural networks seek to imitate the human brain. Genetic algorithms mimic chemical reactions in organic compounds. Tabu search derives and exploits a collection of principles of intelligent problem solving. Scatter search follows the principles that underlie the surrogate constraint design. Ant systems simulate the collective contributions of a colony of ants in solving a common problem.

1.4.1. *Explicit memory versus implicit memory*

The adaptive memory classification is an explicit intelligent memory construction. An adaptive memory constructs a new solution by first creating an intermediate solution using information contained in the memory and then this intermediate solution is improved. The improved solution is then used to update the memory and the process is repeated. Ant systems, tabu search, and neural networks clearly have the features of explicit memory.

In an implicit memory structure, solution combination methods pass on features of selected past solutions to current solutions. These past features are not remembered in a constructive way. Loosely speaking, some good solutions of the past are pushed forward to compete with subsequent solutions. There is an analogy here with sports competition. Think of world sports competition. Athletes compete in their states and the winners are declared. These may be two, three or more. Then there is a national competition in each country. The winners from each state congregate for the national competition. This progresses into international competitions, each time some competitors drop out while others are winning. It is a game of winners. The competitors at the international level are identified with their nations through their jerseys and the records kept by the organizers. It is difficult to trace all the competitors that dropped out. This is because only the best competitors are pushed forward to compete at subsequent level. This simple analogy explains what happens in optimization techniques that have implicit memory structure. Solution combination methods pass on features of selected past solutions to current solution. Genetic algorithm is an example of combinatorial optimization technique that has implicit memory structure.

1.4.2. *Neural networks*

A first adaptive memory programming approach is artificial neural networks (ANNs). Maren *et al.* (1990) define a neural network "as a computational system, either hardware or software, which mimics the computational abilities of biological systems by using a large number of simple, interconnected artificial neurons". Artificial neurons are simple emulation of biological neurons; they take information from sensor(s) or other artificial neurons, perform very simple operations on this data, and pass the results on to other artificial neurons. The combination of simple processing units and complex connections result in a computing tool that is able to perform high non-linear, multi-input, multi-output function.

ANNs correspond to a massively parallel-distributed processor with a natural propensity for storing experiential knowledge and making it available for future use. ANNs resemble the brain with respect to the capacity of acquiring knowledge through a learning process and storing it in the interneuron connections (Haykin, 1994). Many architectures and configurations of ANNs can be built in order to perform different applications, such as classification, clustering, regression and optimization (Kung, 1993). Artificial neural networks are known to be very powerful in scientific and engineering applications when used to predict, classify or recognize patterns due to the inherent data classification capabilities and massively parallel processing power. Incidentally, they have not been as successful as the best adaptive memory programming techniques from the operations research literature when applied to combinatorial optimization problems. The interest in using neural networks for combinatorial optimization problems began in the mid-1980s with the work of Hopfield and Tank (1985, 1986) on the traveling salesman problem. There are two basic approaches to using neural networks for solving optimization problems.

The first approach is based on statistical physics and includes the Hopfield-Tank and the Mean-Field annealing neural networks algorithms. In the Hopfield-Tank approach the set of all possible solutions of a combinatorial optimization problem is mapped onto the set of all possible states of the network. The network is normally composed of one-layered neurons with fully-weighted connections. An energy function is defined over the set of all possible states and it depends on the weights of the connections of the network. The neural networks algorithm iterates towards a ground state of minimum energy that corresponds to an optimal solution satisfying all constraints of the combinatorial optimizations problem. Due to the feedback characteristics, the Hopfield-Tank model converges to a local minimum. Various modifications have been proposed to alleviate this problem such as the incorporation of simulated annealing in a Boltzman machine and other stochastic models.

The second approach is based on competitive neural networks and includes the Kohonen self-organizing network and deformable (elastic) nets in that neurons are allowed to compete to become active under certain conditions. The aim of the Kohonen model is to find a mapping from a high dimensional input space onto a lower dimensional discrete lattice (usually two) of formal neurons. The mapping tries to reflect the topological neighbourhood relationships among the inputs in the arrangement of the corresponding neurons in the lattice. The above two approaches suggest neural networks as an alternative for solving certain optimization problems as compared to classical optimization techniques and other novel approaches. For information on combinatorial optimization problems and neural networks, refer to Udo and Gupta (1994), and Garetti and Taisch (1999).

1.4.3. *Genetic algorithms*

A second adaptive memory programming approach is known as Genetic Algorithms (GAs) (Goldberg, 1989; Davis, 1991). Simply, GAs reach the solution of a problem by generating and modifying a certain number of tentative solutions so as to find the individual (i.e. the solution) best suited to a given environment (i.e. the problem domain). A solution is represented by a string of numbers or characters forming a so-called *string*. A set of strings is referred to as a *solution space*. After a random initialization of a solution space, the algorithm proceeds in an iterative way. It proceeds through: (i) an evaluation of strings by the way of a score function with respect to an objective function of the problem; (ii) a selection of candidates and recombination (iii) a generation of a new solution space from the old one through genetic operators of partial-bits-exchange and partial-bits-exchange. The algorithm stops when a given criterion is reached. Due to the particular way GAs represent the problem solution (sequence of bits), their use in combinatorial problems, such as scheduling, is quite natural (Goldberg, 1989).

1.4.4. *Tabu Search*

A third adaptive memory programming approach is represented by Tabu Search (TS), proposed by (Glover, 1989, 1990) and (Hansen, 1986) as an effective technique to solve combinatorial optimization problems. The TS method is a adaptive memory programming that shares with the genetic algorithms and ant systems the ability to guide the local search descent method to avoid bad local optima. It uses deterministic acceptance criterion. The method works this way: starting from an initial feasible solution (also a trivial one), a modification of the current solution, which results in another feasible solution, is defined as a *move*. For example, an interchange of two jobs in a sequence can be referred to as a move, even if any reasonable method can be used depending on the problem being studied. All the solutions that can be obtained by a single move from a certain solution are referred to as the neighbourhood of the solution itself. Starting from a feasible solution, all the neighbouring solutions are evaluated and the best one is chosen. Anyway, as a result of the move a worse solution can be chosen. On the next iteration, a typical greedy method would choose the previous current solution, thus resulting in cycling. Avoidance of such undesirable effect in tabu search is featured by the tabu list. All solutions not allowed to be selected at the current iteration, are inserted in this list. During each iteration, the past current solution becomes tabu and the oldest move in the tabu-list leaves the list itself. A neighbouring solution is considered forbidden and deemed not admissible if it has attributes on the tabu list. Storing attributes rather than the complete solutions may cause some non-tabu solutions to be wrongly prevented and aspiration criteria are normally used to correct such errors. Therefore a move remains in the tabu list as many moves as the tabu list size. Sometimes, a tabu move can assume the highest value of the objective function ever obtained. In that case, TS recalls an aspiration criteria to override the tabu restrictions and selects anyway the move as the current solution. To produce good results, intensification and diversification strategies are also employed. The former strategy is used to perform a thorough search in good regions while the latter strategy is attempted to consider solutions in a broad area of the search. Different stopping criteria can be developed, even if the most common is a fixed number of iterations without any improvement in the best move.

1.4.5. *Ant systems*

A fourth adaptive memory programming approach is known as ant colony optimization (ACO) introduced by Dorigo, Maniezzo and Colorni (1996). They studied artificial systems that take inspiration from the behaviour of real ant colonies and which are used to solve discrete optimization problems. The first ACO system was called Ant System (AS), which is the result of a research on computational intelligence approaches to combinatorial optimization problems. In AS, an ant represents a single artificial agent and the set of artificial ants cooperate in the search for an optimal solution by exchanging information in the form of a pheromone trail. Since the introduction of AS, there have been a number of variants such AS-Q that uses the Q-learning methology, hybrid ant system (HAS) and fast ant (FANT) system.

Neural networks, genetic algorithms, ant systems and tabu search all belong to the class of Adaptive memory programming. In fact, starting from a solution or set of solutions, they proceed applying special algorithms that make a limited change in the solution itself. Therefore, the new solution is not too far (local) from the previous one. Such a way to operate has been utilized by many authors to refine a first optimal solution found using a different method. Adaptive memory programming have been found to have the power to solve highly constrained, practical, real-life, problems (Dorn *et al.*, 1996).

1.5. Hybrid Systems

Adaptive memory programming approaches have proven powerful in finding high quality solutions to many real world intractable problems. Therefore, over the years, researchers have attempted to combine the best features from different adaptive memory programming approaches to derive more powerful hybrid implementations. It is only natural to do this. Combining the best features of different systems will give a new system that is superior to the individual systems from which these features are derived. Hence, adaptive memory programming continue to prove superior to other known heuristics. It is expected that more serial and parallel implementations based on hybrids of adaptive memory programming with each other or with classical operational research techniques will be on the increase.

In this section, we describe how hybrid adaptive memory programming works, and then we give a table of some known hybrid systems that have been implemented by researchers. Let us describe the Genetic-Tabu Search Hybrid (GTSH) proposed by Fleurent and Ferland (1994). By so doing, readers will easily see the benefit of hybridization.

The GTSH method can be presented as a genetic algorithm, using a special partial bits exchange operator specially designed for permutations (due to Tate and Smith, 1995) and a random bits exchange operator that consists in performing $4n$ steps of a tabu search (due to Taillard, 1991). Let us now describe the memory, its implementation, constructive generation of a provisory solution, improvement via tabu search, and memory update in GTSH.

The memory of GTSH consists of 100 solutions or permutations. Initially, these solutions are obtained by 100 tabu search runs performing $4n$ iterations each and starting with random initial solutions.

The provisory solution is built by first selecting two permutations from the solution space. The probability of selecting the kth worst solution of the solution space is $2k/s(s-1)$, where s is the size of the solution space. The partial bits-exchange of Tate and Smith is then applied to the selected solutions to produce the provisory solution. The elements common in both selected permutations are copied in the new solution simultaneously. Other elements are chosen randomly from the selected permutations, at the same position, if not already chosen. The unfilled positions are chosen randomly in order to complete the solution.

The improvement procedure is the tabu search of Taillard (1991) that runs for $4n$ iterations, starting from the provisory solution. The new solution is inserted in the memory and the worst solution is removed from the memory.

It becomes evident that the description of GTSH is different from the standard genetic algorithm description. But genetic algorithms and tabu search can be joined in a more fundamental way. The introduction of tabu search as an improvement procedure makes GTSH a hybrid system. In traditional genetic algorithm implementation, there is no such improvement procedure other than the generic operators such as partial bits-exchange and random

Table 1.3. List of some hybrid systems.

Year	Investigators	Description
	GENETIC ALGORITHM AND LOCAL SEARCH	
1983	Goldberg	Suggested hybridizing genetic algorithm with problem-specific search techniques.
1983	Goldberg	Hybridized genetic algorithm with local search known as G-bit improvement (gradient-like-bitwise improvement)
1997	Taillard and Gambardella	Genetic Algorithm-Descent Method hybridization (GDH)
	GENETIC ALGORITHM AND TABU SEARCH	
1996	Fluerent and Ferland	Tabu search and genetic algorithm hybrid. Details already discussed in this chapter
	NEURAL NETWORKS AND TABU SEARCH	
1989	de Werra and Hertz	Applied hybrid neural network and tabu search to visual pattern recognition, with increased speed.
1991	Beyer and Ogier	Applied hybrid neural network and tabu search to non convex optimization.

bits-exchange that are embedded in genetic algorithms. The introduction of explicit memory, constructive generation of a provisory solution, improvement via tabu search, and memory update, qualify GTSH as a hybrid system. Normally, a hybrid system will yield better solution than the subsystems that constitute it. Therefore, one expects GTSH to perform better than genetic algorithm and tabu search, separately. Table 1.3 gives a list of some of numerous implementations of hybrid system. For readers who are interested in pursuing the matter further, the bibliography compilation of Osman and Laporte (1996) may be consulted.

Hybridization has taken several forms in recent years. For example, researchers now hybridize adaptive memory programming approaches such as genetic algorithm, neural networks, tabu search, and ant systems, with other good heuristics to generate initial solutions. By so doing, the global solutions obtained by adaptive memory programming are drastically improved. How do we explain this improvement? It is simple. Most adaptive memory programming approaches generate initial solution randomly. Random initial solution may not satisfy the constraint that the performance measure has to be optimized. Therefore, utilizing a good heuristic for generating initial

Table 1.4. Some current hybrid extensions to scheduling problems.

Year	Investigators	Description
		TABU SEARCH
1989	Widmer and Hertz	Hybridized tabu search and an analogy to TSP for makespan performance measure.
1990	Taillard	Hybridized tabu search and an improved NEH algorithm for makespan performance measure.
		GENETIC ALGORITHMS
1995	Chen, Vempati, and Aljaber	Hybridized genetic algorithm and CDS along with rapid access approach for makespan performance measure.
		ANT SYSTEMS
2000a	Onwubolu	Hybridized ant system and NEH;
2000b	Onwubolu	Hybridized ant system and curtailed insertion method

solution is known to yield better quality solution. Table 1.4 shows some of current hybridization of adaptive memory programming with good heuristics for generating initial solutions for scheduling problems.

1.6. Summary

In this first chapter, both the conventional optimization techniques for solving combinatorial problems, and production planning and control, are presented.

Production planning decisions span the *planning schedule* and the *capacity schedule* spectrum. The planning schedule consists of demand forecasting, aggregate production planning, master production schedule, and material requirement planning. The capacity schedule consists of resource capacity planning, rough-cut capacity planning, and capacity requirement planning. Production control decisions span the *planning execution* and the *capacity execution*. The planning-execution consists of operation sequencing, scheduling, order priorities, stock picking, and capacity control. The capacity execution consists of tool control, machine control, factory order control, inventory control, and capacity control.

The conventional optimization techniques for solving combinatorial problems fall into two classes: optimization algorithms and approximate algorithms otherwise known as heuristics. The methods that fall under the first class are attractive if the problems solved are of small size and if optimality must be achieved. These methods fail to perform for large size problems, typically encountered in real life situations. The reason for this failure is that of computational time involved, for which most computers cannot accommodate. The methods that fall under the second class are also attractive if all that is needed is a quick solution, without attention to very good quality. Between these extremes naturally lies a method that finds near optimal solutions in reasonable computational time for practical, real life problems. Adaptive memory programming fills this gap. We have presented the features of adaptive memory programming. These are generally accepted as the yardstick for determining whether or not a method qualifies as being intelligent. Neural networks, genetic algorithms, tabu search, and ant systems are presented as being adaptive memory programming approaches. They qualify as Emerging Optimization Techniques. These are briefly discussed. Hybrid systems are presented as extensions of adaptive memory programming fundamentals. This chapter therefore, sets the scene for what follows in the remaining parts of the book. The details of each of these are discussed in the following chapters. You will notice that simulated annealing is not classified as belonging to adaptive memory programming. This is because it does not have memory since adaptive memory is paramount in classifying an approach as belonging to adaptive memory programming.

Exercises

Production Planning and Control

1. What activities constitute *planning-schedule* aspect of the production planning and control system?
2. What activities constitute *capacity schedule* aspect of the production planning and control system?
3. What activities constitute *planning-execution* aspect of the production planning and control system?
4. What activities constitute *planning-execution* aspect of the production planning and control system?
5. What is the role of production planning and control in integrated manufacturing?

Adaptive memory programming

1. Discuss the disadvantages of optimization algorithms such as *branch-and-bound* and A^* in solving combinatorial optimization problems?
2. Are constructive methods good compared to adaptive memory programming approaches in terms of solution quality? Discuss.
3. Discuss the interpretation of the six adaptive memory-programming dimensions: adaptive memory, strategic exploration, and intensification/diversification, move type, and acceptance of seed solutions from constructive methods.
4. What are the advantages of memory-based structures over memoryless-based structures in designing schemes for solving combinatorial optimization problems?

References

Production Planning and Control

Filho, O. S. S., 1999, An aggregate production planning model with demand under uncertainty, *Production Planning and Control*, **10**(8) 745–756.

Garetti, M. and Taisch, M., Neural networks in production planning and control, 1999, *Production Planning and Control*, **10**(4) 324–339.

Gelders, L. F. and van Wassenhove, L. N., 1981, Production planning: A review. European Journal of Operational Research, **7** 101–110.

Melnyk, S. A., and Denzler, 1996, *Operations Management: A Value-Driven Approach* (Irwin Publishers: Boston).

Nawaz, M., Enscore, E. E., and Ham, I., 1983, A heuristic algorithm for the m-machine, n-job flow shop sequencing problem, *OMEGA The International Journal of Management Science*, **11**(1) 91–95.

Onwubolu, G. C. and Mhlanga, S., 1997, Development of a computer-aided production management course in industrial engineering curriculum, *Production Planning and Control*, **8**(3) 297–306.

Vollmann, T. E., Berry, W. L., and Whybark, D. C., 1991, *Manufacturing Planning and Control System* (Homewood, IL: Irwin).

Vonderembse, M. A. and White, G. P., 1988, *Operations Management: Concepts, Methods and Strategies*, 1st edition (West Publishing, St. Paul, MN).

Wild, R., 1992, *Essentials of Production and Operations Management*, 3rd edition (Casell Educational, London).

Adaptive Memory Programming

Beyer, D. A. and Ogier, R. G., 1991, Tabu learning. A neural network search method for solving non-convex optimization problems, in *Proceedings of the International Conference on Neural Networks* (IEEE Computer Society Press, Los Alamitos, California), 953.

Chen, C.-L., Vempati, V. S., and Aljaber, N., 1995, An application of genetic algorithms for flow shop problems, *European Journal of Operational Research*, **80** 389–396.

Davis, L., 1991, *Handbook of Genetic Algorithms* (VNR computer Library, NY).

De Werra, D. and Hertz, A., 1989, Tabu search techniques. A tutorial and an application to neural networks, *OR Spectrum*, **11** 131.

Dorigo, M., Maniezzo, V., and Colorni, A., 1996, The ant system. Optimization by a colony of cooperating agents, *IEEE Transaction on Systems, Man and Cybernetics-Part B*, **26**(1) 29–41.

Dorn, J., Girsch, M., Skele, G., and Slany, W., 1996, Comparison of iterative improvement techniques for schedule optimization. *European Journal of Operations Research*, **94** 349–361.

Fleurent C. and Ferland, J., 1994, Genetic hybrids for the quadratic assignment problem, *DIMACS Series in Math. Theoretical Computer Science*, **16** 190–206.

Glover, F., 1989, Tabu search, Part I, *ORSA Journal on Computing*, **1**(3) 190–206.

Glover, F., 1990, Tabu search, Part II, *ORSA Journal on Computing*, **2**(1) 4–32.

Hansen, P., 1986, The steepest ascent mildest descent heuristic for combinatorial programming, Presented at the *Congress on Numerical Method in Combinatorial Optimization*, Capri, Italy.

Goldberg, D. E., 1989, *Genetic Algorithm in Search, Optimization and Machine Learning*, Addison Wesley, Workingham, England.

Haykin, S., 1994, Neural Networks. *A Comprehensive Foundation.* Macmillan College Publishing Company, NY.

Hopfield, J. J. and Tank, D. W., 1985, Neural computation of decision in optimization problems, *Biological Cybernetics*, **52** 141–152.

Hopfield, J. J. and Tank, D. W., 1986, Computing with neural circuits: A model. *Science*, **233** 625–633.

Kung S. Y., 1993, *Digital Neural Networks.* Prentice Hall, Englewood Cliffs, NJ.

Maren, A., Harston, C., and Pap, R., 1990, *Handbook of Neural Computing Applications.* Academic Press, San Diego, CA.

Onwubolu, G. C., 2000a, Ants can schedule, *Industrial Engineering Research Conference*, Cleveland, Ohio, USA, May 21–23, 2000.

Onwubolu, G. C., 2000b, Ant system approach for the flow shop problem with bicriteria of makespan and total flow time minimization, *From Ant Colonies to Artificial Ants: 2nd International Workshop on Ants Algorithms*, Brussels, Belgium, September 8–9, 2000, 96–99.

Osman, I. H. and Laporte, G., Metaheuristics: A bibliography, *Annals of Operations Research*, **63** 513–623.

Taillard, E., 1990, Some efficient heuristic methods for the flow shop sequencing problem, *European Journal of Operational Research*, **47** 65–74.

Taillard, E., 1991, Robust tabu search for the quadratic assignment problem, *Parallel Computing*, **17** 443–455.

Taillard, E. and Gambardella, L. M., 1997, *Adaptive Memories for the Quadratic Assignment Problem*, IDSIA-87-97 Technical Report, Switzerland.

Tate, D. E. and Smith, A. E., 1995, A genetic approach to the quadratic assignment problem, *Computers and Operational Research*, **1** 855–865.

Udo, G. J. and Gupta, Y. P., 1994, Applications of neural networks in manufacturing management systems, *Production Planning and Control*, **5** 258.

Widmer, M. and Hertz, A., 1989, A new heuristic method for the flow shop sequencing problem, *European Journal of Operational Research*, **41** 186–193.

PART 2

Production Planning and Control Decisions

Production planning decisions are discussed in Chapter 2. These decisions comprise of *planning schedule* and *capacity schedule.* The planning schedule consists of demand forecasting, aggregate production planning, master production schedule, and material requirement planning. The capacity schedule consists of resource capacity planning, rough-cut capacity planning, and capacity requirement planning. Production control decisions are discussed in Chapter 3. These decisions comprise of *planning execution* and *capacity execution.* The planning execution aspect consists of operation sequencing, scheduling, order priorities, stock picking, and capacity control. The capacity execution consists of tool control, machine control, factory order control, inventory control, and capacity control.

Chapter 2

Production Planning Systems

2.1. Introduction

Production planning and control activities deal with the planning and control of manufacturing processes, and therefore include materials, machines, operators, suppliers, customers, and products. The operational tasks of an industrial enterprise can be referred to as the production planning and control system (PPC), and their problems encompass a wide range of fields and industries. Their solution strongly depends on many interrelated aspects such as market environment, production environment, production strategy, human resources, and degree of automation. Figure 2.1 outlines the production and control framework discussed in this chapter and follows the one suggested by Vollman *et al.* (1991).

With the present trend in manufacturing, products have to be delivered at competitive cost, at the required time and in an acceptable quality to the customer. Competitive cost, timely delivery of products and attainment of high quality require proper planning and effective control of work through a manufacturing system. Production planning is complex because it takes into consideration all the various aspects that are necessary in order to achieve the business and strategic plans of a manufacturing firm. Typically, input to a production planning system is in numeric form such as the number of products to be produced or assembled per week. From this, planned order release has to be determined, each job has to be scheduled and work centres have to be loaded. In each stage, a check on capacity is necessary in order to

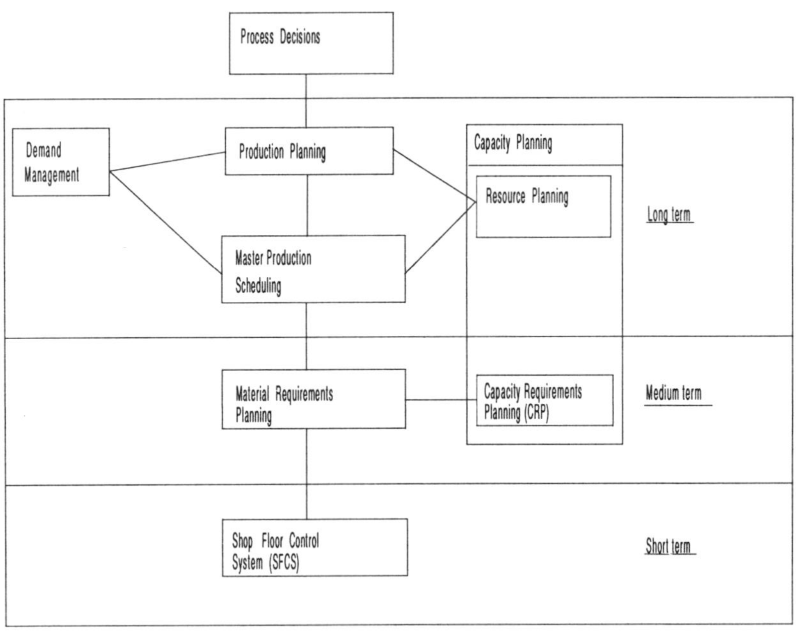

Fig. 2.1. Production planning and control framework.

ensure that equipment and workforce are available to meet production target. Input to a production planning system is composed of forecast from marketing department and also customer orders.

At the top of a production planning and control system is aggregated medium range plan that need to be dis-aggregated into master production schedule. System forecast, customer order, and manual forecast from the marketing/sales department are input into the master production schedule for computing the total demand. The total demand is the net requirement used to drive a material requirement planning system. We now discuss requirement-planning functions in more details.

Production planning and control can be subdivided into planning and execution levels and for each level, we have to consider scheduling and capacity. Schedule determines what is to be produced. The equation has to be balanced by considering capacity, which is the consideration of availability of facilities to meet the production level. Table 2.1 shows one method of classifying production planning and control functions.

Whether at the planning or execution stage, it is important to consider schedule and capacity. It has been observed that most planners ignore planning capacity and execution capacity, and tends to lay emphasis on planning schedule and execution schedule.

The remaining parts of this chapter discuss the various aspects of production planning decisions. We first discuss demand forecasting, aggregate

Table 2.1. Production planning and control components.

Schedule	Capacity	Level
Business plan	Financial planning	Planning
Production planning	Resource requirement plan (RRP)	
Master production schedule (MPS)	Rough cut capacity plan (RCCP)	
Material requirement plan	Capacity requirement plan (CRP)	
Final assembly schedule	Capacity control	
Stock picking schedule	Inventory control	
Order priorities	Factory order control	Execution
Scheduling	Machine (work-centre) control	
Operation sequencing	Tool control	
	Preventive maintenance	

production planning, master production schedule, and material requirement planning. These constitute planning schedule. Then, we discuss resource capacity planning, rough-cut capacity planning, and capacity requirement planning. These constitute capacity schedule.

2.2. Demand Forecasting

Market globalization, increase of quality standards, stronger competition on service and price are driving companies to revise their demand forecasting process. Demand forecasting is the prediction of the future by considering past data. It is an input to production planning, hence accurate methods of forecasting have to be adopted by management in order to achieve a good plan. Forecasting has impact on planning for capacity, schedule and procurement of raw materials. A good sales forecasting system leads to a reduction in production planning disturbances and to an increase in service levels. Traditional quantitative forecasting methods include both time series and causal models. Times series aim at forecasting the future values of a phenomenon based on the historical data of the variable being predicted. Causal models use data from variables other than that being predicted but whose values are linked to that variable by a cause-effect relationship.

Econometric models as well as single and multiple regression methods are examples of causal models. Although these quantitative methods perform well, they suffer from some serious limitations. First, lack of expertise might cause a mis-specification of the functional form linking the independent and dependent variables together, resulting in a peer regression (Hill *et al.*, 1996). Secondly, a large amount of data is often required to guarantee an accurate prediction. Thirdly, non-linear patterns are difficult to capture. Finally, outliers can bias the estimation of the model parameters. Some of these limitations can be overcome by the use of adaptive memory programming (Garetti and Taisch, 1999).

These forecasting techniques are now discussed as important tools for production planning in the remaining part of this section.

2.2.1. *Simple moving average*

When the most recent time periods are used to forecast near future events, significant errors can occur. Simple moving average provides a means of smoothing out undulating data distributions and using this as the forecast for the following period. The forecast for the next period is equal to the average of the most recent number of time periods, which have to be specified. The equation for simple moving average is given as:

$$F_{t+1} = \frac{\sum_{i=0}^{N-1} (X_{t-i})}{N} \tag{1}$$

where F_{t+1} = forecast for time period $(t+1)$, X_{t+1} = actual value at time period, i = time period, and N = number of time periods in the average.

Table 2.2 shows a one-year time period demand and its equivalent three-month simple moving average forecast. The three-month simple moving average forecast is equal to the average of the three previous weeks. For example the forecast of 450 in month 6 is the average of 460, 410 and 480 of months 5, 4, and 3 respectively. The first forecast demand is not available until month 4. Table 2.3 shows the six-month simple moving average for the same observed demand. The forecast of 475 in month 9 is the average of 465, 560, 475, 460, 410, and 480 of months 8, 7, 6, 5, 4, and 3 respectively. The first forecast demand is not available until month 7. The actual demand, three-month, and six-month simple moving averages (SMAs) are plotted in Fig. 2.2.

Table 2.2. Three-month simple moving average.

Week	Observed demand	Forecast demand	Absolute error	Squared error
1	400			
2	420			
3	480			
4	410	433.333	23.333	544.444
5	460	436.666	23.333	544.444
6	475	450	25	625
7	560	448.333	111.66	12469.4
8	465	498.333	33.333	1111.11
9	540	500	40	1600
10	600	521.666	78.333	6136.11
11	700	535	165	27225
12	580	613.333	33.333	1111.11
13		626.666		

Table 2.3. Six-month simple moving average.

Week	Observed demand	Forecast demand	Absolute error	Squared error
1	400			
2	420			
3	480			
4	410			
5	460			
6	475			
7	560	440.833	119.166	14200.7
8	465	467.5	2.5	6.25
9	540	475	65	4225
10	600	485	115	13225
11	700	516.666	183.333	33611.1
12	580	556.666	23.333	544.444
13		574.166		

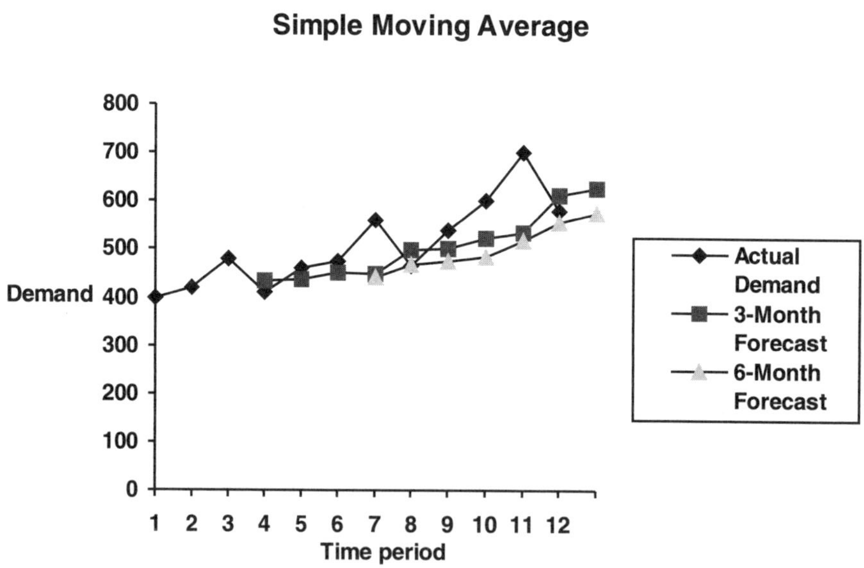

Fig. 2.2. Graphs of actual and forecast demands for SMAs.

2.2.2. *Simple regression*

Simple regression technique is used for predicting future demand based on historical demand. The demand is the dependent variable and the time horizon for planning is the independent variable. The simple-regression is given by the straight-line equation defined as:

$$Y = \alpha + \beta X \tag{2}$$

$$\beta = \frac{n\sum XY - \sum X \sum Y}{n\sum X^2 - (\sum X)^2} \tag{3}$$

$$\alpha = \left(\sum Y - \beta \sum X\right) / n \tag{4}$$

$$r = \frac{n\sum XY - \sum X \sum Y}{[n\sum X^2 - (\sum X)^2][n\sum Y^2 - (\sum Y)^2]} \tag{5}$$

where

X = the independent variable (weeks)
Y = the dependent variable (observed demand)
n = the number of data points
r = the coefficient of correlation.

In the least square regression method, a straight line is drawn through a set of points in such a way as to minimize the square of the deviations of the distance between the observed demand and the forecast demands obtained from the points on the straight line. The degree of deviation is measured using the value of the coefficient of correlation, r.

Table 2.4 shows the observed demand and corresponding forecast demand values using simple regression. The values of the following variables are used to obtain the regression line equation:

$$\sum X = 78$$

$$\sum Y = 6090$$

$$\sum X^2 = 650$$

$$\sum XY = 42630\,.$$

$$\beta = \frac{12(42630) - 78(6090)}{[12(650) - 78^2]} = 21.2937$$

$$\alpha = [6090 - 21.2937(78)]/12 = 369.0909\,.$$

The regression line is: $Y = 369.0909 + 21.2937(X)$.

Using this equation, for the historic data given in Table 2.4, it is now easy to predict the demand of future time periods. For example, the forecast for months 13, and 18 are given by substituting the time periods into the regression line equation. Figure 2.3 shows the graph of the actual and forecast demands for the simple regression data given in Table 2.4.

$$Y_{13} = 369.0909 + 21.2937(13) = 645.9091$$

$$Y_{18} = 369.0909 + 21.2937(18) = 752.3775\,.$$

2.2.3. *Simple exponential smoothing*

The one-period ahead forecast demand for simple exponential smoothing is computed from the combination of the most recent observed demand and the most recent forecast using an associated parameter α with values between 0

Table 2.4. Simple regression.

Week X	Observed demand Y	XY	Forecast demand $Y = \alpha + \beta X$	Absolute error	Squared error
1	400	400	390.3846	9.6154	92.4557
2	420	840	411.6783	8.3217	69.2505
3	480	1440	432.972	47.028	2211.63
4	410	1640	454.2657	44.2658	1959.46
5	460	2300	475.5594	15.5595	242.096
6	475	2850	496.8531	21.8532	477.560
7	560	3920	518.1469	41.8532	1751.69
8	465	3720	539.4406	74.4406	5541.40
9	540	4860	560.7343	20.7343	429.909
10	600	6000	582.028	17.9721	322.994
11	700	7700	603.3217	96.6783	9346.69
12	580	6960	624.6154	44.61542	1990.536

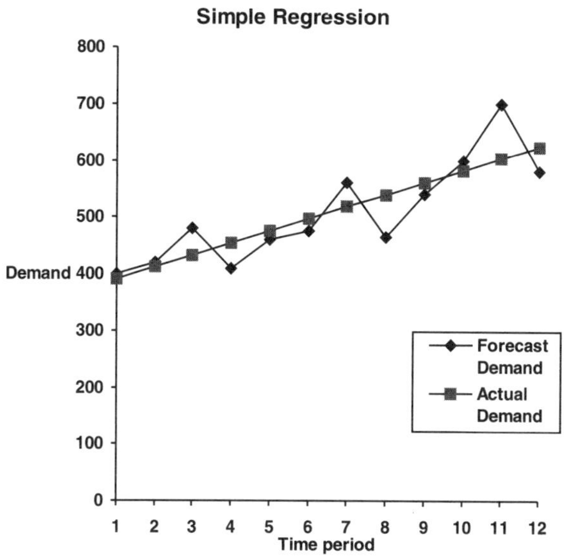

Fig. 2.3. Graphs of actual and forecast demands: SR.

and 1. The equation is given as

$$F_{t+1} = \alpha X_t + (1 - \alpha) F_t \tag{6}$$

where

F_{t+1} = one-period ahead forecast for time $(t + 1)$
F_t = most recent forecast for time t
X_t = observed demand
α = smoothing constant $(0 < \alpha < 1)$.

The formulation for the forecast equation is essentially parametric, with X_t and F_t weighted by the parameter α. For small values of α close to 0, the most recent forecast dominates the one-period ahead forecast. For values close to 1, the most recent observed demand dominates the one-period forecast.

Table 2.5 shows the single exponential smoothing forecast demands for α equal to 0.1, while Table 2.6 shows forecast demands for α equal to 0.9. The graphs for the observed demand and the forecast demands for the two values of

Table 2.5. Simple exponential smoothing: $\alpha = 0.1$.

Week	Observed demand	Forecast demand	Absolute error	Squared error
1	400			
2	420	400	20	400
3	480	402	78	6084
4	410	409.8	0.200	0.0400
5	460	409.82	50.18	2518.04
6	475	414.838	60.162	3619.47
7	560	420.854	139.15	13961.6
8	465	434.768	30.231	913.927
9	540	437.792	102.21	10446.5
10	600	448.013	151.98	23100.1
11	700	463.890	236.78	56068.8
12	580	486.890	93.109	8669.41
13		496.201		

Table 2.6. Simple exponential smoothing: $\alpha = 0.9$.

Week	Observed demand	Forecast demand	Absolute error	Squared error
1	400			
2	420	400	20	400
3	480	418	62	3844
4	410	473.8	63.79	4070.44
5	460	416.38	43.62	1902.704
6	475	455.638	19.362	374.887
7	560	473.064	86.936	7557.901
8	465	551.306	86.306	7448.794
9	540	473.631	66.370	4404.891
10	600	533.363	66.637	4440.485
11	700	593.336	106.66	11377.14
12	580	689.333	109.33	11953.84
13		590.933		

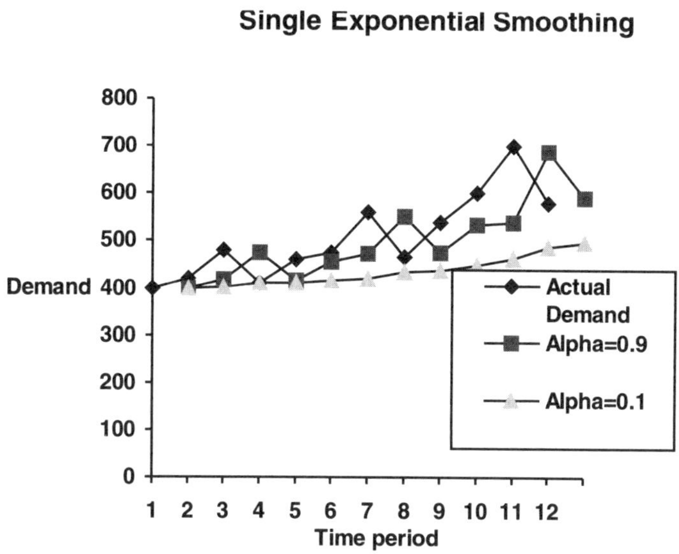

Fig. 2.4. Graphs of actual and forecast demands for SMAs.

α are shown in Fig. 2.4. The graphs show that smaller value of the parameter α gives a smoother forecast, while more variation is observed for larger value. The observed demand for the first week is used to initialize the first forecast in month 2.

2.2.4. *Seasonal exponential smoothing*

As the name implies, seasonal exponential smoothing deals with situations where the observed demand depends on the season of the planning time horizon and some underlying trend in the data. The equations for predicting the seasonal exponential smoothing forecast are:

$$S_{t+1} = \frac{\alpha X_{t+1}}{R_{t+1-L}} + (1-\alpha)(S_t + T_t) \tag{7}$$

$$T_{t+1} = \beta(S_{t+1} - S_t) + (1-\beta)T_t \tag{8}$$

$$R_{t+1} = \frac{\gamma X_{t+1}}{S_{t+1}} + (1-\gamma)R_{t+1-L} \tag{9}$$

$$F_{t+1+m} = (S_{t+1} + T_{t+1}m)R_{t+1-L+m} \tag{10}$$

where

S = smoothed value of the deseasonalized series
T = smoothed value of the trend
R = smoothed value of the randomness seasonal factor
L = length of seasonality
F = forecast demand of m periods after time t.

The forecast demand F is smoothed by three factors R, T, and S, which are associated with randomness, trend, and seasonality respectively. Since the observed demand X_{t+1} depends on the season of the planning horizon, introducing the randomness factor R_{t+1} that parametrically weights the seasonal factor X_t and most recent random factor R_{t+1-L} smoothens this seasonality characteristic by the parameter γ. The trend factor T_{t+1} weights the incremental seaonality factors $S_{t+1} - S_t$ and the recent trend factor using the parameter β. The effect of dividing the observed demand X_{t+1} by R_{t+1-L} is to eliminate

Table 2.7. Seasonal exponential smoothing: $\alpha = 0.30$, $\beta = 0.20$, and $\gamma = 0.10$.

Period (3-months)	Observed demand	Smoothed deseasonalized value S	Smoothed seasonal factor I	Smoothed trend value T	Forecast demand F
1	400		0.9357		
2	420		0.9825		
3	480		1.1228		
4	410	410	0.9421	16.25	
5	460	438.8393	0.9890	18.76786	418.7719
6	475	447.2391	1.1167	16.69424	496.8421
7	560	503.0773	0.9592	24.52304	432.1908
8	465	510.3672	0.9812	21.07641	469.7904
9	540	517.0765	1.1095	18.20298	548.5953
10	600	562.3501	0.9699	23.61711	486.7989
11	700	624.192	0.9953	31.26206	513.9246
12	580	615.641	1.0927	23.30005	599.1265
13					708.9075

Table 2.8. Seasonal exponential smoothing: $\alpha = 0.15$, $\beta = 0.10$, $\gamma = 0.05$.

Period (3-months)	Observed demand	Smoothed deseasonalized value S	Smoothed seasonal factor I	Smoothed trend value T	Forecast demand F
1	400		0.9357		
2	420		0.9825		
3	480		1.1228		
4	410	410	0.9388	16.25	
5	460	432.5446	0.9865	16.87947	418.7719
6	475	445.4675	1.1199	16.48381	496.8421
7	560	482.1261	0.9500	18.50128	430.7154
8	465	496.2373	0.9840	18.06227	468.5909
9	540	509.4772	1.1169	17.58004	550.1909
10	600	542.7335	0.9578	19.14766	482.1354
11	700	584.3026	0.9947	21.3898	515.3879
12	580	592.7272	1.1101	20.09329	603.168
13					684.507

the seasonal fluctuations. The forecast demand is a function of the contributions of the three smoothing factors. Table 2.7 shows the seasonal exponential smoothing forecast demands for $\alpha = 0.3$, $\beta = 0.2$, and $\gamma = 0.1$.

Table 2.8 shows forecast demands for $\alpha = 0.15$, $\beta = 0.10$, and $\gamma = 0.05$. The graphs for the observed demand and the forecast demands for the two sets of values of α, β, and $\gamma =$ are shown in Fig. 2.5. The graphs show that the smaller values of the parameter α, β, and γ give a smoother forecast, while more variation is observed for larger values. For three-month length of seasonality, the observed demand for the first three months are used to initialize the first forecast in month 4.

2.3. Production Planning

Production plan specifies planned rates of production, inventory level, and employee staffing rates and policies to fulfill the charter of a plant. The charter of a plant is the assignment of product families to a specific plant or sub-process within a plant. The environment in which production planning exists comprises of: overall business strategy, human resources policy, financial profile, short-term marketing plan, and product/plant charters. The focus of production

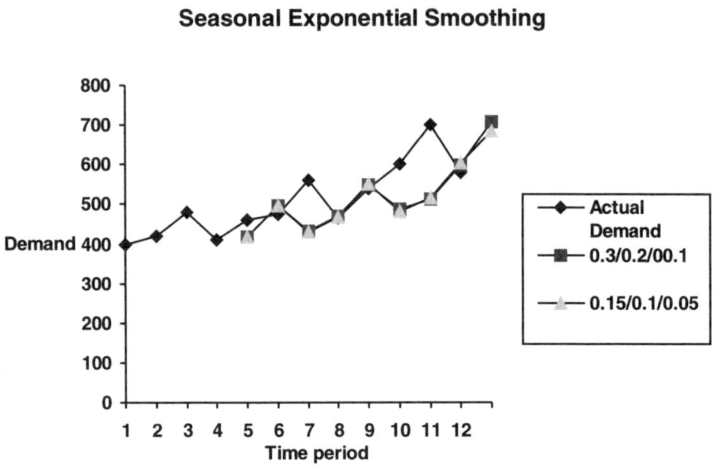

Fig. 2.5. Graphs of actual and forecast demands for SE.

planning is on one product family rather than separate plans for individual combinations of features. For example, production planning might direct the process to make 2000 Toyota Corolla saloon cars rather than 1500 Toyota Corolla 1.3 model and 500 Toyota Corolla 1.5 model. The planning horizon for production planning is usually from 6 months to one and a half years ahead. The production-planning horizon must be sufficiently long to allow decisions, such as the hiring and firing of personnel, to be optimal in the long term and not merely on a short-term basis. The goal of production planning is to define the combination of production rates, employee staffing levels, and inventory patterns that satisfies the needs of the firm, and in addition, reduces the overall costs of manufacturing. Therefore, it is not necessary for planner to waste time searching for optimal arrangements for every resource in an environment subject to change. Emphasis should be placed on searching for a satisfactory plan rather than an ideal one.

For each planning period it is necessary to forecast the sales demand for the aggregated product group and determine the normal production capacity that is available for achieving his demand. The production plan forms the constraints under which detailed scheduling of facilities and personnel then proceeds. Decisions must be made concerning how to correct variances existing between forecast and available manufacturing capacity.

The production planning process is carried out at an aggregated level without the need to provide detailed material and capacity resource requirements for individual products and detailed schedules for facilities and personnel. Consequently, the amount of data used during the planning process is greatly reduced and hence, enables plans to be updated more frequently. The resultant effect is that changes occurring in factors such as forecast sales demand, labour cost rates, production capacity and raw material supply through the supply chain can be readily compensated for. One winning strategy is to focus on bottleneck during production planning. More on this when we deal with resource planning and allocation. If business conditions materially change during the course of a fiscal year, however, updates should incorporate recognized deviations. Planners should communicate any such change to the key individuals in sales, marketing, and finance since major shifts in production planning significantly affect their operations.

The six basic steps involved in the development of a production plan are as follows:

(i) determine relevant company objectives and planning policies;
(ii) determine demand in each planning period;
(iii) determine production capacities for each planning period;
(iv) determine unit costs for regular labour, hiring and firing, over-times, subcontracting, and inventory;
(v) develop alternative plans and choose the minimum;
(vi) select the optimum plan that satisfies the objectives.

Development of the production plan initially commences with the identification of the long-term objectives of the organization and the constraints under which the functional departments within the organization must operate. There are two important issues here: objectives and constraints. Leaving out any of these in planning, defeats the whole purpose of the planning and therefore limits the performance of the organization. Both long-term objectives and constraints have a major effect on the feasibility of a production plan. Fore example, a production plan may require organizations to ensure that customer's orders are always delivered on time, but that inventory levels remain low. Delivery of customer's orders on time is a strategic aim, while keeping low inventory levels is an operational constraint. Often such aim and constraint conflict. For example, in many organizations an overriding constraint when developing production plans is the provision of stable and secure employment for the work force. A stable work-force size would enable stock to be built up during low sales demand periods that may be available during high sales demand periods. However, during the low sales demand periods, there exists a conflict between a low stock level strategy and stable work-force policy.

Sales demand forecasts for each period within the planning horizon for each product-group is determined. Since forecasts of the sales demand for each product-group are inputs to production planning, accurate methods for forecasting should be used by planners. Market globalization, increase of quality standards, stronger competition on service and price are driving companies to revise their demand forecasting process. A good sales forecasting system leads to a reduction in production planning disturbances and to an increase in service levels. Traditional quantitative forecasting methods include both time series and causal models. Times series aim at forecasting the future values of a phenomenon based on the historical data of the variable being predicted. Causal models use data from variables other than that being predicted but whose values are linked to that variable by a cause-effect relationship. Econometric models as well as single and multiple regression methods are examples of causal models. Although these quantitative methods perform well, they suffer from some serious limitations. First, lack of expertise might cause a mis-specification of the functional form linking the independent and dependent variables together, resulting in a peer regression (Hill *et al.*, 1996). Secondly, a large amount of data is often required to guarantee an accurate prediction. Thirdly, non-linear patterns are difficult to capture. Finally, outliers can bias the estimation of the model parameters. Some of these limitations can be overcome by the use of intelligent optimization techniques (Garetti and Taisch, 1999).

For each planning period and each product group it is then necessary to estimate the level of manufacturing resources required to produce the forecast volume of finished goods. In production planning process, individual products are aggregated into groups according to similarities using such criteria as the types of processing operations required, production times required, types of labour skills required or the fact that each product in the group makes use of specific limited resources. The objective is to allow the capacity requirements of a product group to be measured in a common unit, such as man-hours of production time on a specific time of processing equipment. This common resource units is used to estimate the level of manufacturing resources required to produce the forecast volume of finished goods earlier mentioned. When differences exist between required and available capacity then one or more of the following capacity management techniques should be used to achieve a balance:

(1) recruit/fire staff/redundancy;
(2) work overtime/shorter hours;
(3) vary levels of inventory held;
(4) vary lot sizes processed;
(5) subcontract work;
(6) transfer multi-skilled labour between work areas.

The essential problems involved in generating an optimal and feasible production plan include choosing which technique to use, when to commence using the technique, the number of planning periods over which it will be used and the amount of production capacity to add or remove.

Production planning is essentially a stochastic process since the aim is to plan for events that will take place in the near future. Between when the plan is made and when it is carried out, several events would have taken place. For example, worker morale and motivation, level of operator experience and the processing reliability of equipment, which affect the initial estimates of the available resources would have changed. Moreover, the accuracy of sales forecasts would have deteriorated further into the future. The majority of these factors are themselves affected by production planning. For example, equipment availability is greatly influenced by productive maintenance policies, and operator morale is greatly influenced by hiring and firing policies. The stochastic (unpredictable) nature of these variables results in a plant's manager having to consider the trade-off between costs of holding excess inventories and the cost of lost sales. Holding excess inventories occur when demand is lower than forecast, and lost sales occur when demand is greater than forecast. The costs for each of these situations have some consequences, for example obsolescence costs, interest charges, warehousing costs are attributed to holding excess inventories, while lack of revenue and lost customers are attributed to lost sales. The plant manager must therefore, consider these relative costs when deciding to carry stock from one planning period to the next.

The costs must be determined for each alternative method that could be used to vary capacity in order to satisfy the customer. In practice the options available for varying capacity will be limited by the need to meet management policies. Some management policies might be: fulfilling customer's orders within a specific time period, allowing only an occasional stock-out, or never running out of stock. Costs involved in production planning are divided into two categories: (i) quantifiable and (ii) non-quantifiable. Quantifiable costs include: inventory, overtime, subcontracting, and shift premium, recruitment, redundancy, training, and purchasing. Non-quantifiable costs include: stock-outs, loss of orders, low workers' morale, inexperienced personnel, handling complaints, and loss of efficiency

From the several alternative plans, planners should choose the optimum that satisfies the firm's objectives. The search for a satisfactory plan rather than an ideal plan should be emphasized. Planners should not waste resources in a futile search for optimal arrangements for every resource in an environment that is subject to change. Others argue that the goal of minimizing costs should guide production planning. This view is in order, except that it should be noted that the value of the goods or services delivered to a customer is perceived in terms of cost, quality, time and flexibility. Therefore, making choice of the best production plan solely on the basis of cost is questionable. It is acceptable as long as it does not jeopardize the other elements of the value equation as already explained.

Three pure strategies for accommodating mismatch between market demand and production capabilities include:

Chase strategy: This strategy sets the production rate equal to the market demand rate. It then increases and decreases the plant's pool of human resources as needed, usually through hiring and firing permanent workers, and may be augmented by temporary workers.

Level-production strategy: This strategy sets the production rate equal to the market demand rate. While human resources remain at a constant level, the strategy then accumulates inventory during slack demand period and distributes goods from inventory during peak demand periods.

Variable-hour strategy: This strategy sets the production rate equal to the market demand rate, while keeping staffing level stable. It then compensates market demand variations by adjusting the work hours of stable groups of employees.

Example 1

A manufacturing company that produces mining equipment has sales forecast for 6 months period as shown in Table 2.9 and costs were obtained from labour

Table 2.9. Demand forecast for half-year period.

Period	Forecast demand	Number of working days
January	3000	21
February	3200	19
March	3000	22
April	2000	21
May	2800	21
June	3600	22

intensive study as shown. The firm's management decides to keep 40 workers during low demand in April (flexible). As an operations-manager, decide on the strategy to adopt using chase, labour production, and flexible working hour strategies for aggregate production planning.

Further labour-intensive study has yielded the following costs estimates:

Labour hrs/unit	5 hours
Labour cost	$10 per hour
Overtime premium	25%
Hiring Cost	$400 per hour
Firing costs	$250 per fire
Beginning inventory	1000 units
Beginning employment level	75 employees
Inventory holding costs	$8 per period
Material cost per unit	$95

2.3.1. *Chase strategy*

The following steps are used when planning using this strategy:

(1) each period has specified (given) beginning inventory
(2) set the forecast demand rate equal to the production rate
(3) hours required = production rate * given labour hours required per unit
(4) workers required = hours required/(workdays * number of hours/day)

Step by step working for the month of January:

(1) Beginning inventory is given. All periods have this inventory

Table 2.10. The chase strategy worksheet.

Period	Start inv.	Demand	Prod.	Hrs. req.	Work days	Inv. cost	Works req.	Reg. time work	O/T work	Labour cost	Hiring cost	Firing cost
Jan	1000	3000	3000	15 000	21	$8000	89.3	89	0	149 520	5600	0
Feb	1000	3200	3200	16 000	19	$8000	105.2	105	0	159 600	6400	0
Mar	1000	3000	3000	15 000	22	$8000	85.2	82	0	144 320	0	5750
April	1000	2000	2000	10 000	21	$8000	59.5	60	0	100 800	0	5500
May	1000	28 000	2800	14 000	21	$8000	83.3	83	0	139 440	9200	0
June	1000	36 000	3600	18 000	22	$8000	102.2	102	0	179 520	7600	0
						$48 000				$873 200	$28 800	$11 250

(2) Demand is given
(3) Make production = demand
(4) Hours required = production × labour hrs/unit
= 3000*5 = 15 000
(5) Number of working days is given
(6) Inventory cost = beginning inventory * inventory holding cost
= 1000*$8 = $8000
(7) Workers required = hours required/(working days) hrs/day
= 15 000/21*8 = 89.3
(8) Regular time worker = round up solution in Step (7) = 89
(9) No overtime
(10) Labour cost = Regular time workers * Work days * Hrs/day * cost of labour
= 89*21*8*$10 = $149 520
(11) Hiring cost = $C_{\text{hire}}(W_t - W_{t-1})$ = $400*(89 – 75)
= $5600
(12) Firing Cost = $C_{\text{fire}}(W_{t-1} - W_t)$ = $250*(0 – 89) = 0

Total cost = inventory cost + production cost + hiring cost + firing cost
Total cost = $(48 000 + 873 200 + 28 800 + 11 250) = $961 250.

2.3.2. *Level-production strategy*

The following steps are used when planning using this strategy:

(1) set production rate = average of demand rate for all periods
(2) current inventory = (production + previous inventory – demand)
(3) workers required as in Chase strategy
(4) regular time workers = sum of all hours required/sum of workdays for all periods

Step by step working for the month of January:

(1) Beginning inventory for January is given = 1000 units
(otherwise it would have been, 2933 – 3000 = 67)
For February, inventory = 1000 + 2933 – 3000 = 933
(2) Demand is given
(3) Production = $\frac{1}{n}\sum$ Demand = 17600/6 = 2933 unit for all period

Table 2.11. The level-production strategy worksheet.

Period	Start inv.	Demand	Prod.	Hrs req.	Work days	Inv. cost	Works req.	Reg. time work	O/T work	Labour cost	Hiring cost	Firing cost
Jan	1000	3000	2933	14 665	21	$8000	87.3	87	0.3	$146 790	$4800	0
Feb	933	3200	2933	14 665	19	$7464	96.5	87	9.5	$150 290	0	0
Mar	666	3000	2933	14 665	22	$5325	83.3	87	0	$146 160	0	0
April	599	2000	2933	14 665	21	$4792	87.3	87	0.3	$146 790	0	0
May	1532	2800	2933	14 665	22	$12 256	83.3	87	0	$146 160	0	0
June	1665	3600	2933	14 665	22	$13 320	83.3	87	0	$146 160	0	0
					127	$51 160				$882 350	$4800	$0

(4) Hours required = production × \$5 hrs per unit
= 2933*5 = \$14 665 for all work days given
(5) Inventory cost = beginning inventory × inventory holding cost
= 1000*\$8 = \$8000
(6) Workers required = Hrs required/(Work days *8 hrs/day)
= 14 665/(21*8) = 87.3
(7) Regular time workers = (hrs required/(work days
= 14 665*6/127 = 87 for all periods
(8) Overtime workers required = Workers required-regular time workers
= 87.3 − 87 = 0.3
(9) Labour cost = [Regular worker required * daily rate + overtime workers required * daily rate × premium] * Workday * hrs/day = [87*10 + 0.3*10*125%]*21*8 = \$146 790
(10) Hiring cost = $C_{\text{hire}}(W_t - W_{t-1})$ = \$400*(87 − 75) = \$4800 for January, all others zero
(11) Firing cost = $C_{\text{fire}}(W_{t-1} - W_t)$ = \$250*(75 − 87) = 0 for January and all others

Total cost = [inventory + labour + hiring + firing] cost = \$[51 160 + 882 350 + 4800 + 0] = \$938 310.

2.3.3. *Flexible working hours strategy*

The following steps are used when planning using this strategy:

(1) production = beginning inventory × working days × hours perday/ (labour hours per unit)
(2) current inventory = previous (production + inventory − demand)
(3) workers required for period of low demand is decided by management in agreement with employees; some workers may decide to take leave
(4) management decides regular-workers
(5) temporary-workers can then be employed to meet demand

Step by step working for the month of January:

(1) Beginning inventory for January is given
(2) Demand is given
(3) Production is computed 1st in this strategy = Beginning employee level × Weekdays * hrs/day × hrs/unit = 75*21*8/5 = 2520 in January

Table 2.12. The flexible working hours strategy worksheet.

Period	Start inv.	Demand	Prod.	Hrs req.	Work days	Inv. cost	Works req.	Reg. time work	O/T work	Labour cost	Hiring cost	Firing cost
Jan	1000	3000	2520	12 600	21	$8000	75	65	10	$130 200	$0	$2500
Feb	520	3200	2280	11 400	19	$4160	75	65	10	$117 800	$0	$0
Mar	0	3000	2640	13 200	22	$0	75	65	10	$136 400	$0	$0
April	0	2000	1344	16 720	21	$0	40	40	0	$67 200	$0	$0
May	0	2800	2520	12 600	21	$0	75	65	10	$130 200	$0	$0
June	0	3600	2640	13 200	22	$0	75	65	10	$136 400	$0	$0
					127	$12 160				$718 200	$0	$2500

(4) Hrs required = beginning employee level * wk days * hrs/day
= 75*21*8 = 12 600 in January
(5) Workdays are given
(6) Inventory cost = beginning inventory * hold inventory cost = 1000 + 2520 − 3000 = 520 for February
(7) Workers required = hrs. required/(Workdays × 8 hrs/day) = 12 600/ (21*8) = 75
(8) Regular workers = Number that firm decides to keep, while temporary workers are employed (overtime)
(9) Overtime is difference between (7) and (8)
(10) Cost for (9) taken into consideration premium rate for overtime
(11) No hiring available
(12) Fire cost due to (75 − 65) in January

Total cost = $(12 160 + 718 200 + 0 + 2500) = $732 860.

Decision

The total costs are tabulated for the three strategies. If cost is the only basis for making decision, the strategy having the least cost is chosen. Otherwise, other factors must be considered as already discussed.

	Total Costs	Remarks
• Chase strategy	$961 250	Highest
• Level prod. strategy	$938 310	Medium
• Flexible working hr.	$732 860	Lowest

2.4. Master Production Schedule

The master production schedule (MPS) is a definite statement of exactly what the company plans to produce in each time period. It is usually a statement of production within time periods of weeks. Aggregated plan, which is an input to the MPS is disaggregated. Three types of inputs to the MPS are worth discussion: manual forecasts, system forecast and customer order.

Manual forecasts are performed by the sales or marketing department. They are input directly into the MPS system using standard techniques such as simple regression, moving average, single exponential smoothing or seasonal exponential smoothing already discussed.

Customer orders are received by the sales or marketing department. They are input on-line into the MPS system. These orders are the actual customer demand and the time leads are accepted as fixed.

System forecast is a buffer and in particular it is predominant in long range planning horizon. During the short range planning horizon, the effects of customer orders and the forecasts must be taken into account in order to obtain an accurate input into the MPS.

Total demand is computed from the manual forecast, system forecast and customer orders using some rules. Customer-orders take precedence over manual and system forecasts because they represent the actual customer order. Manual forecast takes precedence over system forecast because it gives an indication of the trend of market demand. Based on these criteria, we now set the rules for total demand and formulations for generating the MPS, projected inventory (PI) and available to promise (ATP), all of which are held in a master production schedule record.

Rule for Total Demand (TD)

By defining CD = customer demand; MF = manual forecast; SF = system forecast, the total demand (TD) can be given as:

$$TD = \begin{cases} CD \text{ if } (MF < CD < SF) \\ MF \text{ if } (CD < SF, MF) \text{ and } (MF > 0) \\ SF \text{ if } (MF = 0)\,. \end{cases} \tag{11}$$

The rule encourages planning for excess production because the total demand equals customer demand only if customer demand lies between the extremes of manual forecast by marketing or sales department and system forecast using forecasting techniques. In most cases, there will be excess forecasts and following the rule, excess production is planned for. This procedure is not acceptable in JIT operations where a manufacturing firm produces exactly the customer demand at the right time.

2.4.1. *Calculating the Master Production Schedule (MPS)*

The MPS is a firm production order, which combines the just in case (JIC) stock, the total demand (TD) and projected inventory (PI). For a time bucket i,

$$MPS_i = \max(JIC + TD_i - PI_{i-1};\ 0)\,. \tag{12}$$

Again the just in case (JIC) stock adds to inventory which is unacceptable in JIT philosophy. This means that for JIT operations we must *set* $JIC = 0$. It is easy to compute the MPS for week 1 if we know the projected inventory before then.

Projected Inventory (PI)

The projected inventory is the expected number of completed items held at the end of each time bucket. It combines the previous projected inventory and present MPS less the total demand for a time bucket i,

$$PI_i = PI_{i-1} + MPS_i - TD_i\,. \tag{13}$$

Available To Promise (ATP)

The available to promise represents the uncommitted quantity of products available to a customer. It combines the previous projected inventory and present MPS less the customer order (CO) for a time bucket i,

$$ATP_i = PI_{i-1} + MPS_i - CO_i \tag{14}$$

alternatively, $ATP_i = PI_i + TD_i - CO_i$.

Considering their customer orders, manual forecast and system forecast for a 10-week planning horizon given in Table 2.13, it is now an easy task to generate a complete MPS record containing additional entries such as TD, MPS, PI, ATP and cumulative ATP. The total demand computed is the net requirement of the MPS. The available to promise corresponds to the scheduled receipt and the MPS is the planned order receipt.

Table 2.13. Customer orders and manual/system forecasts.

Item	Part number									
Week number	1	2	3	4	5	6	7	8	9	10
Customer order	60	65	25	22	35	10	15			
Manual forecast	55	65	45	65	65	60				
System forecast	65	65	65	65	65	65	65	65	65	65
Just in case stock = 10										
Initial projected inventory = 65										

Example 2

The customer orders, manual forecast and system forecast for a 10-week planning horizon are given in Table 2.13. Generate a complete MPS record.

For weeks 1 and 2, Eqs. (11)–(14) are used to evaluate the following:,

Total demand	TD_1	$= 60$
	TD_2	$= 65$
Master production	MPS_1	$= \max(10 + 60 - 65, 0) = \max(+5, 0) = 5$
schedule	MPS_2	$= \max(10 + 65 - 10, 0) = \max(65, 0) = 65$
Project Inventory	PI_1	$= 65 + 5 - 60 = 10$
	PI_2	$= 10 + 65 - 65 = 10$
Available to Promise	ATP_1	$= 10 + 60 - 60 = 10$
	ATP_2	$= 10 + 65 - 65 = 10$

Continuing the computation yields the complete MPS record in Table 2.14.

Clearly, manual method will be cumbersome and error-prone. The MPS is the driver of the production planning and control system and any error made here has adverse overall effect on the system. A computer-aided MPS record will be useful in terms of accurate dates.

The information on the MPS record shows that more accurate total demand, MPS, PI, ATP are obtained for short range planning horizon because the effects of customer orders, manual forecast from the sales or marketing

Table 2.14. MPS record for customer orders/forecasts for Table 2.13.

Item					Part number					
Week number	1	2	3	4	5	6	7	8	9	10
Customer order	60	65	25	22	35	10	15			
Manual forecast	55	65	45	65	65	60				
System forecast	65	65	65	65	65	65	65	65	65	65
Total demand TD	60	65	45	65	65	60	65	65	65	65
Master Production Schedule	5	65	45	65	65	60	65	65	65	65
Proj. inven. PI ($PI_0 = 65$)	10	10	10	10	10	10	10	10	10	10
Available to promise ATP	10	10	20	43	30	50	50	65	65	65
Cumulated ATP	10	10	30	73	103	153	203	268	333	398
Just in case stock = 10										

department, and system forecast are taken into consideration. For long-range planning horizon, the system forecast dominates. It is suggested therefore that short-range planning horizon be considered from time to time. This means computing the MPS records after every 4 weeks or so.

2.5. Material Requirement Planning (MRP)

Material requirement planning (MRP) is a process that explodes the master production schedule (MPS) requirements right down through the bill of materials (BOM). It takes into account inventory records and component lead times data in MPS record in order to determine the quantity of orders and when they must be released to replenish inventories of finished parts or raw materials. MRP is a priority planning system because it determines requirement without necessary taking into consideration capacity constraints. An MRP system is driven by the MPS which describes the requirement at top level items. Cascading through BOM, material requirements are determined for assemblies, subassemblies and components that are to be manufactured or purchased at the lower levels.

BOM, is a product structure that can be represented as a tree, with the finished product at the root node (level 0). The immediate subassemblies constitute children-nodes and are linked to the root node. Each subassembly may have their children nodes and this continues until the leave-nodes are reached where components represented by the nodes are either manufactured in-house or purchased. BOM is best understood by example.

The major information needed before material requirements planning (MRP) system can begin processing are the master parts data, the master production schedule (already discussed), the inventory data, and the open order data. These are explained and the respective information for an example will be given.

The *master part data* contains the various levels of each assembly, subassemblies, components, their part numbers, lot size, lead times, description of the items and make/buy code. The *inventory data* holds the current inventory that is the exact amount of material in stock, and allocated which represents the quantity of physical stock already committed for release order but awaiting to be issued. The *open orders data* contain the scheduled receipts that are orders already placed with parts suppliers, but are not yet received. The time of their arrival is known so they are slotted into the time period.

We will define some further terms that are required for MRP system computation and then present a procedure for the computations. MRP record contains gross requirement, scheduled receipt, projected inventory or available, net requirements, planned receipts and planned order release. The terms that have not been already dealt with are now defined.

Gross requirements are entries from the MPS representing the total quantity needed for a portable product in each time bucket of the planning horizon. *Net requirement* is an indication of whether the projected inventory (available) can or cannot cover the gross requirement. If the projected inventory exceeds the gross requirement then no net requirement is needed. If the projected inventory (available) is less than the gross requirement then there is shortage and the net requirements indicates the number that is short which has to be replenished. *Planned receipts* show when orders must arrive in order to avoid shortage problem. *Planned order releases* show when the orders have to be placed to arrive at the correct time. This is obtained by offsetting the planned receipts by the lead-time for the item being considered.

Now that the terminology has been given, we present a procedure for the MRP operation and an example so that readers can follow the preceding discussions. The following variables are used to define terms for the MRP procedure:

$$
\begin{aligned}
i &= \text{time bucket}\\
GR_i &= \text{gross requirement}\\
P_i &= \text{projected inventory}\\
SR_i &= \text{scheduled receipt}\\
PR_i &= \text{planned receipt}\\
NR_i &= \text{net requirement.}
\end{aligned}
$$

The procedure for MRP operation is given in Fig. 2.6. let us now consider an example to see how the MRP logic works.

Example 3

A subassembly is at the next lower level to the product whose master parts data are shown in Table 2.15. There are 4 parts in the subassembly. Lead-time

```
for each level in BOM
for each time bucket i
        Del_i = P_i = GR_i
        if Del_i > 0 then begin
           {P_{i+1} = Del_i + SR_{i+1}
           NR_i = 0
           PR_i = 0 }end;
        if Del_i < 0 then begin
           { NR_i = ABS(Del_i);
           PR_i = NR_i ;(or specified value)
           PR_{i+1} = PR_i - NR_i + SR_{i+1} }end;
end;
end;
```

Fig. 2.6. MRP logic.

Table 2.15. Master parts data.

Week	1	2	3	4	5	6	8
FP-500	50	100	150	50	100	80	75

is 1 week, current inventory is 150, allocated is 50, and scheduled receipt is 500 at the beginning of the first week. Determine the MRP information using the procedure presented in this section.

Solution

Multiplying the master part data by 4, we obtain the gross requirements for the subassembly. Applying the procedure for MRP:

$$\begin{aligned} P_1 &= \text{current inventory} - \text{allocated} + \text{schedule receipt.} \\ &= 150 - 50 + 500 = 600 \\ Del_1 &= 600 - 200 = 400 > 0; \text{ then } P_2 = 400 + 0 = 400 \text{ and } NR_1 = 0 \\ Del_2 &= 400 - 400 = 0; \text{ then } P_3 = 0 + 0 = 0 \text{ and } NR_2 = 0 \end{aligned}$$

$Del_3 = 0 - 600 = -600 < 0$; then $NR_3 = 600$, $PR3 = 600$, $P_4 = 600 - 600 + 0 = 0$

$Del_4 = 0 - 200 = -200 < 0$; then $NR_4 = 200$, $PR4 = 200$, $P_5 = 200 - 200 + 0 = 0$

$Del_5 = 0 - 400 = -400 < 0$; then $NR_5 = 400$, $PR5 = 400$, $P_6 = 400 - 400 + 0 = 0$

$Del_6 = 0 - 320 = -320 < 0$; then $NR_6 = 320$, $PR6 = 320$, $P_7 = 320 - 320 + 0 = 0$

$Del_7 = 0 - 300 = -300 < 0$; then $NR_7 = 300$, $PR7 = 300$, $P_8 = 300 - 300 + 0 = 0$.

The full MRP records are given in Table 2.16.

Table 2.16. MRP information.

Item: sub-assembly	Part number: FP-500						
Week	1	2	3	4	5	6	7
Gross requirement	200	400	600	200	400	320	300
Scheduled receipt	500						
Projected inventory	600	400					
Net requirements			600	200	400	320	300
Planned receipts	600	200	400	320	300		
Planned release		600	200	400	320	300	

Example 4

Consider a simple case of the manufacture of trolley used in a warehouse. The trolley is made up of a handle, a steel body, four wheels and 2 axles. In Fig. 2.7 the warehouse trolley (at level 0) is disassembled into five subassemblies, namely the handle, body, wheels, axles and brackets at level 1. The body is further disassembled in level 2 into the body-base, body-ends and body-sides. Since the body is manufactured in-house, steel sheet will be required from the store. The master parts data, the master production schedule, the inventory data, and the open order data are shown in Tables 2.17, 2.18, 2.19, and 2.20 respectively. Carry out a full MRP record for the warehouse trolley.

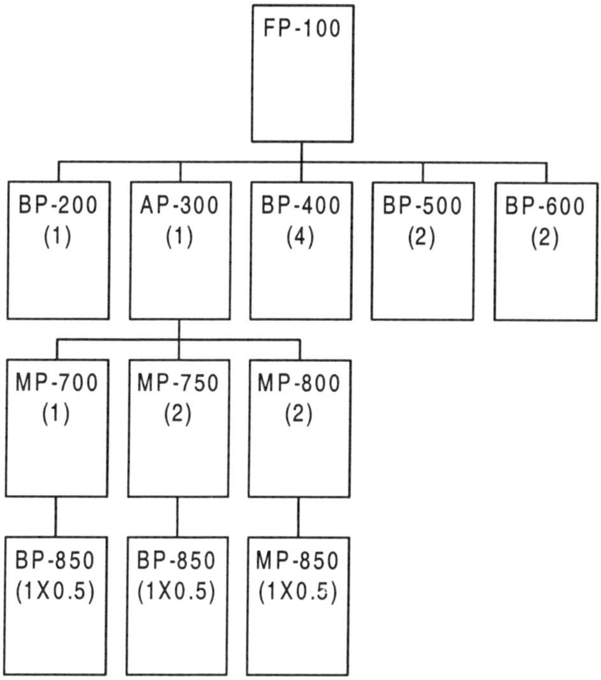

Fig. 2.7. Explosion of warehouse trolley into subassemblies and parts.

Table 2.17. Master parts data for warehouse trolley.

Level	Part number	Lot	Lead size	Description time	Make/ buy
0	FP-100	L	1	warehouse trolley	M
1	BP-200	L	2	handle	B
1	AP-300	L	2	body	M
1	BP-400	L	2	wheels	B
1	BP-500	L	2	axles	B
1	BP-600	L	2	brackets	B
2	MP-700	L	2	body base	M
2	MP-750	L	2	body ends	M
2	MP-800	L	2	body sides	M
2	MP-850	L	2	metal sheets	B

Table 2.18. Master production schedule for warehouse trolley.

Week	1	2	3	4	5	6	7	8
FP-100	—	—	150	50	100	80	75	100

Table 2.19. Inventory data.

Part number	Current inventory	Allocated
FP-100	15	
BP-200	100	
AP-300	200	50
BP-400	150	50
BP-500	600	
BP-600	400	
MP-700	2500	
MP-750	350	
MP-800	100	
MP-850	1500	

Table 2.20. Open order data.

Part number	Scheduled receipt	Due date
FP-100	10	1
BP-200	50	2
AP-300	50	1
BP-400	500	1
BP-500	500	2
BP-600	300	2
MP-700	2000	2
MP-750	500	2
MP-800	400	1
MP-850	1000	1

Solution

We now apply the procedure to obtain the full MRP record for all items in thc BOM for the warehouse trolley. Starting from the level 0 the assembled warehouse trolley is first considered (see Table 2.21). The gross requirement corresponds to the master production schedule already given in Table 2.18.

Table 2.21. Record for warehouse trolley.

Item: warehouse trolley			Part number: FP-100					
Week	1	2	3	4	5	6	7	8
Gross requirement	—	—	150	50	100	80	75	100
Scheduled receipt	10							
Projected inventory 5	25	25	25					
Net requirements			125	50	100	80	75	100
Planned receipts			125	50	100	80	75	100
Planned release	—	125	50	100	80	75	100	

The inventory data in Table 2.19 shows that there are 15 warehouse trolleys held in stock. The open order data in Table 2.20 shows that 10 units scheduled are to be received into stock in week 1. The lead is 1 week (see Table 2.17), and the planned order release is shown in Table 2.21.

Moving to level 1, there are 1 handle, 1 body, 4 wheels, 2 axles and 4 brackets with part numbers: BP-200, P-300, BP-400, BP-500 and BP-600; these are dealt with in turn. The gross requirement for this level is equivalent for the planned order release for level 0 item FP-100. There are four wheels, hence the planned order release of level 0 is multiplied by 4 to obtain the gross requirement for this item.

The gross requirement for the axles are obtained by multiplying the planned order release of level 0 by 2, while those for the brackets are obtained by multiplying the planned order release of level 0 by 4. The gross requirements and the planned order release results are shown for items in level 1 are shown in Tables 2.22, 2.23, 2.24, 2.25, and 2.26.

Table 2.22. Record for handle.

Item: handle			Part number: BP-200					
Week	1	2	3	4	5	6	7	8
Gross requirement	—	125	50	100	80	75	100	
Scheduled receipt		50						
Projected inventory 100	100	150	25					
Net requirements			25	100	80	75	100	
Planned receipts			25	100	80	75	100	
Planned release	25	100	80	75	100			

Table 2.23. Record for body.

Item: body				Part number: BP-300				
Week	1	2	3	4	5	6	7	8
Gross requirement	—	125	50	100	80	75	100	
Scheduled receipt	50							
Projected inventory 150	200	200	75	25				
Net requirements				75	80	75	100	
Planned receipts				75	80	75	100	
Planned release	—	75	80	75	10			
						0		

Table 2.24. Record for wheels.

Item: wheels				Part number: BP-400				
Week	1	2	3	4	5	6	7	8
Gross requirement		500	200	400	320	300	400	
Scheduled receipt	500							
Projected inventory 100	600	600	100					
Net requirements			100	400	320	300	400	
Planned receipts			100	400	320	300	400	
Planned release	100	400	320	300	400			

Table 2.25. Record for axles.

Item: axles				Part number: BP-500				
Week	1	2	3	4	5	6	7	8
Gross requirement		250	100	200	160	150	200	
Scheduled receipt		500						
Projected inventory 600	600	1100	850	750	550	390	240	
Net requirements		0	0	0	0	0	0	
Planned receipts		0	0	0	0	0	0	
Planned release		0	0	0	0	0	0	

Table 2.26. Record for brackets.

Item: brackets				Part number: BP-600				
Week	1	2	3	4	5	6	7	8
Gross requirement		500	200	400	320	300	400	
Scheduled receipt		300						
Projected inventory 400	400	700	200					
Net requirements				400	320	300	400	
Planned receipts				400	320	300	400	
Planned release		400	320	300	400			

Table 2.27. Record for body base.

Item: body base				Part number: MP-700				
Week	1	2	3	4	5	6	7	8
Gross requirement		75	80	75	100			
Scheduled receipt		2000						
Projected inventory 2500	250	4500	4425	4345	4270	4190	4190	4190
Net requirements	0	0	0	0	0	0	0	0
Planned receipts	0	0	0	0	0	0	0	0
Planned release	0	0	0	0	0	0	0	0

Table 2.28. Record for body ends.

Item: body ends				Part number: MP-750				
Week	1	2	3	4	5	6	7	8
Gross requirement		150	160	150	200			
Scheduled receipt								
Projected inventory 350	350	350	200					
Net requirements			110	200				
Planned receipts			110	200				
Planned release		110	200					

Table 2.29. Record for body sides.

Item: body sides		Part number: MP-800						
Week	1	2	3	4	5	6	7	8
Gross requirement		150	160	150	200			
Scheduled receipt	400							
Projected inventory 100	500	500	350	190	40			
Net requirements					160			
Planned receipts					160			
Planned release	—	—	160					

Considering items in level 2, it is found that they are linked to the body (AP-300) of level 1 hence their gross requirements are derived from the planned release order of AP-300 in Table 2.23.

There are one body base, two body ends and two body sides. The gross requirements for the body base is equivalent to the planned order release of item AP-300 while those for the body ends and body sides are equal to twice the planned order release of AP-300. The gross requirements and planned order releases for items in level 2 are shown in Tables 2.27, 2.28, and 2.29.

The last level is the metal sheet (BP-850). The three items in level 2 are constructed from metal sheet, hence its gross requirements is a constituent of the planned order releases of the body base, body ends and body sides. There are steel sheets of 0.5 width in stock. The base is made from 1 m length, the sides from 1 m length and the sides from 0.5 m length. The gross requirements for the steel sheet is therefore obtained by using the result of multiplying the dimension vector and the number of items vector respectively: (1 0.5 1).(1 2 2) = (1 1 2). This means that the planned order release for the body base is added to that of the body ends and the result added to twice the planned order release of the body sides. The planned order release for the steel sheet of 0.5 m standard width is shown in Table 2.30.

The example gives a step by step description of MRP computation. Free issue items such as nuts, bolts, screws are not included because they are easily available.

2.5.1. *Material Resources Planning (MRP II)*

Material Resources Planning (MRP II) is an extension of the closed loop material requirement planning, MRP. We have treated the various functions of

Table 2.30. Record for steel sheets.

Item: steel sheets			Part number: BP-850					
Week	1	2	3	4	5	6	7	8
Gross requirement		130	340	75	100			
Scheduled receipt	1000							
Projected inventory 1500	2500	2370	2030	1955	1855			
Net requirements	0	0	0	0	0			
Planned receipts	0	0	0	0	0			
Planned release	0	0	0	0	0			

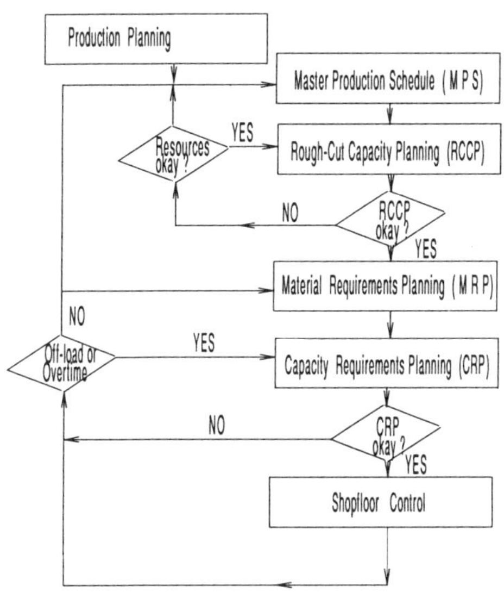

Fig. 2.8. Closed-loop MRP.

MRP, most of which are performed by the computer. A closed loop MRP system determines whether the output from MRP system is feasible or not by checking that there is sufficient capacity. A closed loop MRP is shown in Fig. 2.8. After the capacity requirement planning (CRP) is complete, we must provide an answer to whether capacity is adequate or not in order to determine whether or not to continue to schedule work centres and workforce for production. If capacity is adequate we proceed to scheduling, otherwise we move upstream to determine whether to off-load or plan over time. If any of these actions is planned, we must do a capacity requirement planning. If none of these actions is feasible, then a material requirement planning is done all over, upstream. If the MRP at this stage is not feasible, operation department has no choice but to go back to master production scheduling. The iteration within the closed loop is continued until the rough cut capacity planning confirms that the master production schedule is adequate and the capacity requirement planning confirms that the material requirement planning is adequate.

MRP II includes the business plan at the top and the closed loop MRP at the planning level, together with purchasing, shop flow control and performance measurement at the execution level. But there exists an external framework that affects MRP II actions within the operations function of an organization. This framework includes marketing, finance and human resources. Figure 2.9 shows the functions of an MRP II system and their interrelationship with marketing, finance, human resources and engineering.

It will be useful to know how MRP II interacts with the external framework. Marketing department may carry out a sales promotion, which may result in increase in customer's open orders. Operations department has to be informed of the demand and it must take necessary actions to respond. The actions may

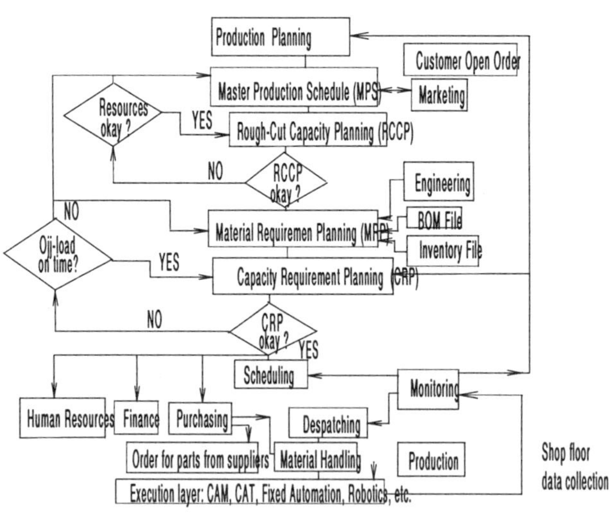

Fig. 2.9. Closed-loop MRP II functions.

include going through the total MRP closed loop and deciding if capacity is available to add the increase in customer demand to existing demand in the factory. It may happen that capacity is not adequate in terms of workforce or equipment. Further decisions are imminent. Human resources department is informed if more temporary or permanent employees are to be hired. There may be need for new equipment to be procured. Finance department has to be informed of expenditures that may result in hiring more employees or procuring more equipment. Materials needed for manufacturing parts have to be procured too and Purchasing department must communicate this need to Finance department so that suppliers have to be contacted for raw materials or bought-out parts.

So far we have discussed the scheduling aspects of production planning. It is now appropriate to discuss the capacity aspects as indicated in the introduction of this chapter where we considered the overall structure of production planning and pointed out that schedule and capacity have to be taken into consideration at planning and execution levels. We now discuss: (1) resource requirement planning (RRP) which is the capacity balance of production planning, (2) rough cut capacity planning (RCCP) as the capacity balance for MPS, and (3) capacity requirement plan (CRP) as the capacity balance for MRP. They act as check to our scheduling. They consider manpower and availability of time on a given work centre or machine, capital, and raw materials within the time bracket of the planning horizon.

2.6. Resource Requirement Planning and Allocation

Resource requirement planning is a typical long-term activity, which provides the capacity required to manufacture products and meet market demand. Using a production plan as an input, resource requirement planning carries out an evaluation of aggregate resources (such as gross labour-hours, capital expansion, machines, warehouse space, etc.) in preparation for future production. In this section, we present a heuristic for resource requirement planning.

The capacity resource planning heuristic schedules n products on m resources. Each product has market demand D_i $(i = 1, 2, \ldots, n)$ and a contribution margin (CM_i) calculated as the product's sales price (S_i) minus its raw material (RM_i) costs. Each resource has capacity CP_j $(j = 1, 2, \ldots, m)$. The heuristic is given here without proof of optimality:

Step 1: Identify the system's constraint(s):

(a) Calculate the difference (d_j) between a resource's capacity and demand on it;

$$d_j = CP_j - \sum_{i=1}^{n} t_{ij} D_i\,, \quad i = 1, 2, \ldots, n \text{ and } j = 1, 2, \ldots, m\,,$$

where t_{ij} is the processing time of product i on resource j.

(b) Determine $CR = \{BN_1, BN_2, \ldots, BN_q\}$, $q \leq m$ where CR is the set of constrained resources with $d_q \leq 0$ and $d_1 \leq d2 \leq \cdots \leq d_q$.

Step 2: Decide how to exploit the system's constraint(s):

(a) Assume that BN_1 is the dominant bottleneck resource and calculate the ratio of the CM to the product's processing time as $R_i = CM_i/t_{i,BN_1}$.

(b) Check that BN_1 is the dominant bottleneck:

(i) If the number of resources (q) in CR equals 1, then the dominant bottleneck constraint is BN_1. Go to (c). GCO

(ii) If $q > 1$ and every product uses every resource then the bottleneck constraint is BN_1 and the ranked constrained resources are $BN_2, BN_3, \ldots, BN_q$. Go to (c).

(iii) Otherwise, determine the dominant bottleneck as follows:

(1) Schedule the products in non-ascending order of their R_i ratio until either one of the following two conditions is met. (Note, if two products to be scheduled have the same R_i, then schedule the product with the higher CM_i first):

(a) The market demand is met (i.e. $Q_i = D_i$);

(b) At least one of the constrained resources does not have sufficient capacity to schedule more of the product.

(2) If the first resource exhausted was not BN_1, make this resource BN_1. The current BN_1 increases rank by 1 to be BN_2. The rank for other bottlenecks is changed in a similar manner. The R_i for the new dominant bottleneck is recalculated. Go to (c).

(c) Schedule all the products requiring the dominant bottleneck resource in non-ascending order of their R_i ratio. For each product P_i schedule the largest quantity possible (i.e. market demand is not exceeded and there is adequate capacity on all constrained resources). If two products to be scheduled have the same R_i, then schedule the product with the largest CM_i first.

(d) When the dominant bottleneck's capacity is exhausted or there is insufficient capacity remaining to produce another unit of product, then treat BN_2 as the new bottleneck. Repeat steps (2c) and (2d) until every resource in CR is exhausted, or there are no products remaining that need these constrained resources.

(e) Schedule the 'free' products (defined here as a product that does not use any of the constrained resources) until their market demand is met. The initial schedule is $P_i, P_2 \ldots, P_i \ldots, P_n$ where P_i is the ith product in the schedule.

(f) Calculate the time left ($t_{\text{left},q}$) on the bottleneck BN_q as follows:

$$t_{\text{left},q} = CP_{BN_q} - \sum_{i=1}^{n} t_{i,BN_{q_i}} Q_i$$

where Q_i is the quantity of product i scheduled.

(g) Examine whether reducing a Q_k and increasing some Q_i (where $i > k$) increases throughput (i.e. examine whether trading-off production of some products for the production of other products increases throughput). A neighbourhood search is conducted to determine which product will be candidates for this interchange. In this search, the neighbourhood of product (k) is defined as the nearest downstream product (i.e. $k + 1$).

Let X be the set of candidate products where reducing some Q_k and increasing some Q_i ($i > k$) may increase throughput. The set X is found as follows:

(i) For $j = 1$ to q
For $i = 1$ to $(n - 1)$
IF $t_{i+1,BN_j} > 0$
THEN IF $R_{i+1}(t_{\text{left},j} + t_{i,BN_j})/CM_i \geq 1$, THEN GO TO (ii)
ELSE $i = i + 1$
ELSE $j = j + 1$

(ii) IF $Q_i < D_i$ OR $Q_{i+1} < D_{i+1}$ THEN set $k = i$ GO TO (iii) ELSE $i = i + 1$ GO TO (i)

(iii) $X = \{P_k, P_{k+1} \dots, P_n\}$. If $X = \phi$, the current master production plan is optimal, so stop.
Otherwise, go to step 2h.

(h) Reduce the current quantity Q_k of product P_k one unit at a time. This creates $(t_{k,BN_q} + t_{\text{left},q})$ of available time on the dominant bottleneck resource. Schedule this available time to process P_{k+1}, $P_{k+2} \dots$, and P_n based on their R_i ratios ($i = k+1, k+2, \dots, n$) in non-ascending order using the procedure in step 2c. Calculate the throughput gain of this change as:

$$\text{Gain} = \Delta Q_k CM_k + \Delta Q_{k+1} CM_{k+1} + \cdots + \Delta Q_n CM_n \,.$$

where ΔQ_k is the change in units for P_k.
Stop if the gain is less than zero, OR if the solution becomes infeasible (i.e. $Q_k > D_k$).

Example 5

Table 2.31 shows the processing times for machines A, B, C, D, E to produce product types R, S, T, U. The master production plan, sales price and material costs are shown on the last three columns. Machine capacities are shown on the last row.

Determine if the master production is realistic. What is the maximum throughput of the system?

Table 2.31. TOC resource requirement planning and allocation data.

(1) Product	(2) A	(3) B	(4) C	(5) D	(6) E	(7) Market demand	(8) Sales price ($)	(9) Raw material cost ($)
R	15	5	10	0	5	70	90	15
S	16	10	5	28	5	60	80	15
T	0	5	10	15	21	50	70	20
U	5	15	7	0	6	150	60	25
Capacity (min)	2400	2400	2400	2400	2400			

Solution

Step 1: (a) This is shown in Table 2.32.

Table 2.32. Identifying the system's constraints.

(1) Product	(2) A	(3) B	(4) C	(5) D	(6) E	(7) Market demand	(8) Sales price ($)	(9) Raw material cost ($)	(10) CM
[1] R	15	5	10	0	5	70	90	15	75
[2] S	16	10	5	28	5	60	80	15	65
[3] T	0	5	10	15	21	50	70	20	50
[4] U	5	15	7	0	6	150	60	25	35
[5] Capacity (min)	2400	2400	2400	2400	2400				
[6] Required	2760	3450	2550	2430	2600				
[7] Difference, D_j	−360	−1050	−150	−30	−200				
Order of BNs	BN_2	BN_1	BN_4	BN_5	BN_3				

(b) CR = {A, B, C, D, E} and $q = 5$.

Step 2: (a) The ratio R_i of each products' CM to its processing times on Resource $B\,(BN_1)$ are shown in column 4 of Table 2.33.

Table 2.33. Original TOC solution.

				B		A		E		C		D	
1	2	3	4	5	6	7	8	9	10	11	12	13	14
	D_i	Q_i	R_i	U	L	U	L	U	L	U	L	U	L
R	70	70	15	350	2050	1050	1350	350	2050	700	1700	0	2400
T	50	50	10	250	1800	0	1350	1050	1000	500	1200	750	1650
S	60	60	6.5	600	1200	960	390	300	700	300	900	168	−30
												0	
U	150	80	2.33	1200	0	400	−10	480	220	560	340	0	−30
					BN_3		BN_2		BN_4		BN_5		BN_1

where $U = Used$, and $L = Left$.

(b) Since $q = 5$ and U does not use every constrained resource in CR, go to step 2b (iii):

(iii) Schedule the products in non-ascending order of R_i to meet their market demand until the resource is exhausted as shown in columns 5 to 14 of Table 3. Since resource D is out of capacity before other resources, resource D is the dominant bottleneck (i.e. BN_1) and others follow as shown. Recalculate R_i for resource D.

(c) Rank the products by their R_i for resource D (column) 3 of Table 2.34). Plan to produce T and S to meet market demand (column 9). Rank product R before U since its CM is larger than U's (column 8) and plan to produce R and U until resources are exhausted. We can produce 50T and 57S.

Table 2.34. Step 2c of the revised TOC resource requirement planning and allocation heuristic.

(1)	(2)	(3) R_{iBN1}	(4) R_{iBN2}	(5) R_{iBN3}	(6) R_{iBN4}	(7) R_{iBN5}	(8) CM	(9) D_i	(10) Q_i	(11) $t_{i,D}$	(12) t_{left}
P_1	T	3.33	—	—	—	—	50	50	50	15	1650
P_2	S	2.32	—	—	—	—	65	60	58	28	26
P_3	R	—	5.0	15.0	15.0	7.5	75	70		0	26
P_4	U	—	7.0	2.33	5.83	5.0	35	80		0	26

(d) The remaining 26 minutes on resource D (column 11) is not adequate to produce another product, so resource A (BN2) is the dominant bottleneck. Plan to produce R.
Since T and S have been previously scheduled, these are dealt with first. Moreover, the values of Q_i for R and U cannot be greater than values already obtained in the previous iteration ($R = 70$, $U = 80$, respectively). Hence, the solution at this step is 70R, 58S, 50T, 80U as shown in Table 2.35:

Table 2.35. Step 2d of the revised TOC resource requirement planning and allocation heuristic.

Product	Q_i	$t_{i,A}$	Used	Left
T	50	0	0	2400
S	58	16	928	1472
R	70	15	1050	422
U	80	5	400	22

(e) There are no ‘free’ products to schedule.

Table 2.36. Step 2e to 2g of the revised TOC resource requirement planning and allocation heuristic.

(1)	(2)	(3) R_{iBN1}	(4) R_{iBN2}	(5) R_{iBN3}	(6) R_{iBN4}	(7) R_{iBN5}	(8) CM	(9) D_i	(10) Q_i	11 t_1	12 t_2	13 Obj	14 Factor
P_1	T	3.3	—	—	—	—	50	50	50	15	1650	2500	$2.3(26 + 15)/50 > 1$
P_2	S	2.3	—	—	—	—	65	65	58	28	26	6270	—
P_3	R	—	5.0	15	15	7.5	75	70	70	0	26	11 520	—
P_4	U	—	7.0	2.3	5.8	5.0	35	80	80	0	26	14 320	—

where $t = t_{i,D}, t_2 = t_{\text{left}}$, obj = throughput, factor = $R_{i+1}(t_{\text{left},j} + t_{i,BN_j})/CM_i$

(f) The time left on resource D is 26 (column 12 of Table 2.36) and the throughput is \$14 320 (column 13).

(g) Determine if decreased production of some products will increase throughput. As shown in column 14, product T is the first product where the ratio $R_{i+1}(t_{\text{left},j} + t_{i,BN_j})/CM_i$ (is greater than or equal to 1. Since the master production plan for S is less than the market demand (column 10), T is the first product in set $X = \{T, S, R, U\}$ considered for interchange to increase throughput.

(h) The impact of increasing or decreasing the order quantity of each number of candidates set X is evaluated in Table 2.37. By not processing one unit of product T, 41 minutes are available on resource D. this is obtained from I1 and column 14 where $(t_{\text{left},j} + t_{i,BN_j}) = 26 + 15 = 41$. This information is used to obtain the gain. We decrease T by one in each iteration and determine how many units of S, R, and U that have to be produced. Therefore, $41/15 = 2.7$. Rounding down gives 2 units that can be produced. This is shared between S, R and U. Assign 2 to S, 0 to

Table 2.37. Step 2h of the revised TOC resource requirement planning and allocation heuristic.

	T	S	R	U	Gain	$Q < D$	D_{remain}	Throughput
(0)								14 320
(1)	−1	2(1)	0	0	15	Yes	13	14 335
(2)	−2	4(2)	0	0	30	Yes	0	14 350
(3)	−3	6(2)	0	0	−20	Yes	15	14 300

R and 0 to U since any further increase for R and U will violate the constraint of demand. The gain is calculated as:

$$\text{Gain} = -1^*50 + 1^*65 + 0^*75 + 0^*35 = 15\,.$$

The remaining time on D becomes $2400 - [(50-1)^*15 + (58+1)^*28] = 13$.

If R and U were incremented then there will not be enough time left in resource D. During the second iteration, T is decremented by 2 units. The possible number of products that can be produced $= 2^*41/15 = 4$ approximately. Again, we cannot add to R and U otherwise they will exceed their demand of 70 and 80 respectively. But if 2 units are added to S, the gain can be calculated as $-2^*50 + 2^*65 = 30$. The remaining time on resource $D = 2400 - [(50-2)^*15 + (58+2)^*28] = 0$.

During the third iteration, T is decremented by 3 units and the possible number of products that can be produced equal to $3^*41/15 = 6$ approximately. Only S can be considered. Only 3 units can be added to S. Remaining time on resource $D = 2400 - [(50-3)^*15 + (58+2)^*28] = 15$. Gain $= -3^*50 + 2^*65 = -20$. The heuristic stops once the gain is negative. The largest throughput of 14 335 gives the optimum solution.

2.7. Rough Cut Capacity Planning (RCCP)

Strictly related to resource planning is rough-capacity planning which, even though on shorter term, provides information on how resource levels should be changed to ensure the execution of the master production schedule (Vollmann *et al.*, 1991). The master production schedule gives the number of end items to be produced each week of the planning horizon. By using the routing sheet which gives the sequence of operations required to be followed, together with the setup time and processing time for predicting a part, it is possible to ascertain whether the master production schedule is possible. We still consider the warehouse trolley example. The operations required are cutting, welding, assembly and painting. Table 2.18 for the MPS is reproduced and the information in the routing sheet is shown in Table 2.38.

It is identified that welding takes the largest time, so it is decided that the resource requirement be estimated in terms of this maximum time to ensure

Schedule for Warehouse trolley: FP-100					Master (Previous)		Production
Week	1	2	3	4	5	6	7	8
FP-100	—	—	150	50	100	80	75	100

Table 2.38. Routing sheet for warehouse trolley.

Operation	Setup time (Hours)	Processing time (Hours)
Fabrication	1	0.6
Assembly	1	0.4
Painting	1	0.2

Table 2.39. RCCR (hours) for warehouse trolley.

Week	1	2	3	4	5	6	7	8
FP-100	—	—	91	31	61	49	46	61

that the MPS is feasible. For week 3, capacity $= 1 + 150 \times 0.6 = 91$ hours. The rough-cut capacity requirement is depicted in Table 2.39.

However, the manufacturers of the warehouse trolley can accommodate maximum of 80 hours of cutting. This means that there is an overshoot in week 3. A solution must be sought at this stage, either by planning an over time in week 3 or redistributing the excess hours to week 4 where under-loading exists.

2.8. Capacity Resources Planning (CRP)

The planned order release which is the output of an MRP system indicates the quantity of materials to be ordered each time bucket over the planning horizon with lead time offsetting, in order to ensure that the parts are available at the appropriate time. Thus more information is provided at this level of planning more than at the MPS level. Lead time offsetting for the warehouse trolley is shown in Fig. 2.10.

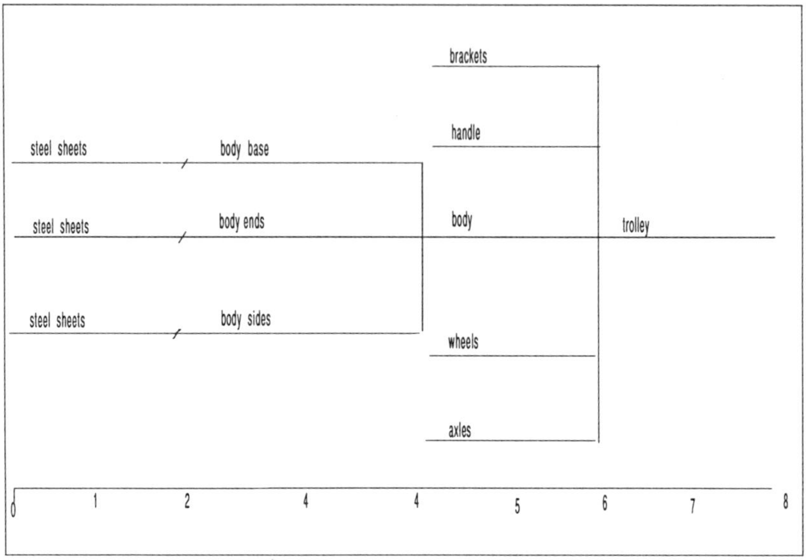

Fig. 2.10. Lead-time offsetting for warehouse trolley.

The same procedure as for MPS is followed for estimating the capacity resources planning for MRP. The order releases obtained from an MRP system and the routing sheets for the operations involved in producing parts or assembly are used to determine the capacity resources. We now carry out a capacity resource planning for the body of the warehouse trolley. The order release for the trolley body (Table 2.23) and routing sheet (Table 2.17) are used. Considering fabrication operation for the warehouse trolley, the capacity in week $2 = 1 + 0.6^{*}125 = 76$ hours. Table 2.40 shows the capacity requirement (hours) for the planning horizon.

This result represents only one part. The normal practice is to consider all the contributions of all parts for a particular jot type in order to obtain the load profile or load report for each time bucket with in planning horizon.

Planned order release of warehouse trolley body (from Table 2.23).

Week	1	2	3	4	5	6	7	8
FP-100	0	125	50	100	80	75	100	0

Table 2.40. RCCR (hours) for warehouse trolley.

Week	1	2	3	4	5	6	7	8
FP-100	—	—	76	31	61	49	46	61

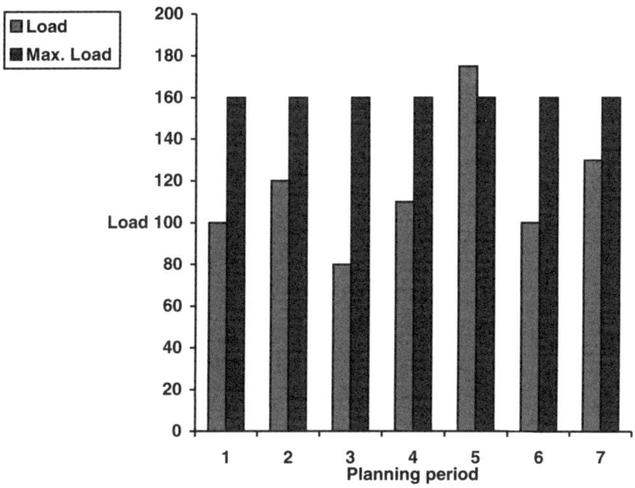

Fig. 2.11. Load profile for fabrication.

Figure 2.11 shows the fabrication work-centre load profile. The maximum load allocated to fabrication is 160 hours/week, thus there is excess load in week 5. Again the excess can be rescheduled for an under-load week or overtime may be planned.

2.9. Summary

At production planning and control level, the planning time horizon becomes tighter, putting much pressure on the operations manager. For proper production planning, an extrapolation of the present to the future is necessary. This is achieved by appropriate demand forecasting techniques. Some of the commonly used techniques are discussed in detail in this chapter. For dependent demand, material requirement planning (MRP) is preferred for production scheduling and inventory. In order for MRP to be functional, information for master production schedule, BOM, and accurate inventory and purchase records, and lead times must be available to management. All these are discussed in this chapter.

Exercises

1. What do you understand by production planning and control?
2. What is the role of demand forecasting in production planning?
3. What external environments affect production planning and control?
4. The marketing department of a chocolate factory submits the observed demand (number of cartons) for one year shown in Table 2.41 to management, for business planning. If 4-month moving average is used, deduce the forecast demand up to the first month of the following year, the absolute error, and the squared error.
5. Simple regression is used as an alternative forecasting method for the demand of Table 2.41. What is the regression line equation? Draw the graph.
6. Use the simple exponential smoothing technique for $\alpha = 0.15$ for the forecasting of demand for Table 2.41. Repeat for $\alpha = 0.2$. Draw the graphs of the forecast demands and the actual demands.
7. What do you understand by seasonal exponential smoothing? Use this technique for Table 2.41 for $\alpha = 0.2$, $\beta = 0.15$, and $\gamma = 0.1$. Repeat for $\alpha = 0.4$, $\beta = 0.3$, and $\gamma = 0.2$. Draw the graphs of the forecast demands and the actual demands. Comment on the difference.
8. How useful would you consider a computer program for each of the forecasting technique?
9. Write a computer program that gives a user the option to choose all the forecasting techniques discussed in this chapter.
10. What are the roles of aggregate production planning in production planning and control?

Table 2.41. Observed demand for chocolate for 1 year.

Jan	Feb	Mar	Apr	May	Jun	Jul	Aug	Sep	Oct	Nov	Dec
345	350	380	300	340	355	400	370	420	430	440	500

11. Define master production schedule (MPS) and discuss its role in production planning and control.
12. Define the following: projected inventory, available to promise, and total demand, as functions of MPS.
13. Write a computer program for a master production schedule based on manual forecasts, system forecasts, and customer demand. Hint: use the definitions in this chapter.
14. Fill in the remaining spaces marked with asterisk in the master production schedule record shown in Table 2.42. The just in case stock is 5 item and the initial projected inventory is 30 items.
15. A product designated by part number AP-10 is made up of two subassemblies BA-10 and BA-20. In the bill of material AP-10 is at level 0, while BA-10 (5 in quantity) and BA-20 (4 in quantity) are at level 1. The master production schedule for AP-10, the master parts data, inventory and open order data are given in Tables 2.43, 2.44, 2.45, and 2.46 respectively. Generate the planned order releases for AP-10, BA-10, and BA-20.
16. What is the significance of rough-cut capacity planning?
17. Discuss capacity resource planning.
18. Which data given in question 15 are relevant for rough cut capacity planning?
19. What data do you need for capacity resources planning for question 15?
20. Clearly distinguish between production planning and production control.

Table 2.42. MPS record for customer orders/forecasts.

Item	Part number									
Week number	1	2	3	4	5	6	7	8	9	10
Customer order	40	35	25	30	20	20	25			
Manual forecast	35	40	25	40	45	35				
System forecast	45	40	45	50	40	40	45	45	45	45
Total demand TD	40	*	*	40	*	35	45	45	45	45
Master Production Schedule	*	*	*	*	45	35	45	45	45	45
Proj. inven. PI ($PI_0 = 30$)	*	*	*	5	5	5	5	5	5	5
Available to promise ATP	*	*	*	*	25	15	20	45	45	45
Cumulated ATP	*	*	*	20	45	60	80	125	170	215
Just in case stock = 5										

Table 2.43. Master parts data.

Level	Part number	Lot	Lead size	Description time	Make/buy
0	AP-10	L	1	final product	M
1	BA-10	L	2	assembly 1	B
1	BA-20	L	1	assembly 2	M

Table 2.44. Master production schedule for final product.

Week	1	2	3	4	5	6	7	8
AP-10	—	50	100	70	120	90	65	90

Table 2.45. Inventory data.

Part number	Current inventory	Allocated
AP-10	10	
BA-10	20	
BA-20	25	5

Table 2.46. Open order data.

Part number	Scheduled receipt	Due date
AP-10	10	1
BA-10	10	2
BA-20	15	1

21. Develop a computer program for material resource planning (MRP).
22. What functions would you include in the MPR software of question 21 in order to upgrade it to material resources planning (MRPII).

References

Bertrand, J. W. M. and Worthman, J. C., 1982, *Production Control and Information Systems for Component Manufacturing Shops* (Amsterdam: Elsevier).

Campbell, H. G., Dudek, R. A., and Smith, M. L., 1970, A heuristic for the n job, m machine sequencing problem, *Management Science*, **16**(10) 30–637.

Corbett, J., Dooner, M., Meleka, J., and Pym, C., 1991, *Design for Manufacture: Strategies, Principles and Techniques* (Addison-Wesley).

Garetti, M. and Taisch, M., 1999, Neural networks in production planning and control, *Production Planning & Control*, **10** 324–339.

Hill, T., O'Conor, M., and Remus, W., 1996, Neural network models for time series forecasts, *Management Science*, **42** 1082–1092.

Johnson, S. M., 1954, Optimal two and three stage production schedules with setup times included, *Naval Research Logistics Quarterly*, **1**(1).

McMahon, C. and Browne, J., 1993, *CAD/CAM: From Principles to Practice* (Addison-Wesley).

Onwubolu, G. C., 1995, A flow-shop manufacturing scheduling system with interactive computer graphics, *International Journal of Operation and Production Management*, **16**(9) 74–84.

Stoll, H. W., 1986, Product design for efficient manufacture, *Workshop notes, Indust. Tech. Inst.*, Ann Arbor, MI.

Vollmann, T. E., Berry, W. L., and Whybark, D. C., 1991, *Manufacturing Planning and Control System* (Homewood, IL: Irwin).

Wild, R., 1990, Essentials for Production and Operations Management, 3rd edition (Cassell Educational Ltd., London).

Chapter 3
Production Control Systems

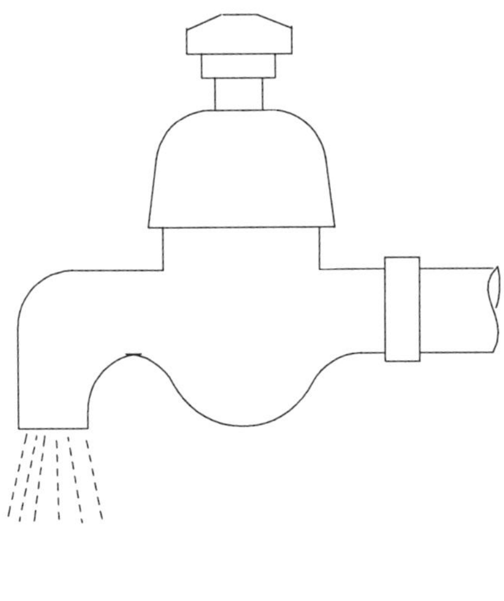

3.1. Introduction

Production control systems (PCS), also referred to as shop floor control systems (SFCS) are used to plan, schedule and execute the tasks required to process an order. Part selection, loading, sequencing and dispatching are just some of these tasks. Other minor but nevertheless important activities deal with tool allocation, machine breakdowns, maintenance and inspection of in-process and/or finished goods (Geretti and Taisch, 1999, Gunasekaran *et al.*, 1993). Production planning commences from intermediate strategic plan towards operational plan, which entails more detailed plans for execution within very short planning horizon. Production control is the last step in production planning and control system, with the main objectives being to allocate different jobs, employee time and machine time within the very near future. There are different criteria for deciding which schedule is optimum, but in general, the following criteria are used:

- minimising the makespan: requires short set-up times; may result in large capacity and WIP inventory;
- minimising the work in progress (WIP): may require excess capacity, with the possibility of high machine or employee idling;
- minimising idle time for equipment: low capacity; may lead to high inventory and overloading of equipment;
- minimising idle time for employees: low work force; may lead to high inventory;
- minimising costs.

Production scheduling exists at the operational level of the production planning and control system. It deals with very near future, hence has very short time horizon ranging from real-time to a few days. The various components of production control are scheduling, dispatching, monitoring, material handling and production and their relationship are shown in Fig. 3.1. These are briefly

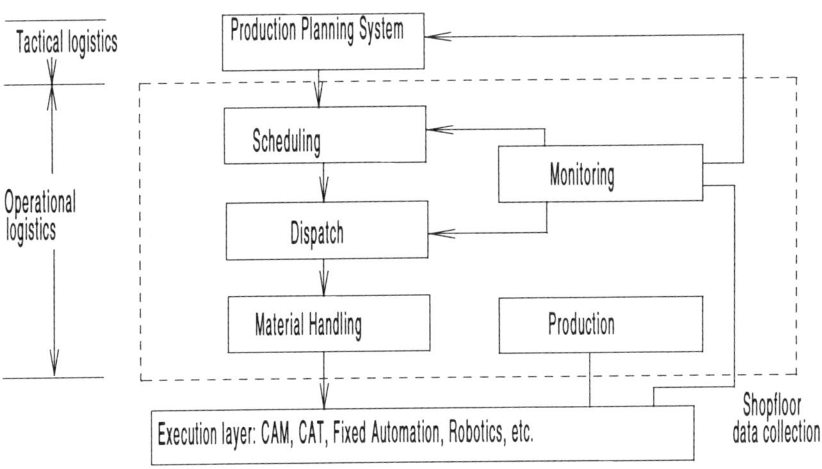

Fig. 3.1. Production control architecture.

discussed. Production scheduling lies between the production planning system and the actual execution at the work centres. It bridges the gap between production planning and the devices at the execution layer. Scheduling is an attempt to allocate scarce resources such as machine time and employee time. Production schedule process varies for different manufacturing process organisations. The way scheduling is considered in assembling line environment differs from scheduling in flow shop or job shop environment. Despatching is the issuing of commands based on the schedule to the material handling and production modules in order to produce the required products at the shop floor. Material handling is the actual movement of material to and from machine centres. Material handling devices include hand operated trucks, fork lift, or automated guided vehicle (AGV). Production is responsible for the execution of various operations at the various work centres. It could be manual by an operator or automatic. Feedback is normally in form of data from the devices showing the status of work at the shop floor. These data may be fed back to the scheduling, dispatching and top level planning modules.

In this chapter, we consider scheduling job shop, batch, line-flow, and assembly line manufacturing systems. We keep the treatment in that order because most of the techniques used for job shop scheduling apply to line-flow scheduling. Techniques of batch and assembly line scheduling vary significantly. Then we also consider other aspects of production control such as material management, inventory control and quality control.

3.2. Scheduling in Job Shop Production

Scheduling is an attempt to allocate machine time to processing of different jobs, or to allocate employee time to different activities to optimise a given measure of performance. Measures of performance include average and maximum completion time, flow time, lateness, and tardiness. The following terms are used to define measures of performance:

d_i — due date of job i
p_{ij} — processing time of job i on machine j
w_i — waiting time of job i
C_i — completion time of job i
F_i — the time that a job i spends in the shop (flow time)
L_i — lateness of job $i = (C_i - d_i)$

T_i — tardiness of job $i = \max(0, L_i)$
E_i — earliness of job $i = \max(0, -L_i)$
M — number of machines
N — number of jobs to be schedules
N_u — number of unfinished job
N_w — number of jobs waiting between machines or workstations
N_p — number of jobs actually being processed

The measures of performance that may be used in evaluating different possible schedules include:

1. minimising average flow time: $\bar{F} = \sum_i F_i/N$
2. minimising the time required to complete all jobs: $C_{\max}$ (makespan)
3. minimising average tardiness: $\bar{T} = \sum_i T_i/N$
4. minimising maximum tardiness: $T_{\max}$
5. minimising the number of tardy jobs: $\sum N_T$

$$N_T = \begin{cases} 1 & \text{if } T_1 > 0 \\ 0 & \text{if otherwise.} \end{cases}$$

Scheduling problems are often represented as $N/M/A/B$. N is the number of jobs, M is the number of machine, A is the job flow pattern and B is the performance measure which is to be minimised or maximised.

The job flow pattern $A(F, P, G)$ depends on the manufacturing process organisation and it takes one of the following forms:

- F is the flow shop situation where all N jobs must visit machines in the same sequence.
- P is the permutation flow shop situation where all M machines process jobs in the same order.
- G is the general job shop problem where N jobs are scheduled on M machines.

Dispatching rules, algorithms, branch and bound, operations research (OR) approaches, Gantt charts (for reverse / forward scheduling) are techniques for scheduling flow shop and job shop cases (see Wild, 1992 for detailed clarification of techniques for scheduling). These are now dealt with in the subsequent sections.

3.2.1. *Dispatching rules*

Dispatching rules are heuristics and are often used in shop scheduling. Rules can be classified as static or dynamic. The difference is that static rules have priority indices that may remain constant throughout the jobs flow time, while dynamic rules are functions of time and queue characteristics. Thirteen common dispatching rules are now discussed and classified as either static or dynamic:

1. *Shortest Processing Time* (*SPT*): select a job with minimum processing time. This is a static rule.
2. *Earliest Due Date* (*EDD*): select a job due first. It is a static rule.
3. *First Come, First Served* (*FCFS*): select a job having the longest queue time in front of a machine. The rule is a static one.
4. *First in System, First Served* (*FSFS*): select a job that has spent the longest time on the shop floor. Again this is a static rule.
5. *Slack per Remaining Operation* (*SRO*): select a job with the smallest rate of slack to number of operations remaining to be performed. Slack-based rules are dynamic. Slack = due date − current time − remaining processing time
6. *Covert Rule* (*CR*): select a job with the highest ratio of delay cost to operation processing time. This is a dynamic rule with definition:

$$\text{Delay } \cos t = \begin{cases} 1 \text{ if slack} < 0 \\ 0 \text{ if slack} > 0 \\ \text{E (wait)} - \text{slack / E (wait) if otherwise}. \end{cases}$$

7. *Critical Ratio* (*CR*): select a job with the least ratio of the current time until due date to processing time remaining. This is a dynamic rule with definition:
8. *Least total work* (*LTWK*): select a job with smallest total processing time. This is a static rule.
9. *Least Work Remaining* (*LWKR*): select a job with smallest total processing time for unfinished operations. The rule is dynamic.
10. *Most Work Remaining* (*MWKR*): select a job with largest total processing time for unfinished operations. This rule is opposite to the LWKR and is dynamic.

11. *Most Operations Remaining* (*MOPNR*): select a job with the most operations remaining in its processing sequence. This is a dynamic rule and utilises the number of unfinished operations, N_u.
12. *Work in Next Queue* (*WINQ*): select a job whose subsequent machine currently has the shortest queue. As queue characteristic is dynamic so also is WINQ
13. *Random* (*RANDOM*): select a job at random. A random number table is used and is not fixed.

Let us consider an example to clarify these common dispatching rules.

Example 1

Table 3.1 presents information for four jobs to be carried out in a shop. Machine B has just finished a job, and the task is to select the next job. The current time is 20. By considering the various dispatching rules, obtain the job sequence.

Table 3.1. Jobs and machine operation times.

Job	Arrival to system	Arrival at B	Due date	Machine/process time 1	2	3
1	20	20	60	(B, 10)	(A, 2)	(D, 12)
2	0	10	40	(A, 10)	(B, 6)	(C, 4)
3	0	18	20	(C, 6)	(D, 4)	(B, 4)
4	0	16	50	(E, 12)	(B, 8)	(C, 8)

The various dispatching rules are now considered:

1. *Shortest Processing Time* (*SPT*): For machine B the processing times are (10, 6, 4, 8) for jobs (1, 2, 3, 4). Using SPT rules, the job sequence is {3, 2, 4, 1}.
2. *Earliest Due Date* (*EDD*): Applying the rule to due dates (60, 40, 20, 50) for jobs (1, 2, 3, 4), and job sequence is {3, 2, 4, 1}.
3. *First Come, First Served* (*FCFS*): Arrival times for jobs (1, 2, 3, 4) at machine B are (20, 10, 18, 16). The required sequence using this rule is {2, 4, 3, 1}.

4. *First in System, First Served* (*FSFS*): Jobs (1, 2, 3, 4) arrive at the system at times (20, 0, 0, 0). Applying the rule and arbitrarily choosing from the second, third and fourth times, the required sequences is {2, 3, 4, 1}.
5. *Slack per Remaining Operation* (*SRO*):

   Slack = due date − current time − remaining processing time

   for job 1, Slack = 60 − 20− processing times (B, A, D) = 60 − 20 − (10 + 2 + 12) = 16

   for job 2, Slack = 40 − 20− processing time (B, C) = 40 − 20 − (6 + 4) = 10

   for job 3, Slack = 20 − 20− processing time (B) = 20 − 20 − 4 = −4

   for job 4, Slack = 50 − 20− processing time (B, C) = 50 − 20 − (8 + 8) = 14

   the unfinished operations, N_u = (3, 2, 1, 2)

   Slack = (16, 10, −4, 14). Slack/N_u = (16/3, 10/2, −4/1, 14/2) = (5.3, 5, −4, 7). The required sequence is {3, 2, 1, 4}
6. *Covert Rule* (*CR*): $E(wait)$ = a percentage of remaining or unfinished operations N_u. Assuming $E(wait) = 0.5\ N_u$, then $E(wait) = 0.5^*(3, 2, 1, 2)$ = (1.5, 1, 0.5, 1). We have already calculated the slack. Slack = (16, 10, −4, 14). Using the rule, Delay cost = $E(wait)$ − slack/$E(wait)$ = (−9.16, −9, 8.5, −13). Processing time for machine = (10, 6, 4, 8), hence CR = (−9.16/10, −9/6, 8.5/4, −13/8) = (−0.916, −1.5, 2.125, −1.625). The sequence is {3, 1, 2, 4}
7. *Critical Ratio* (*CR*):

   Delay cost = Due date − current time. CR = Delay Cost/(time for remaining work)

   Time until due date for: job 1 = 60 − 20 = 40
   job 2 = 40 − 20 = 20
   job 3 = 20 − 20 = 0
   job 4 = 50 − 20 = 30.

   Processing time remaining for: job 1 = 10 + 2 + 12 = 24
   job 2 = 6 + 4 = 10
   job 3 = 4
   job 4 = 8 + 8 = 16.

   CR = $(40/24, 20/10, 0/4, 30/16) = (1.67, 2, 0, 1.88)$. The sequence is {3, 1, 4, 2}

8. *Least total work* (*LTWK*):
   Total processing time for: job 1 = 10 + 2 + 12 = 24
   job 2 = 10 + 6 + 4 = 20
   job 3 = 6 + 4 + 4 = 14
   job 4 = 12 + 8 + 8 = 28.

   Total processing time for jobs (1, 2, 3, 4) = (24, 20, 14, 28). From the rule, sequence = {3, 2, 1, 4}
9. *Least Work Remaining* (*LWKR*): Total processing time for unfinished operations for B, is obtained for
   job 1 = 10 + 2 + 12 = 24
   job 2 = 6 + 4 = 10
   job 3 = 4
   job 4 = 8 + 8 = 16.

   Unfinished processing time = (24, 10, 4, 16). The rule gives sequence {3, 2, 4, 1}
10. *Most Work Remaining* (*MWKR*): We simply reverse the LWKR sequence to obtain {1, 4, 2, 3}
11. *Most Operations Remaining* (*MOPNR*): $N_u = (3, 2, 1, 2)$ and applying the rule yields, sequence = {1, 2, 4, 3}
12. *Work in Next Queue* (*WINQ*): It is best understood by example:
    for job 1, machine B precedes machine A
    for job 2, machine B precedes machine C
    for job 3, machine B precedes no machine
    for job 4, machine B precedes machine C.

    Assume queue times for machines A and C are 20 and 8 respectively, then queue times for machines subsequent to B are (20, 8, 0, 8). The rule gives sequence {3, 2, 4, 1}
13. *Random* (*RANDOM*): Depending on the random number generation a sequence {3, 2, 1, 4} may be acceptable.

3.2.2. *Nondelay schedules*

In nondelay scheduling, a job is started if it can be started before the next job is scheduled to start. Although the policy is logical it does not necessarily lead to optimum solution. The number of such schedules can be very computationally expensive. The nonedelay schedule presented here will allow generation of as many of the nondelay schedules as desired.

Let us define:

m = number of machines

n = number of jobs

A_t = set of the operations scheduled at state t

e_k = the earliest time that operation $k \in A_t$ can be scheduled

m^* = machine needed by $e^* = \min_{k \in A_t} e_k$.

The nondelay schedule algorithm can be stated as:

Step 0: Initialise. Let $t = 1$. A, contains the first operation of each ready job.
Step 1: Select operation. Find e^*. If several e^* exist, break tie orbitrarily.
Step 2: Increment. Remove k from A_t and add the next operation for its job unless that job is completed; this creates A_{t+1}. Let $t = t + 1$. If $t = mxn$ stop. Otherwise go to Step 1.

Let us consider an example to see how the nondelay schedule algorithm works.

Example 2

Table 3.2 shows the data for a 5-job, 4-machine general job shop problem.

(i) Generate a nondelay scheme.
(ii) Compute the makespan for each machine.

Table 3.2. Job shop operation data.

	Operation (machine, processing time)			
Job	1	2	3	4
1	(A, 4)	(C, 12)	(B, 3)	(D, 7)
2	(C, 11)	(D, 5)	(A, 3)	(B, 4)
3	(D, 1)	(C, 4)	(A, 7)	(B, 12)
4	(A, 4)	(C, 8)	(B, 9)	(D, 6)
5	(D, 7)	(C, 14)	(A, 3)	(B, 8)

Solution

Table 3.2 is equivalent to:

$$\begin{bmatrix} \theta_{11} & \theta_{12} & \theta_{13} & \theta_{14} \\ \theta_{21} & \theta_{22} & \theta_{23} & \theta_{24} \\ \theta_{31} & \theta_{32} & \theta_{33} & \theta_{34} \\ \theta_{41} & \theta_{42} & \theta_{43} & \theta_{44} \\ \theta_{51} & \theta_{52} & \theta_{53} & \theta_{54} \end{bmatrix}.$$

Step 0: Initialise $t = 1$ and $A_1 = \{\theta_{11}, \theta_{21}, \theta_{31}, \theta_{41}, \theta_{51}\}$.

Step 1: Select job. The early time for all first operations is O. Let us give priority to job 1 when choice between operations exists. Otherwise use SPT to choose. Hence select operation θ_{11} so that $m^* = A$. Operations θ_{11} and θ_{41} are eligible for selection but choose 0_{11} due to priority.

Step 2: Increment $t = 2$. $A_2 = \{\theta_{12}, \theta_{21}, \theta_{31}, \theta_{41}, \theta_{51}\}$. Since $t \neq 5^*4(20)$, go to Step 1.

Step 3: Choose between θ_{21}, θ_{31} and θ_{51}. θ_{21} is eligible as none of them is from job 1.

The process continues until all 20 operations are assigned as shown in Table 3.3. The final schedule is shown in Fig. 3.2. Key guides to selection of job is as follows:

- The same row is affected if there is a similar machine type with *lower value.*
- If next advance has the same machine type available, it must take a *higher value.*
- The same labels of previous row should have the same value.

In iteration = 1, the earliest starting time = 1 for all jobs on machines A, C, D, A, and D. Since job 1 is priority, we choose θ_{12}. The advancement is machine C. Since machine A are in positions 1 and 4, we have $e_t = \{4, 0, 0, 4, 0\}$.

In iteration = 2, we scan the current e_t for a minimum value which is zero, corresponding to the second position, and corresponding to machine C or θ_{12}. The start time for θ_{12} is 11. The entries to consider for advancement are those that are minimum (0) or entries 2, 3 and 5. Hence, we advance θ_{21} to θ_{22} for earliest starting time = max(11, min(1,7)) = max (11, 1) = 11 for C and D. So we advance θ_{22} by this value.

Table 3.3. Nondelay schedule construction for Example 3.2.

Step	A_t	e_k	m^*	k	
1.	$\theta_{11}^*, \theta_{21}, \theta_{31}, \theta_{41}, \theta_{51}$ A C D A D	$\{0, 0, 0, 0, 0\}$	A	$\theta_{12} = \max(0, \min(4, 4))$ $= \max(0, 4) = 4$	1/1
2.	$\theta_{12}, \theta_{21}^*, \theta_{31}, \theta_{41}, \theta_{51}$ C C D A D	$\{4, \underline{0}, \underline{0}, 4, \underline{0}\}$	C	$\theta_{22} = \max(11, \min(1, 7))$ $= \max(11, 1) = 11$	2/2
3.	$\theta_{12}, \theta_{22}, \theta_{31}^*, \theta_{41}, \theta_{51}$ C D D A D	$\{4, 11, \underline{0}, 4, \underline{0}\}$	D	$\theta_{32} = \max(11, \min(1, 7))$ $= \max(11, 1) = 11$	3/3
4.	$\theta_{12}^*, \theta_{22}, \theta_{32}, \theta_{41}, \theta_{51}$ C D C A D	$\{\underline{4}, 11, 11, \underline{4}, 11\}$	C	$\theta_{13} = \max(12, 4)$ $= 12 + 4 = 16$	1/4
5.	$\theta_{13}, \theta_{22}, \theta_{32}, \theta_{41}^*, \theta_{51}$ B D C A D	$\{16, 11, 11, \underline{4}, 11\}$	A	$\theta_{42} = \theta_{41}(4) + 4 = 8$	4/5
6.	$\theta_{13}, \theta_{22}, \theta_{32}, \theta_{42}^*, \theta_{51}$ B D C C D	$\{16, 11, 11, \underline{8}, 11\}$	C	$\theta_{43} = \theta_{42}(8) + 8 = 16$	4/6
7.	$\theta_{13}, \theta_{22}^*, \theta_{32}, \theta_{43}, \theta_{51}$ B D C B D	$\{16, \underline{11}, \underline{11}, 16, \underline{11}\}$	D	$\theta_{23} = \max(4, \min(5, 7)) + 11$ $= 5 + 11 = 16$	2/7
8.	$\theta_{13}, \theta_{23}, \theta_{32}, \theta_{43}, \theta_{51}^*$ B A C B D	$\{16, 16, \underline{11}, 16, \underline{11}\}$	D	$\theta_{52} = \max(4, 7) + \theta_{51}$ $= 7 + 7 = 14$	5/8
9.	$\theta_{13}, \theta_{23}, \theta_{32}^*, \theta_{43}, \theta_{52}$ B A C B C	$\{16, 16, \underline{11}, 16, 14\}$	C	$\theta_{33} = \max(4) + 11 = 15$	3/9
10.	$\theta_{13}, \theta_{23}, \theta_{32}, \theta_{43}, \theta_{52}^*$ B A A B C	$\{16, 16, 15, 16, \underline{14}\}$	C	$\theta_{53} = \max(14) + 14 = 28$	5/10
11.	$\theta_{13}, \theta_{23}, \theta_{33}^*, \theta_{43}, \theta_{53}$ B A A B A	$\{16, 16, \underline{15}, 16, 28\}$	A	$\theta_{34} = \max(15) + 7 = 22$	3/11
12.	$\theta_{13}, \theta_{23}^*, \theta_{34}, \theta_{43}, \theta_{53}$ B A B B A	$\{\underline{16}, \underline{16}, 22, \underline{16}, 28\}$	A	$\theta_{24} = \max(3, \min(3, 9))$ $+ \theta_{23} = 19$	2/12
13.	$\theta_{13}^*, \theta_{24}, \theta_{34}, \theta_{43}, \theta_{53}$ B B B B A	$\{\underline{16}, 19, 22, \underline{16}, 28\}$	B	$\theta_{14} = \min(3, 9) + \theta_{23}$ $= 3 + 16 = 19$	1/13
14.	$\theta_{14}, \theta_{24}, \theta_{34}, \theta_{43}^*, \theta_{53}$ D B B B A	$\{19, 19, 22, \underline{16}, 28\}$	B	$\theta_{44} = \min(\theta_{43}) + 16$ $= 9 + 16 = 25$	4/14
15.	$\theta_{14}^*, \theta_{24}, \theta_{34}, \theta_{44}, \theta_{53}$ D B B D A	$\{\underline{19}, \underline{19}, 22, 25, 28\}$	D	$\theta_{15} = \min(7, 4) + 19 = 26$	1/15
16.	$-, \theta_{24}^*, \theta_{34}, \theta_{44}, \theta_{53}$ B B D A	$\{-, \underline{19}, 22, 25, 28\}$	B	$\theta_{25} = \min(4) + 19 = 23$	2/16
17.	$-, -, \theta_{34}^*, \theta_{44}, \theta_{53}$ B D A	$\{-, -, \underline{22}, 25, 28\}$	B	$\theta_{35} = \min(12, 6) + 22 = 34$	3/17
18.	$-, -, -, \theta_{44}^*, \theta_{53}$ D A	$\{-, -, -, \underline{25}, 28\}$	D	$\theta_{45} = \min(6) + 22 = 28$	4/18
19.	$-, -, -, -, \theta_{53}^*$ A	$\{-, -, -, -, \underline{28}\}$	A	$\theta_{54} = \min(6, 13) + 28 = 41$	5/19
20.	$-, -, -, -, \theta_{54}^*$ B	$\{-, -, -, -, -\}$	B		5/20

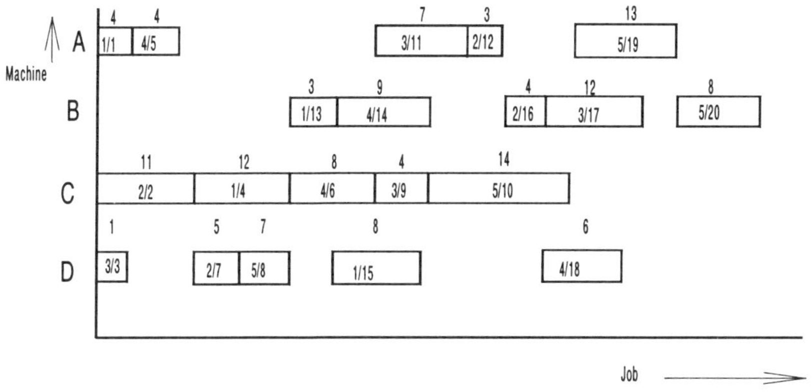

Fig. 3.2. Nondelay schedule.

In iteration = 3, we scan the current $e_t = \{4, 11, 0, 4, 0\}$ for a minimum value, which is zero, corresponding to machine D or θ_{31}. The choice is between positions 3 and 5. Now D is in the three positions, i.e. positions 2, 3, and 5 in e_t. But already, $\theta_{22} = 11$ so the minimum starting time = 11. Se we advace θ_{32} by this value for postions 2, 3, and 5.

In iteration = 4, we scan the current $e_t = \{4, 11, 11, 4, 11\}$ and choose the minimum value, which is 4, corresponding to machine θ_{12} or machine C. The choice is between positions 1 and 5. Since the machines are different, C and A, we add their starting times, 12 + 4 (or θ_{16}) for advancing θ_{13}. The new $e_t = \{16, 11, 11, 4, 11\}$.

In iteration = 5, we scan the current $e_t = \{4, 11, 11, 4, 11\}$ for minimum value, and choose the minimum value, which is 4, corresponding to machine θ_{41} or machine A. We merely advance θ_{42} by 8 (4 + 4). So the new $e_t = \{16, 11, 11, 8, 11\}$.

In iteration = 6, we scan the current $e_t = \{4, 11, 11, 4, 11\}$ for minimum value, and choose the minimum value, which is 4, corresponding to machine θ_{42} or machine C. We advance θ_{42} to θ_{43} by the current value, 8 and value fom the given processing time matrix so that the value becomes 16 (8 + 8). So the new $e_t = \{16, 11, 11, 16, 11\}$.

In iteration = 7, we scan the current $e_t = \{16, 11, 11, 16, 11\}$ for minimum value, and choose the minimum value, which is 11. We are to choose from all entries having value of 11. These correspond to machines D, C, and D so we use max(θ_{32}, min(θ_{22}, θ_{51})) + current value of 11. This leads to a value, max(4, min(5, 7)) + 11 = 5 + 11 = 16. So the new e_t = {16, 16, 11, 16, 11}.

In iteration = 8, we scan the current e_t = {16, 16, 11, 16, 11} for minimum value, and choose the minimum value, which is 11. This corresponds to machines C and D. Machine D is chosen, i.e. θ_{51}. Since the machines are different we advance θ_{52} by max(4, 7), $+\theta_{51}$ = 7 + 7 = 14. So the new e_t = {16, 16, 11, 16, 14}.

In iteration = 9, we scan the current e_t = {16, 16, 11, 16, 14} for minimum value, and choose the minimum value, which is 11, corresponding to machine θ_{32} or machine C. We advance θ_{33} by the current value of θ_{32}, which is 11 and its correponding value given in the processing time matrix. So we advance by 15 (11 + 4). Hence, new e_t = {16, 16, 15, 16, 14}.

In iteration = 10, we scan the current e_t = e_t = {16, 16, 15, 16, 14} for minimum value, and choose the minimum value, which is 14. This corresponds to machines C or θ_{52}. So we advance θ_{53} by θ_{52} + 14, or 28 (14 + 14). Since the machines are different, we advance θ_{52} by max(4, 7), + $\theta_{51} = 7 + 7 = 14$. So the new e_t = {16, 16, 15, 16, 28}.

In iteration = 11, we scan the current e_t = e_t = {16, 16, 15, 16, 28} for minimum value, and choose the minimum value, which is 15. This corresponds to machines A or θ_{33}. So we advance θ_{34} by θ_{33} + 15, or 22 (7 + 15). So the new e_t = {16, 16, 22, 16, 28}.

In iteration = 12, we scan the current e_t = {16, 16, 22, 16, 28} for minimum value, and choose the minimum value, which is 16. This value occurs in postions 1, 2, and 4 or machines B, A, and B. So we choose machine θ_{52} or A and advance θ_{24} by max(3, min(3,9)) + current value of θ_{23}. We obtain = 3 + 16 = 19. So the new e_t = {16, 19, 22, 16, 28}.

In iteration = 13, we scan the current e_t = {16, 19, 22, 16, 28} for minimum value, and choose the minimum value, which is 16. This value occurs in postions 1, and 4 corresponding to machine B. So we choose machine θ_{13} or machine B and advance θ_{13} to θ_{14} by min(Bs) + current value of θ_{13}. We obtain $\theta_{13} = \max(3, 9) + 16 = 19$. So the new e_t = {19, 19, 22, 16, 28}.

In iteration = 14, we scan the current e_t = {19, 19, 22, 16, 28} for minimum value, and choose the minimum value, which is 16. So we advance θ_{44} by θ_{43} + current value of θ_{43}. We obtain $\theta_{43} = 9 + 16 = 25$. So the new e_t = {19, 19, 22, 25, 28}.

In iteration = 15, we scan the current e_t = {19, 19, 22, 25, 28} for minimum value, and choose the minimum value, which is 19. This value occurs in postions 1, and 2 corresponding to machines D and B. So we choose machine θ_{14} or machine D and advance θ_{15} by max(D, B) + current value of θ_{14}. This is equal to max(7, 4) + 19 = 26. So the new $e_t = \{-, 19, 22, 25, 28\}$.

In iteration = 16, minimum is 19, i.e. θ_{24} or machine B and new $e_t = \{-, -, 22, 25, 28\}$.

In iteration = 17, minimum is 22, i.e. θ_{34} or machine B and new $e_t = \{-, -, -, 25, 28\}$.

In iteration = 18, minimum is 25, i.e. θ_{44} or machine D and new $e_t = \{-, -, -, -, 28\}$.

In iteration = 19, minimum is 28, i.e. θ_{53} or machine A and new $e_t = \{-, -, -, -, -\}$, with advancement to θ_{54} or B.

In iteration = 20, B is chosen and the algorithm stops. The makespan value for the schedule is 70.

3.3. Scheduling in Batch Production

Batch production refers to production of a product in small batches or lots by a series of operations with each operation being carried out on the whole batch before a new operation is commenced. It is a medium volume production run of a medium range of products, lying between jobbing shop production and mass production. Large batch size leads to work-in-progress inventory which is prohibited in just-in-time (JIT) environment. JIT encourages batch size that approaches one. Batch or lot size is a function of variable carrying or holding costs and variable ordering costs. When part or raw material is replenished from outside suppliers, then this is an ordering cost. On the other hand, a manufacturing company may produce parts, or finished goods in-house. In this case, the shop floor orders the parts, subassembly, or finished products direct from the store and the ordering cost is actually represented by the costs associated with changing equipment from producing one item to producing another. This cost is referred to as setup cost, and is used in determining batch sizes. The economic order quantity (EOQ) model shows that large batch size results in large set-up time. Low work-in-progress is therefore achieved by reducing the setup costs. Operations managers would achieve this by increasing the processing similarity of items being produced using group technology or by improving the inherent flexibility of processing equipment. Increasing processing similarity means that a tool can be used for a group of parts that belong to a family. This means reduced setup times, and consequently, reduced setup cost, leading to low work-in-progress. Several techniques have been proposed for reduction of setup times (Moden, 1983, Shingo, 1981, 1985) to include:

- reduction of setup processes which renders a machine inoperative during the period of setup.
- aim at achieving single set-up (when set-up time is reduced as small as 10 minutes or less).
- eliminate adjustment as much as possible during set-up process.
- if possible, avoid set-up.
- use standard tooling
- effect changes in equipment design, or use more sophisticated equipment which provides for automated set-up or equipment re-configuration.

Batch processing requires the consideration of three aspects:

1. batch sizing;
2. batch sequencing;
3. batch scheduling.

3.3.1. *Batch sizing*

Batch quantities, which are too small mean frequent processing of small batches. This leads to high costs associated with set-up. When batch quantities are too large, the problem faced by a company is that of increased inventory level and associated costs of holding the inventory. Therefore, an optimum level of batch size must be determined to overcome the unduly high set-up and holding costs. For manufacturing firms where parts, subassemblies, or finished products are produced in-house, ordering cost is actually represented by the set-up cost. Fortunately, the economic order quantity (EOQ) model can be applied to determining the optimum batch size, by replacing the ordering cost by the set-up cost, and the demand by consumption rate per unit of time. The minimum cost or economic batch size (EBS) is given as

$$EBS = \sqrt{\frac{2rC_s}{C_h}}$$

where C_s = Set-up cost / batch
C_h = Holding cost / item / unit of time
r = Consumption rate / unit of time

The cases of finite replenishment rate, and inclusion of safety stock, consider replacement of ordering cost by the set-up cost, and the demand by consumption rate per unit of time.

3.3.2. *Batch sequencing*

Set-up cost may depend not entirely on the type of items to be processed in a batch, but also on the type of items which was previously processed on the machine or workcentre. To fix ideas, consider two similar types of products produced in batches on the same machine. If they are very similar, the set-up cost incurred in changing over tools from one product to the other will be more than if they are of distinctively different types. Therefore, when similar products are to manufactured in batch production system, it will be advantageous to apply the concept of group technology (GT) in grouping them together in order to reduce set-up costs. In order to understand the usefulness of parts family grouping, we consider set-up costs associated with producing six items in batches as shown in Table 3.4. Items 2 and 5 can be grouped together because they have smaller and similar setup costs, while items 1, 3, 4, and 6 can be grouped together because they have larger and similar setup costs. Low setup cost will be incurred by completing processing of items in a group before proceeding items in another group. More of the discussion of this topic in Chapter 4.

Table 3.4. Set-up costs for different parts family groupings.

	Succeeding item					
Preceding item	1	2	3	4	5	6
1	0	20	100	105	50	95
2	25	0	90	75	60	120
3	105	95	0	100	70	100
4	110	80	105	0	75	0
5	80	75	70	85	0	85
6	95	60	105	90	70	0

3.3.3. *Line of balance for batch production scheduling*

Line of balance (LOB) is a technique used to monitor and control progress in a production cycle in batch production system. The technique identifies delays and bottlenecks in the production schedule in order that management can take corrective measures in order to meet delivery deadlines at the customer end. LOB concentrates on bottleneck operations in a job, establishes schedule for them, and draws management attention to those jobs which do not conform to the established schedule. The objectives of LOB are:

1. study the progress of jobs at regular intervals;
2. compare progress made on each operation with the required progress needed to meet the delivery requirements to customers;
3. identify those operations in which progress are below expectation.

The assumptions used in LOB which are often justified in practice are:

1. Constant lead times;
2. Type of operations is independent of production quantities;
3. Sequence of operations is independent of production quantities.

LOB relates the actual status of the elements of a production plan to planned progress. It aims at identifying elements that are lagging prior to delay in delivery of the end item and provides an indication of how well the various phases of production are synchronised. The technique of LOB requires four stages:

1. completion schedule;
2. operation programme;
3. programme progress chart;
4. progress analysis.

Completion schedule

The due dates for job completion are used to construct the completion schedule. The completion schedule is essentially the cumulative completion requirements as shown in Table 3.5 or in Fig. 3.3.

Table 3.5. Scheduled requirements record.

Week	Quantity of finished Items to be delivered	Cumulative quantity
0	0	0
1	10	10
2	15	25
3	10	35
4	10	45
5	15	60
6	20	80
7	25	105
8	30	135
9	25	160
10	30	190
11	35	225
12	40	265

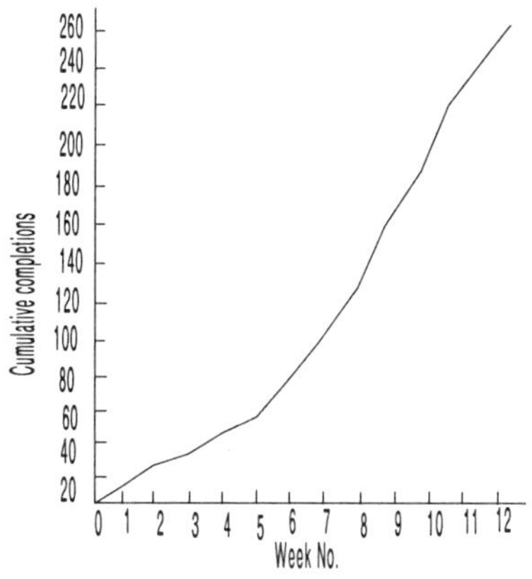

Fig. 3.3. Cumulative requirements.

Operation programme

The operation programme shows the lead time, which is the time interval between an operation and total completion of the product. This is normally measured in working days. The operation programme is best displayed

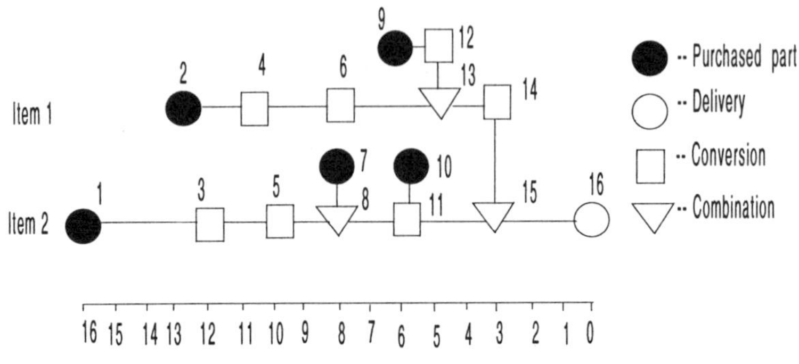

Fig. 3.4. Operation programme.

graphically on a time scale showing the activities to be performed on the product prior to delivery. Four symbols are normally used: a shaded circled is used to depict purchased part; a circle is used to depict delivery; a square is used for conversion; and an inverted triangle depicts combination. Figure 3.4 shows an operation programme for two items produced in batches. The final completion date is taken to be zero and earlier operations have larger lead times. As operations proceed towards the point of deliverying the final product, which in this case is represented by operation 15, the lead times diminish and becomes zero.

Programme progress chart

In production control, it will be useful to know at any point in time the number of items which have been finished at each of the important operations at a given date. This enables the operations manager to know if there are enough items at each operation in order to meet the overall scheduled requirements. Assuming that it is decided to review production level week 5. The number of finished items which have successfully gone through the necessary operations up to the final operation, which is operation 15 can easily be known by checking the inventory level of finished items produced on the shopfloor. This information constitutes what is known as *programme progress record* (PPR) shown in Table 3.6.

In order to assess how close the actual progress made is to the scheduled progress, the *line of balance* (LOB) is superimposed on the PPR so that at a

Table 3.6. LOB record at week 5.

Operation number	Lead time (days)	LOB
1	15	135
2	13	123
3	12	117
4	11	111
5	10	105
6,7,8	8	95
9,10,11	6	85
12,13	5	80
14,15	3	72
16	0	60

glance, the operations controller is able to detect where operation is lagging behind, and where operation is ahead of scheduled requirements. The scheduled requirements (Table 3.5) and lead times (Table 3.6) are used to generate the LOB for operations programme (Fig. 3.4).

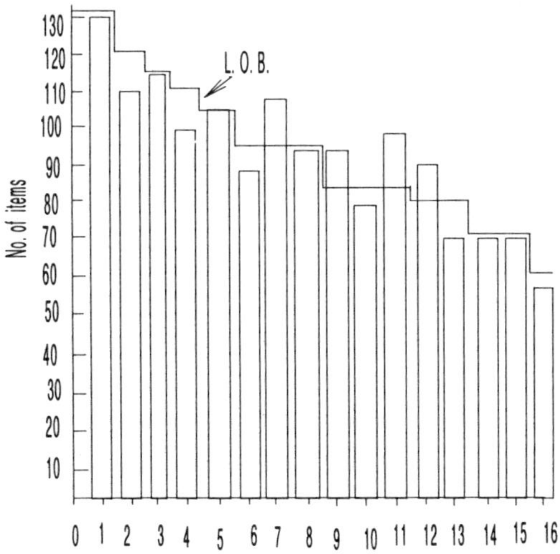

Fig. 3.5. Programme progress chart as at week 5.

Considering week 5 as the review week, the scheduled requirement is equal to 60 items. This value represents the finished items corresponding to operation 16. LOB answers one basic question: if 60 finished items are expected in week 5 during review, how many items should have completed subsequent operations? To answer this question, we use the lead times and scheduled or planned requirements, taking one working week to be 5 days. For operation 1, lead time = 15 days = 15/5 or 3 weeks. Starting from week 5 which is the review week and advancing 3 weeks ahead, gives cumulative requirements of 135 items. The LOB value for operation 1 is therefore 135. For operation 2, the lead time is 13 days (13/5 or 2.6 weeks). Starting from week 5 and advancing 2.6 weeks ahead, the cumulative requirement is computed by interpolation so that, LOB value $= 60 + 0.6(135 - 105) = 123$. The complete LOB record is shown in Table 3.6, for which progress program chart (PPC) is drawn as at week 5 as shown in Fig. 3.5.

Progress analysis

The programme progress chart is analysed to detect problem areas. For example, all production levels that are below the LOB need to be attended to by the operations manager, especially if the operations concerned are near the completion operation. Figure 3.5 shows that operations 13–15 are below the LOB. This means that if the situation is not rectified, they may be problems of achieving the scheduled production requirements at operation 16, because these operations must be completed before the last operation is carried out. Similar earlier operations that are below LOB include operations 1, 2, 3, 4, 6 and 10. It may also be useful to consider production reports on operations–type basis. For example, the operation programme shows that there are three major types of operations. Table 3.7 shows the operations classification and the corresponding operations numbers. These three types of operations have their analysis programme progress charts shown in Fig. 3.6.

Table 3.7. Operations classification.

Operation	Operation No.
Purchasing	1, 2, 7, 9, 10
Conversion	3, 4, 5, 6, 11, 12, 14
Combination	8, 13, 15

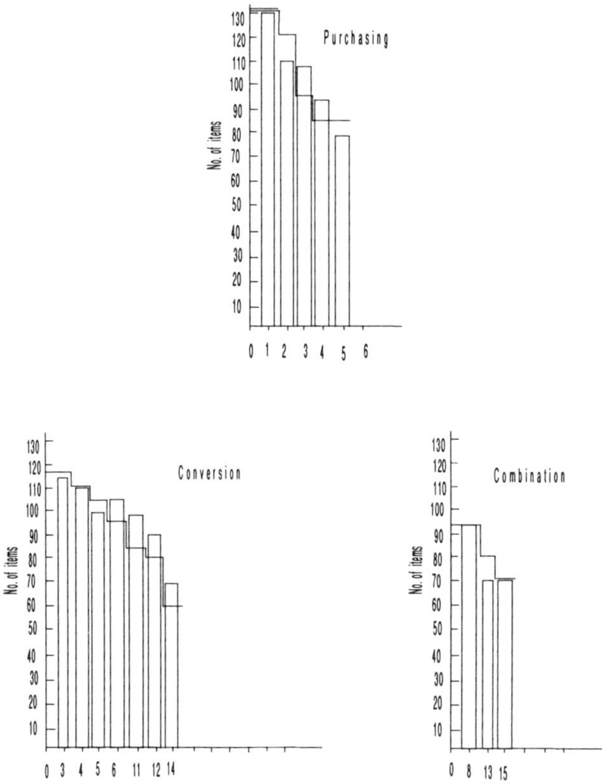

Fig. 3.6. Analysis programme progress charts.

This kind of classification, enables the operations manager to address problems exactly where they occur within a short time.

The different methods discussed so far for scheduling batch production can be made very easy for the operations controller to use, if computerised.

3.4. Scheduling in Line Flow Production

A flow-shop schedule is one in which all jobs must visit machines or work centres in the same sequence. Processing of a job must be completed on a current machine before processing of the job is started on a succeeding machine. This means that initially all jobs are available and that each machine is restricted to processing only one job at any particular time. Since the first machine in the facility arrangement is the first to be visited by each job, the other machines are idle and the other jobs are queued. Although queuing of jobs is prohibited in just-in-time manufacturing environments, flow-shop manufacturing continues to find applications in electronics manufacturing (Wittrock, 1985) and space shuttle processing (Brah, and Hunsucker, 1987), and has attracted much research work (Gupta, 1972, for example). Scheduling an n-job, m-machine with the constraint that all machines process jobs in the same order lead to the permutation situation in which there are $n!$ possible sequences to be considered. For large numbers of jobs this easily leads to combinatorial explosion. One-machine and two-machine flow-shop scheduling have been studied extensively using techniques such as branch and bound (Hariri, 1981, McMahon and Burton, 1967) and others have resorted to heuristic techniques (Kochar, 1987, and Park *et al.*, 1984) for seeking optimal solutions. An important aspect of scheduling is sequencing. Sequencing is the process of defining the order in which jobs visit a machine. Johnson's algorithm (1954) is well suited for two-machine problems and can be extended to three-machine case by splitting machines into two pseudo machines which have processing times equal to the sum of the processing times on the actual machines. A generalisation of Johnson's algorithm is that proposed by Campbell *et al.* (1970) for solving general n-jobs m-machine problems in which $m-1$ two-machines problems are solved and the sequence having the least makespan is selected. The curtailed-enumeration algorithm of NEH (Nawaz, Enscore, and Ham, 1983) for scheduling n jobs on m machines is simple but powerful for solving combinatorial optimization problems.

Scheduling outputs are usually graphically represented by Gantt charts drawn using machine processing times for each job and ensuring that delay times are taken into consideration. This process is repeated for different sequences until a minimum makespan is found which represents the minimum time taken to complete all the jobs. The corresponding sequence is considered to be optimal. It becomes immediately obvious that manual method for scheduling is tedious and prone to error. Clearly, an interactive graphical system which displays the work centres and their loading is very useful to a scheduler. The advantages are that the processing of the load pattern for each machine is sped up, the scheduler can easily visualise the output, the makespan computed by the system can be compared with the lower bound for each sequence and the scheduler is then easily guided to consider other sequences until the minimisation of the makespan is virtually optimal. In the next section we present an interactive graphical system for scheduling flow-shop environment for any number of job and machines. There are two modes in which the scheduler inputs data into the system. Sequences of operations are either input by the user or generated by the system. The processing times of jobs on machines are input interactively and using this information, the system computes the maximum value of the lower bounds so that the scheduler can be guided during the interactive session to seek for the optimal schedule. The sequences and the processing time information are used by the system to display the Gantt charts for each set of sequence. The ability of the system to accept user suggestions as well as utilise system-generated sequences makes it flexible for scheduling environment. First, schedulers may not like to depend entirely on sequences suggested by the system. Secondly, even when sequences are suggested by the system, schedulers may want to decide which sequences to use as input in order to proceed with the scheduling. Computer aided versions for Gantt charts are very useful tools (see Pruett and Schartner, 1993 and Onwubolu, 1996) for scheduling. Three computerised approaches are identified: the successive approach, the interactive approach, and the semi-interactive approach.

3.4.1. *Curtailed-enumeration algorithm of NEH*

The concept of constructive method is best understood using the curtailed-enumeration algorithm of NEH (Nawaz, Enscore, and Ham, 1983) for scheduling n jobs on m machines, which is a combinatorial optimization problem. This is now illustrated using an example.

Table 3.8. Operation time matrix.

	Machines (m)				
Jobs(n)	1	2	3	4	5
1	5	9	8	10	1
2	9	3	10	1	8
3	9	4	5	8	6
4	4	8	8	7	2

Example 3

Suppose we have four jobs, five-machine flow-shop problem given in Table 3.8. The objective is to find the best sequence of jobs using the constructive method of NEH.

Solution

Step 1: Compute the total processing time on all the machines.

$$T_1 = 5 + 9 + 8 + 10 + 1 = 33$$

$$T_2 = 9 + 3 + 10 + 1 + 8 = 31$$

$$T_3 = 9 + 4 + 5 + 8 + 6 = 32$$

$$T_4 = 4 + 8 + 8 + 7 + 2 = 29\,.$$

Step 2: Prioritize jobs based on higher total processing times {1, 3, 2, 4}.

Step 3: The two jobs with the highest total processing times are selected from n jobs. The best partial sequence for these two jobs is found by on "exhaustive search", i.e. considering the two possible partial schedules.

Step 3(a): Pick job 1 and job 3, and in their optimal partial sequence for these two jobs.

Step 3(b): Makespan for partial sequence $1 - 3 = 46$ *Makespan for partial sequence* $3 - 1 = 42$.

Choose best sequence as $3 - 1$ since it has least makespan and fix the positions of the jobs. $i = 3$, so continue.

Step 4: Take the job in the third position of the list of Step 2 (job 2) and find the optimal sequence by placing job 2 at all three possible positions in the partial sequence $3 - 1$ obtained in the last step.

Step 4(a): Makespan for partial sequence $3 - 1 - 2 = 50$

Step 4(b): Makespan for partial sequence $3 - 2 - 1 = 51$

Step 4(c): Makespan for partial sequence $2 - 3 - 1 = 51$.

Choose best sequence as $3 - 1 - 2$ since it has the least makespan.

Step 5: $i = 4(\neq n)$, so continue.
Pick the job in the fourth position of the list of Step 2 (job 4) and find the optimal schedule by placing job 4 in the four possible positions of the partial sequence $3-1-2$ found in Step 4.

Step 5(a): Makespan for partial sequence $3-1-2-4 = 58$
Makespan for partial sequence $3-1-4-2 = 58$
Makespan for partial sequence $3-4-2-1 = 57$
Makespan for partial sequence $4-3-1-2 = 54$.

Stop, because $i = 4+1 = 5 = n$.

Sequence $4-3-1-2$ yields the minimum makespan of 54 corresponds to the optimum solution. This approximate algorithm just described is a good example of constructive method. This problem was solved for the optimal makespan by evaluating all $n! = 24$ sequences and the sequence $4-3-1-2$ is optimal. The constructive method made nine enumeration of which four were complete sequences and five were partial sequences.

3.4.2. *Algorithms*

An algorithm optimises a function based on some rules subject to certain conditions. We present one algorithm that is an extension of the EDD for one machine consideration, another that considers two machines problems, and one for multi-machine case.

Moore's algorithm: $N/1/A/N_T$ problem

Jobs are first sequenced in earliest due date EDD. Jobs which are unable to be finished by their due date are known as tardy jobs. When a tardy job exists, the job having the largest processing time is identified, removed from the sequence and placed at the end. This action enables the tardy job to be ready by the due date.

As an example, Table 3.9 shows the due dates and processing times for 5 jobs. The problem is essentially designated as $5/1/A/N_T$.

First, the EDD sequence is obtained as {3, 2, 1, 4, 5} as shown in Table 3.10 based on the due dates. The processing times are summed until the tardy job is identified. For the first summation job 2 is the tardy job because $C_1 > DD(9 > 8)$ and it has the highest processing time. Hence it is deleted from

Table 3.9. The $5/1/A/N_T$ problem.

Job	1	2	3	4	5
Due date	10	8	5	11	13
Processing	4	5	4	3	1

Table 3.10. Tardy jobs identification and re-sequencing.

EDD sequence	3	2	1	4	5
Due date	5	8	10	11	13
Processing time	4	5	4	3	1
Completion time: C_1	4	9			
C_2	4	*	8	11	
C_3	4	*	8	*	9

the list and placed at the end. Resuming the next summation for C_2, job 4 is deleted from the new list and pushed toward the end. Repeating the process for the last time shows there is no tardy. Hence the sequence is {3, 1, 5, 4, 2}.

Johnson's algorithm: $N/2/F/F_{\max}$ problem

The algorithm considers the processing times for each job for 2 machines. If the minimum processing time is on machine 1, the corresponding job is placed as early as possible in the sequence. If the minimum processing time is on machine 2, the corresponding job is placed as late as possible in the sequence. The algorithm minimises the total flow time.

Consider 6 jobs that have to be processed on two machines with the processing times shown in Table 3.11.

The problem is $6/2/F/F_{\max}$. Apply the algorithm: the minimum processing time is at job 4 ($P_{52} = 4$) on machine 2. This job is pushed to the end of the sequence { , , , , , 4} and it is deleted from the parent list. The next minimum processing time is at job 5 ($P_{42} = 4$) on machine 2, and again the job is pushed towards the end of the sequence { , , , , 5, 4} and deleted from the parent list. The next minimum processing time is at jobs 1 and 3 (P_{12}, $P_{31} = 6$). There is a tie, and job 3 is placed earliest while job 1 is pushed towards the end of the sequence {3, 2, 1, 5, 4} and are deleted from the parent list. The next search yields $P_{61} = 7$ and job 6 is placed early in the sequence 3,

Table 3.11. Processing times for 6 jobs on 2 machines.

	Processing time (min)	
Jobs	Machine 1	Machine 2
1	7	6
2	15	10
3	6	12
4	9	5
5	8	4
6	7	8

6, 2, 1, 5, 4. Finally, $P_{22} = 10$ is minimum and job 2 is placed in the remaining slot in the sequence {3, 6, 2, 1, 5, 4}.

Campbell, Dudek and Smith (C–D–S) algorithm: $N/M/F/F_{\max}$

C–D–S algorithm is an extension of Johnson's algorithm to M-machines. A number of fictitious two-machines schedules are generated from a given number of machines and the Johnson's two-machines algorithm is applied to each of the machines pains. For the critical N/M processing times, a pair of fictitious machines M_1, and M_2 are combined such that:
for the first pair,

processing time on M_1 = processing time on first machine
processing time on M_2 = processing time on last machine

for the second pair,

processing time on M_1 = processing times on first two machines
processing time on M_2 = processing times on last two machines

for the kth pair,

processing time on M_1 = processing times on first k machines
processing time on M_2 = processing times on last k machines

for the $(m-1)$ pair,

processing time on M_1 = processing time on first $(M-1)$ machines
processing time on M_2 = processing time on last $(M-1)$ machines.

The algorithm minimises $F_{\max}$ for both flow shop and permutation cases. For $M > 2$, sets of fictitious machines are obtained since machines are grouped up to $(M - 1)$ machine. Consider a 5-job problem processed on four machines with the processing times shown in Table 3.12. The task is to find the least throughput time.

Since there are 4 machines, there will be 3 pairs of fictitious machines M_1 and M_2 (Tables 3.13, 3.14, and 3.15).

Table 3.12. Processing times for 6 jobs on 4 machines.

	Processing time (min)			
Jobs	Machine 1	Machine 2	Machine 3	Machine 4
1	9	11	8	14
2	12	9	16	20
3	8	12	18	14
4	16	11	14	9
5	12	9	11	15

Table 3.13. First pair of fictitious machines.

	Machine combination	
Jobs	$M_1 = 1$	$M_2 = 4$
1	9	14
2	12	20
3	8	14
4	16	9
5	12	9

Table 3.14. Second pair of fictitious machines.

	Machine combination	
Jobs	$M_1 = 1, 2$	$M_2 = 3, 4$
1	20	22
2	21	36
3	20	32
4	27	23
5	21	26

Table 3.15. Third pair of fictitious machines.

	Machine combination	
Jobs	$M_1 = 1, 2, 3$	$M_2 = 2, 3, 4$
1	28	33
2	37	45
3	38	44
4	41	34
5	32	35

By using Johnson's algorithm, the sequence is {3, 1, 2, 5, 4}.

Johnson's algorithm gives the sequence {1, 3, 2, 5, 4}.

Johnson's algorithm for the machine combination M_1, M_2 gives sequence {1, 5, 2, 3, 4}. From the three sets of sequences, the best is obtained for the objective function under consideration.

3.4.3. *Branch and bound technique*

Branch and bound problem solving technique in the field of operations research (OR) is a very powerful tool for scheduling. The *branch-and-bound* technique commences with a partial solution at the root node. As branches progress into more complete partial solutions, new terminal nodes replace old ones. Although these optimization algorithms guarantee optimum solutions, they are less attractive in solving many large-scale combination optimization problems. This is because this class of problems is solvable by an amount of computational effort that is bounded by a polynomial function of the size of the problem.

To illustrate the *branch-and-bound* concept, let us assume that the partial assignment of a six-department facility layout problem is {2, 1} as shown in Fig. 3.7, and that the lower bound for the cost of assignment is 800. The details of how to obtain the lower bound is omitted here as we are only interested in grasping the concepts involved. The unassigned departments at this step is represented as $u = \{3, 4, 5, 6\}$. We now branch from node {2, 1} and create the four possible extensions by considering all the unassigned departments. Assuming the lower bounds of 910, 890, 880, and 810 are for nodes {2, 1, 3}, {2, 1, 4}, {2, 1, 5}, and {2, 1, 6} respectively. Since the minimum of all these bounds is 810, we choose the partial assignment {2, 1, 6} that corresponds to the best bound, and we branch from this point. There are three unassigned departments at this stage, designated as $u = \{3, 4, 5\}$. To bound node {2, 1, 6}, for instance, we evaluate the cost of assignments {2, 1, 6, 3}, {2, 1, 6, 4}, {2, 1, 6, 5}. Suppose their costs are 990, 810, and 940 respectively. Since node {2, 1, 6, 4} has the least cost of 810, we choose this node. The unassigned departments at this stage is $u = \{3, 5\}$. It should be noted that *branch-and-bound* carries out $n - 2$ assignment before reaching the optimum solution, where n is the problem size. As no terminal node exists with lower bounds less than 810, we stop the *branch-and-bound* algorithm with optimal assignment {2, 1, 6, 4, 3, 5}. Results for these analyses are shown in Fig. 3.7.

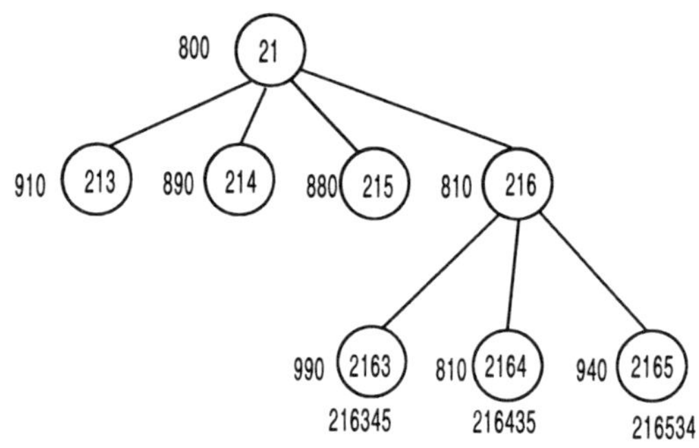

Fig. 3.7. Search trees for branch-and-bound algorithm.

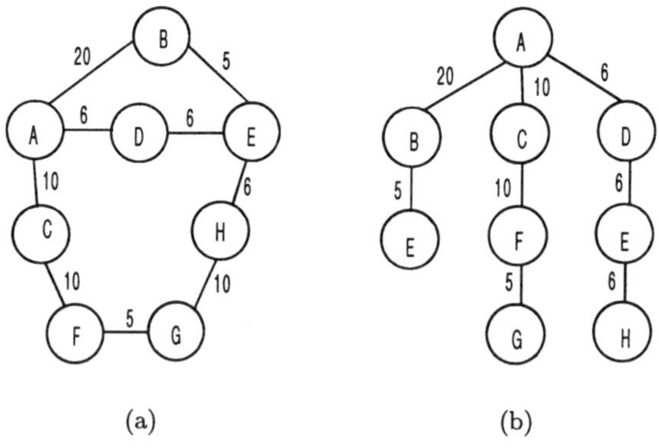

Fig. 3.8. A* algorithm for warehouse logistics.

Another algorithm is the A* algorithm. To illustrate the A* algorithm, let us assume that a material handling system has to travel from the factory warehouse (A), to a depot (H) as shown in Fig. 3.8. The material handling system can travel via several routes which go via B, C, D, E, F, and G. The travel distances between depots are known and depicted on the arcs as shown in Fig. 3.8(a).

Using A* algorithm, all the existing paths that have been found so far are evaluated and the shortest one is chosen and advances. This is illustrated as follows:

Step 1: Starting from the warehouse, A, all the paths from A are set out as branches in the tree structure shown in Fig. 3.8(b) and the shortest path to another depot is chosen. Paths A–B, A–C, A–D have lengths 20, 10,and 6 respectively. Therefore, path A–D is chosen.

Step 2: From A–D, the next path is A–D–E, which has a length of 12. So far, the three existing paths are A–B, A–C, and A–D–E having lengths 20, 10, and 12 respectively. Therefore, A–C is chosen.

Step 3: From A–C, the next path is A–C–F that has a length of 20, 20, and 12, resulting in A–D–E being chosen.

Step 4: From A–D–E, the next path is A–D–E–H that has a length of 18. The existing paths are A–D–E–H, A–C–F, and A–B having lengths 18, 20, and 20 respectively. Therefore, A–D–E–H is chosen and becomes the solution because the algorithm can find no shorter path. Both branch-and-bound and A*, are very useful for small problems when the objective is to find optimum solutions. However, these approaches are limited in solving practical, real life problems. Thus, one is compelled to use approximation algorithms or heuristics for which optimum solution is not guaranteed, but for which polynomial bounds on the computation time can be given.

Since the concept of branch-and bound is now clear, let us apply it to the scheduling problem. An N-job is represented in form of a tree and the priority job is placed at the root. This problem can be stated as: schedule x jobs subject to the constraint that the kth job must be completed first where $k = [1, 2, \dots, i]$. For $i = 4$, we have jobs {1, 2, 3, 4} and there are $(i - 1)!$ combinations. So there are $3! = 6$ combinations which are shown in tree form in Fig. 3.9. These combinations are: {2, 3, 4, 1}; {2, 3, 1, 4}; {2, 4, 3, 1}; {2, 4, 1, 3}; {2, 1, 3, 4}; {2, 1, 4, 3}.

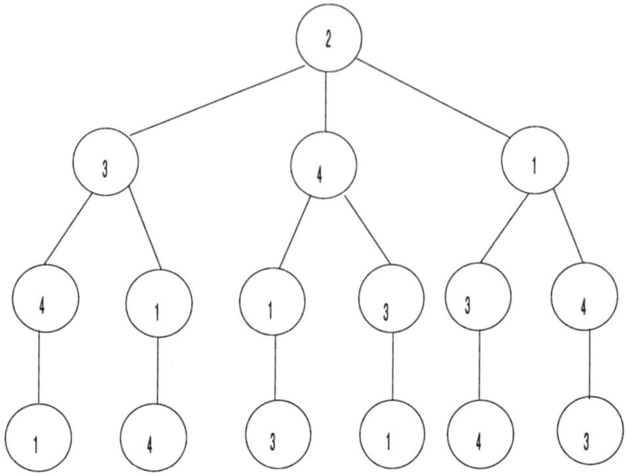

Fig. 3.9. Tree for four jobs on a simple machine.

A propagation cost rule is a very powerful tool for determining *node-propagation direction* in a search tree. Computational time usually associated with goal search for complex problems is drastically reduced to a minimal, using this method. The propagation cost rule must be particular to a problem and the user must know how to set up heuristic for a particular problem domain. The solution to the problem is the route having the least heuristic value. Heuristic value (H) is a combination of (a) the total of cumulative operation cost (C) involved from the root to a given mode in a particular level, and (b) the estimated operational cost (E) from the mode being considered to the goal. By definition, $H = E + C$.

The heuristic value for the goal is least among all nodes in the level where the goal exists. As we do not know the goal in advance, we initially aim at searching every node but we are guided by the heuristic value at each node.

Example 4

Consider the single-machine problem of Table 3.16 with four jobs to be scheduled. The task is to use branch and bound search technique to find the optimum sequence if job 2 must be completed first.

Table 3.16. Processing times for 4 jobs.

Job	1	2	3	4
Processing time (minutes)	3	5	4	2

A heuristic based on the operational cost from the node under consideration to the goal is used to search for the optimum sequence in the tree.

Step 1: start with the given processing time information.

Table 3.17. 4-job single machine process times.

Job	1	2	3	4
Processing time (minutes)	3	5	4	2

Step 2: obtained the total processing time (makespan),
Makespan $= 3 + 5 + 4 + 2 = 14$

Step 3: by using the possible sequences of jobs represented by the tree in Fig. 3.9, generate the possible job-processing times (Tables 3.18–3.23).

Table 3.18. A possible sequence {2, 3, 4, 1}.

Job	2	3	4	1
Processing time (minutes)	5	4	2	3

Table 3.19. A possible sequence {2, 3, 1, 4}.

Job	2	3	1	4
Processing time (minutes)	5	4	3	2

Table 3.20. A possible sequence {2, 4, 3, 1}.

Job	2	4	3	1
Processing time (minutes)	5	2	4	3

Table 3.21. A possible sequence {2, 4, 1, 3}.

Job	2	4	1	3
Processing time (minutes)	5	2	3	4

Table 3.22. A possible sequence {2, 1, 3, 4}.

Job	2	1	3	4
Processing time (minutes)	5	3	4	2

Table 3.23. A possible sequence {2, 1, 4, 3}.

Job	2	1	4	3
Processing time (minutes)	5	3	2	4

Step 4: generate the estimated operational cost from the node being considered to the goal. The goal exists in the path from the root node to the leaf node which gives the minimum value. The completion time is obtained by summing the processing times. The estimate operational cost (E) is the measure of remaining time before getting to the finishing time of all the jobs. The estimated operational cost is given as: E = makespan — current completion time. The makespan is already calculated in Step 2. The cumulative operational cost is the same for each level and hence it may be left out in computing the heuristic cost H. In this case we can without loss of accuracy, define $H = E$. For sequence {2, 3, 4, 1}, using Table 3.18, the completion time for job 2 is 5 minutes, for job 3 is 9 minutes, for job 1 is 12 minutes and for job 4 is 14 minutes. These values are obtained by summing up the *sequenced processing times.* The heuristic cost for job 2 is 14 (makespan) — 5 (completion time) = 9 minutes, for job 3 is 5 minutes, for job 1 is 2 minutes and for job 4 is zero. These results are shown in Table 3.24. Tables 3.25, 3.26, 3.27, 3.28 and 3.29 show the result of the heuristic costs for {2, 3, 1, 4}, {2, 4, 3, 1}, {2, 4, 1, 3}, {2, 1, 3, 4}, {2, 1, 4, 3} respectively.

Table 3.24. Heuristic cost for sequence {2, 3, 4, 1}.

Job	2	3	4	1
Processing time	5	4	2	3
Completion time	5	9	11	14
Heuristic cost	9	5	3	0

Table 3.25. Heuristic cost for sequence {2, 3, 1, 4}.

Job	2	3	1	4
Processing time	5	4	3	2
Completion time	5	9	12	14
Heuristic cost	9	5	2	0

Table 3.26. Heuristic cost for sequence {2, 4, 3, 1}.

Job	2	4	3	1
Processing time	5	2	4	3
Completion time	5	7	11	14
Heuristic cost	9	7	3	0

Table 3.27. Heuristic cost for sequence {2, 4, 1, 3}.

Job	2	4	1	3
Processing time	5	2	3	4
Completion time	5	7	10	14
Heuristic cost	9	7	4	0

Table 3.28. Heuristic cost for sequence {2, 1, 3, 4}.

Job	2	1	3	4
Processing time	5	3	4	2
Completion time	5	8	12	14
Heuristic cost	9	6	2	0

Table 3.29. Heuristic cost for sequence {2, 1, 4, 3}.

Job	2	1	4	3
Processing time	5	3	2	4
Completion time	5	8	10	14
Heuristic cost	9	6	4	0

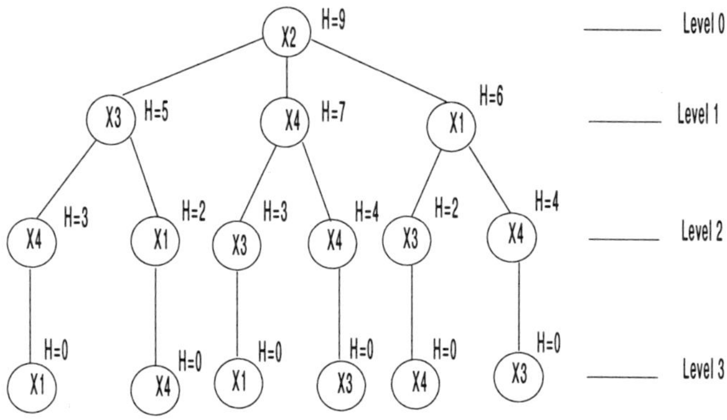

Fig. 3.10. Search tree for optimal sequencing using Heuristic cost.

Step 5: append the heuristic costs on the tree as shown in Fig. 3.10.
Step 6: search for optimum solution.

The heuristic formulation conserves search time because search is carried out level by level. Once a least heuristic cost is encountered in a level, all other branches are abandoned and search concentrates in the direction of least heuristic value. Starting the search from the root node 2, and considering all nodes at the level 1, minimum of heuristic cost, H is 5 for node 3. This means that node 3 is chosen and the left route is retained while all others are abandoned. We now have nodes 2 and 3 as part of the solution path. At node 3, level 1, we are left with nodes 4 and 1 at level 2, each having $H = 3$ and 2 respectively. Node 1 is chosen because minimum heuristic cost at this level is $H = 2$. Our solution is updated to {2, 3, 1}. Since there is no other decision to make at node 1, the leaf-node 4 is automatically chosen so that the goal state and consequently the optimum sequence is {2, 3, 1, 4}.

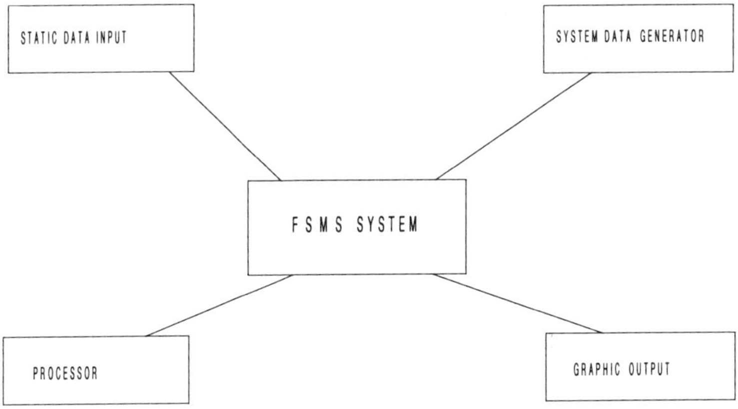

Fig. 3.11. Components of the graphical flow shop manufacturing system.

3.4.4. *A graphical flow-shop manufacturing scheduling system: FSMS*

The overall structure of the flow-shop manufacturing scheduling (FSMS) system described in this section is shown in Fig. 3.11. The system is menu driven and comprises of the data entry, system generator, processor, and graphic output functions. The goals of the FSMS system in an n-jobs m-machines flow-shop environment are:

(1) minimising makespan;
(2) performing the scheduling function;
(3) providing a tool for graphical scheduling.

The different components of the overall structure of the FSMS system are discussed in order to give an insight into various techniques used for the implementation.

The data-entry function

There are three types of data input into the FSMS system: the *primary*, *static* and *system* data. *Primary* data input from the scheduler is made up of the number of machines, number of jobs, and the processing times with inclusion of set-up and transport times. The *static data* are the sequences input to the system by the scheduler, if the scheduler chooses this option. If there are m-machines and the user chooses to utilise sequences generated by the system, the system uses the *primary* data to generate $m - 1$ sequences, based on the algorithm of Campbell *et al.* (1970). This is a generalised form of Johnson's algorithm (1954). The collection of the $m-1$ sequences is referred to as *system data*. The *system* or *static* data is used to drive the processor.

The editor

When the primary data are input in the data-entry mode, it is possible for the scheduler to make mistake. An editor function is available so that primary data are verified before any processing starts. If the primary data is wrong, the system data will also be wrong as well as other results which the processor outputs.

The processor

The processor performs the scheduling functions. It determines the delay times for each job on the various machines. Computation of the delay time is very important because it determines the loading pattern on the Gantt chart. Delay time is either zero or positive. The value determines whether a succeeding machine will have to idle or commence as soon as the subsequent machine finishes operation. The processor also computes the makespan and the maximum lower bound (MLB). The MLB gives the scheduler an indication of the optimal makespan, which in turn gives the optimum sequence. The makespan is the total time required to complete all jobs. This information is important to the scheduler because it helps to indicate whether or not meeting due-date is possible.

The graphic output function

The processor utilises the input data and the delay times generated by the system to output the Gantt chart. The Gantt chart is made up of time scale on one axis of a graph and the machine centres on the other axis. Blocks having lengths equivalent to their processing times are displayed on the correct machine axis for the particular job being scheduled. The correct sequencing is maintained and placed in the blocks for proper identification. The Gantt chart is a powerful tool when used by a supervisor on the factory shop-floor because it graphically displays the work status of the various machines at any time interval within the makespan. It also shows the makespan, which in this case is the time the last job in the sequence is completed on the last machine. The advantages of the FSMS system are:

(1) shortens scheduling time because only limited data are initially input and processing is fast;
(2) offers flexibility because by using the initial input data various sequences can be used by the system to offer alternative scheduling solutions from which the optimum can be obtained;
(3) conveys scheduling in graphical forms to scheduler and shop-floor operators.

Example 5

The 4-machine 4-job processing times shown in Table 3.30 are input data used to demonstrate the FSMS scheduling procedure.

Table 3.30. Processing times (minutes).

	Machines			
Jobs(n)	1	2	3	4
1	2	3.5	1.5	2
2	4.5	3	2.5	1
3	1.5	1.5	5	0.5
4	4	1	2.5	0.5

Lower bounds

Based on the input processing times, the system computes the lower bounds on schedule makespan for each machine. The lower bound is the summation of the minimum of processing time before the machine can begin processing, the total processing time on the machine, and the minimum of the remaining processing time for the last job after it leaves the machine. Machine 1 is the first machine and so no machine starts processing before it. The processing time for all jobs on machine 1 is 12 minutes. The minimum remaining processing times on machines 2, 3, 4 sums up to be 4 minutes. Therefore, the lower bound for machine1 is 0 + 12 + 4 = 16 minutes. Similarly, machine 2 has a lower bound of 13.5 minutes, machine 3 has a value of 15 minutes while machine 4 has a value of 11 minutes. The maximum value of 16 minutes becomes the lower bound for the 4-stage scheduling system. This maximum lower bound helps the scheduler to judge if a schedule is near optimal solution.

System sequences

The system now uses the processing times input by the scheduler to generate three possible sequences based on Campbell *et al.* algorithm (1970). The sequences are {3, 1, 2, 4}, {1, 3, 2, 4}, and {1, 2, 4, 3}. These form the initial options available to the scheduler as *system data* for input.

Schedule optimisation

So far, the lower bound and the system sequences generation have been system-driven. This means that the scheduler is not in control. We prefer a situation in which the scheduler can direct the system towards an optimal solution through

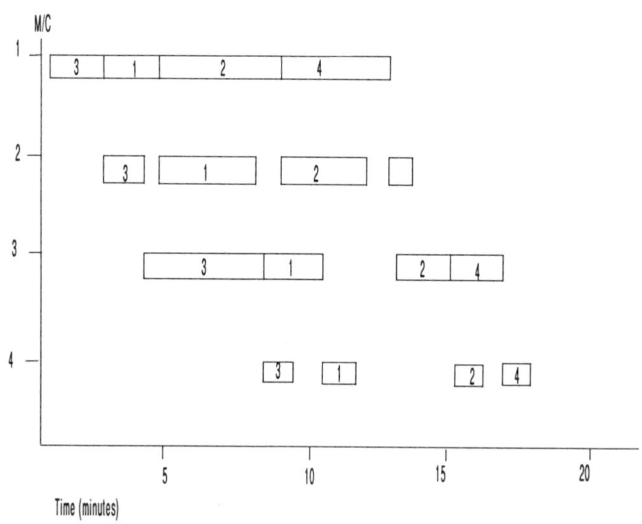

Fig. 3.12. Gantt chart for sequence {3, 1, 2, 4}.

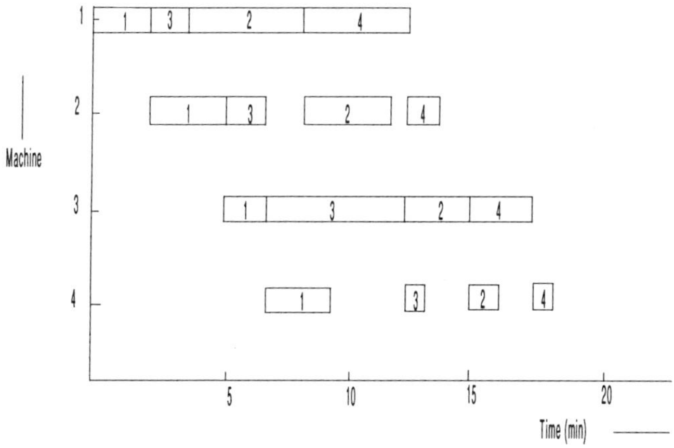

Fig. 3.13. Gantt chart for sequence {1, 3, 2, 4,}.

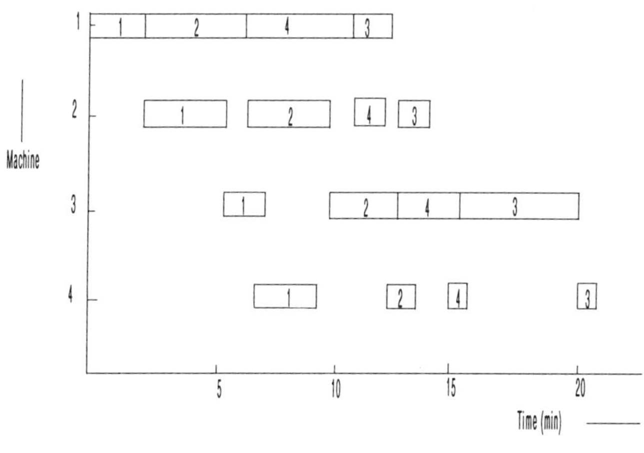

Fig. 3.14. Gantt chart for sequence {1, 2, 4, 3}.

the utilisation of *system* and *static* data. We refer to this as schedule optimisation. First, the system sequences are input by the scheduler for use by the processor. Figures 3.12, 3.13, and 3.14 show the Gantt charts for the sequences {3, 1, 2, 4}, {1, 3, 2, 4}, and {1, 2, 4, 3} respectively. Their makespans are also computed as {17.5 minutes, 17.5 minutes and 20 minutes}. Since the lower bound is 16 minutes, sequence {1, 2, 4, 3} need no further consideration by the scheduler because it has a deviation of 25%. Sequences {3, 1, 2, 4} and {1, 3, 2, 4} each has deviation of 9.38%.

For sequence {3, 1, 2, 4}, we start by assigning job 3 to machine 1 at time zero. The job lasts for 1.5 minutes on machine 1. Since all jobs must visit machine 1 first, while job 3 is being processed by machine 1, jobs 1, 2, and 4 are queued and machines 2, 3, and 4 are idle. At the end of 1.5 minutes, job 3 visits machine 2 while job 1 goes on to machine 1. Jobs 2 and 4 are queued for machine 1 while machines 3 and 4 idle. Meanwhile, job 3 is completed on machine 2 at 1.5 minutes while job 1 is still being processed on machine 1. This means that machine 2 idles for 0.5 minutes, waiting for machine 1 to complete job 1 before it commences. Job 1 then commences on machine 2 and completes 3.5 minutes later while job 2 is still being processed on machine 1. Machine 2 idles for 1 minute, waiting for machine 1 to complete to complete job 2. Job 2 takes 3 minutes on machine 2 and is completed while job 4 is still being processed on machine 1, so that machine 2 idles for 1 minute before job 4 commences on machine 2. Job 3 takes 5 minutes on machine 3 while job 1 on machine 2 is already completed 1 minute earlier on machine 2. Job 1 is completed on machine 3 while job 2 is still being processed on machine 2. Machine 3 therefore idles for 2.5 minutes before job 2 commences. Job 4 takes

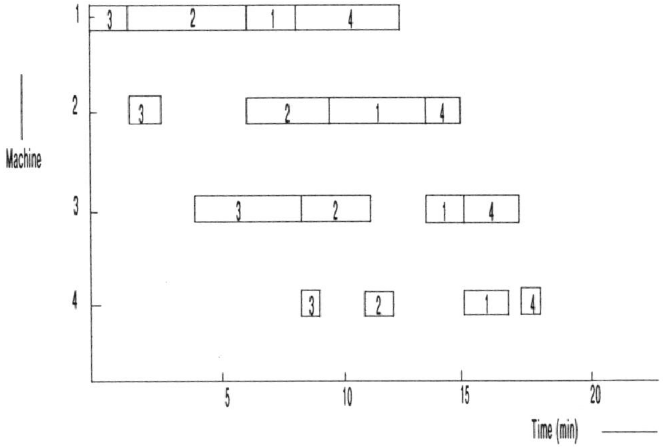

Fig. 3.15. Gantt chart for sequence {3, 2, 1, 4}.

only 1 minute on machine 2 while job 2 takes 2.5 minutes to be completed on machine 3. This means that job 4 starts as soon as job 2 is completed on machine 3 and therefore there is no idling. On machine 4, job 3, job 1, job 2 and job 4 commence when they are completed on machine 3. This means that machine 4 idles for 1 minute before job 1 starts, idles for 3 minutes before job 2 commences, and idles for 1.5 minutes before job 4 starts. The processor unit of the FSMS keeps track of the idle times and the output module uses the idle times and sequenced processing times to draw the Gantt chart. The makespan is computed and displayed so that the scheduler can decide whether to accept the schedule or not. The deviation of the makespan from the lower bound helps the scheduler to take an appropriate decision.

The scheduler may decide to try a few combinations of the system sequences that gave the least makespan. By swapping 1, and 2 in the sequence {3, 1, 2, 4}, it is observed in Fig. 3.15 that the sequence {3, 2, 1, 4} gives the least makespan of 17 minutes, with a deviation of 6.25% form the lower bound.

3.5. Scheduling in Assembly Line Production

An assembly line is made up of work stations where employees or automated machines arrange various components into a finished product. The assembly line has precedence relationships which describe the order in which each assembly task should be performed. Usually the item required for each assembly task is determined and used together with the precedence relationships to estimate the total production rate. The objectives of assembly line layout are:

(1) to minimise material handling cost by minimising distance between adjacent work stations;
(2) to achieve a balanced assembly line so that no workstation is overloaded.

Line balancing forms the basis for scheduling in assembly lines because by assigning tasks to work stations, the line balance specifies the job assignment and material flow.

3.5.1. *Line balancing*

The technique for line balancing is to split the total assembly procedure into sub-assemblies. Precedence relationships are determined, showing which assembly process is performed before another commences. The time taken for each assembly procedure is also determined. Two manual techniques for line balancing which are easily computerised are: largest candidate rule and ranked positional weighted method. Computerising these techniques removes errors and drastically shortens time for assigning workstations for balanced line. Before going into details of the methods used for line balancing, it is worthwhile to define some useful terms:

C = cycle time for the balance
C_a = actual time for the balance after balance is achieved
t_i = time for completing task for element i; $i = 1, \ldots, n$ tasks
T = total time for completing all tasks
N = theoretical minimum number of workstations
N_a = actual number of workstations in the balanced line
B_d = balanced delay

$$N = \sum_i t_i / C = T/C$$

$$B_d = 1 - \frac{T}{N_a C_a}\,.$$

We now discuss two methods which have been found useful for line balancing.

Lowest Candidate Rule (LCR)

The steps for LCR are as follows:

(1) Arrange the elements in the precedence diagram in decreasing order of the time for each task. The immediate predecessor tasks are also listed along side;
(2) Commence assigning tasks to station 1;
(3) Start from the top of the list and find a feasible task and calculate the time remaining at the station. Once a feasible task is found, it is deleted from the list. A feasible task must not have predecessors. Continue to find tasks which are feasible with respect to the remaining time at the workstation;
(4) If no more feasible task is found for remaining time for a workstation, start another workstation assignment by using the cycle time for the line balancing;

(5) When all feasible tasks are assigned, the total number of stations represent the actual number of workstations required for the balanced line and the maximum cumulative time obtained by considering each station is the true actual cycle time used to compute the balance delay.

Example 6

Consider the precedence diagram in Fig. 3.16 and the work elements information in Table 3.31. The master production schedule indicates that 9600 units are required per month. The company works 20 days per month and 8 hours per day. Using the largest candidate rule, the workstations balance delay and efficiency are obtained.

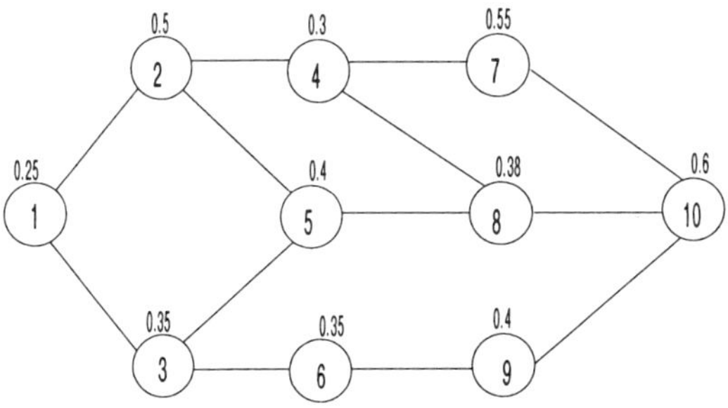

Fig. 3.16. Precedence diagram for example 1.

Table 3.31. Work element time and intermediate predecessor.

Element	Time (min.)	Intermediate predecessor
1	0.25	—
2	0.50	1
3	0.35	1
4	0.30	2
5	0.40	2, 3
6	0.65	3
7	0.55	4
8	0.38	4, 5
9	0.40	6
10	0.60	7, 8, 9

Solution

The cycle time is first calculated. From the master production schedule, production rate per day (PR) = 9600/(20 × 8) = 60 units/hr = 1 unit/min.

$$\text{Cycle time} = 1/\text{PR} = 1\ \text{min/unit}$$

$$\text{Total time } T = \sum_i t_i = 4.38\ \text{min};\ N = T/C = 4.38 \sim 5.$$

The elements are arranged in decreasing order of the task time and shown in Table 3.32.

Table 3.32. Elements arranged in descending order of time.

Element	Time (min.)	Intermediate predecessor
6	0.65	3
10	0.60	7,8,9
7	0.55	4
2	0.50	1
5	0.40	2,3
9	0.40	6
8	0.38	4,5
3	0.35	1
4	0.30	2
1	0.25	—
	$T = 4.38$ min	

Starting from the top of the list and seeking for the first feasible task and noting that it must not have a predecessor and that it has to be equal to or less than the cycle time of 1 min/unit, we find that element 1 qualifies so t_i enters the list of station 1. The time remaining at the station is the cycle time less the time for task 1 which is $1 - 0.25 = 0.75$ min. Task 1 is deleted from the list and we go to the top and move down seeking for the next feasible task. Since task 2 has no predecessor because task 1 has already been deleted and the remaining time is $0.75 - 0.5 = 0.25$, it passes the test. No feasible task exists for station 1 and cumulative time is $0.25 + 0.5 = 0.75$. Station 2 starts, now with the cycle time = 1 min/unit. Task 3 is feasible because task 1 predecessor has been completed and remaining time is $1 - 0.35 = 0.65$. So we delete task 3 and return to the top of the list and moving downwards we find task 6 is eligible because precedence task 3 has been completed and remaining time = 0.65 − 0.65 = 0. This station is complete and station 3 commences with task 5 qualifying since tasks 2, 3 are already completed and remaining time = 1 − 0.4 = 0.6. We return to the top of the list and move downwards to find task 9 qualifying because task 6 has been completed and remaining time = 0.6 − 0.4 = 0.2. Station 3 is complete and we commence station 4. Moving from the top, task 4 qualifies since task 2, he predecessor is completed and remaining time = 1 − 0.3 = 0.7. Task 4 is deleted. We return to the top and continue search, with task 7 qualifying because its predecessor 4 has been completed and remaining time = 0.7 − 0.55 = 0.15. Task 7 is deleted. Station 5 commences with cycle time = 1. Searching from the top, task 8 qualifies with predecessors 4, 5 completed and remaining time = 1 − 0.38 = 0.62. It is deleted from the list and we return to top of the list. Move downwards we find task 10 qualifying because predecessors 7, 8, 9 have been completed and remaining time = 0.62 − 0.6 = 0.02.

Table 3.33. Station assignment.

Station	Element t	Time remaining at the station		Time (min.)	Cumulative time (min.)
1	1	0.75	1	0.25	0
	2	0.25		0.50	0.75
2	3	0.65	1	0.35	0
	6	0.0		0.65	1.00
3	5	0.60	1	0.40	0
	9	0.20		0.40	0.80
4	4	0.70	1	0.30	0
	7	0.15		0.55	0.85
5	8	0.62	1	0.38	0
	10	0.02		0.6	0.98

The complete stations assignment is depicted in Table 3.33 and station 2 has the maximum cumulative time which is the actual cycle time, $C_a = 1$. Actual number of workstation $N_a = 5$.

The balance delay, $B_a = 1 - 4.38/(5^*1) = 0.124$.
Efficiency of balance $= (1 - B_d)^*100 = 87.6\%$.

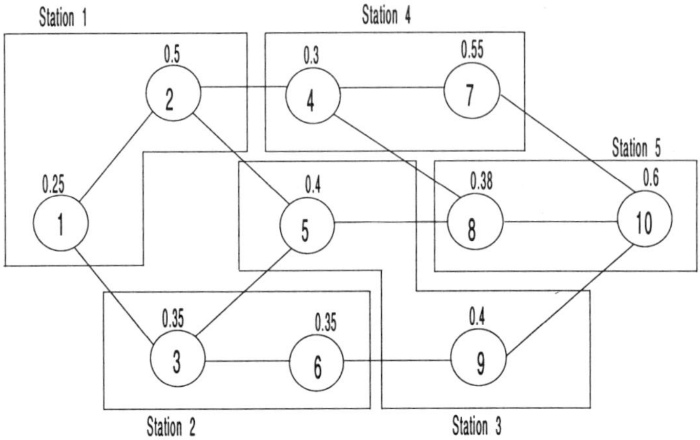

Fig. 3.17. Solution to example 1 showing stations configuration.

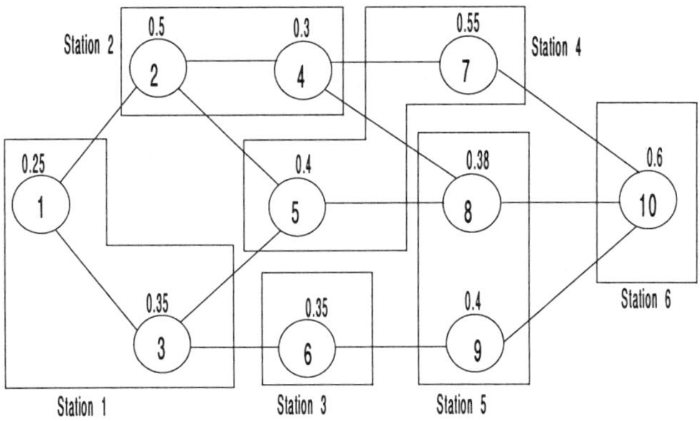

Fig. 3.18. Station configuration of example 1 using RWR method.

The final solution is graphically shown by assigning the station tasks in blocks as shown in Fig. 3.17.

Ranked Positional Weights (RPW) method

The summation of the time from any task to the last task gives the weight of that task. The weight for task 1 is the total time for the precedence diagram.

Referring to Fig. 3.18, the weight for task 5 is the summation of the time for tasks 5, 8 and 10 (i.e. 0.4 + 0.38 + 0.60 = 1.38). The tasks are then ranked according to the weights in a descending order. The ordering of elements for the example depicted by precedence relationships of Fig. 3.15 is shown in Table 3.34.

Similar steps are taken as in largest candidate rule by using the time for each task and the cycle time and concept of remaining time to obtain the stations. The stations assignment are shown in Table 3.35

The actual cycle time, $C_a = 0.95$.
The actual number of stations, $N_a = 6$.

Table 3.34. Ranked positional weights for tasks in descending order.

Element	RWW	Time (min.)	Immediate predecessor
1	4.38	0.25	—
3	2.78	0.35	1
2	2.73	0.50	1
6	1.65	0.65	3
4	1.45	0.30	2
5	1.38	0.40	2, 3
7	1.15	0.55	4
9	1.00	0.40	6
8	0.98	0.38	4, 5
10	0.6	0.60	7,8,9

Table 3.35. Stations assignment

Station	Element	Remaining time (min.)	Time (min.)	Cumulative time (min.)
1	1	0.75	0.25	
	3	0.40	0.35	0.60
2	2	0.50	0.50	
	4	0.20	0.30	0.80
3	6	0.35	0.65	0.65
4	5	0.60	0.40	
	7	0.10	0.55	0.95
5	9	0.6	0.40	
	8	0.22	0.38	0.78
6	10	0.4	0.6	0.60

The balance delay, $B_d = 1 - 4.38/(6^*0.95) = 0.2316$ min.
Efficiency of the balanced line $= (1 - 0.2316)^*(100) = 76.84\%$.

The complete graphical solution of the station are shown in Fig. 3.18.

3.6. Material Management

Material management includes decisions regarding purchasing, inventory control, material handling and storage, and distribution system. We briefly discuss these components and dedicate the next subsection to inventory control.

3.6.1. *Purchasing*

Purchasing involved the activities of procuring goods or services from a supplier or vendor for an organization. The organization is referred to as the customer. Various decsions are reauired in purchasing. The aspects covered in purchasing are as follows:

1. *Defining specification for purchased goods/service.* This begins with the design of the product, up to product delivery. The specifications should be checked to be met by current suppliers, checked if they are consistent with accepted commercial standards.
2. *Sourcing price quotations*, through:
   (i) published price list;
   (ii) competitive bidding (by considering several suppler);
   (iii) two-step bidding (competitive bidding and negotiation);
   (iv) negotiation (usually for specialised products).
3. *Developing criteria for selecting suppliers*, such as:
   (i) quality goods/services;
   (ii) timely deliveries;
   (iii) affordable price;
   (iv) meeting customer requirements.
4. *Classification of suppliers.* This is done by clasifying suppliers using performance measures.

5. Make-or-buy decision. This decision is made through determining:
   (i) whether suppliers can deliver a quality product;
   (ii) whether suppliers can deliver on time;
   (iii) associated costs with each alternative;
   (iv) who is technically qualified to supply.
6. *Awarding contract.* This is done through comparison of suppliers with low price and slow delivery with suppliers with higher prices and better delivery history. Quality should be taken very seriously and without compromise.
7. Expediting. This is the monitoring of supplier deliveries of materials which are critical to the customer

The make-or-buy decision can be made by means of process selection, which is now treated in some details in the next subsection.

3.6.2. *Process selection*

Process selection is a series of decisions that includes technical issues and volume (scale) issues. While product design determines what should be made, and capacity determines volume to be made, process selection determines how a product should be made. The formulation of process selection criteria utilises the cost associated with production on one side and the selling price on the other side:

$$\text{Total cost production} = \text{fixed costs } (F) + \text{variable costs } (V)\,.$$

The variable cost affect the number of units, X, to be produced.

Selling cost, S is also dependent on the number of units to be produced. Hence:

$$T = F + X \bullet V$$

$$R = S \bullet X$$

where R is revenue generated, T is total production cost.

$$\text{Profit, } P = \text{Revenue} - \text{Total production costs}\,.$$

We define:

$$P = S \bullet X - (F + V \bullet X) = X \bullet (S - V) - F\,.$$

Consequently,

$$X = \frac{P+F}{S-V}\,.$$

For the case of mixed model where each model has a weight m associated with it,

$$X = \frac{P+7}{\sum_{i=1}^{n} m_i(S_i - V_i)}\,.$$

Simple as these equations are, they are effective for process selection. Let us solve some examples in order to clarify the use of these process selection equations.

Example 7

Table 3.36 shows the costs associated with 2 methods of production and their production rate. Which option is preferable for producing 2000 of each unit per year?

Table 3.36. Costs data.

Production Type	Annual Fixed Costs	Variable Costs	Production Rate
Method I	\$2000	\$10/hr	30/hr
Method II	\$7000	\$4/hr	400/hr

Solution

Method I $\quad T = F + X.V = 2000 + (\$\underline{10}/\text{hr})/(30/\text{hr})^{*}(2000)$

$= \$2666.7.$

$\text{Unit Cost} = \dfrac{2667}{2000} = \$1.3/\text{unit} \quad \text{volume of 2000.}$

Method II $\quad T = 7000 + \dfrac{4/\text{hr}}{400/\text{hr}} \times (2000)$

$= 7020.$

$\text{Unit cost} = \dfrac{7020}{2000} = \$3.5/\text{unit} \quad \text{volume of 2000.}$

From the analysis, method I is preferred because \$1.3/unit < \$3.5/unit.

Example 8

A city bakery sells high-quality bread. The loaves of bread sell for \$5.00 each and cost \$1.50 per loaf to produce. Fixed costs total \$30 000 per annum.

(i) How many loaves of bread must the bakery sell to cover fixed costs?
(ii) How many loaves of bread must the bakery sell to make a \$60 000 profit per year?
(iii) If fixed costs increase to \$50 000 per year, what are the answers to (i) and (ii)?

Solution

Using $X = \dfrac{P+F}{S-V}$.

(i) to cover fixed costs, profit, $P = 0$.

$$X = \frac{0 + 30\,000}{5 - 1.50} = 8571 \text{ loaves}.$$

(ii) $P = 60\,000$,

$$X = \frac{60\,000 + 30\,000}{5 - 1.50} = 25\,714 \text{ loaves}.$$

(iii) for $F = 60\,000$,

$$X = \frac{0 + 50\,000}{5 - 1.50} = 14\,286 \text{ loaves}, \text{ for problem (i)}, \text{ and}$$

$$X = \frac{60\,000 + 50\,000}{5 - 1.50} = 31\,429 \text{ loaves for problem (ii)}.$$

Example 9

A small company produces resistors, capacitors and transistors in the same facility and costs involved are shown in Table 3.37:

Table 3.37. Costs for producing various product mixes.

	Transistors	Capacitors	Resistors
Produce mix	0.4	0.4	0.2
Selling price/unit	\$0.09	\$0.075	\$0.03
Variable price/unit	\$0.06	\$0.04	\$0.029

Annual fixed costs are estimated at \$2 750 000.

(i) At what volume will the company cover its costs, given the present mix?
(ii) At what volume will it report a \$2 000 000 profit, given the present mix?
(iii) If the price of a transistor is raised by \$0,02, what happens to the volume required to make a \$2 000 000 profit ?

Solution

(i) $$\sum_{i=1}^{3} M_i(S_i - V_i) = 0.4(0.09 - 0.06) + 0.4(0.075 - 0.04) + 0.2(0.03 - 0.02)$$
$$= 0.012 + 0.014 + 0.002 = \$0.028/\text{composite units}$$

$$X = \frac{P + F}{\sum_{i=1}^{3}(S_i - V_1)} = \frac{0 + 2\,750\,000}{0.028}$$

$$= 98 \text{ million composite units}\,.$$

(ii) $$X = \frac{2\,000\,000 + 2\,750\,000}{0.028} = 169 \text{ million composite units.}$$

(iii) If transistor selling price is raised by \$0.02, it becomes \$0.11/unit.

$$\sum_{i=1}^{3} M_i(S_i - V_i) = 0.4(0.11 - 0.06) + 0.4(0.075 - 0.04) + 0.2(0.03 - 0.02)$$
$$= 0.02 + 0.014 + 0.002$$
$$= 0.036\,.$$

$$X = \frac{2\,000\,000 + 2\,750\,000}{0.036} = 132 \text{ million composite units}\,.$$

3.6.3. *Inventory control*

Inventory is material that is meant to be turned into a useful product, but which is yet to be transformed into a useful form. Three forms of inventories are common in the manufacturing sector: raw material, work-in-progress, and finished product inventories. The goal of manufacturing is to eliminate inventories. This means that raw material, work-in-progress, and finished product inventories should be reduced. More of inventory control in a later part of this chapter.

3.6.4. *Material handling and storage*

Although work-in-progress inventory is reduced through flexible automation, set-up time reduction programs and point of use storage, it is still necessary to provide warehouses or storage systems, particularly for distribution and spare parts storage. Storage system or warehouse may range from manually controlled fork-lift-based operations to fully automated storage and retrieval system (AS/RS) with narrow aisle. Storage systems require material handling systems for effective and efficient operations. Material handling and storage includes receiving/shipping, material handling devices (AGVs, forklifts, conveyors, overhead cranes, etc.), storage devices, and intelligent warehouse systems.

3.6.5. *Distribution and logistics*

A distribution system links suppliers and customers, from suppliers facility. Logistics is the process of strategically managing the acquisition, movement and storage of materials, parts and finished inventory (and related information flows) through an organisation and its marketing channels to fulfill orders most cost-effective. The significance of logistics in value chain management include: (i) 2 out of 7 (30%) activities in the integrated materials value pipeline are attributed to logistics, (ii) logistics expenditures for individual firms typically range from 5 to 35 percent of sales logistics costs in the United States was just under 10% of the 1994 gross national product.

The components of logistics are typically:

(i) network design, which concerns the determination of the size, and geographical relationship of facilities used to perform logistical operations.

Logistic facilities include manufacturing plants, warehouses, cross-docks operations , and retail stores.

(ii) information, which includes electronic data interchange (EDI), application of recently developed information technology for control concepts such as just-in-time (JIT), quick response (QR), and continuous replenishment (CR).

(iii) inventory, which entail achieving the desired consumer service with minimum inventory commitment, consistent with lowest total cost.

(iv) warehousing, material handling, and packaging, which represent a mini-system within the logistical process since major activities such as sorting, sequencing, order selection, transportation consolidation are done while products are warehoused; material handling deals with receiving, moving, sorting and assembling products in the warehouse to meet customer order requirement.

Integrated logistic involves:

(i) logistics is the competency that links an enterprise with its customer and supply chains.

(ii) information from and about customer flows enterprise in forms of sales activities, forecasts, and orders.

(iii) value-added inventory flow is initiated as products and materials are procured that ultimately results in ownership of finished products to customers.

Barriers to internal logistic integration include:

(i) organizational structure — traditional structure may inhibit cross-functional process being implemented.

(ii) measurement systems — traditional measurement systems reflect organisation structure and hence inhibit cross-functional coordination.

(iii) inventory ownership — cost-benefit relationship and the risks associated with incorrectly located inventory. Decision on availability of inventory in firm or forward commitment of inventory to local market.

(iv) information technology, which is the key resource to achieve integration.

(v) knowledge transfer — inability of a firm to develop procedures and systems for transferring cross-functional knowledge is a limiting factor of integration.

3.7. Inventory Control

When supplier performance or customer demand is uncertain, a company may be forced to hold some inventories. This is referred to as just-in-case inventory control system and the uncertain demand situation is known as *independent demand*. However, when the customer demand can be anticipated or when supplier performance/customer demand is certain as in MRP or JIT, a different situation arises, known as *dependent demand*.

Materials management encompasses the subject of independent demand, dependent demand, and materials inventory control systems. Other areas that are currently receiving attention include supplier chain and distribution logistics. Techniques for lot sizing or order quantity differ for independent demand and dependent demand items. These techniques for lot sizing are now discussed in detail.

3.7.1. *Lot sizing for independent demand items*

Independent demand items have been defined as those in which it is difficult to predict when they will arrive from suppliers or when they will be ordered

Table 3.38. Holding and replenishing inventory costs.

Inventory type	Fixed cost	Variable cost
Ordering (replenishing)	• Employee cost (payroll, benefits, etc.) • Facility, furniture and equipment	• Dispatch Costs • Cost of placing and order (order form, phone, postage) • Cost of lost production during setup • Cost of material wasted during start after changeover
Holding (carrying)	• Warehouse cost (capital, operating costs) • Taxes on company property • Personnel Costs	• Cost of capital inventory • Insurance on inventory value • Taxes on inventory values • Losses: pilfering, spoilage

by customers. This kind of uncertainty compels a company to manufacture or place order in batches in order to protect against possible shutdowns resulting from shortage in raw-materials or parts. Holding (carrying) or replenishing (ordering) inventory is associated with some costs shown in Table 3.38. Some are fixed, others are variable. These cost are used to formulate the various models used to determine lot sizes.

Variable costs have more effect than fixed cost on short term planning, hence they are considered in formulating the economic order quantity (EOQ) model.

3.7.2. *Economic order quantity*

The more the quantity of materials ordered at a time, the less the ordering or replenishing cost and the more the carrying or holding cost, and vice versa. This results in a trade-off between the annual ordering costs and carrying costs as shown in Fig. 3.19. The EOQ model is based on minimising the total annual cost by varying the order quantity or the lot size.

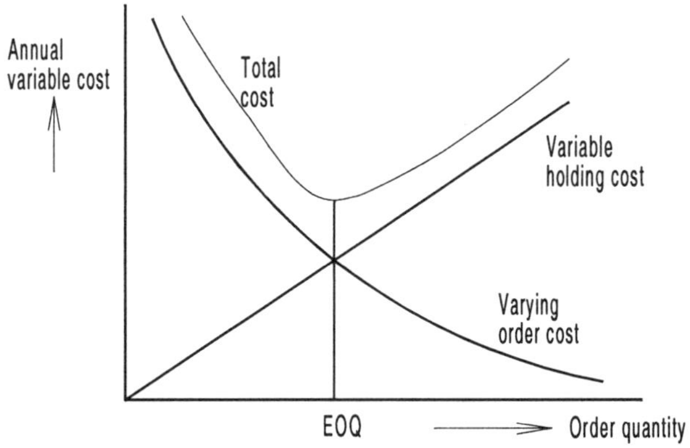

Fig. 3.19. Total annual cost: EOQ model.

The assumptions in EOQ are:

(i) constant demand;
(ii) purchase price is independent of order quantity;
(iii) order delivery is instantaneous.

The first and last assumptions lead to the following EOQ inventory pattern shown in Fig. 3.20:

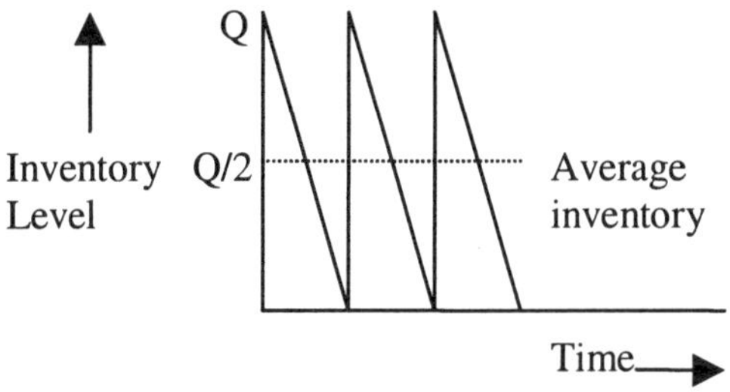

Fig. 3.20. EOQ inventory pattern.

We define: D = annual demand (units/yr)
Q = order quantity or lot size (units)
C_o = variable ordering cost (cost/order)
P = holding cost: % of unit cost, U
C_h = variable holding cost (cost/unit/yr).

The annual ordering cost is $\frac{D}{Q} * C_o$ (Cost/order/yr).

The annual holding cost is $\frac{Q}{2} * C_h$ (Cost/yr).

The total annual variable cost is given by

$$T = \frac{D}{Q} * C_o + \frac{Q}{2} * C_h \,.$$

When the total annual variable cost is differentiated with respect to Q we have,

$$\frac{\partial T}{\partial Q} = \frac{-DC_o}{Q^2} + \frac{C_h}{2} = 0 \,.$$

$$\therefore \quad \text{EOQ} = \sqrt{\frac{2DC_o}{C_h}} \,.$$

Example 10

A fabrication shop uses flat sheets for manufacturing metal boxes. The following information relate to the flat metal sheets:

$$\begin{aligned}&\text{demand} = 1000/\text{annum}\\&\text{cost price} = \$15/\text{unit standard size}\\&\text{order cost} = \$10/\text{order}\\&\text{holding cost} = 25\%\text{ of unit cost/annum.}\end{aligned}$$

Determine the quantity of metal sheets to be ordered each time in order to minimise total annual variable costs.

Solutions

$$\begin{aligned}&D = 1000;\quad U = \$15/\text{unit};\\&C_o = \$10;\quad P = 25\%;\\&C_h = U^*P = 25\%^*15 = \$3.75.\end{aligned}$$

$$\therefore \qquad \text{EOQ} = \sqrt{\frac{2 \times 1000 \times 10}{3.75}} = 73.03.$$

The company has to order 73 standard flat sheet sizes at a time.

3.7.3. *Economic Production Quantity (EPQ)*

The EPQ model assumes that delivery is done at one time. Known as *infinite or instantaneous replenishment rate*, this is difficult to achieve. For finite replenishment rate, the inventory does not instantaneously jump up to the order quantity level. The reason is that as some items are being removed from inventory, other items are also being added as finished products from the shop floor especially for a company that manufactures the product it sells.

Let us define the following terms:

$$\begin{aligned}&p = \text{replenishment or production rate (unit produced/day)}\\&d = \text{daily demand rate (unit draw up/day)}\\&t_p = \text{replenishment period (days)}\\&U_p = \text{usage during replenishment period (units/day).}\end{aligned}$$

The time taken up to produce the quantity ordered, Q is:

$$t_p = \frac{Q}{p} \quad \text{(days)}$$

$$U_p = (t_p) \bullet d \quad \text{(units)}.$$

The maximum inventory level is the difference between Q and U_p i.e. $Q - (\frac{Q}{p})d$. This affects only annual inventory holding costs. The total annual variable costs T is given as:

$$T = \frac{DC_o}{Q} + \frac{Q(1 - \frac{d}{p})C_h}{2}$$

$$\frac{\partial T}{\partial Q} = \frac{-DC_o}{Q^2} + \frac{(1 - \frac{d}{p})C_h}{2} = 0$$

and hence we obtain:

$$\text{EPQ} = \sqrt{\frac{2DC_o}{C_h(1 - \frac{d}{p})}} = \frac{EOQ}{\sqrt{(1 - \frac{d}{p})}}.$$

Example 11

A company which produces corn-flour purchases a sack of corn for \$20. Variable costs of storing it in inventory is 15% of the unit cost per year. It costs the company \$8 per order. The company requires 7000 sacks of corn per annum. The production rate is 90 sack of corn-flour per day. Determine the economic production quantity (EPQ). Assume 250 working days/annum.

Solution

$$D = 7000 \text{ sacks/year};$$
$$d = \frac{7000}{250} = 28/\text{day}$$
$$C_o = \$8$$
$$U = \$20/\text{unit}$$

$$
\begin{aligned}
P &= 15\% \\
C_h &= U^*P = 0.15^*\$20 = \$3 \text{ per unit/yr} \\
\text{EPQ} &= \sqrt{\frac{2 \times 7000 \times 8}{3(1 - 28/90)}} \\
\text{EPQ} &= 232.8.
\end{aligned}
$$

3.7.4. *Economic Discounts Quantity (EQD)*

The EQD model assumes that the purchase price is not dependent of order quantity. In practice, discount is given for purchasing larger quantity of items at a time. The variation of purchase cost with order quantity is stepwise as shown in Fig. 3.21.

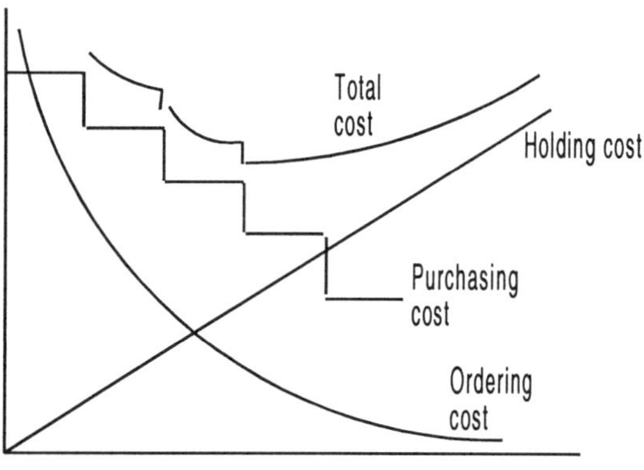

Fig. 3.21. Economic discount quantity total annual cost.

The steps involved in calculating the lot size are:

Step 1: Begin with the lowest unit price. Determine the holding cost C_h and EOQ. If EOQ is within range of order quantity, the optimum order quantity exists. Otherwise continue.

Step 2: For the next higher unit price, determine C_h and EOQ. If EOQ is within range of order quantities, then compute its annual variable costs and compare with total annual variable costs of Step 1. Select lower value. Otherwise continue.

Example 12

A cement factory is purchasing cement bags from manufacturer of polythene bags at discount rates shown in Table 3.39.

Table 3.39. Discounts per unit price.

Bags ordered	Unit price
24 or less	\$3.00
25–99	\$2.80
100–149	\$2.50
150 or more	\$2.40

Annual demand of cement bags is 6000; Order cost for each bag is \$7.5; variable cost of storing bag is 30% of the unit cost per year. Determine the EOQ.

Solution

Step 1: Using lowest unit price of \$2.40 holding cost of 30%*\$2.40 = \$0.72/bag/year.

$$\text{EOQ} = \sqrt{\frac{2 \times 6000 \times 7.5}{0.72}} = 353.5 \text{ or } 354 \text{ bags}.$$

Since 354 > 150, we calculate the total annual variable costs for ordering 150 each time as

$$T_{150} = \frac{Q}{2}C_h + \frac{DC_o}{Q} + DU$$

$$= \frac{150}{2} \times 0.72 + \frac{6000}{150} \times 7.5 + 6000^*2.4 = \$14\,655\,.$$

Step 2: For next higher unit price of \$2.50, $C_h = 30\%^*\$2.50 = \$0.75/\text{bag/year}$, and

$$\text{EOQ} = \sqrt{\frac{2^*6000^*7.5}{0.75}} = 346.4 \text{ or } 346\,.$$

Since 346 > 149, we calculate for 149:

$$T_{149} = \frac{149}{2}{*}0.75 + \frac{6000}{149}{*}7.5 + 6000{*}2.5 = \$15357.90\,.$$

Consider the next unit price since previous EOQ was outside range. For unit price \$2.8, $C_h = 30\%{*}\$2.80 = 0.84/\text{bag/yr}$, and

$$\text{EOQ} = \sqrt{\frac{2 \times 6000 \times 7.5}{0.84}} = 327.3 \text{ or } 327\,.$$

Since 327 > 99 we calculate for 25:

$$T_{99} = \frac{99}{2}{*}0.84 + \frac{6000}{99}{*}7.5 + 6000{*}2.8 = \$17296.1\,.$$

Consider the next case, since previous EOQ was outside the range. For unit price of \$3.00, $C_h = 30\% \times \$3.0 = 0.9/\text{bag/year}$, hence

$$\text{EOQ} = \sqrt{\frac{2 \times 6000 \times 7.5}{0.9}} = 316.2 = 316\,.$$

We then calculate for 24 since 316 > 24:

$$T_{24} = \frac{24}{2}{*}0.9 + \frac{6000 \times 7.5}{24} + 6000 \times 3.0 = \$19\,885.8\,.$$

Costs are shown in Table 3.40 in discounted form.

Table 3.40. Discounts per unit price.

Order quantity	Total Annual Variable Costs
150 bags or more	\$14 754.0
100–149 bags	\$15 375.9
25–99 bags	\$17 296.1
25 or less bags	\$19 885.8

3.8. Inventory Control Systems

Inventory control concerns dealing with inventory levels review which means an accurate count of materials, parts, or finished products held in stock by an organisation. Some concepts are important in formulating the two popular control systems.

3.8.1. *Order point and stockout*

The simplified EOQ model assumes that stock is replenished just at the point when inventory is zero. This zero inventory before replenishment is known as *stockout.* To avoid stock out, lead time for order should be taken into account. Management has to establish an order point which corresponds to the level of inventory at which an *order* should be placed. It is a function of the daily demand rate, leadtime and safety stock (an extra inventory). There are two cases involved:

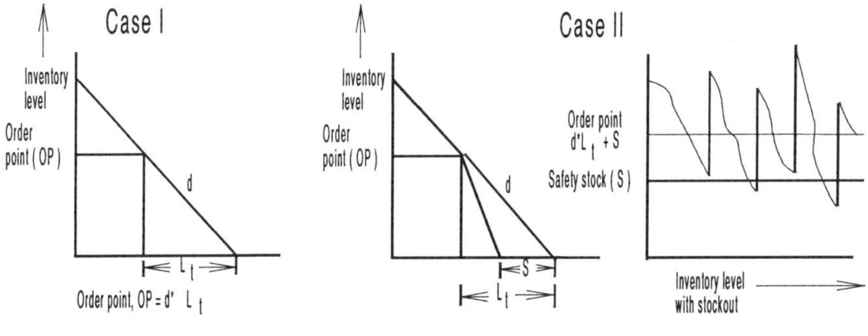

Fig. 3.22. Stockout concept.

Case I is when order point is a function of daily demand and leadtime. Case II is when it is a function of daily demand, leadtime and safety stock.

Service level and normal distribution of inventory

Service level is the percentage of orders that are received before a stockout occurs in inventory control. When historic data exists, then the expected demand can follow the normal distribution pattern so that

$$z = (x - \mu)/\sigma .$$

In this case we define:

$$\text{mean} = \text{expected demand} = \mu$$
$$\text{safety stock, } S = z\sigma$$
$$\text{order point, } OP = x.$$

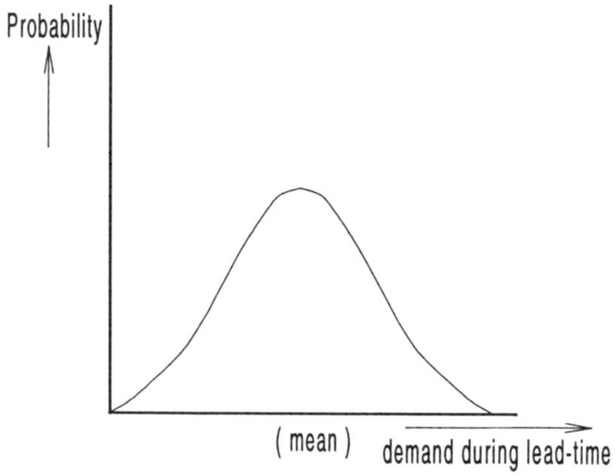

Fig. 3.23. Normal distribution pattern.

The table of normal distributions is based on a variable z, which is normally distributed with a mean of 0 and standard deviation of 1. The values are given in the Appendix at the end of this book. We now deal with the two inventory control systems.

3.8.2. *Continuous review system*

This review system uses all the ingredients discussed so far — order point, based on safety stock and lead time.

Example 13

A company maintains a continuous review system for its inventory. The following data are available:

Price of purchasing material	= \$7.5
% of price/unit attributed	
to inventory	= 20%
Ordering cost	= \$5
Annual demand	= 3000
Working weeks/year	= 50
Service level	= 99.4%
Lead time (weeks)	= 1 week
Standard deviation	= 20
period of standard deviation (wks)	= 1 week.

What reorder rule should the company adopt?

Solution

For a service level of 99.4%, $z = 2.51$ (from chart in the Appendix)
Expected demand (μ) in 1 wk $= \frac{3000}{50} = 60$

$$\text{Safety stock, } S = z\sigma = 2.51{*}20.$$

$$\text{Order point, } OP = z\sigma + m = 2.51{*}20 + 60.$$
$$= 110.2\ (110).$$

$$\text{EOQ} = \sqrt{\frac{2 \times 3000 \times 5}{20\% \times 7.5}} = 141.4(141).$$

Rule: plan order of 141 when inventory falls to 110 units.

3.8.3. *Periodic review system*

Reviewing the inventory on continous basis is very expensive. Therefore we utilise another method here based on "order-up-to" level. The review interval, R, is based on the EOQ and the annual demand as shown in Fig. 3.24. Again, two cases exist as shown in Fig. 3.25.

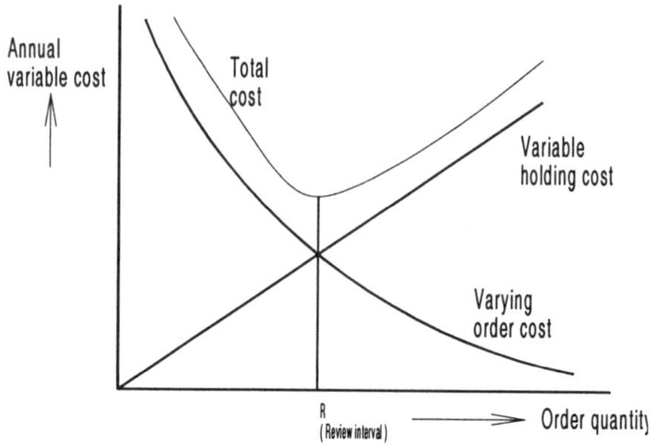

Fig. 3.24. Periodic review system: total annual cost.

The "order-upto-level", is given as $UTL = D(R + L_t)$ or $UTL = D(R + L_t) + S$.

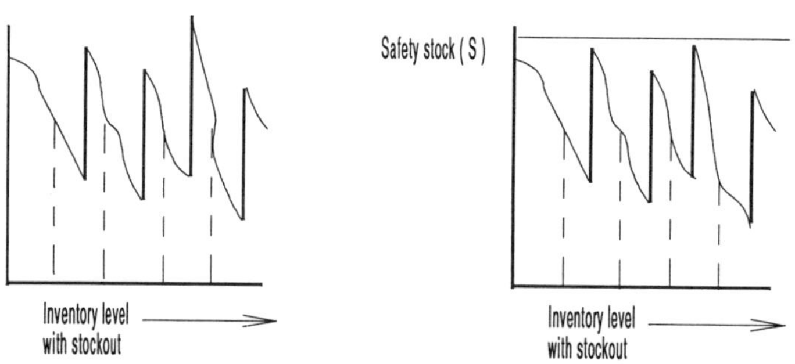

Fig. 3.25. "Order-upto-level" in periodic review system.

The safety stock is given as:

$$\text{Safety stock} = z\sigma\sqrt{\frac{(R + L_T)}{L_T}}.$$

Example 14

The same company discussed in the previous example contemplates a periodic review system and seeks the advantage. What periodic review rule should be adopted?

Solution

$z = 2.51$.
Expected demand, $\mu = 3000/50 = 60$.
EOQ $= 141$ (already computed previous example).
Review interval, $R =$ EOQ/D $= 141/3000 = 0.047$ yrs.
$= 0.047{*}50$ or 2.35 wks (2 weeks).
Review interval + lead time, $R + L_t = 2 + 1 = 3$ wks.
Safety stock $= 2.51 \times 20^{*}\sqrt{\frac{3}{1}} = 86.9$ or 87.
Demand during lead time review $= \frac{3000\times 3}{50} = 180$.
Order point $= \mu + S = 180 + 87 = 267$.
Rule: review inventory every 2 weeks and up to 267 units.

3.8.4. *Periodic Order Quantity (POQ)*

The economic order quantity can be converted into an equivalent time period for MRP in a similar way as in the case of periodic review system. For MRP the demand is known in advance from the master production schedule and the time period corresponds to the time period. The net requirements are combined in one planned order release over a certain time interval. The demand is taken to be the sum of the gross requirement, and knowing the ordering cost and handling cost, the periodic order quantity (POQ) can be computed in a similar manner as in the case of EOQ as follows:

$$POQ = \sqrt{\frac{2DC_o}{C_h}}$$

$$\text{Order period} = \frac{POQ}{D}\,.$$

Example 15

A subassembly is at the next lower level to the product whose master parts data are shown in Table 3.41. There are 4 parts in the subassembly. Lead-time is 1 week, current inventory is 150, allocated is 50, and scheduled receipt is 500 at the beginning of the first week. The following information relate to the product: cost price = \$1.6/unit standard size; order cost = \$10/order; holding cost = 25% of unit cost/annum. The company operates 50 working weeks per annum. Determine the periodic order quantity and period suitable for the product.

Table 3.41. Master parts data.

Week	1	2	3	4	5	6	8
FP-500	50	100	150	50	100	80	75

Solution

Multiplying the master part data by 4, we obtain the gross requirements for the subassembly.

$$\text{Demand} = \text{sum of the gross requirements} = 2420 \text{ for 8 weeks;}$$
$$= 2420^{*}50/8 = 17\,286 \text{ per annum;}$$

$$U = \$1.6/\text{unit};$$
$$C_o = \$10;$$
$$P = 25\%;$$
$$C_h = U \times P = 25\% \times 1.6 = \$0.4.$$

$$\therefore \quad \text{POQ} = \sqrt{\frac{2 \times 17\,286 \times 10}{0.4}} = 929.68 = 930.$$

$$\text{Order period} = 930/17\,286 = 0.0538 \text{ years}$$
$$= 0.0538(50) = 2.69 \text{ weeks or 3 weeks}.$$

Applying the procedure for MRP, the full MRP records are given in Table 3.42. The planned order releases are then lumped for every three weeks to obtain the periodic order quantity (POQ) in Table 3.43.

Table 3.42. MRP information.

Item: sub-assembly			Part number: FP-500				
Week	1	2	3	4	5	6	7
Gross requirement	200	400	600	200	400	320	300
Scheduled receipt	500						
Projected inventory	600	400					
Net requirements			600	200	400	320	300
Planned receipts			600	200	400	320	300
Planned release		600	200	400	320	300	

Table 3.43. POQ information.

Item: sub-assembly			Part number: FP-500				
Week	1	2	3	4	5	6	7
Gross requirement	200	400	600	200	400	320	300
Scheduled receipt	500						
Projected inventory	600	400					
Net requirements			600	200	400	320	300
Planned receipts			600	200	400	320	300
Planned release		1200			620		

3.9. Quality Control

To deliver competitive products, manufacturers must change their strategy from solving quality problem by inspecting products to designing quality into products. World class manufacturing can be attained only by designing products and manufacturing processes to resolve problems before they occur during usage by customers. Poorly designed product puts responsibility on customers to incur losses from initial design stage through to when delivery is made.

The traditional way of considering quality in products and processes is that quality is only given attention during the actual production at the shop floor and during the receipt of incoming raw materials. Product delivered to customer merely passes some acceptance level, with no guarantee that it will meet customer requirements. There is no feedback to the designer who initiated the product, and manufacturing department merely uses statistical failure data to seek for possible causes. Operators in most cases are not often courageous to report defects noticed during production. The result is excessive loss to manufacturers and high price by customers because they are paying for losses incurred throughout the product development stream. More recently, quality culture has changed and manufacturers have realised that if they are to remain in business and compete effectively in the market place, they must begin to consider quality right from when the customer places order for a product to when the final product is delivered.

Cost incurred during the design cycle affects the concept development, demonstration validation, full-scale development, production and operations support. It has been realised by manufacturers that about 70% of product development cost is incurred at the conception development stage, and changes made in later stages very close to production affect only about 30%. Since changes made at the design stage does not cost much as compared to high costs that are involved in attempting to make changes at the production stage due to detailed planning that has taken place, it is therefore useful to consider quality early enough at the design stage. Once the customer's voice has been captured using quality deployment function, building quality into product has the effect of satisfying customer requirement at low price, and just in time. This section outlines the steps, which are taken in order to design quality into products and manufacturing processes.

Total quality management tools are mainly: (i) design tools, and (i) process analysis and control tools. Design tools are used to design quality into products at the design stage of the product life cycle, and are made up of the following:

- Quality function deployment;
- Concurrent engineering;
- Hoshin management;
- Taguchi method.

In this section of this chapter, we consider only Taguchi method, which has been given much attention by researchers and practitioners in the area of designing quality into products.

3.9.1. *Taguchi method for quality control*

The contribution of Taguchi (1987) to the field of quality engineering is one of the most significant in the last few decades. Taguchi method is currently one of the most acclaimed techniques for building quality into product. Following his contribution towards solving Japanese telephone system problems, he developed new methods which are totally different from traditional procedures for optimising engineering experimentation. These methods known as Taguchi methods have become invaluable tools for designing quality into products early enough in the product development. The fundamental concepts of Taguchi method are quite simple: (1) quality should be designed, and not inspected into the product (2) quality is best achieved by minimising deviation from functional target (3) quality cost should be measured as a function of the target and the losses incurred should be viewed as being systemic. These fundamental concepts have far reaching effects in manufacturing industries and many companies in Western countries have applied them in order to change their quality culture and solve quality problems which were almost, if not impossible to solve. Taguchi method answers these three questions: when do we start considering quality, how do we effect it, and what cost is involved? The observations of Deming (1986) that 85% of poor quality is attributed to the manufacturing process and only 15% to the worker was instrumental to Taguchi's development of robust manufacturing systems.

In the first concept, Taguchi states that quality improvement should be started at the design phase of a product or process and it should cascade into the production phase. Taguchi proposed an *off-line* strategy (as far as possible away from the production phase) of building quality improvement into the product rather than the traditional approach of inspection during production.

In the second concept, Taguchi views quality as directly related to deviation to a design parameter from a target value. By developing manufacturing processes to satisfy a specified target value, with little deviation, it is possible to improve the product life cycle. The second concept is concerned with how to effect quality.

In the third concept, designers are instructed to measure deviations from a given design parameter in relation to the overall life cycle cost of the product. Costs include rejects during production, warranty service call, replacement, etc. Tremendous cost improvements can be made at the design stage using Taguchi method, as we shall demonstrate later in this section.

Off-line quality improvement techniques take place away from the actual production point, normally at the product or manufacturing systems design stage. The Taguchi *off-line* quality improvement strategy is a preventive measure at the product or manufacturing systems design stage which has the capability of making the product design robust, hence becoming immune to external uncontrollable environmental noise factors such as humidity, ambient temperature, machinery deterioration, tool wear, and batch to batch variation operators. The strategy guarantees that the product delivered will meet customer requirements because the potential problems which might occur during usage have been considered early enough in the design stage. Quality improvement techniques at the point of production are termed on-line techniques. If quality is only considered down-stream very close to the point of manufacturing, much cost will be incurred in re-work, scraps, and customer warranty. Disney and Bendell (1990) show the relationship between off-line and on-line quality controls. Various sources of literature on Taguchi philosophies and methodologies include Taguchi and Konishi (1987), Addelman (1962), (1985), Juran *et al.* (1979), Lucas Engineering and Systems (1992) and Ross (1988). For a full discussion on the use of Taguchi in on-line quality control see Taguchi (1981). Onwubolu (1997) developed a computer software, known as QUALITY to study the Taguchi method for optimising product and process design.

Taguchi has proposed three stages in the design process to achieve *off-line* product quality as:

(1) Systems design in which the overall factors affecting the system are determined;
(2) Parameter design in which the factors affecting the performance of the product or process are specified;
(3) Tolerance design in which allowable range of deviation in the parameter values that affect the product are defined.

The five steps to applying Taguchi method are: (1) define the problem; (2) experiment design; (3) conduct the experiment; (4) analysis of variance; and (5) confirm the improvement by determining the loss function.

Quality factors

Two types of factors are commonly used in Taguchi method: design or control factors, and noise or uncontrollable factors. For *design factors*, as much as possible, all factors which may affect the design have to be exhaustively considered, then prioritized in terms of their relative importance. The factors with higher weighting of importance are then filtered and used in the experiment. The number of levels have to decided too. In the planning stage it is important to decide what constitutes *noise factors* in the experiment, and what steps are to be taken to design a product that is insensitive to variations in these noise factors. Similarly, the number of factors and their levels have to be determined. Another important point to consider is the factors which are likely to *interact*; a decision has to be taken whether to include interactions among factors or replace them with additional factors.

Quality design

The traditional method of laying out the conditions or designs of experiments is to identify all possible combinations for a set of factors. These factors may be design factors (off-line) or manufacturing process factors (on-line). Known as factorial design of experiments, it has been used extensively by scientists since the 1920s when it was first proposed by Fisher (1951). For example, an experiment that involves three factors at two levels, requires 8 (2^3) combinations; and in general, an experiment that involves n-factors at m-levels, requires m^n combinations. A factorial experiment of 3 factors, at 2 levels each, is shown in Fig. 3.27 (a) with all the 8 combinations. We represent the factors as A, B, and C. A_1, B_1, and C_1 represent the values of the factors at level 1, while A_2, B_2, and C_2 represent the values of the factors at level 2.

		A_1	A_2
B1	C_1	*	
	C_2		*
B2	C_1		*
	C_2	*	

(a)

Factors			
Trials	1	2	3
1	1	1	1
2	1	2	2
3	2	1	2
4	2	2	1

(b)

Fig. 3.26. Taguchi design experiment using L4 (2/3) array.

A special set of orthogonal arrays (OAs) were devised by Taguchi for designing an experiment. The cells in Fig. 3.26(a) marked with asterisk represent the Taguchi partial factorial orthogonal array for the experiment. The Taguchi orthogonal array reduces the number of factors to be considered, to 4 instead of 8 in the factorial method. Figure 3.26(b) shows the Taguchi orthogonal array, in which factors affecting a design are assigned to the top row, while trial numbers are placed on the first column. The relationship between the design factors and trial numbers is described by the trial conditions with factor levels indicated by integer numbers. The Taguchi experimental technique is useful in a wide range of situations where solutions are not straight forward, as in the case where there are multi-reasons for defect or deficiency.

Control and noise orthogonal arrays

In *control* orthogonal arrays, the control parameters are assigned to the columns of the orthogonal array, and the integers are translated into actual tests values of the assigned parameters. The control orthogonal array is known as *inner* array. The noise array is known as *outer* array and it is used to introduce real life factors which are not under the control of the engineer. For example, the engineer cannot directly control the humidity under which the

	Outer Array					
Inner Array						
	Control Factors			Repetitions	(Results)	
Column						
Exp No.	1	2	3	1	2	3
1	1	1	1	40	45	38
2	1	2	2	30	35	32
3	2	1	2	45	48	40
4	2	2	1	35	40	36

Fig. 3.27. Inner and outer orthogonal arrays describing control and noise factors.

product or process design will operate. Therefore humidity is considered to be noise variable. The normal method in Taguchi experiment is to carry out the experiment using the control factors, under different conditions of a noise variable such as humidity. These uncontrollable factors as they are called are also treated like the control factors using orthogonal arrays. However, noise (outer) array is placed perpendicular to control (inner) array as shown in Fig. 3.27. In practice, the outer array describes the noise conditions for the repetitions.

Designing with interactions

There are cases in practice when the effect of changes in the level of a factor determines the influence of another factor, and vice versa. When this happens, the two factors are said to interact. The commonest example which most people can easily relate to, is the effect of temperature and humidity on human comfort. Lowering temperature alone causes some degree of discomfort but this is heightened by lowering humidity in addition. Temperature and humidity have interaction on each other, when considering human comfort.

Designing using interactions needs special attention. Not all orthogonal arrays accommodate interaction. Typically, arrays which accommodate interactions are those which can be described as 2^{k+1}-type (L_4, L_8, L_{16}, L_{32} etc. for level 2), 3^{k+1}-type (L_9, L_{27}, L_{81}, etc. for level 3), $\ldots, m^{k+1}$-type where $k = 1, 2, \ldots, n$. Figures 3.28(a)–(d) show linear graphs for L_4, L_8, L_{16}, and L_{32}. The nodes represent control parameters, while the line segments represent interactions. The effect of interaction is usually reserved for control parameters while repetitions are considered for noise parameters. There are several methods of constructing linear graphs but one which is easy to understand using level 2 orthogonal arrays as example, is described in the following steps.

Step 1: Assign node 1, then connect a line segment to node 2.
Step 2: Continuing other m-line segments to obtain other nodes, where 2^m corresponds to L_{2m}.
Step 3: Connect all nodes.

For example in constructing an L_{16} array, we assign node 1, connect line segment to node 2. There are remaining 4-segments to form a polygon since $2^4 = 16$. Therefore in the design, the control parameters are assigned to nodes 1, 2, 4, 8, and 15. It is observed that intermediate nodes between nodes 1 and

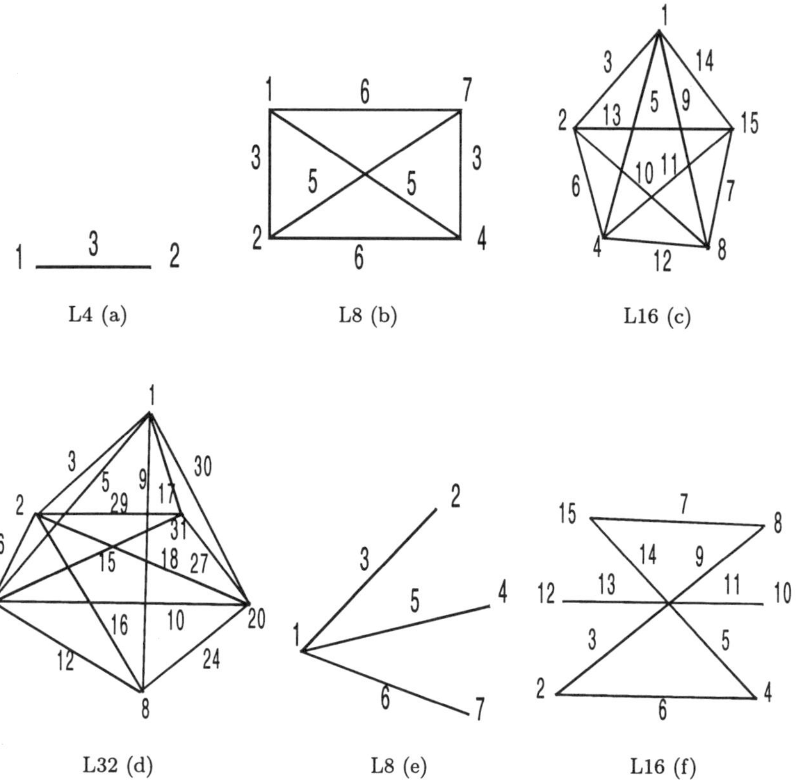

Fig. 3.28. Linear graphs showing interactions.

the last nodes are multiples of 2. All nodes are then connected as shown in Fig. 3.28(c), with line segments designated 3, 5, 6, 7, 9, 10, 11, 12, 13 and 14. The interaction between columns 1 and 2 is column 3, between columns 2 and 4 is column 6, etc.

There are rules for assigning columns of the orthogonal array when interactions exist. Using the linear graphs, the control parameters are assigned to the orthogonal array columns corresponding to the nodes of the linear graph. When nodes are used up, the line segments are assigned. The columns for interaction should be filled according to the labels in the linear graph.

Upgrading and collapsing levels in mixed factors design

Mixed factors exist in practical design for optimisation of products and processes. The details of this technique is not discussed in this chapter.

Quality analysis

A methodical way of analysing the Taguchi design is the use of analysis of variance techniques (ANOVA) in statistics. ANOVA easily indicates which design factors are important so that we concentrate on their effects and by so doing, we filter the less important factors. The basic computations which are covered in ANOVA are as follows:

- Degrees of freedom (f)
- Sum of squares (s)
- Mean square (variance) (V)
- Variance ratio (F)
- Pure sum of squares (S′)
- Percentage contribution (p).

Repetition of the experiments for several number of times is normally recommended. But when only one set of experimental readings are taken, only one column is utilised. This simple procedure for design applies to any complicated problem, once the necessary factors have been determined.

The solution of optimum performance may be obtained from the ANOVA result by considering *pooling effect.* Pooling means adding the effect of factors that are not very significant to other significant factors. The optimum performance value gives indication of the **signal to noise ratio** (S/N) which is an input to the **Taguchi loss function** (TLF).

Quality cost

Two kinds of losses are taken seriously. The first, is loss incurred by manufacturers from their supplier, and the second and more important is the loss incurred by customers when they purchase products which have poor quality. Taguchi aims at minimising losses throughout the product development stream so that customers at the front end do not have to incur unnecessary costs which are measured in monetary terms. Taguchi defined a loss function which is proportional to the deviation from the target quality characteristic, given in parabolic form as shown in Fig. 3.29.

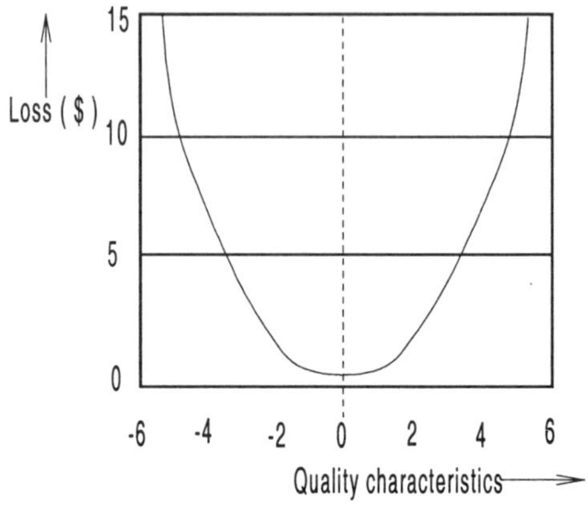

Fig. 3.29. Taguchi loss function.

Case Study

A contracting firm involved in making automobile body parts for major motor industries is concerned with the cost incurred in re-work of parts. The firm uses mainly mechanical presses for piercing operations for sheet-metal work. Progressive die is used for producing the parts, with flat sheet-metal being progressively released from the stock reel, and it passes through the air feeder for straightening and finally through the die where the press pierces out the required part shape. As the sheet-metal passes through the air feeder, it is cooled using a coolant. There is also a sag between the stock reel and the air feeder.

A Taguchi experiment was carried out to: identify the cause of the variation of the manufactured parts from the target; use the knowledge of the cause of variation to either eliminate the cause of variation or redesign the product/process in order to make the product robust against the effects of the variation.

In order to proceed with the task, the factors which affected production were first identified as: the stock reel speed, press punch speed, angle of deflection of the sheet-metal, the feeder release rate, lubricant flow rate. These factors are designated A, B, C, D, E. It is found that two interactions exist: between A and C, and between B and C. Management therefore considers that

Table 3.44(a). Design factors and their levels.

Column	Factors	Level 1	Level 2
1(A)	Reel speed	Low	High
2(C)	Reel angle	Small	Large
3(A × C)			
4(B)	Press speed	Low	High
5(D)	Air feeder release	Slow	Fast
6(B × C)			
7(E)	Lubricant flow rate	Low	High

Table 3.44(b). Taguchi experiment data with interactions included.

	A 1	C 2	A × C 3	B 4	D 5	B × C 6	E 7	Results (y)
1	1	1	1	1	1	1	1	50
2	1	1	1	2	2	2	2	55
3	1	2	2	1	1	2	2	45
4	1	2	2	2	2	1	1	52
5	2	1	2	1	2	1	2	60
6	2	1	2	2	1	2	1	42
7	2	2	1	1	2	2	1	58
8	2	2	1	2	1	1	2	48

there are seven factors which are considered to have significant impact on the quality of product and that they are two levels for each factor, hence an L8 (2/7) Taguchi orthogonal array is needed to investigate the factors A, B, C, D, E which are assumed to be causing high production scraps. The operations manager of the company receives results of scraps taken at a machine centre by an operator at the interval of 2 hours (see Table 3.44(a)). There are 8 readings shown in Table 1(b), representing the averages of readings taken at the specified times. We use Taguchi method for investigating this problem, and carry out a complete analysis of variance (ANOVA) for the problem. The monthly saving derived from the experiment is also to be computed, based on the following information:

Target value = 40

Tolerance = 2

Reject cost per unit = \$1.00

Production rate per month = 10 000.

Quality factors

The design factors are A, C, A × C, B, D, B × C, and E given in the problem formulation.

Quality design

The Taguchi orthogonal array of Table 3.44(b) shows the design. The linear graph of Fig. 5(b) shows that interaction between nodes 1 and 2 is at line segment 3, while that between nodes 2 and 4 is at line segment 6. These information are used to assign the factors at the top of Table 3.44(b).

Quality analysis

The results in Tables 3.45 and 3.46 constitute the quality analysis results which were using the software QUALITY. Estimate of the Optimum Condition of Design/Manufacturing Process based on 'smaller the better characteristic' is shown in Table 3.46.

The result shows 3.47 significant factors-A × C (6.6%), B (11.7%), D (73%), B × C (4.5%). The pooled factors are A (1.6%), B (0.7%), E (1.6%). Hence

Table 3.45. ANOVA results.

Source	F	S	V	F	S	P
A	1	4.5	4.5	0	0	1.645338
C	1	2	2	0	0	0.7312614
A × C	1	18	18	0	0	6.581353
B	1	32	32	0	0	11.70018
D	1	200	200	0	0	73.12614
B × C	1	12.5	12.5	0	0	4.570384
E	1	4.5	4.5	0	0	1.645338
Error (before pooling)	1	0	0	0	0	0
Total	7	273.5	273.5	0	0	100

Table 3.46. Optimum solution.

Factor description	Contribution
Contribution from all factors (total)	−9.75
Current grand average of performance	51.25
Optimum result	41.50

the firm is informed that the most important factor is D with as much as 73% of the overall effect. The firm is informed not to too concerned with the effects of A, C, and E, and so spends more time and on the important factors. Non-value activities are greatly reduced and important factors which affect the quality of the product or process are highlighted. Upper and lower bounds are determined by the analysis as 42 and 61 respectively, beyond which management cannot tolerate scraps.

Quality cost

Taguchi loss function is the deviation from the target quality characteristic. It is an excellent tool for determining the magnitude of the process and supplier tolerances, based upon quality as perceived by the customer. By determining the loss before and after the Taguchi experiments, it is possible to estimate the savings to the firm. The following results were obtained for the press investigation:

Loss before experiment = $40.18.
Loss after experiment = $32.64.
Monthly savings:
Based on production rate of 10 000/month = $75 400.

We considered Taguchi method in details because it is one of the most powerful tools available to the designer for building quality into the product and diagnosing the cause of variation during production for the purpose of either eliminating the cause of variation or redesigning the product/process to make the product robust in order to satisfy customer requirement. Operations managers and engineers should rigorously seek for methods which are capable of solving problems in products and processes before they occur.

3.9.2. *Process analysis and control tools*

Process anaylis and control tools are on-line tools for ensuring that products are manufactured to the right quality for the customer. The following tools are commonly used:

- Pareto charts;
- Cause and effect (CE) diagrams;
- Process flow charts;
- Histograms;

- Check sheets;
- Scatter diagram;
- Plan-do-check-act (PDCA);
- Brain storming;
- Statistical process control charts;
- Process capability measure.

Some of these tools are very basic. Let us discuss some of these tools in detail. Since statistical process control charts are widely used in the industry, we devote a subsection to it.

Pareto charts

Pareto chart shown in Figure 3.30 is based on the concept that 20% of the items in a group will account for 80% of the value, cost, cost, problem, etc. Enables the operations management system to separate the *important few* from the *trivial many.*

Pareto chart shows that inventory, packaging, and return of purchased goods from customers dominate in contributing to costs. Therefore, the operations management system has to consider these factors more than others.

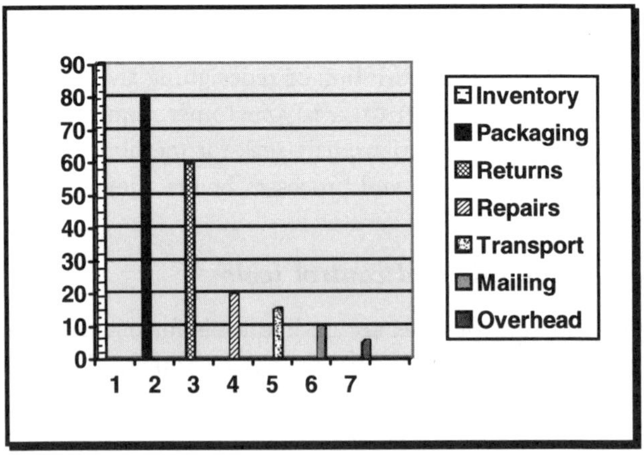

Fig. 3.30. Pareto analysis.

Cause and Effect (CE) diagrams

The cause and effect, Ishikawa or fishbone diagram (see Fig. 3.31) is used to analyze a problem. It is a tool to indicate how various causes can operate independently as well as simultaneously in order to produce an eventual effect on the manufactured product. It is also applicable in service sector.

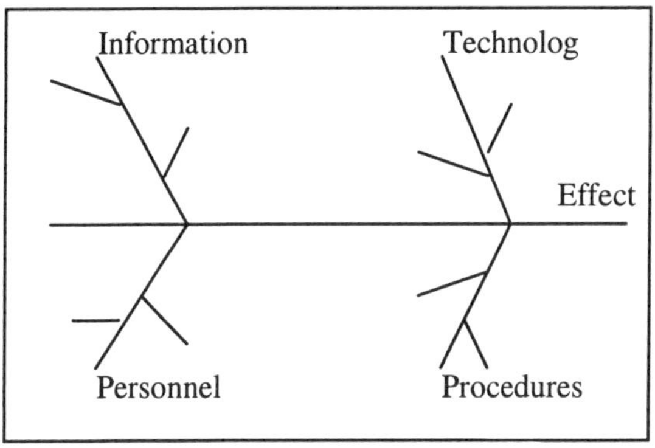

Fig. 3.31. Cause-and-effect (Ishikawa) diagram.

3.9.3. *Statistical Process Control (SPC) tools*

Statistical Process Control (SPC) tools are subdivided into the following categories:

- Process control for variables (tangible measurements)
  - X-bar charts
  - R-charts.
- Process control for attributes (accept/reject situations)
  - P-chart (% of defect plotted)
  - C-chart (no. of defects plotted).
- Process capability measures
  - process variability
  - process capability.

3.9.4. *Process control for variables*

In process control for variable, tangible measurements are made. The central limit theory of probability states that whenever samples are taken from a population, irrespective of its distribution, those sample means will approximately follow a normal distribution pattern. This means that about 99% of all values fall within plus or minus three standard deviation of the mean. This distribution helps the operations management system to determine whether a process is *in control or out of control.*

- **X-bar charts**

X-bar charts are used when enough information are available to calculate the process standard deviation.

$$\text{Upper limit } (\mathrm{UCL}_{\bar{x}}) = \overline{\overline{X}} + 3\sigma\bar{x}\,.$$

$$\text{Lower limit } (\mathrm{LCL}_{\bar{x}}) = \overline{\overline{X}} - 3\sigma\bar{x}\,; \quad \text{where } \sigma\bar{x} = \frac{\sigma}{\sqrt{n}}\,.$$

When information on standard deviation is not available, sample range factors shown in Fig. 3.47 may be applied.

Example 16

Ten mean values of boxes of soap packaged in a soap-making factory for sample size 5 are 95, 97, 95, 99, 97, 102, 104, 109, 106, 108. The process standard deviation is 5. Grand mean is 101. Analyse the SPC problem.

Solution

$\sigma\bar{x} = 5/\sqrt{5} = 2.236.$

Then $\mathrm{UCL}_{\bar{x}} = 101 + 3(2.236) = 108.$

$\mathrm{LCL}_{\bar{x}} = 101 - 3(2.236) = 94.$

The raw data, upper control limit and lower control limit are plotted as shown in Fig. 3.32. Clearly the operations management system detects the 3 last boxes as being out of control. Something has to be done!

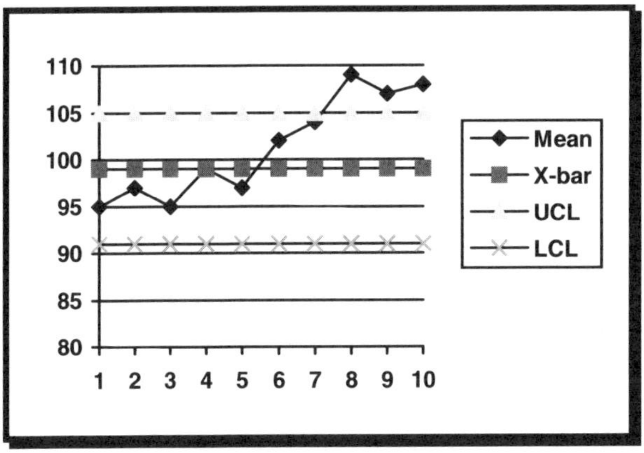

Fig. 3.32. Control chart.

Table 3.47. Factors for sample ranges.

n	A_2	D_3	D_4
2	1.880	0	3.268
3	1.023	0	2.574
4	0.729	0	2.282
5	0.577	0	2.114
6	0.483	0	2.004
7	0.419	0.706	1.924
8	0.373	0.136	1.864
9	0.337	0.184	1.816
10	0.308	0.223	1.777

Let us use Table 3.47 to solve the same problem. Since $n = 5$, maximum value of data = 109, and minimum value of data = 95. This means that

$$\text{Range} = (109 - 95) = 14.$$
$$\text{Mean} = 101.$$
$$A_2 = 0.577.$$

Hence,

$$UCL_{\bar{x}} = 101 + 0.577(109 - 95) = 109.$$
$$LCL_{\bar{x}} = 101 - 0.577(109 - 95) = 93.$$

- **R-charts**

R-charts are used to solve the problem of calculating standard deviation. Instead of using discrete values, sample range is used (difference between maximum and minimum values in a sample). Factors are applied depending on the sample size, n. The upper control limit and lower control limit factors are obtained from Table 3.47. these limits are obtained emperically as follows:

$$\text{Upper limit } (\text{UCL}_R) = \text{D}_4\ \overline{\overline{R}}\,.$$

$$\text{Lower limit } (\text{LCL}_{\bar{x}}) = \text{D}_3\ \overline{\overline{R}}\,.$$

For the previous example, since $n = 5$, $\text{D}_3 = 0$, $\text{D}_4 = 2.114$. The maximum value of data $= 109$, and minimum value of data $= 95$. This means that range $= (109 - 95) = 14$. Hence,

$$\text{UCL}_R = 2.144{*}14 = 29.596\,,$$

and

$$\text{LCL}_{\bar{x}} = .0.14 = 0\,.$$

Process control for attributes

In process control for attributes, intangible measurements are made such as accept/reject.

- ***P*-charts**

P-charts are used when percentage defects are known. The standard deviation for this class of problems is given as,

$$\sigma_p = \sqrt{\frac{(1-\bar{p})\bar{p}}{n}}\,.$$

The upper control limit and lower control limits are computed as follows:

$$\text{Upper control limit } (\text{UCL}_p) = \bar{p} + 3\sigma_p.$$

$$\text{Lower control limit } (\text{LCL}_p) = \bar{p} - 3\sigma_p.$$

Example 17

An airline records an average delay of 5 percent in 100 recent flights. The airline is determined to offer customers quality service hence management decides to determine the upper and lower control limits for a p-chart using this information. Analyse the SPC problem.

Solution

Percentage delay $\bar{p} = 0.05$. Therefore, $\sigma_p = \sqrt{0.05^*(1-0.05)/500} = 0.0097$. Then,

$$\text{UCL}_p = 0.05 + 3(0.0097) = 0.079.$$
$$\text{LCL}_p = 0.05 - 3(0.0097) = 0.021.$$

Clearly the operations management system will investigate operations procedure if the percentage of fight delays exceed 8 percent for every 500 scheduled flights. How would you interpret the lower limit for this problem?

- **c-charts**

C-charts are used when the track of only defects are known. The standard deviation for this class of problems is given as,

$$\sigma_c = \sqrt{\bar{c}}.$$

The upper control limit and lower control limits are computed as follows:

$$\text{Upper control limit } (\text{UCL}_c) = \bar{c} + 3\sigma_c.$$
$$\text{Lower control limit } (\text{LCL}_c) = \bar{c} - 3\sigma_c.$$

Example 18

A retail shop keeps record of the number of customer transactions returned during a particular period of business operation. The number has been averaging 10 while the process has been in control. The shop manager is interested in offering customers quality service and so decides to develop a c-chart for the business operation. Analyse the SPC problem.

Solution

Average return $\bar{c} = 10$. Therefore, $\sigma_c = \sqrt{10} = 3.16$. Then,

$$\text{UCL}_c = 10 + 3^*3.16 = 19.48.$$

$$\text{LCL}_c = 10 - 3^*3.16 = 0.52.$$

The shop's operations management system will investigate operations procedure if the returns exceed 19 for similar period. How would you interpret the lower limit for this problem?

3.10. Summary

Scheduling presents many day-to-day problems and challenges to production and operations mangers due to last minute changes that may occur in customer order. The way jobs are scheduled affect the operations output of a manufacturing firm, therelore it is important that jobs are properly scheduled. Scheduling in job shop, batch production, flow shop and line balancing are discussed in this chapter. Several priority rules for job shop are used for scheduling. We have examined batch production techniques in details and in a very practical way. For flow shop scheduling, some heuristics, algorithms, and a graphical method are presented. Due to the complexity of most scheduling problems, there are few really satisfactory techniques which have been developed. Graphical scheduling systems which make scheduling flexible and which drastically cut time for flow-shop environment are very helpful tools for decision makers such as production and operation managers because they can be easy way of communicating the scheduling situation to them. Additionally, inventory an quality control, which are important elements of production control systems, are discussed.

We also considered material management, inventory control, and quality control. Both Taguchi method and statistical quality control are discussed in details because they are important quality-tools available to the designer.

Taguchi method is for building quality into the product and diagnosing the cause of variation during production for the purpose of either eliminating the cause of variation or redesigning the product/process to make the product robust in order to satisfy customer requirement. SPC is an on-line quality-tool. Operations managers and engineers should rigorously seek for methods which are capable of solving problems in products and processes before they occur.

Exercises

1. Distinguish between flow-shop and job-shop manufacturing.
2. Are dispatching rules for scheduling more appropriate for flow-shop or job-shop?
3. Table 3.48 shows information for four jobs to be carried out in a fabrication shop. Machine B has just finished a job, and the task is to select the next job. The current time is 25. By using slack per remaining operation dispatching rule, obtain the job sequence.

Table 3.48. Jobs and machine operation times.

Job	Arrival to system	Arrival at B	Due date	Machine/process time 1	2	3
1	25	20	58	(B, 11)	(A, 2)	(D, 12)
2	0	12	40	(A, 11)	(B, 6)	(C, 5)
3	0	17	22	(C, 5)	(D, 4)	(B, 4)
4	0	15	52	(E, 13)	(B, 9)	(C, 9)

4. Use (i) Covert rule and (ii) Critical ratio rule to determine the operation sequence.
5. Write a computer program for scheduling using the dispatching rules given in Sec. 12.2.1.
6. Distinguish between Moore's and Johnson's algorithms for scheduling
7. Table 3.49 shows the due dates and processing times for 6 jobs. Use Moore's algorithm to determin the sequence of the $5/1/A/N_T$ problem.
8. Apply the Campbell, *et al.* algorithm for sequencing the q6 jobs that have to be processed on two machines with the processing times shown in Table 3.50.
9. Does the Campbell, *et al.* algorithm for scheduling have advantages over Johnson's algorithm? Justify your answer.

Table 3.49. The $5/1/A/N_T$ problem.

Job	1	2	3	4	5	6
Due date	12	8	6	11	13	7
Processing time	4	5	5	3	12	9

Table 3.50. Processing times for 6 jobs on 2 machines.

Job	Processing time (min) Machine 1	Machine 2
1	8	7
2	17	11
3	7	14
4	10	6
5	9	5
6	8	9

10. Write a computer program for Campbell, *et al.* algorithm for scheduling.
11. How many different 2-machine flow shop problems must be solved for Table 3.35 information? Draw the Gantt charts and determine their makespans.
12. Is the schedule for problem 11 optimal?
13. Discuss the advantages of computerised Gantt chart for scheduling in a manufacturing industry.
14. Consider the single-machine problem of Table 3.51 with four jobs to be scheduled. Find the optimum sequence using branch and bound technique if job 2 must be completed first.

Table 3.51. Processing times for job 4.

Job	1	2	3	4
Processing time	5	8	6	4

15. Outline the procedure for batch production scheduling.
16. Discuss the usefulness of line of balance in batch production scheduling.
17. Differentiate between flow shop production and assembly line.
18. Consider the work elements information in Table 3.52. The master production schedule indicates that 480 000 units are required per annum. The company works 50 months in a year, 20 days per month,and 8 hours per day.

Table 3.52. Work element time and intermediate predecessor.

Element	Time (min.)	Intermediate predecessor
1	0.20	—
2	0.50	1
3	0.3	1
4	0.30	2
5	0.40	2, 3
6	0.6	3
7	0.5	4
8	0.40	4, 5
9	0.40	6
10	0.60	7, 8, 9

(i) Draw the precedence diagram
(ii) Using the largest candidate rule, determine the workstations balance delay and efficiency.

19. Use the ranked positional weights method for the same problem of Table 3.52. Compare answer with that obtained in question 19.
20. Write a computer code for Moore's algorithm.
21. Write a computer code for Johnson's algorithm.
22. Write a computer code for CDS algorithm.
23. Write a computer code for the four stages of line of balance (LOB) technique for batch scheduling.
24. Write a computer code for lowest candidate rule method for assembly balancing.
25. Write a computer code for ranked positional weight method for assembly balancing.

Appendix

Z	.00	.01	.02	.03	.04	.05	.06	.07	.08	.09
.0	.5000	.5040	.5080	.5120	.5160	.5199	.5239	.5279	.5319	.5359
.1	.5398	.5438	.5478	.5517	.5557	.5596	.5636	.5675	.5714	.5753
.2	.5793	.5832	.5871	.5910	.5948	.5987	.6026	.6064	.6103	.6141
.3	.6179	.6217	.6255	.6293	.6331	.6368	.6406	.6443	.6480	.6517
.4	.6554	.6591	.6628	.6664	.6700	.6736	.6772	.6808	.6844	.6879

Z	.00	.01	.02	.03	.04	.05	.06	.07	.08	.09
.5	.6915	.6950	.6985	.7019	.7054	.7088	.7123	.7157	.7190	.7224
.6	.7257	.7291	.7324	.7357	.7389	.7422	.7454	.7486	.7517	.7549
.7	.7580	.7611	.7642	.7673	.7704	.7734	.7764	.7794	.7823	.7852
.8	.7881	.7910	.7939	.7967	.7995	.8023	.8051	.8078	.8106	.8133
.9	.8159	.8186	.8212	.8238	.8264	.8289	.8315	.8340	.8365	.8389
1.0	.8413	.8438	.8461	.8485	.8508	.8531	.8554	.8577	.8599	.8621
1.1	.8643	.8665	.8686	.8708	.8729	.8749	.8770	.8790	.8810	.8830
1.2	.8849	.8869	.8888	.8907	.8925	.8944	.8962	.8980	.8997	.9015
1.3	.9032	.9049	.9066	.9082	.9099	.9115	.9131	.9147	.9162	.9177
1.4	.9192	.9207	.9222	.9236	.9251	.9265	.9279	.9292	.9306	.9319
1.5	.9332	.9345	.9357	.9370	.9382	.9394	.9406	.9418	.9429	.9441
1.6	.9452	.9463	.9474	.9484	.9495	.9505	.9515	.9525	.9535	.9545
1.7	.9554	.9564	.9573	.9582	.9591	.9599	.9608	.9616	.9625	.9633
1.8	.9641	.9649	.9656	.9664	.9671	.9678	.9686	.9693	.9699	.9706
1.9	.9713	.9719	.9726	.9732	.9738	.9744	.9750	.9756	.9761	.9767
2.0	.9772	.9778	.9783	.9788	.9793	.9798	.9803	.9808	.9812	.9817
2.1	.9821	.9826	.9830	.9834	.9838	.9842	.9846	.9850	.9854	.9857
2.2	.9861	.9864	.9868	.9871	.9875	.9878	.9881	.9884	.9887	.9890
2.3	.9893	.9896	.9898	.9901	.9904	.9906	.9909	.9911	.9913	.9916
2.4	.9918	.9920	.9922	.9925	.9927	.9929	.9931	.9932	.9934	.9936
2.5	.9938	.9940	.9941	.9943	.9945	.9946	.9948	.9949	.9951	.9952
2.6	.9953	.9955	.9956	.9957	.9959	.9960	.9961	.9962	.9963	.9964
2.7	.9965	.9966	.9967	.9968	.9969	.9970	.9971	.9972	.9973	.9974
2.8	.9974	.9975	.9976	.9977	.9977	.9978	.9979	.9979	.9980	.9981
2.9	.9981	.9982	.9982	.9983	.9984	.9984	.9985	.9985	.9986	.9986
3.0	.9987	.9987	.9987	.9988	.9988	.9989	.9989	.9989	.9990	.9990

References

Addelman, S., 1962, Orthogonal main-effect plans for asymmetrical factorial experiments, *Technometrics*, **4**(1) 21–46.

Brah, S. A. and Hunsucker, J. L., 1987, *Optimal scheduling in an m-stage flow shop with multiple processors*, Technical Report, University of Houston, TX.

Campbell, H. G., Dudek, R. A., and Smith, M. L., 1970, A heuristic for the n job, m machine sequencing problem, *Management Science*, **16**(10) 630–637.

Deming, W. E., 1986, *Out of Crisis* (Cambridge, M.A.: Massachusetts Institute of Technology, Centre for Advanced Engineering Study).

Disney, J. and Bendell, A., 1990, *The Potential for the Application of Taguchi Method of Quality Control in British Industry*, In Managing Quality (Eds: Dale, B. G. and Plunkett, J. J.), Philip Allan Publishers.

Garetti, M. and Taisch, M., 1999, Neural networks in production planning and control, *Production Planning & Control*, **10** 324–339.

Gunasekaran, A., Goyal, S. K., Virtanen, I., and YL1_OLL1, 1993, Production planning and control in automated manufacturing systems, *Production Planning & Control*, **4** 2–16.

Gupta, N. D., 1972, Heuristic algorithms for multistage flow shop scheduling problem, *AIIE Transaction*, **4**(1) 11–18.

Gupta, J. N. D. and Darrow, W. P., 1971, The two-machine sequence dependent flow shop scheduling problem, *AIIE Transaction*, **3** 199–205.

Fisher, R. A., 1951, *Design of Experiments* (Edinburgh: Oliver and Boyd).

Hariri, A. M. A., 1981, *Scheduling using branch and bound techniques*, Ph.D. thesis, University of Keele, UK.

Ishikawa, K., 1985, *What is Total Quality Control? The Japanese Way* (Englewood Cliffs, N J: Prentice-Hall, Inc).

Juran, J. M., Gryna, F. M. Jr., and Bingham, R. S. Jr., 1979, *Quality Control Handbook*, 3rd ed. (New York: McGraw-Hill).

Johnson, S. M., 1954, Optimal two and three stage production schedules with setup times included, *Naval Research Logistics Quarterly*, **1**(1).

Kochhar, S. and Morris, R. J. T., 1987, Heuristic methods for flow lines scheduling, *Journal of Manufacturing Systems*, **6** 299–314.

Lucas Engineering and Systems, 1992, *The Lucas Manufacturing Systems Handbook: Mini Guides*, Lucas Engineering and Systems Ltd.

Monden, Y., 1981, *Toyota Production System: Practical Approach to Production Management*, American Institute of Industrial Engineers.

McMahon, G. B. and Burton, P. G., 1967, Flow-shop scheduling with the branch and bound method, *Operations Research*, **15** 475–482.

Nawaz, M., Enscore, E. E., and Ham, I., 1983, A heuristic algorithm for m-machine, n-job flow-shop sequencing problem, OMEGA, *The International Journal of Management Science*, **11**(1) 91–95.

Onwubolu, G. C., 1996, A flow-shop manufacturing scheduling system with interactive computer graphics, *International Journal of Operations and Production Management*, **9** 74–84.

Onwubolu, G. C., 1997, On and off-line quality control for product design and manufacturing processes, *South African Journal of Industrial Engineering*, **8** 61–71.

Park, Y. B., Pegden, C. D., and Enscore, E. E., 1984, A survey and evaluation of static flow shop scheduling heuristics, *International Journal of Production Research*, **22** 127–141.

Pruett, J. M. and Schartner, A., 1993, JOB: An intrustive job shop scheduling environment, *International Journal of Operation and Production Management*, MBC University Press Ltd., Bradford, **13**(11).

Ross, P. J., 1988, *Taguchi Techniques for Quality Engineering* (McGraw Hill Book Co., New York).

Shingo, S., 1981, *Study of Toyota Production System from Industrial Engineering Viewpoint*, Japanese Management Association, 352.

Shingo, S., 1985, *A revolution in Manufacturing: The SMED System* (SA: The Productivity Press).

Taguchi, G., 1981, On-line Quality Control During Production, *Japanese Standard Association.*

Taguchi, G., 1987, *System of Experimental Design* (New York: UNIPUB, Kraus International Publications).

Taguchi, G. and Konishi, S., 1987, *Orthogonal Arrays and Linear Graph — Tools for Quality Engineering* (Dearborn, Michigan: American Supplier Institute, Inc).

Wittrock, R. J., Scheduling algorithms for flexible lines, 1985, *IBM Journal Res. Dev.*, **29** 401–412.

PART 3

Emerging Optimization Techniques

Part 3 of this book presents four memory-based optimization techniques and one memoryless optimization technique. Chapter 4 deals with several types of neural networks such as back-propagation network (BPN), self-organizing map (SOM), adaptive resonance theory (ART1 and ART2) and Hopfield network. Hopfield network is then applied to combinatorial optimization problems in production planning and control, such as machine tooling and production sequencing in manufacturing cell, as well as flow shop and job shop scheduling. Chapter 5 discusses genetic algorithm (GA) and its application to aggregate production planning. Chapter 6 discusses tabu search (TS) and its application to resource planning and allocation based on constraint management principles. Chapter 7 discusses ant systems (AS) and its application to scheduling bi-criterion flow shop problems. Chapter 8 presents one memoryless optimization technique, simulated annealing (SA) and its application to the cell formation problem. Chapter 9 discusses programming techniques, and presents the code of one of the memory-based optimization techniques presented in this book.

Chapter 4

Artificial Neural Networks

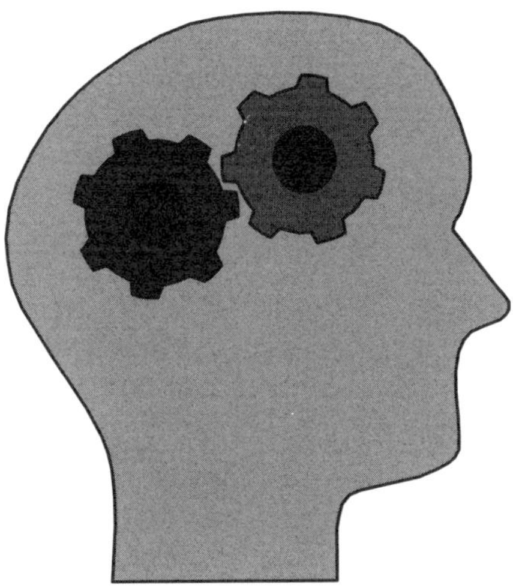

4.1. Background to Neural Networks

The way that biological neurons work has intrigued neurologists for several years. It is known that neurons are connected together via synapses, and a synapse produces a chemical response to input. The biological neuron fires if the sum of all the reactions from the synapses is sufficiently large. For many years, scientists and engineers have been interested in the actions of biological neurons in order to define new models of parallel problem solving. The McCulloch–Pitts theory (1943) that treats the human brain as a computational organism is the foundation for all activities in the central nervous system and forms the basis for most neural-network models. In their work, McCulloch and Pitts model the central nervous system as neural circuits that have computational power. Each neuron sends impulses to many other neurons in a process known as *divergence*, receives impulses from many other neurons in a process known as *convergence*, and also receives impulses from *feedback paths.*

Let us briefly examine the activity that occurs at the connection between two neurons called the synaptic junction or synapse. Communication between two neurons occurs as a result of postsynaptic cell absorbing chemical substances called neurotransmitter by the presynaptic cell as shown in Fig. 4.1. As the action potential arrives at the presynaptic membrane, the permeability of the membrane changes, resulting in influx of calcium irons. These irons

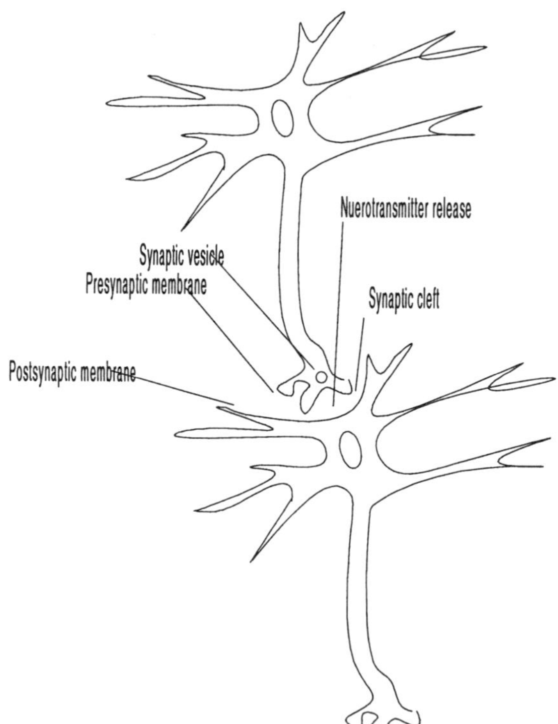

Fig. 4.1. Neurons connections at synapses.

cause the vesicles containing the neurotransmitters to fuse with the presynaptic membrane, resulting in their releasing neurotransmitters into the synaptic cleft. Consequently, the neurotransmitters diffuse across the synaptic cleft into the membrane of the postsynaptic membrane at certain receptor sites. The chemical-action at the receptor sites influences the permeability of the postsynaptic membrane. When positive ions enter the receptor sites, the action results in depolarisation, an effect referred to as *excitatory*. On the other hand, when negative ions enter the receptor sites, the action results in hyperpolarisation, an effect referred to as *inhibition*. Both, the excitatory and inhibition actions are local actions that take place within a finite distance into the cell body and are summed up at the axon hillock. If the sum is greater than a certain threshold, an action potential is generated. If the sum is less than this threshold, an action potential is not generated.

The artificial neural network (ANN) is an adaptive algorithm that takes its roots in the way that biological neurons work. Neural networks are massively parallel interconnected networks of simple elements which are usually adaptive and their hierarchical organizations which are intended to interact with the objects of the real world in the same way as biological nervous systems do. The basic components of a neural network are nodes that correspond

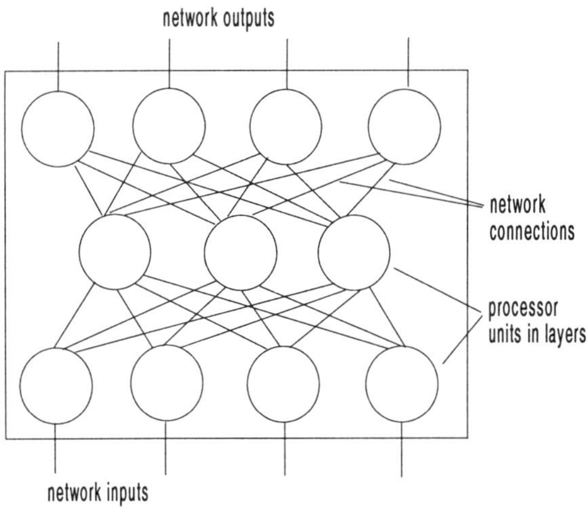

Fig. 4.2. Neural network architecture.

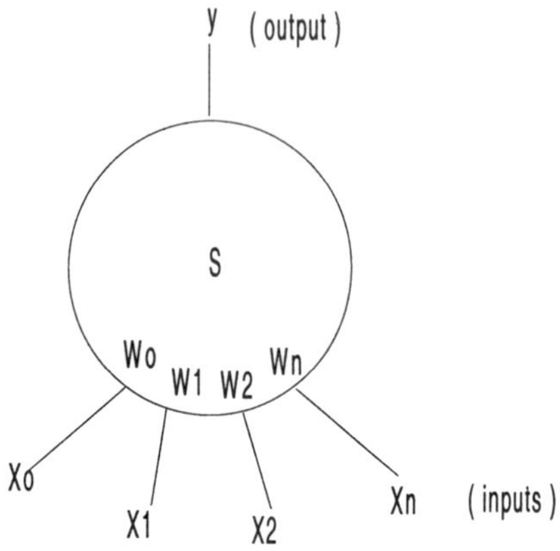

Fig. 4.3. A single processing unit.

to biological synapses. The weighted-inputs to a neuron are accumulated and then passed on, to an activation function that determines the nervous response. A positive weight represents an excitatory connection while a negative weight represents an inhibitory connection. In fact, the units were originally invented as an attempt to model biological neurons, hence the use of the term neural networks. A neural network is divided into layers: input layer, hidden layer(s) and the output layer as (see Fig. 4.2). Figure 4.3 shows one of these units.

Neural network may be classified on the basis of the directions in which, signals flow. In the feed-forward network, signals propagate in one direction from the input neurons through intermediate neurons in the hidden layer(s) to the neurons in the output layer. In the recurrent network, signals may propagate from the output of any neuron to the input of any neuron.

Another way in which neural networks may be classified is the extent to which the user (teacher) guides the learning-process. A supervised learning neural network adjusts weights of nodes of the hidden layer(s) and output on the basis of the difference between the values of the output units and the expected values assigned by the teacher, for a given input pattern. An unsupervised learning neural network adjusts weights of the nodes and classifies the input into sets without being guided. In artificial intelligence application to most engineering problems, it is useful to implement unsupervised feed-forward or unsupervised recurrent neural networks. There are basically two major classifications of neural networks: feed forward and recurrent. A neural network is either supervised or unsupervised. Supervised, feed forward neural networks include: perceptron, Hamming network, counter-propagation network (CPN), linear associative memory (LAM), and Boltzman machine. Unsupervised, feed forward neural networks include: clustering-network and self-organising feature maps (SOM). Supervised, recurrent neural networks include: bi-directional associative memory (BAM), auto-associative memory, and Hopfield network. Unsupervised recurrent neural networks include; adaptive resonance theory (ART1 and ART2). An excellent summary of the taxonomy of the most important network models may be found in Huang and Zhang (1994).

An artificial neural network, as we use the term here, is a computational paradigm inspired by the way that biological neurons work in responding to external stimuli. A similar process can be transposed to combinatorial optimisation: solutions of the problem are built using the parallel-processing actions of the neural network units. The remaining sections of this chapter discuss some of the common neural networks. Thereafter, an application-area in production planning and control is discussed, followed by the application of one of the neural networks.

4.2. Learning in Supervised Neural Networks: Delta Rule

A feed forward network consists of units in different layers that are interconnected. The units are associated with a set of weights that produce the correct response. Finding this appropriate set of weights requires us to search for appropriate solution, which leads us to a combinatorial optimisation problem. In fact, it is the search for optimum solutions for these weights that bring about inclusion of local search techniques, leading to hybrid-systems. We do not want to consider hybrid-system this early in our discussion. In feed forward networks, the technique often used for optimising the weights, is the popular gradient-descent method. The gradient-descent method is referred to as the delta rule when it is applied to training feed forward neural networks.

In the multi-layer feed forward network, how good a solution is measured by the mean squared error, $\bar{E}$. The error, e, is the difference between the desired output, d and the actual output, y, as given in Eq. (1). The error, e, is squared and the average is taken over all examples in a training set, as given in Eq. (2). Here, the training set refers to all the known N-input and output pairs for a particular problem.

$$e_j = d_j - y_j \tag{1}$$

$$\bar{E} = \frac{1}{N}\sum_{j=1}^{N} e_j^2 \,. \tag{2}$$

The mean squared-error, $\bar{E}$, gives a two-dimensional surface. Applying the gradient descent approach for training feed forward neural network requires that the gradient of the mean-squared error, $\bar{E}$ be changed by inclusion of a negative sign so that search is directed downward of the surface toward the grand minimum. This representation is given as follows:

$$\Delta w_i = -\alpha \frac{\partial \bar{E}}{\partial w_i} \,. \tag{3}$$

The derivative of the mean squared-error $\bar{E}$ is done using chain rule. Differentiating $\bar{E}$ with respect to w_i is done in two steps: first we differentiate $\bar{E}$ with respect to y, and then we differentiate y with respect to w_i.

$$\frac{\partial \bar{E}}{\partial w_i} = \frac{\partial E}{\partial y} \bullet \frac{\partial y}{\partial w_i} = \frac{\partial}{\partial y}\left(\frac{1}{N}\sum_{j=1}^{N}(d_j - y_i)^2\right) \bullet \frac{\partial y}{\partial w_i} \,. \tag{4}$$

The result of this differentiation is:

$$\frac{\partial \bar{E}}{\partial w_i} = \frac{-2}{N}\sum_{j=1}^{N} e \bullet \frac{\partial y}{\partial w_i} \,. \tag{5}$$

The result of this exercise shows that differentiating $\bar{E}$ with respect to w_i requires that the output y have to be differentiated with respect to w_i. This means that the output should be represented by a function that is differentiable. The function usually employed in representing feed forward neural network output is the *sigmoid* function. The sum of the products of the inputs and corresponding weights for a processing unit is given as:

$$S = \sum_{i=0}^{N} w_i x_i \,. \tag{6}$$

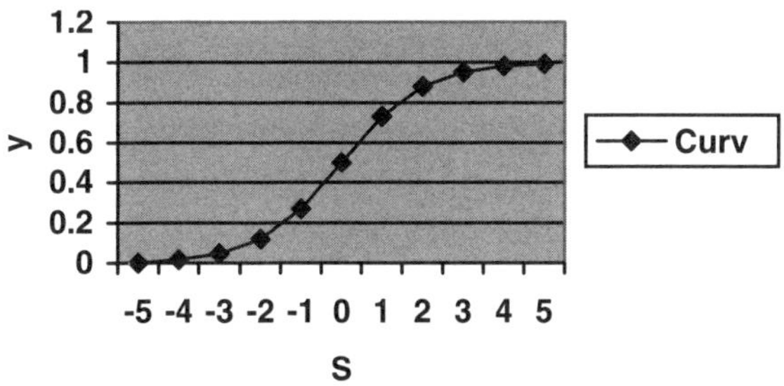

Fig. 4.4. Sigmoid function.

The output of a processing unit is obtained using sigmoid function, shown in Fig. 4.4 and given as follows:

$$y = \frac{1}{1 + e^{-S}} . \tag{7}$$

The *sigmoid* function has several interesting features. As the variable $S \to -\infty$, the output y tends to zero. As the variable $S \to \infty$, the output y tends to 1. When the variable $S = 0$, the output y, is 0.5.

The differentiation of the sigmoid function $y = \frac{1}{1+e^{-S}}$, with respect to the sum of the products of the inputs and corresponding weights for a processing unit, S is as follows:

$$\frac{\partial y}{\partial S} = \frac{e^{-S}}{(1 + e^{-S})^2} = y(1 - y) . \tag{8}$$

Differentiating the output, y, with respect to w_i is done using chain rule. It is carried out in two steps: first we differentiate the output, y with respect to S, and then we differentiate S with respect to w_i, as follows:

$$\frac{\partial y}{\partial w_i} = \frac{\partial y}{\partial S} \bullet \frac{\partial S}{\partial w_i} . \tag{9}$$

Since $S = \sum_{i=0}^{n} w_i x_i$, differentiating S with respect to w_i, is easy and given as follows:

$$\frac{\partial}{\partial w_i} \sum_{k=0}^{n} w_k x_k = \sum_{k=0}^{n} \frac{\partial w_k}{\partial w_k} w_k = x_k . \tag{10}$$

Therefore, the result of differentiating the output, y with respect to w_i is obtained from Eqs. (8), (9) and (10):

$$\frac{\partial y}{\partial w_i} = y(1-y)x_i\,. \tag{11}$$

Substituting these equations into Eq. (5) for the derivative of the mean squared-error, $\bar{E}$ is given as:

$$\frac{\partial \bar{E}}{\partial w_i} = \frac{-2}{N}\sum_{j=1}^{N} e \bullet y \bullet (1-y) \bullet x_i\,. \tag{12}$$

Let

$$\delta = ey(1-y) = y(1-y)(d-y)\,. \tag{13}$$

Then,

$$\frac{\partial \bar{E}}{\partial w_i} = \frac{-2}{N}\sum_{j=1}^{N} \delta x_i\,. \tag{14}$$

Equation (3) then becomes:

$$\Delta w_i = -\alpha\frac{\partial \bar{E}}{\partial w_i} = \frac{2\alpha}{N}\sum_{j=1}^{N} \delta x_i \tag{15}$$

$$\Delta w_i = \eta \sum_{j=1}^{N} \delta x_i \tag{16}$$

where $\eta = \frac{2\alpha}{N}$ is a constant and Eq. (16) then becomes:

$$\Delta w_i = \sum_{j=1}^{N} \eta\delta x_i \tag{17}$$

$$\Delta w_i = \eta\delta x_i \tag{18}$$

where $\delta = y(1-y)(d-y)$ and $\eta \in [0,1]$ is the *learning rate*.

$$w_i(t+1) = w_i(t) + \Delta w_i\,. \tag{19}$$

Equation (18) is the *delta rule* formula for feed forward neural networks, and Eq. (19) is the *learning rule.* The concepts discussed in this section are now applied to back-propagation neural network in the following section.

4.3. Backpropagation Neural Network (BPN)

The basic components of a neural network are nodes, called neurons and weights, called connections. A neuron (processing element) is an information-processing unit that is fundamental to the operation of a neural network. There are three basic operations of the neuron (Fig. 4.5):

(1) Synaptic weights (w_{ij}), with a positive value representing an excitatory connection, and a negative value representing an inhibitory connection;
(2) An accumulator for summing the input signal (x_i), weighted by the respective synapses of the neuron (w_{ij});
(3) An activation function $f_a(\bullet)$ which determines the neuron's response (y_j).

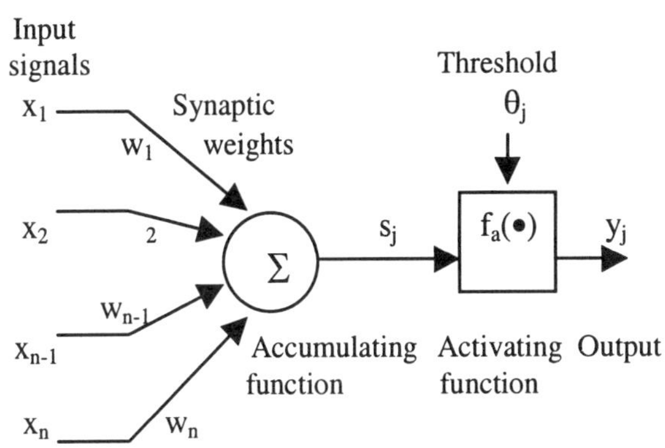

Fig. 4.5. Basic functions of a neuron.

When presented with an input vector, the neuron responds by generating output signals, which are based on the dot products of the input vector and weight vector. The same input vector is sent to each neuron, and the dot product ensures that the neuron whose weight vector is most nearly in the same direction as the input vector is the winner.

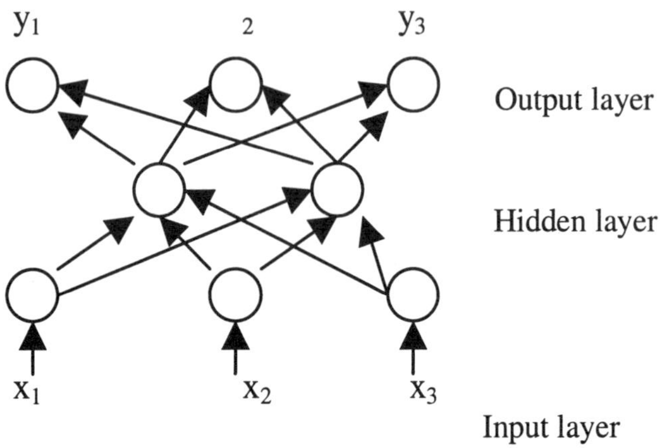

Fig. 4.6. A typical neural network.

Backpropagation neural networks (BPN) are multi-layered feed-forward neural networks that apply the error backpropagation-procedure for learning (Rumelhart *et al.*, 1986). An n-layer BPN has $n-2$ hidden layers as shown in Fig. 4.6. The BPN procedure uses a gradient descent technique to minimize a cost function equal to the mean square difference between the desired and the actual network outputs as already discussed in the preceding section.

BPN is trained by initially selecting small weights in a random manner and internal thresholds and then presenting all training data repeatedly. After every trial, weights are adjusted using side information specifying the correct class until weights converge and the cost function is reduced to an acceptable value. In BPN, the learning proceeds by a "forward activation" and by a "back-propagation" of error signals. Typically, at least three layers of neurons are involved, and the input nodes are presented with the input vector. The outputs of the input layer are transmitted to the middle layer and subsequently, to the outer layer. This is referred to as the forward activation. The output signals of the output layer are then compared with the desired outputs, and the error computed. The weights are then adjusted according to the least mean square training law, also referred to as delta rule (see Sec. 4.2). Parts of the errors are attributed to misalignment of the weight vectors of the hidden and input layers, and these error components are computed subsequently, and their weights adjusted. This is referred to as back-propagation of errors. The training algorithm therefore involves adjusting the weights sequentially from the output layer toward the input layer.

The BPN network assumes a sigmoid logistic non-linearity function and the algorithm step for its implementation are as follows:

Step 1: Initialization of weights and thresholds
Start with small random values of synoptic weights (w_{ij}) and threshold (θ) levels of the network.

Step 2: Presentation of training exemplars
Present the network with a continuous value input vector (x_i) together with the desired output vector (d_i). The training exemplars should be presented cyclically to the network until weights reach equilibrium. The number of times exemplars are presented to the network is known as *epoch* and when equilibrium weights are obtained at sufficiently high epochs, the weights change towards the shape of the input patterns.

Step 3: Forward output computation
Compute the output of neurons and the output values of the activation function layer by layer through the network, from the input layer to the output layer:
For the input layer, compute

$$s_j = \sum w_{ji} x_i \,, \quad \forall i \tag{20}$$

$$y_j = \max(s_1, \ldots\ldots, s_p) \tag{21}$$

where x_i is the input to the neuron i, s_j is the output of the neuron j, w_{ji} is the weight of the connection between the input layer and the first hidden layer, and y_j is the output value of the activation function. For net internal activity level, s_j for neuron j is

$$s_k = \sum w_{kj} y_j \,, \quad \forall j \tag{22}$$

$$y_k = f_a(s_k) = \frac{1}{a + \exp(-s_j + \theta_j)} \tag{23}$$

where y_k is the function signal of neuron k in the previous layer, w_{kj} is the synaptic weight of the neuron k, and θ_k is the threshold applied to neuron k. If neuron k is in the output layer, we set $z_k = y_k$, hence the error signal at the output layer is

$$\varepsilon_k = d_k - z_k \tag{24}$$

where d_k is the kth element of the desired response vector d_n. A BPN can thus be thought of as a mapping procedure, where m-dimensional input space is transformed into a corresponding p-dimensional output space. It requires a training set $T : T = \{(x_1, d_z), (x_2, d_2), \ldots, (x_n, d_n)\}$, where $d_i = \Phi(\bullet)$ is the mapping function $\Phi : \Re^m \rightarrow \Re^p$. Here, x_i is the input vector and d_i is the desired output vector.

Step 4: Backward weights adjustment

Starting at the output nodes and working back to the first hidden layer recursively adjust weights by computing the local gradients, ε_s, according to the generalized delta rule:

$$w_{ij}(t+1) = w_{ij}(t) + \eta\varepsilon_j y_j + \alpha\{w_{ij}(t) - w_{ij}(t-1)\}\,. \tag{25}$$

In the generalized delta rule, $w_{ij}(t)$ is the weight from hidden node j or from an input to node j at time t, y_j is either the output of node j or is an input, η is the learning-rate parameter, $\alpha(0 < \alpha < 1)$ is the momentum constant to smooth out the weight change and to accelerate convergence of the network, and ε_j is an error term for node j given as

$$\varepsilon_j = \begin{cases} z_j(1-z_j)(d_j - z_j) & \text{if node } j \text{ is an output node} \\ y_j(1-y_j)\sum_k \varepsilon_k w_{jk} & \text{if node } k \text{ is an int ernal hidden node} \end{cases} \tag{26}$$

where k is over all nodes in the layers above node j and y_j is either the output of node j or is input.

The generally good performance of the BPN algorithm is somewhat surprising considering that it is a gradient search technique that may find a local minimum in the least mean square cost function instead of the desired global minimum. Three steps to improve performance and reduce the occurrence of local minima include: lowering the learning-weight parameter η; allow extra hidden units; and making many training runs starting with different sets of random weights.

4.4. Self-Organising Map (SOM) Neural Network

Kohonen (1984) developed an unsupervised forward network known as self-organizing map (SOM). The primary use of the SOM network is for applications such as clustering, pattern recognition, and various optimization problems. SOM works for both binary and analog input pattern, hence, it preferable to the adaptive resonance theory (ART-1) developed by Carpenter and Grossberg (1988). The SOM network typically has two layers of nodes, the input layer and the Kohonen layer as shown in Fig. 4.7. SOM utilizes the concept of topology-preserving map of the input data to the competitive or Kohonen units. In a topology-preserving map, units located physically next to each other will respond to classes of input vectors that are likewise next to each other. Large-dimensional input vectors are projected down on the two-dimensional map such that the natural order of the input vectors are maintained. The input-layer of SOM is fully connected to a two-dimensional Kohonen layer. The Kohonen layer behaves similar to the biological systems because it can compress the representation of sparse data and spread out the representation of dense data using a two-dimensional map.

The values of the input pattern are each fed to the input nodes of the SOM network. There are as many input nodes in an SOM-network as there are values in the input pattern. An m-valued, input pattern is represented by a

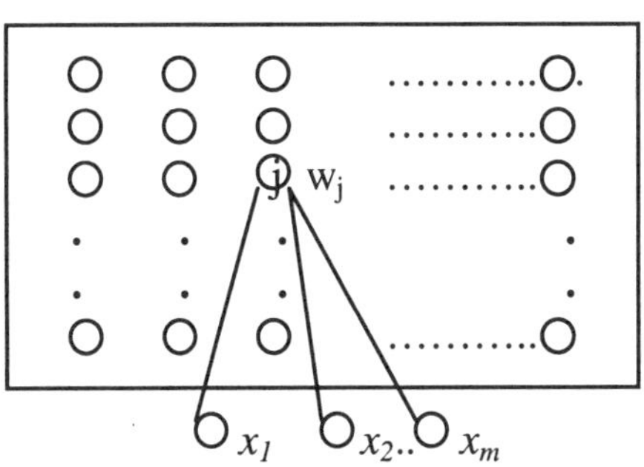

Kohonen layer with $N = nxn$ nodes
and weight vector $w_j = w_{1j}, w_{2j}, \ldots, m_j$.

Fig. 4.7. The self-organizing map (SOM) network.

vector of order m given as:

$$\mathbf{x} = (x_1, x_2, \dots, x_m)\,. \tag{27}$$

The m nodes in the input layer are fully connected to the nodes in the Kohonen layer that is a two-dimensional layer of nodes, usually arranged as a square matrix of size $N = nxn$, where n is the number of nodes in one of the directions of the two-dimensional Kohonen layer. Each node in the Kohonen layer has a weight vector with m-weight values contributed by the m-valued input pattern, so that the weight vector of a node j is represented as:

$$w_j = (w_{1j}, w_{2j}, \dots, w_{mj})\,, \quad j = 1, 2, \dots, N \tag{28}$$

where w_{ij} is the weight associated with Kohonen node j corresponding to the input node i.

Lateral interactions exist among processing elements in the Kohonen layer. During the learning process, the positive feedback will extend from the winning unit to other units in some finite neighbourhood. In this way, all the units in the neighbourhood that receive positive feedback from the winning unit participate in the learning process.

A self-organizing map algorithm generally commences the training cycle with random values assigned to the weight vectors in the Kohonen layer. The training process consists of repeatedly presenting the input data to the network through the nodes in the input layer. As the training process continues, the Kohonen nodes adjust their weight values according to the topological relations in the input data. Once training has progressed sufficiently, the weight vector of each unit in the Kohonen layer will converge to a value that is representative of the coordinates of the points near the physical location of the unit. For each training cycle, the winning node is determined in the Kohonen layer by finding the node, which has the minimum distance between it and all the nodes at the input layer. All the nodes within a defined neighbourhood around the winning node are updated and are considered to participate in the victory. As training proceeds, the size of the neighbourhood is diminished until it surrounds only a single node. An outline for the training process of a SOM algorithm is given as follows:

Step 1: Generate random weight values for each Kohonen node.
Step 2: **While** not end of each training cycle.
Step 3: **While** not end of input patterns.

3.1 Present the input vector x to the input layer.

3.2 Compute the distance d_j between each Kohonen node weight vector w_j and the input vector x_i:

$$d_j = \sum_{i=1}^{m} \|x_i(t) - w_{ij}(t)\| . \tag{29}$$

3.3 Determine the winning Kohonen node according to the winning criteria:

$$Z_j^* = \begin{cases} 1 & \text{if } \|x_i - w_{ijj}^*\| > \|x_i - w_{ik}\| \quad \forall_{i,j \neq k} \\ 0 & \text{if otherwise} . \end{cases} \tag{30}$$

3.4 Adjust the weight vector of the winning node j^* and those of the neighbourhood $N_{bj}{}^*(t)$ to be close to the input vector according to the training process criteria:

$$w_{ij}(t+1) = \begin{cases} w_{ij}(t) + \alpha(t)(x_i - w_{ij}(t)) & \text{if } j \in N_{bj}^*(t) \\ 0 & \text{if otherwise} . \end{cases} \tag{31}$$

Until termination (*end of input patterns and training cycle*).

In order to use the SOM algorithm, we need to define three important parameters: neighbourhood, size of Kohonen layer, and training coefficients α.

4.4.1. *Neighbourhood*

The algorithm that forms feature maps requires a neighbourhood to be defined around the winning node and all the nodes within the neighbourhood participate in the weight vector update process. The neighbourhood size is used to model the "*mexican hat*" function for lateral inhibition. The effect of this is to set up a competitive environment for training as only winning neurons and their neighbours participate in training for a given input pattern. Nodes that are outside the neighbourhood do not participate in the training process. As training proceeds, the size of the neighbourhood is diminished until it encompasses only a single node. This process adopts the winner-take-all strategy because as the neighbourhood size decreases, a single winner node is isolated for a given input pattern.

4.4.2. *Size of Kohonen layer*

The size of the Kohonen layer is the number of nodes in the layer which are all connected to each input node of the input layer. The Kohonen region is a square region that depends on the size of the Kohonen layer. We define the size of the Kohonen layer L_s as:

$$L_s = \frac{problem\ size}{neighbourhood}\,. \tag{32}$$

The square two-dimensional Kohonen region is usually reduced to a one-dimensional region. The reduction is achieved by reforming the two-dimensional layer of nodes into a single layer of nodes organized as a long vector composed of the concatenation of the original row vectors. The advantage of restructuring the network in this manner is that it leads to efficient locating and updating of the neighbourhood encompassing the winning node in the competition process.

4.4.3. *Training coefficient*

The training coefficient, $\alpha(t)$ is effectively a Gaussian function of the distance of a Kohonen node to the input node with a width that decreases with time. The significance of the training factor is that the lower its value, the lower the tendency for the network to be trapped in a local optimum state, but longer is the time for the network to converge. Conversely, a higher value of the training coefficient increases the tendency for the network to be trapped in a local optimum state, accompanied with a faster rate of convergence. Since we want to avoid being caught up in a local optimum state, experimentation usually commences with a high value of the training coefficient ($0.1 < \alpha < 0.4$) and gradually reduced as the training process continues. Progressive reduction of α continues until the network converges to similar configurations in successive tests.

4.5. Adaptive Resonance Theory

Adaptive resonance theory (ART), developed by Carpenter and Grossberg (1988), solves the problem of *stability-plasticity dilemma.* The stability-plasticity dilemma in neural network systems can be explained as follows.

When developing a neural network system to perform a particular pattern-classification operation, what is normally done is that training patterns or a set of exemplars are gathered for training the system. During the training process, information is encoded in the neural network system by the adjustment of the weight values. After training, the neural network system is ready for use, and no additional weight modification is permitted. In practice, this is not what really happens. We want a system that is dynamic. We want to add new information to the previous information used for training the neural network system. Think of a robot that is used for object recognition. Training data have been fed into the robot's memory for recognising a class of objects. It will be useful that if these objects change in shape, these changes should be used in updating previous geometry information. If this is not possible, then it means that each time we have such changes, the neural network system has to be retrained. Training is an expensive exercise, and such retraining due to changes in previous information stored in the memory makes the neural network system not to be adaptive (plastic). What we need is a neural network system that is adaptive and at the same time stable. This means that we need a neural network system that can retain previously learned information while continuing to learn new things. Counter propagation networks (CPN) are competitive systems in which nodes compete with one another, based on some specified criteria, and the winner is said to classify input pattern. This type of neural network system is excellent for pattern-classification. However, in CPN, there is the possibility of different nodes responding to the same input pattern, on different occasions, and previous learning may be overridden by a later learning.

Adaptive resonance theory (ART) solves the problem of stability-plasticity dilemma by incorporating a feedback mechanism between the competitive layer and the input layer of a network. This feedback mechanism allows the neural network system to learn new information that is added, without destroying the previous information used for training the neural network system. In physics, resonance is said to occur when small-amplitude vibration results in frequency that matches the frequency of a mechanical or electrical system. The same phenomenon is applied in ART where unit outputs travel to and from one layer to the other. Since they are being processed during the trips, there is the possibility of their matching the inputs to the network. When this happens, the system is said to have stable oscillation that is known as the resonance of the neural network.

4.5.1. *Basic features of ART network*

The basic features of ART network are shown in Fig. 4.8. Essentially, there are two main subsystems in the ART architecture: the *attentional* subsystem (A) and the *orienting* subsystem. The attentional subsystem consists of short-term memory (STM) and the long-term memory (LTM). The pattern of activation in the initial pre-processing neuron layer, F_1 and the output pattern in the competitive neuron layer, F_2 constitute the short-term memory of the attentional subsystem. The weights associated with the bottom-up and top-down connections between F_1 and F_2 constitute the short-term memory of the attentional subsystem respectively. In treating ART, the following notation, shown in Fig. 4.8, is useful:

- X pattern of activation across F_1;
- S output pattern from F_1;
- T net input received from layer F_1;
- V net input received from layer F_2;
- U output pattern from F_2;
- V pattern of activity from F_2 (in response to inputs from F_1).

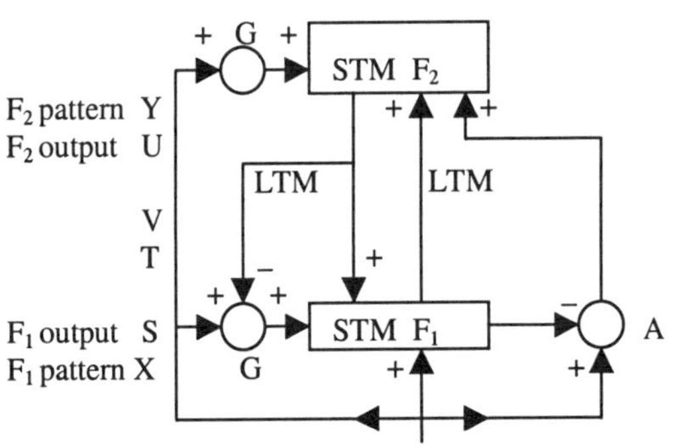

Fig. 4.8. Adaptive resonance theory architecture.

Pattern Matching

The output pattern from F_1, which is S, sends an inhibitory signal to the orienting subsystem, A, which exactly cancels the excitatory effect of the signal from the input I (see Fig. 4.9(a)). The output pattern from F_2, which is U,

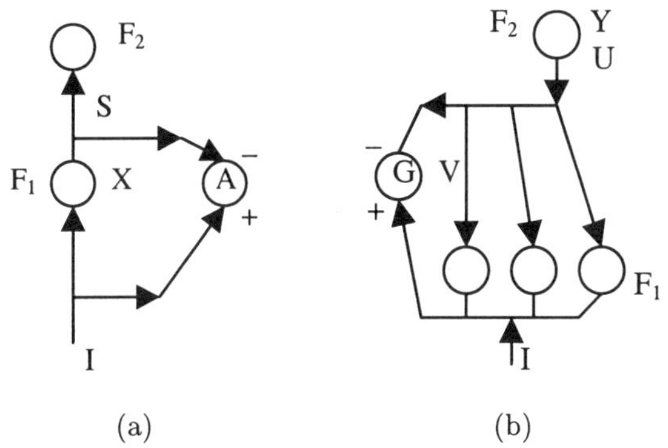

Fig. 4.9. Output patterns activities in F_1 and F_2.

sends an inhibitory signal to the gain control system, G, which is so configured that when this signal from F_2 (i.e. U) is received, the gain control system cease activity (see Fig. 4.9(b)). The output pattern for F_2 which is U, also becomes input pattern for F_1-units. The output pattern from F_2, which is U, is transformed by LTM traces on the top-down connections from F_2 to F_1. This transformed pattern is V. Consequently; there are three possible sources of inputs to F_1 (i.e. V, U, and I). However, the units on F_1 and F_2 layers are so constructed that they are active when V and U are active. These new sources of input to F_1 cause a mismatch, because a new activity pattern, X^* will be obtained, resulting in a new output S^*. This new output is different from the previous output S. Hence, the new signal $I \cap V$ reaching the attentional subsystem A does not cancel the incoming excitation, I in F_1. The mismatch in F_1 due to $I \cap V$ causes A to be active and sends signal to all units resulting in a new output Y^*. The cycle of pattern matching will continue until a match is found by the system.

Gain Control

The gain control mechanism plays significant role when the ART network of Fig. 4.8 is part of yet another larger network. In this case, F_2 layer might receive signals from the much larger network and from F_1 layer. When the F_2 layer is stimulated by the much larger-network, then top-down signal is sent to layer F_1. Problem could arise from this, because this top-down signal may arrive at F_1 before the incoming signal I, resulting in a phenomenon known as *expectation.*

The expectation phenomenon causes some problems because of the way the gain control-mechanism, G operates. Any signal from F_2 causes inhibition in G. Any signal from F_1 causes G to arouse F_1 (see Fig. 4.8), but this is only possible if an expectation signal in not received. The gain control-mechanism, together with the 2/3 rule enables the F_1 layer to discriminate between expectation signal, I_e and input signal, I. The remaining sections deal with the variants of ART, known as ART1 and ART2.

4.5.2. *ART*1

The ART1 architecture is essentially the same as the ART architecture of Fig. 4.8. The input to ART1 is binary, i.e. a set of {0, 1}. This constraint limits the application of ART1 in practice. In this section, we discuss the attentional and gain control subsystems, and the processing that takes place at the F_1 and F_2 layers. Specifically, for the F_2 processing, we discuss the bottom-up traces and the top-down traces.

The attentional subsystem

The governing differential equation for processing units on layers F_1 and F_2 is given by

$$\varepsilon \ddot{x}_k = -x_k + (1 - Ax_k)J_k^+ - (B + Cx_k)J_k^- \, . \tag{33}$$

The total excitatory input, J_i^+ and inhibitory input, J_i^-, are given respectively by

$$J_i^+ = I_i + B_1 G + D_1 V_i \tag{34}$$

$$J_i^- = 1 \, . \tag{35}$$

The unit net-input from F_2, gain G, and output S, are given respectively as:

$$V_i = \sum_j u_j z_{ij} \tag{36}$$

$$G = \begin{cases} 1 & \text{if } I \neq 0 \text{ and } U = 0 \\ 0 & \text{if } I = 0 \text{ and } U \neq 0 \, . \end{cases} \tag{37}$$

$$S_i = h(x_{1i}) = \begin{cases} 1 & \text{if } x_{1i} > 0 \\ 0 & \text{if } x_{1i} \le 0\,. \end{cases} \tag{38}$$

Processing on F_1

There are now three possible sources of inputs to F_1. These inputs include I, V, and gain control system G, shown in Fig. 4.10. Let us consider the three possible cases.

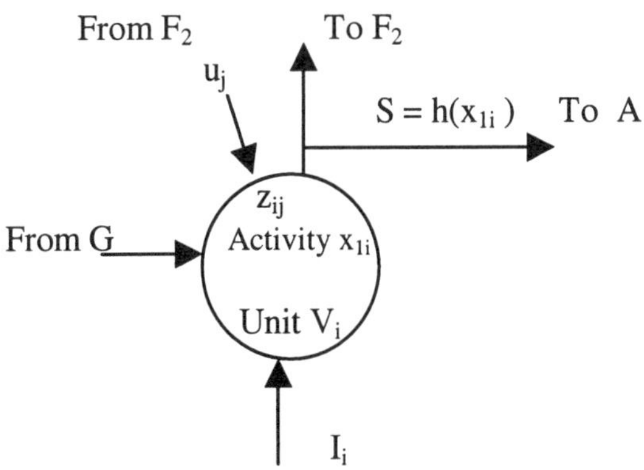

Fig. 4.10. Inputs to F_1.

Case I: $I_i = 0$ and F_2 is not active (i.e. $V_i = 0$); In this case, $G = 0$ since $I_i = 0$. This means that $J_i^+ = 0$.

Using the governing differential equation for processing units on layers F_1 and F_2 (Eq. (33)), together with the conditions: $J_i^+ = 0$, $\ddot{x}_k = \ddot{x}_{1i} = 0$ at steady state condition, yields

$$0 = -x_{1i} - (B_1 + C_1 x_{1i})\,.$$

Leading to:

$$x_{1i} = \frac{-B_1}{1 + C_1}\,. \tag{39}$$

Case II: I_i is applied (i.e. $I_i \neq 0$) and F_2 is not active (i.e. $V_i = 0$); In this case, $G = 1$ since $I_i \neq 0$. This means that $J_i^+ = I_i + B_1$.

Using the governing differential equation for processing units on layers F_1 and F_2 (Eq. (33)), together with the conditions: $J_i^+ = I_i + B_1$, $\ddot{x}_k = \ddot{x}_{1i} = 0$ at steady state condition, yields

$$0 = -x_{1i} + (1 - A_1 x_{1i})(I_i + B_1) - (B_1 + C_1 x_{1i}) .$$

Leading to:

$$x_{1i} = \frac{I_i}{1 + A_1(I_i + B_1) + C_1} . \tag{40}$$

Case III: I_i is applied (i.e. $I_i \neq 0$) and F_2 is active (i.e. $V_i \neq 0$); In this case, $G = 0$ since $I_i \neq 0$. This means that $J_i^+ = I_i + D_1 V_i$.

Using the governing differential equation for processing units on layers F_1 and F_2 (Eq. (33)), together with the conditions: $J_i^+ = I_i + D_1 V_i$, $\ddot{x}_k = \ddot{x}_{1i} = 0$ at steady state condition, yields

$$0 = -x_{1i} + (1 - A_1 x_{1i})(I_i + D_1 V_i) - (B_1 + C_1 x_{1i}) .$$

Leading to:

$$x_{1i} = \frac{I_i + D_1 V_i - B_1}{1 + A_1(I_i + D_1 V_i) + C_1} . \tag{41}$$

ART1 parameters setting

Let us examine several conditions that ART1 parameters must satisfy:

- For large positive net input, using Eq. (41)

  $$I_i + D_1 V_i - B_1 > 0 .$$

  The maximum output of an F_2-unit is 1 (only one node of $F_2 \neq 0$) and the maximum weight on any top-down connection is also 1. Since $V_i = \sum_j u_j z_{ij}$, and only one unit is non-zero, it means that $V_i \leq 1$. The maximum value of $V_i = 1$ and the maximum value of $I_i = 1$. Hence $I_i + D_1 - B_1 > 0$, or

  $$B_1 < D_1 + 1 . \tag{42}$$

- For no top-down signal from F_2 on any unit, a negative activity results, i.e. $I_i + 0 - B_1 < 0$, and since maximum of $I_i = 1$, it means that $1 - B_1 > 0$, leading to

$$B_1 > 1\,. \tag{43}$$

- For top-down signal from F_2 but there is not yet an input vector, $I_i(I_i = 0)$, using Eq. (41) gives

$$x_{1i} = \frac{D_1 V_i - B_1}{1 + A_1 D_1 V_i + C_1}\,.$$

- If $V_i > 0$, then the unit's activity reaches some value, but must remain negative to avoid the unit having non-zero output based on top-down inputs alone (i.e. $(D_1 \bullet \max(V_1) - B_1 < 0)$), or

$$B_1 > D_1\,. \tag{44}$$

Combining Eqs. (42), (43), and (44), we have

$$B_1 > 1, D_1 \tag{45}$$

$$B_1 < D_1 + 1\,. \tag{46}$$

Leading to:

$$\max(D_1, 1) < B_1 < D_1 + 1\,. \tag{47}$$

Equation (47) must be satisfied by ART1 parameters. These parameters must be carefully chosen so that the specified conditions are net before running an ART1 program.

For a unit to have positive value of activation,

$$I_i + D_1 V_i - B_1 > 0\,.$$

And for maximum of $I_i = 1$, we have $1 + D_1 V_i - B_1 > 0$, leading to the condition that must be satisfied for a top-down signal to be activated:

$$V_i > \frac{B_1 - 1}{D_1}\,. \tag{48}$$

Processing on F_2

The unit v_j of layer F_2 receives inputs from layer F_1, the gain control system, G and the orienting subsystem, A as shown in Fig. 4.11. The activity of the unit is x_{2y}. Outputs, u_j are sent back to F_1. Each unit receives a positive feedback term, $g(x_{2y})$ from itself and sends identical signal through an inhibitory

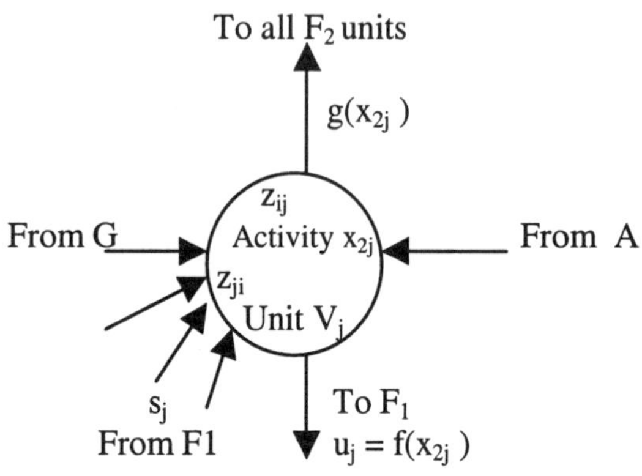

Fig. 4.11. Inputs to F_2.

connection to all other units on the layer. This is because ART1 is a recurrent ANN system.

The bottom-up signals, s_i from F_1 are gated by the weights, z_{ji} and are used to calculate the net input as

$$T_j = \sum_i s_i z_{ji} \,. \tag{49}$$

The total excitatory input to v_j is

$$J_j^+ = D_2 T_j + g(z_{2j}) \,. \tag{50}$$

The inhibitory input to each unit is given by

$$J_j^- = \sum_{k \neq j} +g(x_{2j}) \,. \tag{51}$$

Substituting Eqs. (49), (50), and (51) into Eq. (33) yields

$$\varepsilon \dot{x}_{2j} + (1 - A_2 x_{2j})(D_2 T_j + g(x_{2j})) - (B_2 + C_2 x_{2j}) \sum_{k \neq j} g(x_{2k}) \,. \tag{52}$$

Outputs of F_2 units are given as

$$u_j = f(x_{2j}) = \begin{cases} 1 & T_j = \max_k \{T_k\} \quad \forall\, k \\ 0 & \text{otherwise} \,. \end{cases} \tag{53}$$

Top-down LTM traces

Top-down LTM traces represent weights on connections form F_2 units to F_1 units. The mechanics of ART1 is based on the counter propagation network (CPN) theory. We use the CPN learning law, which states that

$$\dot{w} = (-cw + dI)U(net)\,.$$

Applying this equation, we have

$$\dot{z}_{ij} = (-z_{ij} + h(x_{1i}))f(x_{2j})\,. \tag{54}$$

Since $f(x_{2j})$ is only none-zero for only one value of j (for the wining unit) and if the jth F_2 unit is active and the F_1 unit is also active, then

$$\dot{z}_{ij} = \begin{cases} -z_{ij} + 1\,; & v_j \text{ active and } v_i \text{ active} \\ -z_{ij}\,; & v_j \text{ active and } v_i \text{ in active} \\ 0\,; & v_j \text{ and } v_i \text{ both in active}\,. \end{cases} \tag{55}$$

At equilibrium,

$$0 = \begin{cases} -z_{ij} + 1 \\ -z_{ij}\,. \end{cases}$$

What this means is that:

$$0 = \begin{cases} 1 & \text{if } V_i \text{ is active} \\ 0 & \text{if } V_i \text{ is in active}\,. \end{cases}$$

Since only one F_2-unit is active (i.e. the wining unit) at a time, this means that $V_i = \sum_j u_j z_{ij} = z_{ij}$. This leads to:

$$z_{ij} > \frac{B_1 - 1}{D_1}\,. \tag{56}$$

Bottom-up LTM traces

Bottom-up LTM-traces represent weights on connections from F_1 units to F_2 units. The weights on the connections from v_i on layer F_1 to v_j on layer F_2 are determined by

$$\dot{z}_{ji} = K\left[- z_{ij} \sum_{k \neq i} h(x_{1k}) + (1 - z_{ji})Lh(x_{1l})\right] \bullet f(x_{2j}) \tag{57}$$

where K and L are constants. Although the bottom-up LTM-traces are more complicated than those of the top-down LTM traces, Eq. (57) for determining the weights is essentially of the form, $y = \mu x + (1-\mu)x$, which is the parametric representation of a straight-line. As was the case with the top-down LTM traces, the factor $f(x_{2j})$ ensures that only weights on the winning unit are allowed to change. The individual weights compete among one another, rather than individual units competing among themselves.

We define a few notations. The input pattern is I and its magnitude is: $|I| = \sum_i h(x_{1i})$. The output of layer F_1 is S and its magnitude is: $|S| = \sum_i h(x_{1i})$. Accordingly, we define:

$$S = \begin{cases} I & \text{if } F_2 \text{ is in active} \\ I \cap V^j & \text{if } F_2 \text{ is active}\,. \end{cases} \tag{58}$$

Since $|S| = \sum_i h(x_{1i})$, then $\sum_{k \neq i} h(x_{1k}) = \sum_k h(x_{1k}) - h(x_{1i}) = |S| - 1$ or $|S|$, depending on whether v_i is active. Hence,

$$\dot{z}_{ji} = \begin{cases} K[-z_{ij}(|S|-1) + (1-z_{ij})L] & \text{if } v_i \text{ and } v_j \text{ are active} \\ -K[z_{ij}|S|] & \text{if } v_i \text{ is in active and } v_j \text{ is active} \\ 0 & \text{if } v_j \text{ is in active}\,. \end{cases} \tag{59}$$

It can be seen that v_i can remain active only if it receives an input, I_i from layer F_1 and a sufficiently large input, V_i from layer F_2. At equilibrium, $\dot{z}_{j1} = 0$, leading to

$$0 = K[-z_{ji}(|S|-1) + (1-z_{ji})L]\,.$$

Hence,

$$z_{ij} = \begin{cases} \dfrac{L}{L-1+|S|} & \text{if } v_i \text{ is active} \\ 0 & \text{if } v_i \text{ is in active}\,. \end{cases} \tag{60}$$

During initialisation, we have to keep all initial weights below a certain value. Since all patterns are a subset of the pattern containing all ones, we should keep all initial weights with the range

$$0 \leq z_{ji}(0) \leq \frac{L}{L-1+M} \tag{61}$$

where M is the number of units on F_1.

The orienting subsystem

The orienting subsystem in an ART network is responsible for sensing mismatches between bottom-up and top-down patterns on the F_1 layer. The inputs to the orienting subsystem, A are the outputs of the F_1 units, S, and the input vector, I. The weights on the connection from the input vector are all equal to P. Those on the connection from F_1 are all equal to the value Q. The net-input to the orienting subsystem, $A = P|I| - QS$. The output of A switches on, if the net input becomes non-zero. This means that

$$P|I| - Q|S| > 0$$

or

$$P|I| > Q|S| \,. \tag{62}$$

We can define the ratio:

$$\frac{P}{Q} > \frac{|S|}{|I|}, \quad \text{or } \rho > \frac{|S|}{|I|} \tag{63}$$

where ρ is the *vigilance-parameter*. The value ρ is usually less than unity. Consequently, we have two cases: (i) If $\frac{|S|}{|I|} < \rho$, then the winning unit is marked inactive; (ii) If $\frac{|S|}{|I|} > \rho$, then the winning unit is declared active. All the equations in this section are now put together to form an ART1 algorithm.

ART1 algorithm

Step 1: Initialisation

$$\begin{aligned}
&A_1 \geq 0 \\
&C_1 \geq 0 \\
&D_1 \geq 0 \\
&\max\{D_1, 1\} < B_1 < D_1 + 1 \\
&L > 0 \\
&0 < \rho \leq 1 \\
&x_{1i}(0) = \frac{-B_1}{1 + C_1}
\end{aligned}$$

$$z_{ij}(0) > \frac{B_1 - 1}{D_1}$$

$$0 < x_{ji}(0) < \frac{L}{L - 1 + M}.$$

Step 2: Apply an input vector I to F_1 and calculate the activities according to

$$x_{1i} = \frac{I_i}{1 + A_1(I_i + B_1) + C_1}.$$

Step 3: Compute the output vector for F_1,

$$S_i = h(x_{1i}) = \begin{cases} 1 & \text{if } x_{1i} > 0 \\ 0 & \text{if } x_{1i} \leq 0. \end{cases}$$

Step 4: Forward-propagate S to F_2 and calculate the activities according to

$$T_j = \sum_i s_i z_{ji}.$$

Step 5: Only the winner-unit in F_2 has a non-zero output

$$u_j = f(x_{2j}) = \begin{cases} 1 \; T_j = \max_k\{T_k\} & \forall k \\ 0 & \text{otherwise}. \end{cases}$$

Step 6: Back-propagate the output from F_2 to F_1 and calculate the activities according to

$$V_i = \sum_j u_j z_{ij}.$$

Step 7: Calculate the new activities according to

$$x_{1i} = \frac{I_i + D_1 V_i - B_1}{1 + A_1(I_i + D_1 V_i) + C_1}.$$

Step 8: Determine the new output values, s_i exactly as in Step 3.

Step 9: Determine the degree of mismatch the top-down and input patterns

$$\frac{|S|}{|I|} = \frac{\sum_{i=1}^{M} s_i}{\sum_{i=1}^{M} I_i}.$$

Step 10: Compare the degree of mismatch with the vigilance parameter. If $\frac{|S|}{|I|} < \rho$, label v_J as being active, zero the outputs of F_2, and return to Step 1 using the original input pattern, I.

Step 11. Update bottom-up weights on v_J (winning F_2 unit) only

$$z_{Ji} = \begin{cases} \dfrac{L}{L-1+|S|} & \text{if } v_i \text{ is active} \\ 0 & \text{if } v_i \text{ is in active}. \end{cases}$$

Step 12. Update top-down weights coming from v_J (winning F_2 unit) only to all F_1 units

$$z_{iJ} = \begin{cases} 1 & \text{if } V_i \text{ is active} \\ 0 & \text{if } V_i \text{ is in active}. \end{cases}$$

Step 13. Remove the current input pattern. Restore all disabled F_2 units. Pick a new input pattern and return to Step 1.

Example 1

We solve the problem for the following ART1 parameters setting: $M = 6$; $N = 7$; $A_1 = 1$; $B_1 = 1.3$; $C_1 = 6$; $D_1 = 0.7$; $L = 4$; $\rho = 0.85$, $I_1 = (0,0,0,0,1,0)^t$.

Solution

Step 1: Initialisation:

$$z_{ij}(0) > \frac{B_1 - 1}{D_1} = \frac{1.3 - 1}{0.7} + 0.3 = 0.729$$

$$z_i = (0.729, 0.729, 0.729, 0.729, 0.729, 0.729, 0.729)^t$$

$$0 < z_{ji}(0) < \frac{L}{L-1+M} = \frac{4}{4-1+6} - 0.095 = 0.349$$

$$z_j = (0.349, 0.349, 0.349, 0.349, 0.349, 0.349)^t$$

$$x_{1i}(0) = \frac{-B}{1+C_1} = \frac{-1.3}{1+6} = -0.186$$

$$X(0) = (-0.816, -0.816, -0.816, -0.816, -0.816, -0.816)^t$$

Step 2: $x_{1i} = \frac{I_i}{1+A_1(I_i+B_1)+C} = \frac{I_i}{1+I_i+1.3+6} = (0,0,0,0,0.108,0)^t$

Step 3: $S = (0,0,0,0,1,0)^t$

Step 4: $T_j = \sum_i s_i z_{ji} = (0.349, 0.349, 0.349, 0.349, 0.349, 0.349, 0.349)^t$

Step 5: $U = (1,0,0,0,0,0,0)^t$

Step 6: $V_i = \sum_i z_i U = (0.729, 0.729, 0.729, 0.729, 0.729, 0.729)^t$

Step 7: $x_{1i} = \frac{I_i+D_1V_i-B_1}{1+A_1(I_i+D_1V_i)+C_1} = \frac{I_i+0.7*0.729-1.3}{1+1*(I_i+0.7*0.729)+1.3}$

$X = (-0.0105, -0.0105, -0.0105, -0.0105, 0.025, -0.0105)^t$

Step 8: $S = (0,0,0,0,1,0)^t$

Step 9: $|S|/|I| = 1 > \rho$

Step 10: No reset is required. Resonance is attained.

Step 11: $z_1 = \frac{L}{L-1+|S|} = \frac{4}{4-1+1}$, z_1 on F_2 becomes $z_{Ji} = (0,0,0,0,1,0)^t$

Step 12: $$z_{ij} = \begin{bmatrix} 0 & 0.729 & 0.729 & 0.729 & 0.729 & 0.729 & 0.729 \\ 0 & 0.729 & 0.729 & 0.729 & 0.729 & 0.729 & 0.729 \\ 0 & 0.729 & 0.729 & 0.729 & 0.729 & 0.729 & 0.729 \\ 0 & 0.729 & 0.729 & 0.729 & 0.729 & 0.729 & 0.729 \\ 1 & 0.729 & 0.729 & 0.729 & 0.729 & 0.729 & 0.729 \\ 0 & 0.729 & 0.729 & 0.729 & 0.729 & 0.729 & 0.729 \end{bmatrix}$$

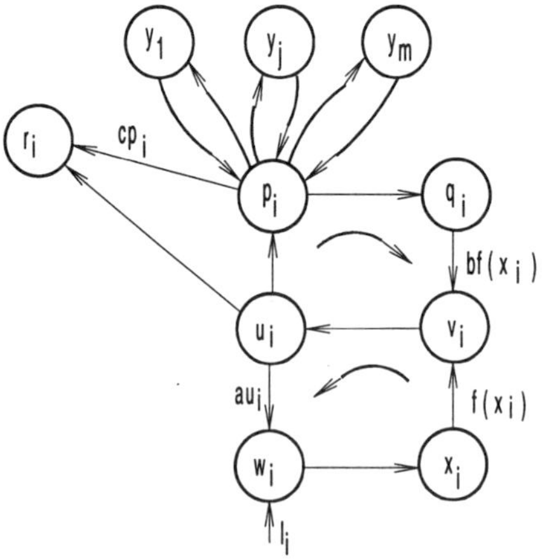

Fig. 4.12. ART2 architecture.

4.5.3. *ART2*

The ART2 architecture is essentially the same as the ART architecture of Fig. 4.8. Unlike ART1, the input to ART2 is an analogue pattern, i.e. a set of {1.5, 15.3, etc.}. Like ART1, ART2 has attentional and gain control subsystems, and the processing that takes place at the F_1 and F_2 layers. In order to deal with analogue patterns, Carpenter and Grossberg (1988) divided the F_1 layer into sub-layers containing feed-forward and feedback connections as shown in Fig. 4.12.

In this section, we discuss the attentional and gain control subsystems, and the processing that takes place at the F_1 and F_2 layers. Specifically, for the F_2 processing, we discuss the bottom-up traces and the top-down traces.

The attentional subsystem

The governing differential equation for processing units on layers F_1 and F_2 is given by

$$\varepsilon \ddot{x}_k = -Ax_k + (1 - Bx_k)J_k^+ - (C + Dx_k)J_k^- \,. \tag{64}$$

For ART2 model, the constants, B and C are taken be identically zero. The total excitatory input is J_i^+ and inhibitory input is J_i^-, and for asymptotic solution,

$$x_k = \frac{J_k^+}{A + DJ_k^-} \,. \tag{65}$$

The values of individual quantities in Eq. (65) vary with different layers, and these values are shown in Table 4.1.

Table 4.1.

Layer	A	D	J_i^+	J_i^-
w	1	1	$I_i + au_i$	0
x	e	1	w_i	$\|w\|$
u	e	1	v_i	$\|v\|$
v	1	1	$f(x_i) + bf(q_i)$	0
p	1	1	$u_i + \sum g(y_j)z_{ij}$	0
q	e	1	p_i	$\|p\|$
r	e	1	$u_i + cp_i$	$\|u\| + \|cp\|$

Processing on F_1

The activities of on each of the six sub-layers on F_1 are defined as follows, based on the values in Table 4.1:

$$W_i = I_i + au_i \tag{66}$$

$$x_i = \frac{w_i}{e + \|w\|} \tag{67}$$

$$u_i = \frac{v_i}{e + \|v\|} \tag{68}$$

$$v_i = f(x_i) + bf(q_i) \tag{69}$$

$$p_i = u_i + \sum_j g(y_j) z_{ij} \tag{70}$$

$$q_i = \frac{p_i}{e + \|p\|} \tag{71}$$

$$r_i = \frac{u_i + cp_i}{e + \|u\| + \|cp\|} \approx \frac{u_i + cp_i}{\|u\| + \|cp\|}, \quad (e \ll 1)\,. \tag{72}$$

Processing on F_2

The processing on layer F_2 for ART2 and ART1 are identical. The bottom-up signals, s_i from F_1 are gated by the weights, z_{ji} and are used to calculate the net input as

$$T_j = \sum_i p_i z_{ji}\,. \tag{73}$$

Outputs of F_2 units are given as

$$g(y_j) = f(x_{2j}) = \begin{cases} d \; T_j = \max\limits_k \{T_k\} & \forall\, k \\ \quad 0 \;\; \text{otherwise}\,. \end{cases} \tag{74}$$

The equation for processing on the $\boldsymbol{p}$ sub-layer of F_1 is now written as

$$p_i = \begin{cases} u_i & \text{if } F_2 \text{ is in active} \\ u_i + dz_{iJ} & \text{if the } J\text{th node on } F_2 \text{ is active}\,. \end{cases}$$

Top-down and bottom-up LTM traces

Both the top-down and bottom-up LTM traces have the same form and are simpler than those for ART1:

$$\dot{z}_{ij} = g(y_i)(p_i - z_{ij})\,. \tag{75}$$

If v_J is the wining node, then

$$\dot{z}_{iJ} = d(u_i + dz_{iJ} - z_{iJ})\,. \tag{76}$$

Similarly

$$\dot{z}_{Ji} = d(u_i + dz_{Ji} - z_{Ji})\,. \tag{77}$$

At equilibrium, $\dot{z}_{Ji} = \dot{z}_{iJ} = 0$ for $j \neq J$:

$$z_{Ji} = z_{iJ} = \frac{u_i}{1-d}\,. \tag{78}$$

The orienting subsystem

The orienting subsystem in an ART network is responsible for sensing mismatches between bottom-up and top-down patterns on the F_1 layer. We make use of our earlier definition for the r layer. This means that

$$r_i = \frac{u_i + cp_i}{e + \|u\| + \|cp\|} \approx \frac{u_i + cp_i}{\|u\| + \|cp\|}\,, (e \ll 1)\,.$$

The condition for reset is

$$\frac{\rho}{\|r\|} > 1 \tag{79}$$

where ρ is the *vigilance-parameter*. Defining $\|r\| = (r \bullet r)^{1/2}$, the expression becomes

$$\|r\| = \frac{[1 + 2\|cp\| \cos(u,p) + \|cp\|^2]1/2}{1 + \|cp\|}\,. \tag{80}$$

All the equations in this section are now put together to form an ART2 algorithm.

ART2 algorithm

Step 1: Initialisation

$$a, b > 0$$

$$0 \leq d \leq 1$$

$$\frac{cd}{1-d} \leq 1$$

$$0 \leq \theta \leq 1$$

$$0 \leq \rho \leq 1$$

$$e \ll 1$$

$z_{ij}(0) = 0$ for top-down weights-initialisation

$z_{ji}(0) \leq \dfrac{1}{(1-d)\sqrt{M}}$ for bottom-up weights-initialisation.

Step 2: Apply an input vector I to layer F_1 and calculate the output according to

$$w_i = I_i + au_i\,.$$

Step 3: Forward-propagate to the next $\boldsymbol{x}$ sub-layer,

$$x_i = \frac{w_i}{e + \|w\|}\,.$$

Step 4: Forward-propagate to the next $\boldsymbol{v}$ sub-layer,

$$v_i = f(x_i) + bf(q_i)\,.$$

Step 5: Propagate to the next $\boldsymbol{u}$ sub-layer,

$$u_i = \frac{v_i}{e + \|v\|}\,.$$

Step 6: Forward-propagate to the next $\boldsymbol{p}$ sub-layer,

$$p_i = u_i + \sum_j g(y_j) z_{ij} = u_i + dz_{iJ}$$

where the Jth node on F_2 is the winner-node.

Step 7: Propagate to the next $\boldsymbol{q}$ sub-layer,

$$q_i = \frac{p_i}{e + \|p\|}\,.$$

Step 8: Repeat Steps 2 through 7 in order to stabilise the values on sub-layer F_1.

Step 9: Determine the output of $\boldsymbol{r}$ sub-layer,

$$r_i = \frac{u_i + cp_i}{e + \|u\| + \|cp\|}\,.$$

Step 10: Determine whether a reset is required using

$$\frac{\rho}{(e + \|r\|)} > 1\,.$$

Step 11: Propagate the output of $\boldsymbol{p}$ sub-layer to the F_2. Then calculate the net input to F_2.

$$T_j = \sum_i p_i z_{ji}\,.$$

Step 12: Update top-down weights coming from v_J (winning F_2 unit) only to all F_1 units

$$g(T_j) = \begin{cases} d\ T_j = \max\limits_k\{T_k\} & \forall\, k \\ \quad 0 \quad \text{otherwise}\,. \end{cases}$$

Step 13: Repeat Steps 6 to 10.

Step 14: Modify bottom-up weights on the winning F_2 unit according to

$$z_{Ji} = \frac{u_i}{1-d}\,.$$

Step 15: Modify top-down weights on the winning F_2 unit according to

$$z_{iJ} = \frac{u_i}{1-d}\,.$$

Step 16: Remove the input pattern. Restore all disabled F_2 units. Pick a new input-pattern and return to Step 1.

Let us consider an example to see how ART2 works.

Example 2

Analyse an ART2 for $M = 5$; $N = 6$; $a = 10$; $b = 10$; $\theta = 0.2$; $d = 0.8$; $c = 0.2$; $\rho = 0.85$; $I_i = (0.4, 0.5, 0.1, 0.6, 0.2)$.

Solution

Step 1: The top-down weights and the bottom-up weights are respectively initialised as follows:

$$z_{ij}(0) = 0$$

$$z_{ji}(0) \leq \frac{1}{(1-d)\sqrt{M}} = \frac{0.5}{(1-0.8)\sqrt{5}} = 1.118\,.$$

Step 2: Since there is no feedback from $\boldsymbol{u}, \boldsymbol{w}$ is a copy of the input vector:

$$w_i = I_i + au_i = I_i = (0.4, 0.5, 0.1, 0.8, 0.2)\,.$$

Step 3: $\boldsymbol{x}$ is the normalised version of the same vector,

$$\begin{aligned} x_i &= \frac{w_i}{e + \|w\|} \\ &= \frac{(0.4, 0.5, 0.1, 0.6, 0.2)}{(0.4^2 + 0.5^2 + 0.1^2 + 0.6^2 + 0.2^2)^{1/2}} = \frac{(0.4, 0.5, 0.1, 0.6, 0.2)}{(0.82)^{1/2}} \\ &= (0.441, 0.552, 0.110, 0.663, 0.221)^t\,. \end{aligned}$$

Step 4: Since there is no feedback from $\boldsymbol{q}, \boldsymbol{v}$ is equal to $f(x)$:

$$\begin{aligned} v_i &= f(x_i) + bf(q_i) = (0.441, 0.552, 0.110, 0.663, 0.221)^t \\ &= (0.441, 0.552, 0, 0.663, 0.221)^t\,. \end{aligned}$$

We observe that the third component is zero since the third component is less than the threshold value $\theta = 0.2$. Meanwhile, F_2 is not active hence $\boldsymbol{p}$ and $\boldsymbol{q}$ are copies of v:

Steps 5–7: $\boldsymbol{u} = (0.441, 0.552, 0, 0.663, 0.221)^t$;
$\boldsymbol{p} = (0.441, 0.552, 0, 0.663, 0.221)^t$;
$\boldsymbol{q} = (0.441, 0.552, 0, 0.663, 0.221)^t$;

Step 8: Repeat Steps 2 through 7 in order to stabilise the values on sublayer F_1. Since u and q are now not zero, we proceed as follows:

$$\begin{aligned}
w_i &= I_i + au_i = I_i \\
&= (0.4, 0.5, 0.1, 0.8, 0.2) + 10^*(0.441, 0.552, 0, 0.663, 0.221)^t \\
&= (4.81, 4.02, 0.1, 7.23, 2.41)\,;
\end{aligned}$$

$$\begin{aligned}
x_i &= \frac{w_i}{e + \|w\|} \\
&= \frac{(4.81, 4.02, 0.1, 7.23, 2.41)}{(4.81^2 + 4.02^2 + 0.1^2 + 7.23^2 + 2.41^2)^{1/2}} \\
&= \frac{(4.81, 4.02, 0.1, 7.23, 2.41)}{(97.3875)^{1/2}} \\
&= (0.487, 0.407, 0.100, 0.733, .244)^t\,.
\end{aligned}$$

$$\begin{aligned}
v_i &= f(x_i) + bf(q_i) \\
&= (0.487, 0.407, 0.100, 0.733, 0.244) \\
&\quad + 10^*(0.441, 0.552, 0, 0.663, 0.221)^t \\
&= (4.897, 5.927, 0, 7.363, 2.454)\,.
\end{aligned}$$

$$\begin{aligned}
u_i &= \frac{v_i}{e + \|v\|} \\
&= \frac{(4.897, 5.927, 0, 7.363, 2.454)}{(4.897^2 + 5.927^2 + 7.363^2 + 2.454^2)^{1/2}} \\
&= \frac{(4.897, 5.927, 0, 7.363, 2.454)}{(99.63)^{1/2}} \\
&= (0.491, 0.593, 0.110, 0.737, 0.246)^t\,.
\end{aligned}$$

$$\begin{aligned}
\boldsymbol{p} &= (0.491, 0.593, 0, 0.737, 0.246)^t \\
\boldsymbol{q} &= (0.491, 0.593, 0, 0.737, 0.246)^t\,.
\end{aligned}$$

Let us see the effect of propagating the vector forward to F_2 leading to a vector of activities across the F_2-units of

$$\begin{aligned} T_j &= \sum_i p_i z_{ji} \\ &= (0.491 + 0.593 + 0.737 + 0.246)^*(1.118, 1.118, 1.118, 1.118, 1.118, 1.118) \\ &= (2.01, 2.01, 2.01, 2.01, 2.01, 2.01)^t . \end{aligned}$$

Since all the components are the same the first component becomes the winner so that

$$(2.01, 2.01, 2.01, 2.01, 2.01, 2.01)^t$$

leading to an output vector of

$$(0.8, 0, 0, 0, 0, 0)^t .$$

The bottom-up weight matrix is:

$$z_{Ji} = \frac{u_i}{1-d} = u_i/(1-0.8) = (0.491, 0.593, 0, 0.737, 0.246)^t/0.2 ,$$

given as:

$$\begin{pmatrix} 2.455 & 1.965 & 0.000 & 18.425 & 1.230 \\ 1.118 & 1.118 & 1.118 & 1.118 & 1.118 \\ 1.118 & 1.118 & 1.118 & 1.118 & 1.118 \\ 1.118 & 1.118 & 1.118 & 1.118 & 1.118 \\ 1.118 & 1.118 & 1.118 & 1.118 & 1.118 \\ 1.118 & 1.118 & 1.118 & 1.118 & 1.118 \end{pmatrix} .$$

The top-down weight matrix is:

$$\begin{pmatrix} 2.4550 & 0 & 0 & 0 & 0 & 0 \\ 1.9650 & 0 & 0 & 0 & 0 & 0 \\ 0.0000 & 0 & 0 & 0 & 0 & 0 \\ 18.425 & 0 & 0 & 0 & 0 & 0 \\ 1.2300 & 0 & 0 & 0 & 0 & 0 \end{pmatrix} .$$

4.6. Hopfield Neural Network

The Hopfield neural network is essentially a recurrent network consisting of a single layer of processing elements (PEs) with real valued connection weights. All elements are simply both input and output units as shown in Fig. 4.13. They are non-adaptive networks because their interconnection weights which, correspond to synaptic efficiency in a real biological neuron are not determined as a result of a learning process, but rather derived initially from the energy function. This energy function encodes the problem data in terms of the objective function and the constraints of the original problem. The objective function is known as the soft constraint and the constraints of the original problem, is known as the hard constraints. The Hopfield network, which is partly inspired by real neurons, is an optimiser. Hopfield and Tank (1985) applied this network to the travelling salesman problem (TSP) by mapping the problem into a two-dimensional neuron matrix. This breakthrough aroused the attention of people solving combinatorial optimisation problems to understand the great potential of neural network in combinatorial optimisation applications (Foo and Takefuji 1998, Zurada, 1992).

In the Hopfield neural network, each processing element has an activity value or state. The entire network also has a state at any particular time with an associated energy value that corresponds to the objective function to be

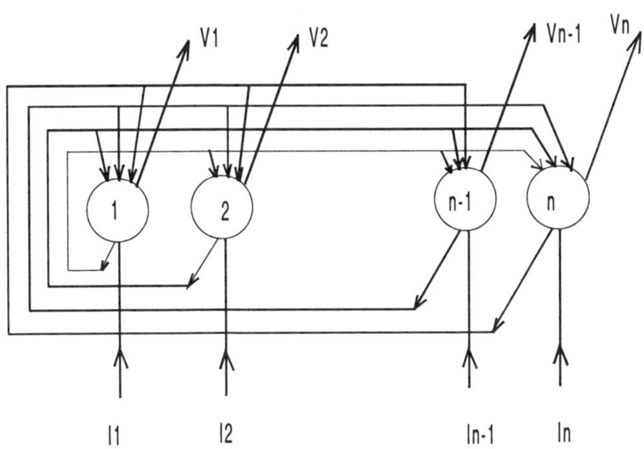

Fig. 4.13. Hopfield neural network.

minimised by the network. There are two versions of Hopfield memory: discrete Hopfield memory, and continuous Hopfield memory. The continuous Hopfield network shares commonality with real neurons which have a continuous, graded output response as a function of their inputs, similar to the characteristics of electronic-circuits using non-linear amplifiers and resistors.

In real neurons, there is a time delay between the appearance of the outputs, v_j of other connected cells, and the resultant net input u_i to a cell. This is due to the resistance (ρ) and capacitance (C) of the cell membrane and the finite conductance of the synapse between the jth and ith cells. In the circuit for the continuous Hopfield network, amplifiers with a sigmoid output characteristic are used as processing units. Each amplifier has an input resistance (ρ), an output capacitance (C), and external signals, I_i that supply each amplifier with a constant current. The voltage-difference between jth unit and the ith unit is $v_j - u_i$, so that the current is $(v_j - u_i)/R_{ij}$. The leakage current across the input resistor is u_i/ρ. The sum of individual current contributions from each other unit is I_i. Hence, the total current from all connecting units and the external input is given as

$$I = \sum_j \frac{(v_j - u_i)}{R_{ij}} - \frac{u_i}{\rho} + I_i$$

$$= \sum_j \frac{v_j}{R_{ij}} - \left(\sum_j \frac{1}{R_{ij}} + \frac{1}{\rho}\right) \bullet u_i + I_i \,. \tag{81}$$

If we define $T_{ij} = \frac{1}{R_{ij}}$ and, $\frac{1}{R_i} = \sum_j \frac{1}{R_{ij}} + \frac{1}{\rho}$ then

$$I = \sum_j v_j T_{ij} - \frac{u_i}{R_i} + I_i \,. \tag{82}$$

As an RC-circuit, we have

$$C\frac{du_i}{dt} = \sum_j v_j T_{ij} - \frac{u_i}{R_i} + I_i \,. \tag{83}$$

The net input and net output signals are given respectively as

$$u_i = \sum_{j=1}^{n} w_{ij} x_j \tag{84}$$

$$v_i = g_i(\lambda\, u_i) = \frac{1}{2}(1 + \tanh(\lambda\, u_i))/2 \,. \tag{85}$$

The input signal is transformed into the output signal through the operator, S, hence $S = g_i(\bullet)$.

It is now an easy matter to formulate the energy equation of the continuous Hopfield network, by defining energy, E, as the sum of the mean of the product of the total current and the output voltage:

$$E = \frac{1}{2}\sum_i Iv_i\,.$$

Using this definition and Eq. (82), we have

$$E = -\frac{1}{2}\sum_i \sum_{j\neq i} u_i v_j T_{ij} + \frac{u_i v_i}{R_i} - \sum_i I_i v_i\,. \tag{86}$$

From Eqs. (85) and (86), we have

$$E = -\frac{1}{2}\sum_i \sum_{j\neq i} u_i v_j T_{ij} + \frac{1}{\lambda}\sum \frac{1}{R_i}\int_0^{v_i} g_i^{-1}(v)dv - \sum_i I_i v_i\,. \tag{87}$$

As has been mentioned, Hopfield and Tank (1985) and Tank and Hopfield (1986) first applied the energy formulation to solve the TSP problem, which is a combinatorial optimisation problem. They formulated the problem as one of minimising energy. Therefore, when using Hopfield neural network for optimisation of combinatorial problem, the objective function value is referred to as the energy level, E, of the network. The goal is to minimise the energy level.

In the remaining of this chapter, we apply the Hopfield neural network to machine tooling and production sequencing in manufacturing cell planning, which is an aspect that has not been emphasized in production planning and control.

4.7. Application of Neural Networks to Machine Tooling and Production Sequencing in Manufacturing Cell Planning

The effectiveness and efficiency of operating a manufacturing cell is affected by the problem of machine tooling and production activities sequencing within the cell (Askin and Standridge, 1993). There are several ways productivity may be maximised: (i) by sequencing batches to minimise tooling changeovers; (ii) by sequencing machining of jobs (a batch of one or more similar parts) to minimise idle time of material handling devices; (iii) by optimising the layout of the work-cell. This chapter addresses the first problem, which is that of sequencing batches to minimise tooling changeovers. For bottleneck work-centres, maximising the production of goods that meet customer requirements is the prime objective of a manufacturing enterprise. Therefore, any strategy that can be adopted to minimise the production time will have much impact on achieving the firm's objectives.

4.7.1. *Complete tooling changeover: a TSP approach*

In group-technology, parts in the same family may require little or no changeover. The reason for this is that the parts may have similar manufacturing process or possess similar geometric shapes. However, parts in different families may require complete breakdown and rebuilding of machine tooling. Individual changeover times depend only on the current and next product. Total *processing time* is fixed by the processing time per unit and batch-size. Total *set-up time* depends on the sequence in which jobs (a batch of one or more similar parts) will be run. Suppose we are given N jobs (a batch of one or more similar parts) to be performed on one machine. After a particular family of parts is processed, the tools are removed and a complete tooling changeover takes place. We want to schedule these jobs so as to minimise production time. This situation corresponds to the travelling salesman problem (TSP). It is therefore important that we start by discussing TSP, then we present one heuristic used for solving this problem, and then we see how Hopfield neural network is applied to the TSP, which is a combinatorial optimisation problem.

Travelling Salesman Problem (TSP)

In the TSP, a salesman must visit each city in his designated area and then return home. In our case, the worker (tool) must perform each job and then return to the starting condition. The problem can be visualised on a graph. Each city (job) becomes a node. Arc lengths correspond to the distance between the attached cities (job changeover times). The salesman wants to find the shortest tour of the graph. A tour is a complete cycle. Starting at a home city, each city must be visited exactly one time before returning home. Each leg of the tour travels on an arc between two cities. The length of the tour is the sum of the lengths of the arcs selected. Figure 4.14 illustrates a five-city

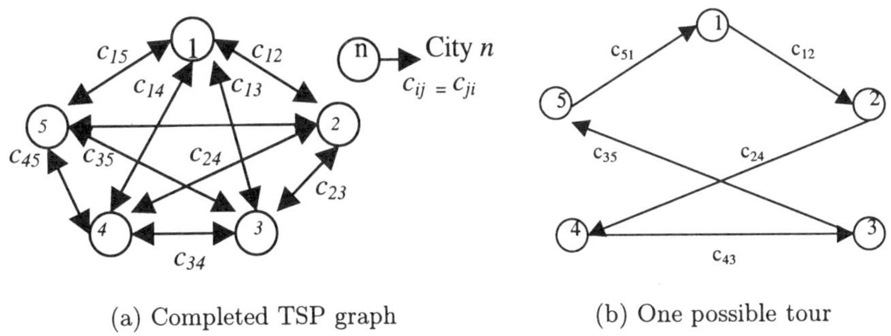

Fig. 4.14. TSP illustrated on a graph.

TSP. Trip lengths are shown on the arcs in Fig. 4.14(a), the distance from city i to j is denoted by c_{ij}. We have assumed in the figure that all paths (arcs) are bi-directional. If arc lengths differ depending on the direction of the arc, the TSP is said to be asymmetric, otherwise it is symmetric. A possible tour is shown in Fig. 4.14(b). The cost of this tour is $c_{12} + c_{24} + c_{43} + c_{35} + c_{51}$.

Several mathematical formulations exist for the TSP. One approach is to let x_{ij} be 1 if city j is visited immediately after i, and be 0 if otherwise. A formal statement of TSP is given as follows:

$$\text{minimise} \sum_{i=1}^{N} \sum_{j=1}^{N} c_{ij} x_{ij} \tag{88}$$

$$\text{subject to} \sum_{j=1}^{N} x_{ij} = 1\,; \quad \forall\, i \tag{89}$$

$$\sum_{i=1}^{N} x_{ij} = 1\,; \quad \forall\, j \tag{90}$$

$$\text{No subtours} \tag{91}$$

$$x_{ij} = 0 \text{ or } 1\,.$$

No subtours mean that there is no need to return to a city prior to visiting all other cities. The objective function accumulates time as we go from city i to j. Constraint 89 ensures that we leave each city. Constraint 90 ensures that we visit (enter) each city. A subtour occurs when we return to a city prior to visiting all other cities. Restriction 91 enables the TSP-formulation, differ from a linear assignment programming (LAP) formulation. Unfortunately, the non-subtour constraint significantly complicates model solution. One reasonable construction procedure for solving TSP is the *closest insertion algorithm.* This is now discussed.

TSP using closest insertion algorithm

The closest insertion algorithm starts by selecting any city. We then proceed through $N-1$ stages, adding a new city to the sequence at each stage. Thus a partial sequence is always maintained, and the sequence grows by one city each stage. At each stage we select the city from those currently unassigned that is closest to any city in the partial sequence. We add the city to the location that causes the smallest increase in the tour length. The closest insertion algorithm can be shown to produce a solution with a cost no worse than twice the optimum when the cost matrix is symmetric and satisfies the triangle inequality. In fact, the closest insertion algorithm may be a useful seed-solution for combinatorial search methods when large problems are solved. Symmetric implies $c_{ij} = c_{ji}$ where c_{ij} is the cost to go from city i directly to city j. Unfortunately, symmetry need not exist in our changeover problem. Normally, the triangular inequality ($c_{ij} \leq c_{ik} + c_{kj}$) will be satisfied, but this alone does not suffice to ensure the construction of a good solution. We may also try repeated application of the algorithm choosing a different starting city each time and then choose the best sequence found. Of course, this increases our workload by a factor of N. Alternatively, a different starting city may be chosen randomly for a specific number of times, less than the total number of cities. This option is preferred for large problem instances.

We now state the algorithm formally. Let S_a be the set of available (unassigned) cities at any stage. S_p will be the partial sequence in existence at any stage and is denoted $S_p = \{s_1, s_2, \ldots, s_n\}$, implying that city s_2 immediately follows s_1. For each unassigned city j, we use $r(j)$ to keep track of the city in the partial sequence that is closest to j. We store $r(j)$ only to avoid repeating calculations at each stage. Last, bracketed subscripts $[i]$ refer to the ith city in the current partial sequence. The steps involved are:

Step 0: Initialize. $N = 1$. $S_p = \{1\}$. $S_a = \{2, \ldots, N\}$. For $j = 2, \ldots, N$. $r(j) = 1$.

Step 1: Select new city. Find $j^* = \arg\min_{j \in S_a}\{c_{j,c(j)}$, or $c_{c(j),j}\}$ Set $n = n+1$.

Step 2: Insert j^*, update $r(j)$. $S_a = S_a - j^*$. Find city $i^* \in S_p$ such that $i^* = \arg\min_{[i]\in S_p}\{c_{[i]j^*} + c_{j^*,[i+1]} - c_{[i],[i+1]}\}$. Update $S_p = \{s_1, \dots, i^*, j^*, i^*+1, \dots, s_n\}$. For all $j \in S_a$, if $\min\{c_{j,j^*}, c_{j^*,j}\} < c_{j,r(j)}$ then $r(j) = j^*$. If $n < N$, go to 2.

As can be seen, the closest insertion algorithms a constructive method. In order to understand the steps involved, let us consider an example related to changeover times for a flexible manufacturing cell (FMC).

Example 3

Table 4.2 shows the changeover times for a flexible manufacturing-cell. A machine is finishing producing batch T1 and other batches are yet to be completed. Use the closest insertion heuristic to find a job sequence, treating the problem as a TSP.

Table 4.2. Changeover times (mins).

From/To	T1	T2	T3	T4	T5
T1	—	8	14	10	12
T2	4	—	15	11	13
T3	12	17	—	1	3
T4	12	17	5	—	3
T5	13	18	15	2	—

Solution

Step 0: $S_p = \{1\}$, $S_a = \{2,3,4,5\}$, $r(j)_{j\in S_a} = 1$; $j = 2, \dots, 5$. This is equivalent to choosing the first city from the partial-list and eliminating this city form the available list.

Step 1: Select the new city: find $j^* = \arg_{j\in S_a}\min\{c_{j,r[j]}\}$ and set $n = n + 1$ $\min_{j\in S_a}\{c_{12}, c_{13}, c_{14}, c_{15}, c_{21}, c_{31}, c_{4,1}, c_{5,1}\} = \min\{8, 14, 10, 12, 0, 0, 0, 0\} = 8$; $j^* = 2$.
But ignore c_{21}, c_{31}, $c_{4,1}$, and $c_{5,1}$ because city 1 is already considered in S_a.

Step 2: Insert city 2, and update $r(j)$ for the remaining jobs 3, 4, and 5.
$S_p = \{1,2\}$; $S_a = \{3,4,5\}$ $c_{12} + c_{21} - c_{11} = 8 + 4 - 0 = 12$

Step 1: Select new city.
$\min\{c_{23}, c_{24}, c_{25}, c_{32}, c_{42}, c_{52}\} = \{15, 11, 13, 17, 17, 18\} = 11$; $j^* = 4$; $n = 3$. So we have job 4 after job 1 or 2

Step 2: Insert job 4.
There are the following possibilities from {1, 2}: {1, 2, 4} or {1, 4, 2}.
For {1, 2, 4}, $c_{12} + c_{24} - c_{14} = 8 + 11 - 10 = 9$.
For {1, 4, 2}, $c_{14} + c_{42} - c_{12} = 10 + 17 - 8 = 19$.
The minimum occurs for inserting job 2 after job 4. Update $r(j)$ for remaining jobs 3, 5 i.e. $r(3) = r(5) = 4$

Step 1: Select new job.
$\{c_{34}, c_{5,4}, c_{4,3}, c_{4,5}\} = \min\{1, 2, 5, 3\}$; $j^* = 3$ min $= C_{34}$, but 4 is already considered. Hence, $j^* = 3$.

Step 2: Insert job 3
There are the following possibilities from {1, 2, and 4}:
$\{1,\ 2,\ 4,\ 3\} : c_{43} + c_{31} - c_{41} = 5 + 12 - 12 = 5$
$\{1,\ 2,\ 3,\ 4\} : c_{23} + c_{34} - c_{24} = 15 + 1 - 11 = 5$
$\{1,\ 3,\ 2,\ 4\} : c_{13} + c_{32} - c_{12} = 14 + 17 - 8 = 25.$
Choosing {1, 2, 4, and 3} breaks the tie. Updating $r[5] = 3$

Step 1: Select new job.
Since job 5 remains, $j^* = 5$

Step 2: Insert job 5.
There are the following possibilities from {1, 2, 4, and 3}
$\{1,\ 2,\ 4,\ 3,\ 5\} : c_{35} + c_{51} - c_{31} = 3 + 13 - 12 = 4$
$\{1,\ 2,\ 4,\ 5,\ 3\} : c_{45} + c_{53} - c_{43} = 3 + 15 - 5 = 13$
$\{1,\ 2,\ 5,\ 4,\ 3\} : c_{25} + c_{54} - c_{24} = 13 + 2 - 11 = 4$
$\{1,\ 5,\ 2,\ 4,\ 3\} : c_{15} + c_{52} - c_{12} = 12 + 18 - 8 = 22.$
Choosing {1, 2, 4, 3, and 5} breaks the tie as the final sequence. The cost $= c_{12} + c_{24} + c_{43} + c_{35} = 8 + 11 + 5 + 3 = 27$.

The TSP construction is shown in Fig. 4.15. The meaning of this solution is that batch 1 is first produced, followed by batch 2, then batch 4, then batch 3, and finally batch 5.

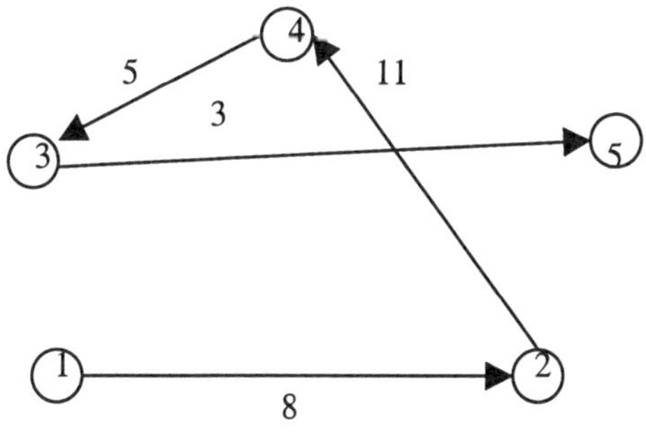

Fig. 4.15. TSP solution for Example 1.

TSP using Hopfield neural network

The most important aspect of applying Hopfield neural network to combinatorial optimization problems, is the formulation of the energy function. For the TSP, the energy function is given as

$$E = \frac{A}{2}\sum_{i=1}^{u}\sum_{k=1}^{n}\sum_{j\neq k}^{n} x_{ik}x_{ij} + \frac{B}{2}\sum_{i=1}^{u}\sum_{k=1}^{n}\sum_{j\neq k}^{n} x_{ki}x_{ji} + \frac{C}{2}\left[\left(\sum_{i=1}^{n}\sum_{k=1}^{n} x_{ik}\right) - n\right]^2$$

$$+ \frac{D}{2}\sum_{k=1}^{u}\sum_{j\neq k}^{n}\sum_{i=1}^{n} x_{kj}x_{ki}(x_{j,i+1} + x_{j,i-1}) \,. \qquad (92)$$

Remember that in our TSP, when a city is visited, the position of that city on the tour is equal to 1 while the positions of other cities on the tour are each 0. To illustrate our problem representation, let us examine a tour from city 2 to city 6 to city 1 to city 3 to city 5 and finally to city 4. In this case, since we have a square matrix in TSP formulation, the information is represented as shown in Fig. 4.16.

The first term of the TSP energy equation becomes zero if there is only one row having 1, representing only one city appearing in the square matrix of the PE. The second term of the TSP energy equation becomes zero if there is only one column having 1, representing only one position associated with a city in the square matrix of the PE. The third term of the TSP energy equation represents the summation of all n^2 output values, which becomes zero if there are n terms, each having a value of 1. In other word, the third term is a

1	2	3	4	5	6	
0	0	**1**	0	0	0	**1**
1	0	0	0	0	0	**2**
0	0	0	**1**	0	0	**3**
0	0	0	0	0	**1**	**4**
0	0	0	0	**1**	0	**5**
0	**1**	0	0	0	0	**6**

(a) Each row of PE is a city; each column is a position on the tour

$$\mathbf{2}\rightarrow\begin{Bmatrix}1\\0\\0\\0\\0\\0\end{Bmatrix}\quad \mathbf{6}\rightarrow\begin{Bmatrix}0\\1\\0\\0\\0\\0\end{Bmatrix}\quad \mathbf{1}\rightarrow\begin{Bmatrix}0\\0\\1\\0\\0\\0\end{Bmatrix}\quad \mathbf{3}\rightarrow\begin{Bmatrix}0\\0\\0\\1\\0\\0\end{Bmatrix}\quad \mathbf{5}\rightarrow\begin{Bmatrix}0\\0\\0\\0\\1\\0\end{Bmatrix}\quad \mathbf{4}\rightarrow\begin{Bmatrix}0\\0\\0\\0\\0\\1\end{Bmatrix}$$

(b) Six-city TSP problem output vectors: 1 show position of city in tour

Fig. 4.16. Hopfield network TSP representations.

penalty function that penalizes any output value not having a 1 as output. The last term of the TSP energy equation computes a value proportional to the distance traveled on a tour. The last term defines the neighbourhood of a unit j to be $j-1$ before it and $j+1$ after it and the objective is to inhibit the selection of these two neighbourhoods on adjacent columns. Minimizing each of these four components has a global effect on minimizing the overall energy function.

A neuron is normally described by two subscripts, one for the city it refers to, and the other, to the order of that city in the tour. Consequently, when we are referring to two neurons that at connected, the weights have to be described by four subscripts. The description of weights in this regard, makes use of the Kronecker delta function,

$$\delta_{ik} = \begin{cases} 1 & \text{if } i = k \\ 0 & \text{if otherwise}. \end{cases}$$

The weight connecting two processing units in Hopfield neural network is given as

$$w_{ik,lj} = -A\delta_{il}(1-\delta_{kj}) - B\delta_{kj}(1-\delta_{il}) - C - D^* d_{il}(\delta_{j,k+1}+\delta_{j,k-1})\,. \tag{93}$$

We now make use of Eq. (83) to obtain the change in the activation u_{ij}. Since the term T_{ij} is an arbitrary constant and I_i can be adjusted to any value, we now have the change in activation expressed as follows:

$$\frac{du_i}{dt} = \sum_j v_j T_{ij} - \frac{u_i}{\tau} + I_i\,. \tag{94}$$

For sufficiently small value of increment in time, Δ_t, Eq. (94) can be written as

$$\Delta u_i = \left(\sum_j v_j T_{ij} - \frac{u_i}{\tau} + I_i\right)\Delta t\,. \tag{95}$$

The change in activation is then written as

$$\Delta u_{ij} = \Delta t\Bigg(- u_{ij}/\tau - A\sum_{k\neq j}^{n} x_{ik} - B\sum_{k\neq i}^{n} x_{kj} - C\Bigg(\sum_{i=1}^{n}\sum_{k=1}^{n} x_{ik} - \beta\Bigg)$$

$$- D\sum_{k\neq i}^{n} d_{ik}(x_{k,j+1}+x_{k,j-1})\Bigg) \tag{96}$$

where τ = time constant and β = a constant to be fixed. An updating rule that is a function of time is given as follows:

$$u_{ij}(t+1) = u_{ij}(t) + \Delta u_i\,. \tag{97}$$

We will see in a moment how Hopfield neural network is applied to combinatorial optimization problems in the subsequent subsections.

4.7.2. *Partial tooling changeover*

In the previous subsection, we considered the case of distinct part families being machined using distinct tools, in which case there is complete rebuilding of machine tooling when changing from one part family to another. This is not always the case. In partial tooling changeover, the situation is more complex when sequencing the jobs. Consider for example an automated machine having

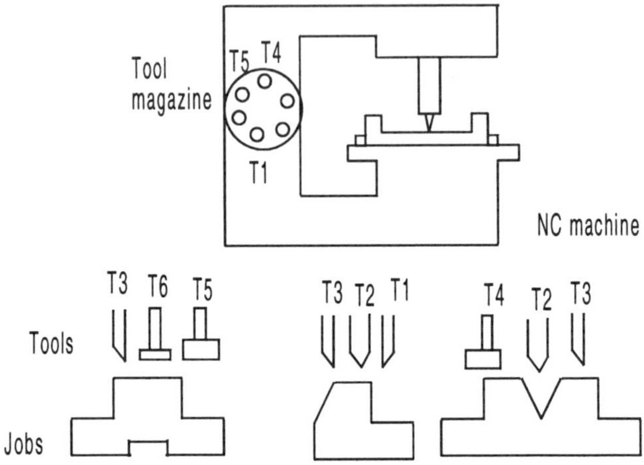

Fig. 4.17. Partial tool changeover problem in NC work-centre.

a tool magazine that can hold only m tools, and is to process n jobs as shown in Fig. 4.17. Tools will normally be required by several jobs and the total number of tools is limited by the maximum, m. It becomes necessary that tool changes will take place. Any job that requires more than m tools cannot be processed. Remember our definition of jobs: a batch of one or more similar parts. The jobs have to be sequenced on the machine. When each job is completed, any tool required by the current job but not required by next job is added to the current job. This is not an easy task. Two issues affect the optimum total number of tools changed on the machine. These are (i) tool changeovers and (ii) jobs ordered. Tool changeover means removing one tool and adding another to the tool magazine as we produce one batch after the other. Some tools that are needed for producing one batch may not needed for another batch, so we have to remove such tool when producing another batch. It may become necessary that some tools are needed which are not currently on the tool magazine. In such a case therefore, new tools are added. This process continues until all the batches are completely produced. The question that we are to answer, is: which tools are required for producing which jobs? How can the jobs be best sequenced on the machine in order to achieve our objective? Our objective is to order jobs and tool changeovers to minimise the total number of tools changed on the machine.

The problem at hand is similar to machine grouping in grout technology. Following the method of using the routing sheet to form the machine-part incidence matrix, we can similarly form a tool-job incidence matrix, such that $a_{ij} = 1$ if a tool i machines a part j and $a_{ij} = 0$ if otherwise. This is a familiar group technology (GT) problem, and we can use any method for solving the GT problem to solve the partial changeover problem.

Tool changeover, which involves removing one tool and adding another to the tool magazine as we produce one batch after the other is best solved using the keep tool needed soonest (KTNS) procedure (Tang and Denardo, 1988). Tools that are needed for producing one batch may not be needed for another batch. When we do not need any tool, we remove such tool as long as it is not needed *until the longest time.* When we produce a batch, we keep tools that are needed.

In this section, two steps for the grouping problem for tools and jobs are discussed (Askin and Standridge, 1993):

Step 1: *Job Ordering.* Any part-family grouping procedure or the TSP can be used.

Step 2: *Tool Set-up Planning.* The keep tool needed soonest (KTNS) procedure gives optimum solution.

Let us see how we go about the partial changeover problem by solving a problem. We apply the two solution-steps. For the first step, we make use of neural network for the binary clustering. We use the self-organising map (SOM) already discussed in this chapter. This way, we get to understand how SOM is applied in solving practical problems. Therefore in this section we present the SOM approach to clustering problems. Neural networks are very good at clustering. Clustering is an aspect of pattern-classification, which is an important area of study in artificial intelligence. It will be in order to present SOM for the clustering problem.

Self-Organizing Map (SOM) for the clustering problem

Recognizing the need for incorporating some form of intelligence in the grouping process, researchers have proposed AI-based approaches that can adapt to changes in their environment. The first effort in this direction is the application of knowledge-based systems (Kusiak 1988). Neural networks, another AI approach, has emerged as a revolutionary technology for solving problems that are difficult for conventional approaches. The applications of neural networks in the industry are diverse and include process planning (Osakada *et al.*, 1990, Osakada and Yang, 1991, Hwang and Henderson, 1992, Knapp and Wang, 1992), scheduling (Foo and Takefuji, 1988, Chen, 1992). Kaparthi and Suresh (1991) proposed a neural network system for shape-based classification and coding of rotational parts in GT. The problem of maintaining consistency in dealing with part family formation during the part classification process was solved using the learning ability of neural networks to automate the process (Kao and Moon, 1990). The group technology, cell-formation problem for binary machine-part matrix has been solved using Adaptive Resonance Theory (ART-1) neural network (Dagli and Huggahalli, 1991). Malave and Ramachandran (1991) and Lee *et al.* (1992) have also studied the application of neural network approaches in the machine-part family formation problem.

The purpose of this subsection is to solve the clustering problem for binary and non-binary input for combinatorial situations. Since the self-organizing map (SOM) approach has the characteristic of performing a dimensional reduction of the input space by transforming the high order-dimensional input data into a two-dimensional space, it has been found to be useful in solving combinatorial problems.

We have already discussed that the SOM network typically has two layers of nodes, the input layer and the Kohonen layer. SOM utilizes the concept of topology-preserving map of the input data to the competitive or Kohonen units. In a topology-preserving map, units located physically next to each other will respond to classes of input vectors that are likewise next to each other. Large-dimensional input vectors are projected down on the two-dimensional map such that the natural order of the input vectors are maintained. The input layer of SOM is fully connected to a two-dimensional Kohonen layer. The Kohonen layer behaves similar to the biological systems because it can compress the representation of sparse data and spread out the representation of dense data using a two-dimensional map.

The values of the input pattern are each fed to the input nodes of the SOM network. There are as many input nodes in an SOM-network as there are values in the input pattern. An m-valued input pattern, is represented by a vector of order m given as:

$$\mathbf{x} = (x_1, x_2, \ldots, x_m).$$

The m nodes in the input layer are fully connected to the nodes in the Kohonen layer that is a two-dimensional layer of nodes. The Kohonen layer is usually arranged as a square matrix of size $N = nxn$, where n is the number of nodes in one of the directions of the two-dimensional Kohonen layer. Each node in the Kohonen layer has a weight vector with m weight values contributed by the m-valued input pattern, so that the weight vector of a node j is represented as:

$$w_j = (w_{1j}, w_{2j}, \ldots, w_{mj}), \quad j = 1, 2, \ldots, N$$

where w_{ij} is the weight associated with Kohonen node j corresponding to the input node i. Lateral interactions exist among processing elements in the Kohonen layer. During the learning process, the positive feedback will extend from the winning unit to other units in some finite neighbourhood. In this way, all the units in the neighbourhood that receive positive feedback from the winning unit participate in the learning process.

A self-organizing map algorithm generally commences the training cycle with random values assigned to the weight vectors in the Kohonen layer. The training process consists of repeatedly presenting the input data to the network through the nodes in the input layer. As the training process continues, the Kohonen nodes adjust their weight values according to the topological relations in the input data. Once training has progressed sufficiently, the weight vector of each unit in the Kohonen layer will converge to a value that is representative of the coordinates of the points near the physical location of the unit. For each training cycle, the winning node is determined in the Kohonen layer by finding the node, which has the minimum distance between it and all the nodes at the input layer. All the nodes within a defined neighbourhood around the winning node are updated and are considered to participate in the victory. As training proceeds, the size of the neighbourhood is diminished until it surrounds only a single node. Working through an example will clarify the issue.

Example 4

A numerical control (NC) machine can hold only four tools (T1, T2, T3, and T4), and the tool magazine is considered empty at the start of operations. The tool-job incidence matrix is given in Table 4.3. Analyze the work-centre for partial tool changeover.

Table 4.3. Tool-job incidence matrix.

	Jobs									
Tools	1	2	3	4	5	6	7	8	9	10
1			1		1			1		
2		1				1	1			1
3		1				1				
4		1				1				1
5	1	1		1		1			1	
6			1		1		1			1
7			1		1			1		
8	1			1		1			1	
9				1					1	
10	1			1						

Solution

SOM clustering approach

Step 1: Job Ordering. We now solve the problem of the 10×10 size shown in Table 4.3 in order to understand the application of the SOM algorithm.

In order to use the SOM algorithm, we need to define three important parameters: neighbourhood, size of Kohonen layer, and training coefficients α.

Neighbourhood

The algorithm that forms feature maps requires a neighbourhood to be defined around the winning node and all the nodes within the neighbourhood participate in the weight vector update process. The neighbourhood size is used to model the "*mexican hat*" function for lateral inhibition. The effect of this is to set up a competitive environment for training as only winning neurons and their neighbours participate in training for a given input pattern. Nodes that are outside the neighbourhood do not participate in the training process. As training proceeds, the size of the neighbourhood is diminished until it encompasses only a single node (see Fig. 4.18). This process adopts the winner-take-all strategy because as the neighbourhood size decreases, a single winner node is isolated for a given input pattern. Figure 4.18 shows the neighbourhood $N_{bj}(t)$ which is the set of nodes considered to be in the

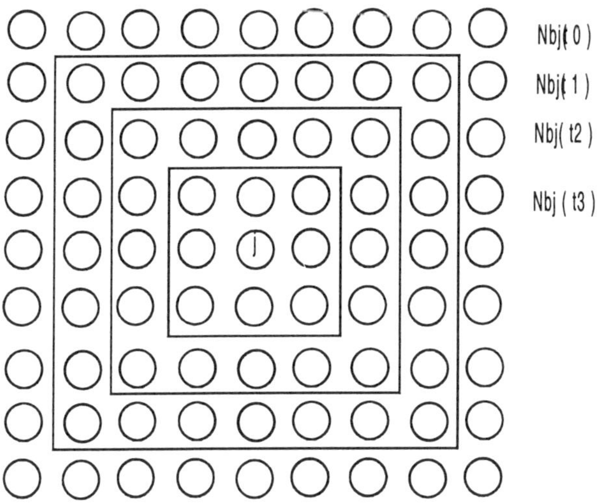

Fig. 4.18. Decreasing neighbourhood at different times as feature maps are formed.

neighbourhood of node j at the time t. The neighbourhood starts with a large size and slowly diminishes in size over time such that $0 < t_1 < t_2 < t_3$ and $N_{bj}(t_3) < N_{bj}(t_2) < N_{bj}(t_1) < N_{bj}(0)$. In the present application of SOM to the clustering problem the following parameter settings have been used: neighbourhood size of 5.

Size of Kohonen layer

The size of the Kohonen layer is the number of nodes in the layer, which are all connected to each input node of the input layer. The Kohonen region is a square region, which depends on the size of the Kohonen layer. For the machine-part families grouping we define the size of the Kohonen layer L_s as:

$$L_s = \frac{problem\ size}{neighbourhood}\,.$$

The square two-dimensional Kohonen region is usually reduced to a one-dimensional region. The reduction is achieved by reforming the two-dimensional layer of nodes into a single layer of nodes organized as a long vector composed of the concatenation of the original row vectors. The advantage of restructuring the network in this manner is that it leads to efficient locating and updating of the neighbourhood encompassing the winning node in the competition process.

The determination of the Kohonen layer size needs explanation. In the present application of SOM to the clustering problem the following parameter settings have been used: neighbourhood size of 5. For a neighbourhood size of 5, the size of the Kohonen layer $L_s = 10 \times 10/5 = 20$. Since the Kohonen region is a square, the size of ${L_r}^2 = 20$ is defined so that $L_r = 4.47$. We choose a value of $L_r = 5$ (an integer) so that the Kohonen layer size $L_s = L_r \times L_r = 5 \times 5 = 25$. This value should be chosen to be greater than the number of input nodes for each pattern ($25 > 10$ in the problem solved).

Training coefficient

The training coefficient $\alpha(t)$ is effectively a Gaussian function of the distance of a Kohonen node to the input node with a width that decreases with time. The significance of the training factor is that the lower its value, the lower the tendency for the network to be trapped in a local optimum state, but longer is the time for the network to converge. Conversely, a higher value of the training coefficient increases the tendency for the network to be trapped in a local optimum state, accompanied with a faster rate of convergence. Since we want to avoid being caught up in a local optimum state, experimentation usually commences with a high value of the training coefficient ($0.1 < \alpha < 0.4$) and gradually reduced as the training process continues. Progressive reduction of α continues until the network converges to similar configurations in successive tests. In the present application of SOM to the clustering problem the following parameter settings have been used: training coefficient of 0.02, and training period of 50. The number of maximum training cycles depends on the size of problem, but in general the range of 100–2000 cycles is adequate for most problems.

An outline for the training process of a SOM algorithm is given as follows:

Step 1: Generate random weight values for each Kohonen node.
Step 2: While not end of each training cycle.
Step 3: While not end of input patterns.

3.1 Present the input vector x to the input layer.
3.2 Compute the distance d_j between each Kohonen node weight vector w_j and the input

$$\text{vector } x_i : d_j = \sum_{i=1}^{m} \|x_i(t) = w_{ij}(t)\|$$

$$i = 1\,.$$

3.3 Determine the winning Kohonen node according to the winning criteria:

$$Z_j{}^* = \begin{cases} 1 & \text{if } \|x_i - w_{ij}^*\| > \|x_i - w_{ik}\| \quad \forall_{ij \neq k} \\ 0 & \text{if otherwise}\,. \end{cases}$$

3.4 Adjust the weight vector of the winning node j^* and those of the neighbourhood $N_{bj}{}^*(t)$ to be close to the input vector according to the training process criteria

$$w_{ij}(t+1) = \begin{cases} w_{ij}(t) + \alpha(t)(w_i - w_{ij}(t)) & \text{if } j \in N_{bj}^*(t) \\ 0 & \text{if otherwise}\,. \end{cases}$$

Until termination (*end of input patterns and training cycle*).

The 10×10-machine clustering problem is essentially represented as a 10-valued input vector so that the SOM network has 10 input nodes in the input-layer. These 10 nodes are fully connected to the 25 nodes in the in the (5×5) Kohonen layer. Each Kohonen node has a 10-dimensional weight vector associated with it. In one iteration of the training process, an input vector of a part is compared with each of the 25 weight vectors in the Kohonen layer. The winning node for the iteration is the one whose weight is closest to the input vector. The weight vector of the winning node and those nodes within its neighbourhood are adjusted to be closer to the input vector.

The next input vector is then considered. Each of the 10 input vectors is considered during one training cycle. After several training cycles have been chosen at the initialization stage, the network converges, resulting in the values of weight vectors stabilizing and each part maps to a certain wining node in the Kohonen layer. Parts which are similar and belong to the same family (similar input vectors) are bundled together in the Kohonen layer, with the possibility of several parts being mapped to the same Kohonen node. It is this characteristic of SOM being able to perform dimensional reduction from any-dimensional input data to a two-dimensional space that makes it a powerful technique for solving clustering problems in group technology. The procedure

Table 4.4. Tool-job clustering.

Tools	Jobs 3	8	5	10	7	2	6	4	1	9
7	1	1	1							
6	1		1	1	1					
1	1	1	1							
2				1	1	1	1			
4				1	1	1	1			
5						1	1	1	1	1
3						1	1			
8								1	1	1
10								1	1	
9								1		1

converges to the ordering of Table 4.4. SOM finds job orders of {3, 8, 5}, {10, 7}, {2, 6}, and {4, 1, 9}. Jobs {3, 8, and 5} are grouped using tools {7, 6, and 1}. Jobs {10, 7} are grouped using tools {6, 2, and 4}. Jobs {2, 6} are grouped using tools {2, 4, 5, and 3}. Finally, jobs {4, 1, and 9} are grouped using tools {5, 8, 10, and 9}.

TSP approach

Step 1: Job Ordering. The four-city TSP is shown in Table 4.5, which is a from-to matrix for the jobs. Notice that in this case, jobs are made up of part families and not single parts. The four jobs are {3, 8, 5}, {10, 7}, {2, 6}, and {4, 1, 9}. S/E represents the start and end position. The cost to return to S/E is 0 since we will not require tools

Table 4.5. TSP costs for Example 4.2.

Tools	To Jobs S/E	{3, 8, 5}	{10, 7}	{2, 6}	{4, 1, 9}
S/E	—	3	3	4	4
{3, 8, 5}	0	—	2	4	4
{10, 7}	0	2	—	2	4
{2, 6}	0	3	1	—	3
{4, 1, 9}	0	3	3	3	—

to be removed when the last job is completed. The cost from S/E to a set of jobs is equal to the number of tools required for producing that particular set of jobs. For example, jobs {3, 8, 5} require tools {7, 6, 1}, hence the cost in this case = 3. From jobs {3, 8, 5} to {10, 7} requires cost = 2 because tool {6} is already available, although jobs {10, 7} requires tools {6, 2, and 4}. Other costs are similarly obtained.

Step 2: Tool Setup Planning. Since the tool magazine can only hold four tools at one time ($m = 4$), the first four tools are loaded on to the tool. This means that tools {7, 6, 1, and 2} are loaded on the tool magazine, and used to run jobs {3, 8, and 5}. Jobs {10, 7} require tools {6, 2, and 4}, so tool 4 is added. If this happens, one tool must be removed. Either tool 1 or 7 should be removed. Tool 1 is removed. Jobs {2, and 6} require tools {2, 4, 5, and 3}. In this case tools 6 and 7 are removed and replaced with tools 3 and 5. Finally, jobs {4, 1, and 9} require tools {5, 8, 10, and 9}. Therefore, tools 2, 3, and 4 are removed and tools 8, 9, and 10 are loaded from the tool magazine. The process of removing and adding tools to tool magazine (setups) is shown in Table 4.6.

Table 4.6. Sequence of setups.

Tools added	Tools removed	Tools on machine	Jobs run
7, 6, 1, 2	—	7, 6 1, 2	3, 8, 5
4	1	7, 6, 4, 2	10, 7
5, 3	7, 6	5, 3, 4, 2	2, 6
8, 10, 9	4, 3, 2	8, 10, 9, 5	4, 1, 9

4.7.3. *Assembly cell assignment and sequencing*

During production of a product, several production activities take place. Parts visit different machines, and finally converge at the assembly point. Human operatives or robots may carry out the assembling exercise. In an assembly cell, important material handling devices include robots and bins. Bins are used as temporal-buffers for part-types that have to be sent to the assembly board for assembly. A robot picks the part-types from bins and then transfer the picked part-type to the assembly board as shown in Fig. 4.19. It is assumed

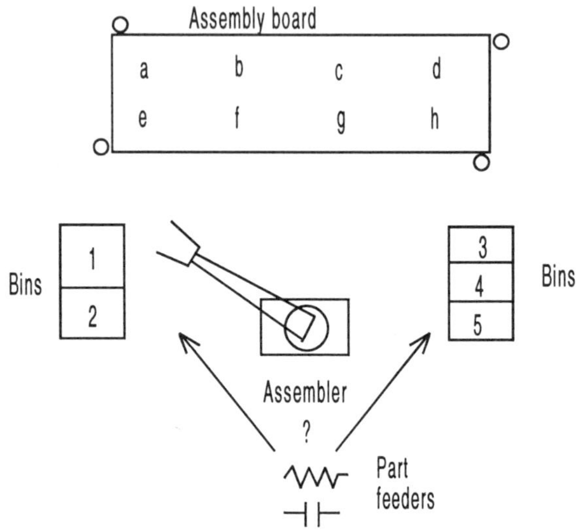

Fig. 4.19. Assembly cell operations.

that a particular part-type is held in a particular bin. This is important to avoid jumbling of different parts-types, which may lead to chaos. Assuming there are N-bins, assignment of part-types to bin is an N-city TSP problem. Again, assuming there are N-assembly board locations, sequencing of part-types to assembly board locations is another N-city TSP problem. Therefore, in effect, we have $2N$-city TSP to deal with. It will be easier to manage this kind of problem by dividing the task into two steps. In the first step, we assign part-types to bins, and in the second step, we add parts to the assembly board location for assembly. In practice, the number of bins may not be equal to the number of locations on the assembly board. To simply our problem, we use assignment method to assign part-types to bins during the first step and then apply TSP in the second step. During the first step, when assignment method is applied, dummy variable may be used if the number of bins is not equal to the number of part-types. For example, if there are five bins and six part-types, then we introduce a dummy part-type, and vice versa. One operational issue that has to be resolved is how to deal with the situation where a part-type is used more than once. This is dealt with simply by summing up its trips to obtain the cost of assigning the part-type to a bin. This is a reasonable assumption since the objective of the assembler is to minimize total loaded travel time, which is represented by the time it take to transfer a part-type from the feeder holding the part-type to the assembly board.

Stated differently, the problem before us is as follows: in assembly cell operation, parts are picked from bins and moved to a product frame where they are assembled. The product frame has N locations where the parts are added. The assembly cell has N bins or locations where part feeders can be placed, and transferred to the product frame. Two problems arise: (i) how do we assign the parts to bins? (ii) And how do we determine the order of transferring the part types from the bins to the frame?

There are therefore $2N$-city TSP problems to solve, but fortunately, we can use assignment method for solving the first part of the problem, and use TSP for the second part of the problem. Someone may be wondering why we made all the effort to discuss TSP. The reason is now obvious, because from now onward, we find that the problems before us are essentially modeled as TSP. Hopfield network is one of the few neural networks that have been identified for solving TSP which is a combinatorial optimization problem. This is one of the motivations for first treating the TSP model in this chapter, before discussing the problem of machine tooling and production sequencing in manufacturing cell planning. Let us now illustrate the two steps involved in assembly cell assignment and sequencing by consider two problems.

Bin assignment

Let us concentrate on the first problem of assigning part types to the bins. The problem is essentially that of minimizing the total loaded time. But here, we replace time with distance. The same solution is obtained. We are assigning part type i to bin location j.

Example 5

Table 4.7 shows the traveling times for a material handling system for eight parts, and there are 2 of type A, 2 of type B, 1 of type C, 2 of type D and 1 of E must be added to the product. Products type A are required in positions a and c on the assembly frame. B is required in positions, b and d on the assembly frame. C is required in position e on the assembly frame. D is required in

Table 4.7. Travel times from bins to frame positions in FMC assembly.

	Frame Locations							
Bins	a	b	c	d	e	f	g	h
1	6	6	4	6	7	7	8	6
2	5	5	7	3	20	9	6	10
3	4	4	4	2	5	6	4	8
4	8	3	6	2	8	5	5	6
5	3	8	3	4	8	7	3	10

positions, f and g on the assembly frame, and E is required in positions h on the assembly frame. How should parts be assigned to the bins?

Solution

We first construct a tableau shown in Table 4.8 such that the costs are as follows:

$$c_{A,i} = t_{a,i} + t_{c,i}\,; \quad i = 1, \ldots, n \text{ (number of bins)}$$

$$c_{B,i} = t_{b,i} + t_{d,i}\,; \quad i = 1, \ldots, n$$

$$c_{C,i} = t_{e,i}\,; \quad i = 1, \ldots, n$$

$$c_{D,i} = t_{f,i} + t_{g,i}\,; \quad i = 1, \ldots, n$$

$$c_{E,i} = t_{h,i}\,; \quad i = 1, \ldots, n\,.$$

The bin assignment is therefore, a linear assignment problem (LAP).

Table 4.8. Part type assignments total travel times.

Part type	Bin Locations				
A	10	12	8	14	6
B	12	8	6	5	12
C	7	20	5	8	8
D	15	15	10	10	10
E	6	10	8	6	10

The solution to the assignment problem of Table 4.7 (i.e. total travel times in Table 4.8) is as follows. Assign product type A to bin 5; assign product type B to bin 2; assign product type C to bin 3; assign product type D to bin 4; and assign product type E to bin 1.

Assembly task sequencing

The bin assignment solved the problem of fixing where the part-types can be found by the robot-arm. It minimized the loaded travel time but did not take into consideration the unloaded travel time from the assembly board to the bin. When a part-type is picked form a bin, it has to be taken to a particular location on the assembly board, so we have to deal with a TSP.

Let us see how to deal with the assembly task-sequencing. Use is made of the assignment solution. For example, in the previous problem, the assignment solution together with assembly board location-visits are summarized as follows:

$$\begin{array}{ccccc} 5 & \rightarrow & A & \rightarrow & a, c \\ 2 & \rightarrow & B & \rightarrow & b, d \\ 3 & \rightarrow & C & \rightarrow & e \\ 4 & \rightarrow & D & \rightarrow & f, g \\ 1 & \rightarrow & E & \rightarrow & h\,. \end{array}$$

What this information means is that we set up a square matrix of the location-visits on the assembly board. Locations a and c have part-type A and assignment to bin 5. Locations b and d have part-type B and assignment to bin 2. Location e has part-type C and assignment to bin 3. Locations f and g have part-type D and assignment to bin 4. Location h has part-type E and assignment to bin 1. Using this information and those of Table 4.7 for the loaded travel distances, the empty travel times are generated as shown in Table 4.9. The information in the second row and the first column of Table 4.9 are used in conjunction with Table 4.7, to obtain the unleaded times. For example, bin 2 on the second row and a on the first column of Table 4.9 show that travel time required is 5 (from Table 4.7). Bin 5 on the second row and a on the first column show that travel time required is 3. Other values are similarly obtained. The last column corresponds to the time the product-type takes from the assembly board to feeder-home. The last row corresponds to

Table 4.9. Empty travel times for Example 3.

Part	A	B	A	B	C	D	D	E	
Bin	5	2	5	2	3	4	4	1	
Sequence	a	b	c	d	e	f	g	h	0
a	—	5	3	5	4	8	8	6	9
b	8	—	8	5	4	3	3	6	10
c	3	7	—	7	4	6	6	4	12
d	4	3	4	—	2	2	2	6	13
e	8	20	8	20	—	8	8	7	11
f	7	9	7	9	6	—	5	7	13
g	3	6	3	6	4	5	—	8	15
h	10	10	10	10	8	6	6	—	16
0	2	1	2	1	3	4	3	2	—

the time the product-type takes from the feeder-home to the bin. The information generated in Table 4.9 is therefore, an input to the TSP approach. We now use the Hopfield neural network to solve our TSP problem. It is noted that the problem we are solving is an asymmetric TSP since the cost matrix is not symmetric, i.e. ($c_{ij} \neq c_{ji}$). Using the Hopfield neural network, the sequence of the tours is {2, 3, 1, 4, 5, 8, 6, 7, and 9}. This constructed sequence has cost (time) of 47.

4.7.4. *Automated turret punch press location and hit sequencing*

The first class of problem that was discussed in this chapter relates to complete tool changeover where distinct part families exist and parts are produced using distinct tools, therefore requiring complete tooling. The next class of problem that was discussed relates to partial tool changeover where distinct part families do not exist and the parts are therefore, produced by adding and removing some tools, but ensuring that the tool magazine maintains the maximum number of tools. The class of problem now treated in this subsection is different from these other two already discussed. Consider an automated punch press that hits holes on a flat sheet. The turret has to be loaded with all the tools required to hit the holes. There is no removing or adding of tools. The machine bed carries the flat plate and moves from home, locating each scheduled hit on the flat plate under the machine turret. Then the turret rotates and aligns the proper tool under the hit. This process continues until all hits are completed and the bed returns to the home position.

There are two problems to be solved here. One is to load tools to the turrets and the other is to locate or sequence hits. The objective is to minimize the cycle time such that the appropriate tools are loaded and the best hits-sequence is obtained. The problem can therefore be divided into two: (i) solve a TSP for the inter-hit sequencing; (ii) solve a quadratic assignment problem (QAP) for the tool loading. Walas and Askin (1984) developed a mathematical formulation to this problem and iterated between the TSP and QAP. Once the hit sequence is known, the sequence of tools to be used is then fixed since each hit requires specific tool. On the other hand, if we know the tool assignment on the turret, we need to know the inter-hit sequence. Connecting each hit in the best sequence is definitely a TSP, where we consider the machine bed home as the home for the TSP, and each hit, a city. Inter-hit travel times and the rotation of the turret are the costs involved and we take the maximum between them, i.e. inter-hit cost = max (inter-hit travel time, turret rotation travel time). The cost to place tool k in position i and tool l in position j is the time it takes the turret to rotate from i to j multiplied by the number of times the turret switches from tool k to l.

The inter-hit travel times are easy to estimate from the geometry of the plate to be punched and the tools required per punch. The inter-hits times are first estimated and then adjusted according to the turret rotation times. This information constitutes the data for solving the TSP. Once the hit sequence is obtained from the TSP, the tools are placed, by solving the QAP. Let us illustrate the TSP-QAP solution procedure by considering an example.

Example 6

A numerically controlled (NC) machine is to punch holes on a flat metal sheet and the hits are shown in Fig. 4.20. The inter-hit times are shown in Table 4.10. There are four tools $\{a, b, c, \text{ and } d\}$ and the hits are $\{1, 2, 3, 4, 5, 6, \text{ and } 7\}$. The machine turret can hold five tools and rotates in clockwise or anti-clockwise direction. When the turret rotates from one tool position to an adjacent position, it takes 60 time units. It takes 75 time units and 90 time units to two locations and three locations respectively. The machine bed home is marked 0. Assign tools to the turret and sequence the hits.

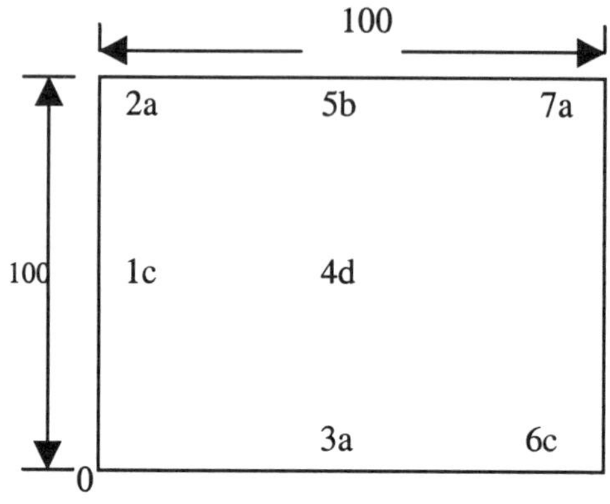

Fig. 4.20. Flat metal sheet to be punched.

Table 4.10. Inter-hit travel times.

	Hit							
Hit	0	1	2	3	4	5	6	7
0	—	50	100	50	100	150	100	200
1	50	—	50	100	50	100	150	150
2	100	50	—	150	100	50	200	100
3	50	100	150	—	50	100	50	150
4	100	50	100	50	—	50	100	100
5	150	100	50	100	50	—	150	50
6	100	150	200	50	100	150	—	100
7	200	150	100	150	100	50	100	—

Solution

From the given inter-hit times, modified inter-hit times have to be calculated using the condition: inter-hit cost = max (inter-hit travel time, turret rotation travel time). For example, for inter-hit between locations 1 and 2, the inter-hit travel time is 50 time units. Now, the tool for hit 1 is c while the tool for hit 2 is a. This means there is change in tools because that the turret will rotate. The cost of rotation is 60 time units, which exceeds the 50 inter-hit time unit. This means that the modified inter-hit time between locations 1 and 2 is 60

Table 4.11. Modified inter-hit travel times (considering turret movements).

	Hit							
Hit	0	1	2	3	4	5	6	7
0	—	50	100	50	100	150	100	200
1	50	—	60	100	60	100	150	150
2	100	60	—	150	100	60	200	100
3	50	100	150	—	60	100	60	150
4	100	60	100	60	—	60	100	100
5	150	100	60	100	60	—	150	60
6	100	150	200	60	100	150	—	100
7	200	150	100	150	100	60	100	—

time units. From the home to any hit is not affected. The modified inter-hit times are shown in Table 4.11. This information is used for TSP. The TSP solution for Table 4.10 is {0, 1, 2, 3, 4, 5, 6, and 7}, with a cost of 830. Next the QAP is solved for assigning tools. We used the "ant" heuristic of chapter 7 to obtain tool sequence of $c \to b \to a \to d$, and the cost is 670. As can be seen a better solution is obtain by the latter. Let us explain how we obtained the tool sequence. Solving the QAP, the sequence obtaineed is {2, 7, 6, 8, 3, 1, 4, and 5} or {1, 6, 5, 7, 2, 0, 3, and 4}. What we do is to refer to Fig. 4.20 and get the labels that corrspond to this sequence as $\{c, c, b, a, a, , d\}$. Hence the optimum sequence is *c–b–a–d*.

4.7.5. *Flow shop sequencing*

The classical problem of scheduling n jobs on m machines in a flow shop is to minimize the throughput time of all the jobs under the assumption that all jobs are processed on all machines at the same sequence. This scheduling problem leads to the permutation situation in which there are $n!$ possible sequences to be considered, and for large number of jobs leads to combinatorial explosion.

The flow shop scheduling-problem is among the combinatorial optimization problems because for large number of jobs, the search for the best job-sequence can be very demanding in terms of computational time. Optimization algorithm such as branch-and-bound, has been employed by some researchers (see, for instance, Ignall and Linus (1965), and Hariri (1981)). Due to the difficulties associated with the computational requirements of optimization algorithms for large-sized problems, many researchers have opted for heuristic methods, which, though do not guarantee optimal solutions, do produce satisfactory and reliable solutions with a reasonably small amount of computational efforts. The min-idle heuristic of Gupta (1968), heuristic scheduling rules and synthetic job technique (Gupta and Maykut, 1973), schedule optimization technique (Onwubolu, 1996), minimum makespan criteria (King and Spachis, 1980; Widmer and Hertz, 1989) are some reported heuristic methods. Flow shop scheduling problem has been solved using flexible-resource allocation method (Daniels and Mazzola 1994). A summary of different methods for scheduling flow shop problem is found in Dudek *et al.* (1992).

In this section of this chapter, the general (mxn) flow shop problem is formulated as a TSP problem, with the input data generated by the heuristic scheduling rule similar to Gupta and Maykut (1973). Their synthetic job definition is less attractive when compared to the TSP model because many cumbersome steps are involved in forming partial sequences until the final sequence is obtained.

Mathematical formulation of flow shop schedules

Machine idleness and/or job waiting are the two situations that constantly arise in a flow shop schedule (see Fig. 4.21). Machine idleness occurs when

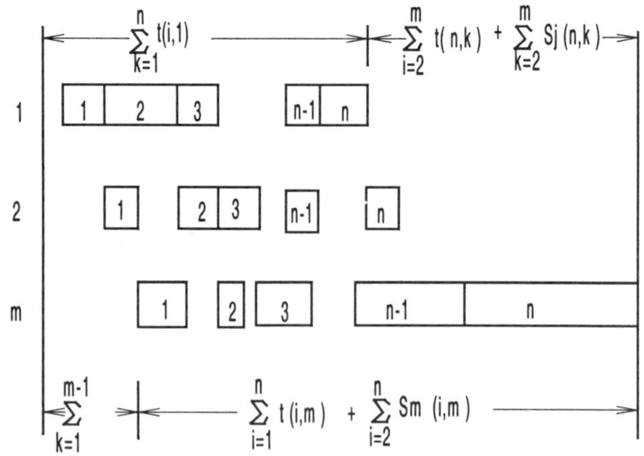

Fig. 4.21. Gantt chart for sequence $\{1, 2, \ldots, n\}$ in flow shop.

the second machine waits for a unit of time (machine slack) during which the second job completes processing on the first machine. Job-waiting can be illustrated by the third job, which after completion of processing on the first machine, waits for unit time on the second machine until the second job completes processing on the second machine.

Let $t(i, m)$ represent the processing time of the ith job on the mth machine, SM (i, m) represent the machine slack on the last machine between the $(i-1)$th and the ith jobs, $S_J(n, m)$ represent the job slack experienced by the nth job on the mth machine, then the total makespan $T_S(n, m)$ can be formulated in two possible ways:

$$T_s(n, m) = \sum_{k=1}^{m-1} t(1, k) + \sum_{i=1}^{n} t(i, m) + \sum_{i=2}^{n} S_M(i, m) \tag{98}$$

$$T_s(n, m) = \sum_{k=1}^{m-1} t(1, k) + \sum_{i=1}^{n} t(i, m) + \sum_{i=2}^{n} S_M(i, m)\,. \tag{99}$$

Scheduling algorithm

The scheduling algorithm consists of a job-pair schedule evaluation scheme, which is based on the analysis of the behaviour of a flow shop schedule described in expression (98) and (99), and the TSP model. Before giving details of the job-pair schedule evaluation scheme, an overview of the generalized schedule algorithm is discussed.

A general schedule evaluation scheme

Arraying the processing times of the jobs on machines in their sequence as an array $T = [t_{ij}]$ is the starting point in the job-pair schedule evaluation scheme:

$$\begin{array}{ll} & \text{Machines} \\ & \begin{array}{lllll} & 1 & 2 & 3 \ldots\ldots\ldots\ldots\ldots & m \\ 1: & t(1,1) & t(1,2) & t(1,3) \ldots\ldots\ldots\ldots\ldots & t(1,m) \\ 2 & t(2,1) & t(2,2) & t(2,3) \ldots\ldots\ldots\ldots\ldots & t(2,m) \\ 3 & t(3,1) & t(3,2) & t(3,3) \ldots\ldots\ldots\ldots\ldots & t(3,m) \\ \vdots & \vdots & & & \\ n & t(n,1) & t(n,2) & t(n,3) \ldots\ldots\ldots\ldots\ldots & t(n,m) \end{array} \end{array}$$

with the rows labelled "Jobs".

The algorithm propagates column by column in the array t_{ij}, comparing processing times $t(i+1,j)$ and $t(i,j+1)$, where $i = 1,2,\ldots.,n-1; j = 1,2,\ldots.,m-1$. If $t(i+1,j) > t(i,j+1)$, the difference is written beside a vertical line drawn between $t(i+1,j)$ and $t(i,j+1)$. This difference represents the machine slack of the second machine $(j+1)$ between the first and second jobs, $S_M(i+1,j+1)$.

If $t(i+1,j) < t(i,j+1)$, the difference is written beside a horizontal line drawn between $t(i+1,j)$ and $t(i+1,j+1)$. This difference represents the job slack of the second job $(j+1)$ between the first and second machine, $S_J(i+1,j+1)$. If $t(i+1,j) = t(i,j+1)$, then the process proceeds without annotating the vertical or horizontal lines.

As the algorithm propagates from the top of each column of the array of processing times to the bottom, and from the left to the right, the computations becomes more involved. When a horizontal line lies to the left of a processing time (job slack) and when a vertical line lies above a processing time (machine slack), these have to be taken into consideration when making the comparison.

Figure 4.22 shows an illustration of the algorithm for a 4×4 problem. For a given sequence, the algorithm solves the machine loading problem because the completion time of any job i on any machine m may be determined by simply summing the processing times in the shortest path between the jobs (1, 1) and $t(i,m)$. It has been proved that the makespan of a given sequence may be expressed as the sum of $(m+n-1)$ processing times along the shortest

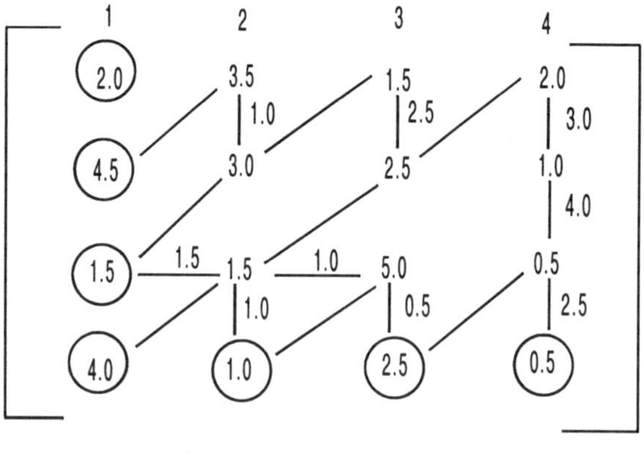

Fig. 4.22. The generalized schedule algorithm.

path (Heller, 1959). This shortest path, known as the *critical path* is obtained by backtracking from the last job on the last machine through the path where neither jobs slack nor machine slack is encountered.

Although we have been able to analyze schedules, this is only possible when the job sequence is known. The flow shop problem can only be said to be completely solved when we can find sequence that gives the optimum or near-optimum schedule.

A job-pair evaluation

The job-pair is the most elementary unit that can generate machine slack and job slack which leads to a makespan that is sequence dependent. The job-pair evaluation scheme used to define the schedule evaluation matrix is similar to the one used to define rules for forming a heuristic algorithm for building a near optimum sequence (Gupta and Maykut, 1973). The following steps are involved in the scheme:

Step 1: Form all possible $n(n-1)$ job pairs among the n jobs.

$$\Im_{ij} = \left(\sum_{k=1}^{n} t_{ik}, \sum_{k=1}^{n} t_{jk} \right), \quad \forall i \neq j . \tag{100}$$

Step 2: Then we can select the job pair that has the minimum of

$$s_{ij} = \sum_{l=1}^{m-1} t(i,l) + S_M(2,m) \quad \text{from } \Im_{ij} \text{ if } n \leq 12 \tag{101}$$

$$s_{ij} = S_M(2,m) \quad \text{from } \Im_{ij} \text{ if } n > 12 \tag{102}$$

$$s_{ij} = 0 \quad \forall i = j . \tag{103}$$

As an illustration of the above algorithm, let us consider the processing times for the 4×4 problem in Fig. 4.22.

Example 7

Use the job-pair evaluation scheme to define the schedule evaluation matrix for the 4×4 problem in Fig. 4.22, and use Hopfield network to determine an appropriate makespan value.

Solution

We consider job pairs 1 and 2, then from expression (100):

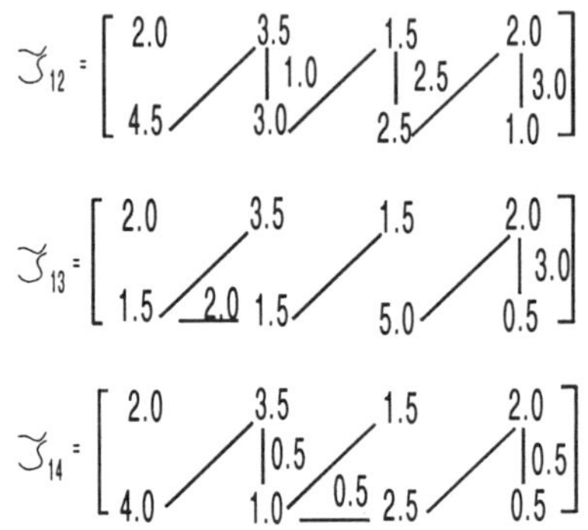

Applying the rules in expressions (101) to (103) for scheduling the pair of jobs, the machine slacks are determined and annotated as shown in the subset array $\Im_{12}$, and

$$s_{12} = 2.0 + 3.5 + 1.5 + 3.0 = 10$$

$$s_{13} = 2.0 + 3.5 + 1.5 + 3.0 = 10$$

$$s_{14} = 2.0 + 3.5 + 1.5 + 0.5 = 7.5\,.$$

Continuing the evaluation for all $n(n-1)$ job pairs leads to the schedule evaluation matrix s_{ij} shown in Fig. 4.23. The schedule evaluation matrix is a square matrix of size (nxn). Using the Hopfield neural network leads to a makespan of 20. The deviation from the lower bound of 16.0 minutes is 25%.

	k			
i	1	2	3	4
1	0	10.0	10.0	7.5
2	11.5	0	14.0	11.5
3	9.0	11.0	0	10.5
4	10.5	13.5	12.0	0

Fig. 4.23. Schedule evaluation matrix for Fig. 4.22 problem.

4.7.6. *Job shop sequencing*

In manufacturing, production scheduling comprises of the important tasks of dispatching jobs and allocating resources. Production scheduling is the key link in carrying out the production plan and process. The larger the production scale or wider the product variety, the more important the action of scheduling becomes. Although the objects to be produced are different in manufacturing industries, the basic attributes of production schedule are the same. The main goal is to seek an optional program for allocating resources under a production goal and related constraint conditions. There are several different production goals, e.g. shortest production time, minimum production cost, JIT mode, which all will seriously effect the production management and organization.

The constraint conditions, e.g. operation time, product priority, equipment implementation, production rate, production batch, due date, etc., are dependent upon the enterprise features including objects, environment, technology and process. The production scheduling includes the pre-process schedule for initiating the production process as static requirement, as well as the post-process schedule to feed new products on the production line for maintaining the continuous running of the production process as a dynamic requirement (Gao *et al.* 2000). The former is usually called the static schedule and the latter the dynamic schedule. Operations research, combinatorial mathematics, especially the branch and bound method and heuristic inference method, have all made a great contribution to production scheduling, it is still problematic to meet the modern production demands of increasing variety, expanding production scale, responding promptly, reforming production mode, etc. There is no unified model to deal with the problems with success.

The application of Hopfield neural network in, 1982 to the solution of the complicated travelling salesman problem (TSP), which is NP-hard aroused researcher's attention and interest to understand the great potential of neural network in applications (Foo and Takefuji, 1998, Zurada, 1992). In other words, the application of computation energy function to solving combinatorial optimization problem was a major breakthrough. A number of neural network approaches applied to the scheduling problem are reported in the literature in the last decade. The approaches include: job-shop scheduling by Foo and Takefuji (1988), real-time scheduling by Liang and Hsu (1996). Willems and Brandts (1995) proposed a Hopfield-based neural network with a built-in rule optimization criteria in its architecture for job-shop problems, and it performs better than the traditional heuristic procedure for small job-shop problems. Liang and Hsu (1996) developed a neural network approach, which has been successfully implemented in scheduling the Taiwan power system. Sabuncouglu (1998) proposed a paralyzed neural network based on competition of a group of neurons for solving the single-machine total tardiness problem and the minimum makespan job-shop scheduling problem.

Huang and Zhang (1994) have made a state-of-the-art survey of the neural network applications in manufacturing, in which eight aspects in manufacturing were involved, of course, the scheduling aspect was one of them. As Huang and Zhang indicated, the scheduling problem is one of immense importance, it is certain that considerable efforts will continue in applying neural networks in scheduling. Also, Sabuncouglu (1998) contributed a wide review of the literature on scheduling problems and their solution methods using neural networks. Special focus was put on neural networks' models and their applications to scheduling problems. He concluded, as one of the research directions, that more research-work is needed to improve the ways of setting the values of some key parameters.

This subsection discusses a unified neural network model and algorithm to deal with different mode of schedule in manufacturing for job shop of Gao *et al.* (2000). A discussion of selecting parameters is also included.

Mathematical model for job shop scheduling

The neural network-scheduling model of Gao *et al.* (2000) mentioned above is derived from the Hopfield linear network (Tank and Hopfield, 1986). The mathematical description of the neural network-scheduling model will cover aspects from the energy function to its state equation as in the following discussion.

First, according he production conditions mentioned previously, the constraint conditions of job-shop schedule can be given as follows:

$$s_{i,k} - s_{i,k+1} + t_{i,k} \leq 0 \tag{104}$$

$$s_{i,1} \geq 0 \tag{105}$$

$$\left.\begin{array}{l} \quad\lfloor (s_{i,k} - s_{j,p} + l_{i,k}), (s_{j,p} - s_{i,k} + t_{j,p}) \rfloor \\ \text{Min} \\ \quad 1 \leq i, j \leq n, 1 \leq k, p \leq k \end{array}\right\} . \tag{106}$$

Secondly, the goal function can be worked out as an enterprise requirement. If a manufacturing firm pursues the maximum productivity as the goal, then the goal function can be derived by the minimum total idle time of all machine tools. So, the following expressions are the proper function to be used as the goal function, i.e.

$$\left.\begin{array}{l}\min \sum_i s_{i,k_i} \\ \text{or} \\ \min \sum_i (s_{i,ki} + t_{i,k_i})\end{array}\right\} \tag{107}$$

and

$$\left.\begin{array}{l}\min \left[\max_i (s_{i,k_i})\right] \\ \text{or} \\ \min \left[\max_i (s_{i,ki} + t_{i,k_i})\right]\end{array}\right\} . \tag{108}$$

Now, the job-shop schedule can be described by the energy function E based on the Hopfield neural network:

$$\begin{aligned} E = D \cdot G(s_{ik_i}) &+ \sum_i \sum_k A \cdot F_1(s_{i,k} - s_{i,k+1} + 1_{i,k}) \\ &+ \sum_i B \cdot F_2(-s_{i,1}) \\ &+ \sum_{i(j)} \sum_{k(p)} C \cdot \min \begin{bmatrix} F_3(s_{i,k} - s_{j,p} + t_{i,k}) \\ F_3(s_{j,p} - s_{i,k} + t_{j,p}) \end{bmatrix} . \end{aligned} \tag{109}$$

$G(x)$ – goal function

$$F_{1,2,3}(x) = \begin{cases} 0.5x^2 , & \text{where } x > 0 \\ 0 , & \text{else} \end{cases}$$

where A, B, C and D are the weight parameters.

Obviously, the first item, function $G(\cdot)$, in Eq. (109) is the goal function, which can be taken from Eq. (108). The other items second, third and fourth are the constraint conditions corresponding to expressions (104)–(106), in which the $F_{1,2,3}(\cdot)$ are the penalty functions, respectively.

Therefore, the general expression of function E for production scheduling can be expressed as follows:

$$
\begin{aligned}
E &= DG(s_{i,k}) \\
&+ \sum_{i(j)}^{n} \left\{ \sum_{k(p)} AF_i(s_{i,k}) + BF_2(s_{i,k}) + \sum_{i(j)} CF_3(s_{i,k}\vee, s_{j,p}) \right\}
\end{aligned}
$$

where $g(\cdot)$ is the goal function, $F_{1,2,3}(\cdot)$ are penalty functions where their intrinsic function meaning has been mentioned before.

Then, the dynamic state equation of function E based on the nature of the neural network can be described as follows:

$$
\frac{\partial E}{\partial s_{i,k}} = \frac{du_{i,k}}{dt}
$$

where u is the neuron internal-state in the neural network at time t.

Therefore, the dynamic state equation of Eq. (109) can be derived from partial differential of $s_{i,k}$ when $k = 1, k, k_i$ with the goal function $G(s_{i,k}) = \min[\max_i(s_{i,ki})]$, and given in the following Eq. (110),

$$
\left.
\begin{aligned}
&\frac{du_{i,k}}{dt} = D(-1)\delta(s_{i,k} - \max(s_{i,k})) \\
&-A[f(s_{i,k} - s_{i,k+1} + t_{i,k}) - f(s_{i,k-1} - s_{i,k} + t_{i,k-1})] \\
&-B \cdot (s_{i,k})\delta(k-1) \\
&- \sum_{m(i,k)=m(j,p)} C \left(\min \begin{pmatrix} f(s_{i,k} - s_{j,p} + t_{i,k}), \\ f(s_{j,p} - s_{i,k} + t_{j,p}) \end{pmatrix} \right) \\
&f(x) \frac{\partial(F(x))}{\partial x} x > 0 \\
&\partial(x) = \begin{cases} 1, & x = 0 \\ 0, & x \neq 0 \end{cases}
\end{aligned}
\right\} . \tag{110}
$$

Equation (110) is called the neural network state equation of the job-shop schedule, which is the basic equation for solving job-shop scheduling. There are two issues that need to be discussed in detail. One is how to determine the parameters A, B, C and D, which have a great effect on the convergence; the other is how to give the constraint items and the goal item in formula (110) according to the different application requirements of the production mode.

Analysis of the parameters

Of course, while solving the neural network state equation, the convergence is a key issue which affects the result of problem solving, in which the parameters A, B, C, D and step Δt play important roles.

Preliminary analysis on parameters A, B, C

First, according to the feature of job scheduling, the physical analysis of Eq. (110) will give us the first operation starting time $s_{i,1} = 0$, and then other starting times can be accumulated successively, i.e. $s_{i,k+1} = s_{ik} + t_{i,k}(k > 1)$. Because item 3 in Eqs. (109) and (110), the first operation item, should be satisfied previously, this infers that the relationship $B > A$ and $B > C$ existed.

Next, in Eqs. (109) and (110), item 4, the constraint for avoiding two work-pieces produced on one machine tool simultaneously, should be satisfied secondly, so that will infer the relationship $C > A$.

Therefore, the qualitative basis for determining parameters can comply with the relationship rule $B > C > A$.

Determination of step Δt

The network for the state Eq. (110) is obtained in the following way:

$$u_{new} = u_{old} + \frac{du_{t,k}}{dt} \times \Delta t \tag{111}$$

where u_{new} and u_{old} indicate the new state and old state of neuron $s_{i,k}$ respectively, the step Δt is constant, and item

$$\frac{du_{j,k}}{dt} \times \Delta t$$

denotes the increment with the state of the neuron varies within step Δt.

Assume that the variable corresponding to the current neuron is $s_{i,k}$, as $F(x) = 0.5x^2$ given in energy function (109) is a quadratic function, therefore E can be expressed in general as:

$$E = \alpha_1(s_{i,k} - \beta)^2 + \gamma$$

where

$$\alpha_1 \le D + A + \sum_{i=1}^{N-1} C + \begin{cases} B\,, & \text{if } k = 1 \\ A\,, & \text{else} \end{cases}\,.$$

To ensure that the network is convergent, it is essential to ensure the network against the energy function E increasing during its evolving. According to the extreme characteristic of quadratic $F(x)$ in Eq. (109), the following derivation must be satisfied:

$$\left|\frac{du_{i,k}}{dt}\right| \times \Delta t |s_{i,k} - \beta|$$

$$\because \left|\frac{\partial E}{\partial s_{i,k}}\right| = \left|\frac{du_{i,k}}{dt}\right| \quad \text{and} \quad \left|\frac{\partial E}{\partial s_{i,k}}\right| = 2\alpha_1 \times |s_{i,k} - \beta|$$

$$\therefore \Delta t \le \frac{1}{2\alpha_1}\,.$$

Contextually, the value of step "Δt" is not only concerned with the constraint items, which include variable $s_{i,k}$, in Eqs. (109) and (110) qualitatively, but also determined quantitatively by the number of constraint items including $s_{i,k}$.

Before reaching the goal function in the course of convergence, only the constraint item 4 in Eqs. (109) or (110), i.e. the number of different work-pieces overlapping on a machine tool simultaneously is possibly involved in the case of disobeying the constraint. Hence, Δt must be concerned with such a situation occurring.

If L is the number of different work-pieces overlapping on a machine tool simultaneously, then the value of L is determined by the specific value corresponding to every iteration in evolving, so usually $L \le N - 1$.

$$\therefore \alpha_2 = D + A + \sum_{i=1}^{L} \begin{cases} B\,, & \text{if } k = 1 \\ A\,, & \text{else} \end{cases}$$

and

$$\Delta t \le \frac{1}{2\alpha_2}\,.$$

Associated selection for parameters A, B, C and D

Because the schedule network is run by means of Eqs. (110) and (111), the associated parameters (PA_0, PB_0, PC_0, PD_0, $\Delta t_0/P$) and (A_0/D_0, B_0/D_0, C_0/D_0, 1, $D_0 \times \Delta t_0$) both will make the following equation set-up as the same result with (A_0, B_0, C_0, D_0, Δt_0):

$$u_{new} - u_{old} = \frac{du_{i,k}}{dt} \times \Delta t = \left(\frac{du_{i,k}}{dt}\right) x \times (\Delta t)/x\,.$$

As mentioned previously, the associated parameters (A_0, B_0, C_0, D_0, Δt_0) can make the schedule network convergent, then associated parameters (PA_0, PB_0, PC_0, PD_0, $\Delta t_0/P$) where $P > 0$, and (A_0/D_0, B_0/D_0, C_0/D_0, 1, $D_0 \times \Delta t_0$) will also make the network convergent with the same solution.

On discussion of the above analysis, we can obtain the following conclusions.

(1) You can make parameters maintain the relationship $B > C > A$ and $D = 1$, then the schedule network equation (110) will be evolved into convergence.

(2) During the adaptation of the schedule network Eq. (110), you can make step

$$\Delta t \leq \frac{1}{2\alpha_2}$$

where

$$\therefore \alpha_2 = 1 + a + \sum_{i=1}^{L} C + \begin{cases} B\,, & \text{if } k = 1 \\ A\,, & \text{else} \end{cases}$$

then step Δt can be controlled automatically by means of variable number L, and the schedule network can be adjusted dynamically to speed up its convergence.

Algorithm for Hopfield neural network-based procedure

The algorithms for the Hopfield neural network-based procedure (Gao *et al.* 2000) is as follows:

Step 1: To initialize the parameters A, B, C and D and the increment step Δt.
Here we assigned $A = 200$, $B = 500$, $C = 250$ and $D = 1$.

Step 2: To set up the starting time for work-piece i: the first operation's starting time $s_{i,l} = 0$ coordinated with the time axis. The operation k starting time $s_{i,k}$ and its successive operations are accumulated: $s_{i,k+1} = s_{i,k} + t_{i,k}(i = 2, \ldots, n, k = 1, 2, \ldots, k-1)$.
Step 3: To compute $E(t_0)$.
Step 4: To compute $du_{i,k}/dt$ according to Eq. (7).
Step 5: To compute new $u_{i,k}(t + \Delta t)$ according to Eq. (8).
Step 6: Let $s_{i,k} = u_{i,k}$.
Step 7: To derive $E(t_0 + \Delta t)$.
Step 8: To derive $\Delta E = E(t_0 + \Delta t) - E(t_0)$.
Step 9: Let $(t_0 + \Delta t) \to t0$.
Step 10: To check $\Delta E < \varepsilon$, else go to Step 3, then output $s_{i,k}$.

Putting together steps 1 to 10, the Hopfield neural network is constructed for the job shop scheduling-problem.

4.8. Summary

The fundamental aspects of neural networks, as well as the different types are discussed. Machine tooling and production sequencing in manufacturing cell planning is an aspect that has not been emphasized in production planning and control. This chapter, bridges the gap by discussing in details the application of neural networks to complete tooling changeover, partial tooling changeover, assembly cell assignment and sequencing, automated turret punch press location and hit sequencing, flow shop scheduling and job shop scheduling. The chapter presents several neural networks and specifically applies the Hopfield neural network model to these production planning and control problems.

Exercises

1. What is the role of the *delta rule* in back-propagation neural network?
2. Describe how back-propagation neural network carries out its learning process.
3. Self-organizing map (SOM) neural network is often used for solving clustering and pattern recognition problems. How can it be adapted to be used for other combinatorial optimization problems?
4. What are the constraints in the application of ART-1 neural network and which areas, is it best used?

5. What are the advantages and disadvantages of ART-2 over ART-1?
6. Hopfield neural network easily converges to a local optimum, for combinatorial optimization problems. How can its performance be improved?
7. Write a computer code for the TSP using the Closest Insertion Algorithm described in this chapter, and compare its results with those obtained using Hopfield neural network.
8. Does it make sense apply the partial tooling changeover when the job-tool groupings, are non-overlapping?
9. An automated machine can hold only four tools (T1, T2, T3, and T4), and the tool magazine is considered empty at the start of operations. The tool-job incidence matrix is given in Table 4.25. Analyze the work-centre using the procedure described in this chapter for partial tool changeover.

Table 4.25. Tool-job incidence matrix.

	Jobs							
Tools	1	2	3	4	5	6	8	9
1				1			1	
2		1			1	1		1
3		1			1			
4		1			1			1
5	1	1	1		1			
6				1		1		1
7				1			1	
8	1		1					
9			1					
10	1		1					

10. For problem 9, use the similarity clustering-algorithm for forming groups, and then analyze the work-centre for partial tool changeover.
11. For problem 9, use the binary clustering order-algorithm for forming groups, and then analyze the work-centre for partial tool changeover. Which clustering approach is preferred and why?
12. Table 4.26 shows the traveling times for a material handling system for eight parts, 2 of type A, 2 of type B, 1 of type C, 2 of type D and 1 of E must be added to the product. Products type A are required in positions a and c on the assembly frame. B is required in positions, b and d on the assembly frame. C is required in position e on the assembly frame. D is required in positions, f and g on the assembly frame, and E is required in positions h on the assembly frame. How should parts be assigned to the bins?

Table 4.26. Travel times from bins to frame positions in FMS assembly.

	Frame Locations							
Bins	*a*	*b*	*c*	*d*	*e*	*f*	*g*	*h*
1	7	5	15	6	7	4	15	6
2	3	8	12	3	20	7	10	10
3	5	3	9	2	5	9	12	8
4	10	7	8	2	8	2	11	6
5	2	4	10	4	8	5	15	10

13. Solve the TSP for the problem described in question 12.
14. Write a computer code for improving the performance of the Hopfield neural network described in this chapter.
15. Write a computer code for solving the job shop problem described in this chapter.

References

Neural Networks

Carpenter, G. A. and Grossberg, S., 1988, The ART of adaptive pattern recognition by a self-organizing neural network, *Computer*, **21**(3) 77–88.

Chen, C. L. P., 1992, Design of a real-time AND/OR assembly scheduler on an optimization neural network, *Journal of Intelligent Manufacturing*, **3**(4) 251–262.

Dagli, C. and Huggahalli, R., 1991, Neural network approach to group technology, *Knowledge-based systems and neural network: Techniques and applications and applications* (New York: Elservier) 213–228.

Foo, Y. P. S. and Takefuji, Y., 1988, Integer linear programming neural networks for job-shop scheduling, *Proceeding* 1988 *International IEEE Conference Neural Networks*, **2** 341–348.

Gao, L., Shen, G., and Wang, S., 2000, Intelligent scheduling model and algorithms for manufacturing, *Production Planning and Control*, **11**(3) 234–243.

Hopfield, J. J. and Tank, D. W., 1985, Neural computation of decisions in optimization problems, *Biological Cybernetics*, **52** 141–152.

Hwang, J. L. and Henderson, M. R., 1992, Applying the perception to three-dimensional feature recognition, *Journal of Design and Manufacturing*, **2**(4) 187–198.

Kao, Y. and Moon, Y. B., 1990, Learning part families by the back-propagation rule of neural networks, *Proceeding 1st International Conference on Automation Technology*, 819–824.

Kaparthi, S. and Suresh, N. C., 1991, A neural network system for shape-based classification and coding of rotational parts, *International Journal of Production Research*, **29**(9) 1771–1784.

Knapp, G. M. and Wang, H. P., 1992, Neural networks in acquisition of manufacturing knowledge, *Intelligent Design and Manufacturing* (New York: John Wiley and Sons) 723–744.

Kohonen, T., 1984, *Self-Organization and Associative Memory* (Berlin: Springer Verlag).

Kusiak, A., 1988, EXTGT-S: A knowledge based system for group technology, *International Journal of Production Research*, **26**(5) 887–904.

Liang, R. H. and Hsu, Y. Y., 1996, Short-term hydro scheduling using Hopfield neural network, *IEEE International Conference on Neural Networks*, San Diego, **2** 275–280.

Lee, H., Maleve, C. O., and Ramachandran, S., 1992, A self-organizing neural network approach for the design of cellular manufacturing systems, *Journal of Intelligent Manufacturing*, **3**(5) 325–332.

Malave, C. O. and Ramachandran, S., 1991, A neural network-based design of cellular manufacturing system, *Journal of Intelligent Manufacturing*, **2** 305–314.

McCullon, W. S. and Pitts, W. A., 1943, A logical calculus of the ideas imminent in nervous activity, *Bulletin of Mathematics and Biophysics*, **5** 115–133.

Osakada, K., Yang, Nakamura, T., and Mori, K., 1990, Expert system for cold forging process based on FEM simulation, *Annals of the CIRP*, **39**(1) 249–252.

Osakada K. and Yang, 1991, Neural networks for process planning of cold forging, *Annals of the CIRP*, **40**(1) 243–246.

Reisig, W., 1985, *Petri Nets* (Berlin, Germany: Springer).

Rumelhart, D. E., Hinton, G. E., and Williams, R. J., 1986, Learning internal representations by error propagation, in *Parallel Distributed Processing: Explorations in the Microstructure of Cognition*, 1 : *Foundations* (Cambridge, MA, MIT Press) 318–362.

Sabuncuoglu, I., 1998, Scheduling with neural networks: A review of the literature and new research direction, *Production Planning and Control*, **9** 2–12.

Shen, G., 1994, *The Research on Neural Network based Intelligent Optimization Techniques Scheduling*, The Dissertation of Beijing University Aeronautics and Astronomy, China.

Tang, C. S. and Denardo, E. V., 1988, Models arising from a flexible manufacturing machine, Part I: Minimization of the number of tool switches, Operations Research, **36**(5) 767–777.

Tank, D. W. and Hopfield, J. J., 1986, Simple "neural" optimization networks: An A/D converter, signal decision circuit and a linear programming circuit, *IEEE Transaction on Circuit Systems*, CAS-33, 533–541.

Walas, R. A. and Askin, R. G., 1984, An algorithm for NC turret punch press tool location and hit sequencing, *IIE Transactions*, **16**(3) 280–287.

Willems, T. M. and Brandts, L. E. M. W., 1995, Implementing heuristics as an optimization criterion in neural networks for job-shop problems, *Journal of Intelligent Manufacturing*, **6** 377–387.

Huang, S. H. and Zhang, H. C., 1995, Application of neural networks in manufacturing a state-of-the-art survey, *International Journal of Production Research*, **33** 705–728.

Zheng, J. L. and Li, X. J., 1992, *Artificial Neural Networks* (Beijing: High Education Publisher) 115–119.

Zurada, J. M., 1992, *Introduction to Artificial Neural System* (West Publishing Company) 456–464.

Applications

Askin, R. G. and Standridge, C. R., 1993, *Modeling and Analysis of Manufacturing Systems* (John Wiley).

Daniels, R. L. and Mazzola, J. B., 1994, Flow shop scheduling with resource flexibility, *Operations Research*, **42**(3) 504–522.

Dudek, R. A., Panwalkar, S. S., and Smith, M. L., 1992, The lessons of flowshop scheduling research, *Operations Research*, **40**(1) 7–13.

Gupta, J. N. D., 1968, Heuristic rules for $n \times M$ flowshop scheduling problem, Opsearch, **5** 165–170.

Gupta, J. N. D. and Maykut, A. R., 1973, Heuristic algorithms for scheduling n jobs in a flowshop, *Journal of the Operations Research Society of Japan*, **16**(3) 131–150.

Hariri, A. M. A., 1981, *Scheduling using branch and bound techniques*, Ph.D. thesis, University of Keele, Keele, 1981.

Heller, J., 1959, Combinatorial, probabilistic and statistical aspects of a $M \times J$ scheduling problem, *NYO-2540, New York: AEC Computing and Applied Mathematics Centre, Institute of Mathematical Sciences, New York University.*

Ignall, E. and Linus, S., 1965, Application of the Branch and Bound technique to some flowshop scheduling problems, *Operations Research*, **13** 400–412.

King, J. L. and Spachis, A. S., 1980, Heuristics for flowshop scheduling, *International Journal of Production Research*, **18** 345–357.

Onwubolu, G. C., 1996, A flowshop manufacturing scheduling system with interactive computer graphics, *International Journal of Operations Management*, **16**(9) 74–84.

Widmer, M. and Hertz, A., 1989. A new heuristic method for the flowshop sequence problem, *European Journal of Operations Research*, **41** 186–193

Chapter 5
Genetic Algorithms

5.1. Introduction

Our earlier emerging optimization techniques classification, which differentiates between population-based strategies and adaptive memory strategies, is often taken to be a fundamental distinction in the literature. Population-based strategies manipulate a collection of solutions rather than a single solution at each stage. Genetic algorithms, scatter search, path re-linking, and ant system approaches belong to this subclass of population-based strategies. The ingenuity of such methods in coordinating the individual solutions contributions is the focus of parallel processing methods. Consequently, several population-based strategies turn out to be a subset of parallel processing methods.

Population-based strategies induce a pattern whose present state depends on the past states, hence incorporate an *implicit memory* structure. The preservation of previous choices is in a sense remembered by current choices, since the present is inherited from the past. Genetic algorithms and scatter search approach are techniques, whose mode of solution combination transmit features of selected past solutions to current solutions. They use no conscious design for keeping record of the past and no purposeful strategy of comparing previous and currently anticipated states. We therefore, conclude that genetic algorithms are based on *implicit memory* rather than adaptive memory. Implicit memory does not take a form normally viewed to be a hallmark of an intelligent memory construction. Does genetic algorithm therefore, qualify as an emerging optimization technique? The answer is that genetic algorithms do not strictly qualify as emerging optimization techniques because they do not exhibit adaptive memory structure, although they can be said to have weak membership since they exhibit implicit memory and since elements of adaptive memory can be included into it. A few proponents of genetic algorithms have gone further to propose the inclusion of elements of adaptive memory as embedded in tabu search, scatter search, and ant systems. These proposals have been labeled hybrid systems since they combine various aspects of different emerging optimization techniques.

5.1.1. *Genetic algorithms versus traditional optimization and search techniques*

Traditional search methods include such methods as calculus-based methods, and enumerative methods. Calculus methods search for extrema (maxima and minima) by equating the first derivative to zero. Enumerative method uses a finite search space or a discretised infinite search space to consider objective function values at every point in the space one at a time. The limitation of this method is that it is only effective in a small search space. Traditional methods lack robustness because

(1) they are local in scope and the optima that they seek are the best in a neighborhood of the current point;
(2) calculus method depends on the existence of derivatives;
(3) they depend on restrictive requirements of continuity, whilst many practical functions are discontinuous in nature;

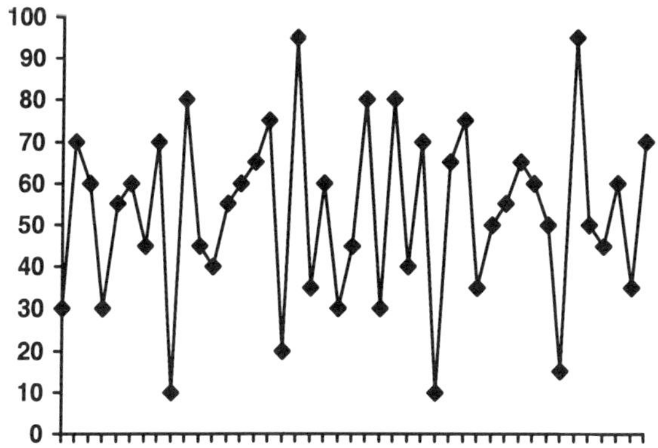

Fig. 5.1. Several discontinuous functions that confound traditional methods.

(4) traditional methods fail in cases where there are several minima and maxima as shown in Fig. 5.1.

Genetic algorithms differ from traditional methods in their very fundamental features. They differ from most optimization and search procedures in four different advantageous ways, in that they:

(1) work with a coding of the parameter set, and not the parameters themselves;
(2) search from a solution space of points, not a single point;
(3) use objective function information, not derivatives or other auxiliary knowledge;
(4) use probabilistic transition rules, not deterministic rules.

Genetic algorithms require the natural parameter set of the optimization problem to be coded as a finite-length string over some finite alphabet. The approach uses that old adage that says "*there is safety in numbers*". This property enables the algorithm to handle discontinuous functions since it uses only objective function values at some points.

All these four differences taken together: direct use of a coding, search from a solution space or solution space of strings, blindness to auxiliary information, random operators, contribute to the robustness of genetic algorithms. Therefore, we can conclusively say genetic algorithms have many advantages over other commonly used search techniques.

Genetic algorithms are search techniques that cautiously exploits historical information in order to speculate strings with higher potential of better performance, i.e. those approaching the optimal solution. Simply, genetic algorithms reach the solution of a problem by generating and modifying a certain number of tentative solutions so as to find the individual (i.e. the solution) best suited to a given environment (i.e. the problem domain). A solution is represented by a string of numbers or characters forming a so-called *string*. A set of strings is referred to as a *solution space*. After a random initialization of a population, the algorithm proceeds in an iterative way through:

(i) an evaluation of strings by the way of a score function with respect to an objective function of the problem;
(ii) a selection of candidates and recombination;
(iii) a generation of a new solution space from the old one through genetic operators of partial bits exchange and random bits exchange.

In every generation of strings (strings of binary numbers, characters, etc.), a new set of strings (of integers, characters, etc.) is generated using pieces/bits of the old strings and are tried for good measure by using a predetermined evaluation criterion. The algorithm stops when a given criterion is reached. Due to the particular way genetic algorithms represent the problem solution (sequence of bits), their use in combinatorial problems, such as scheduling, is quite natural. A strategic decision must be made in terms of suitability of problem representation in genetic algorithms before it is used for solving a combinatorial optimization problem.

Genetic algorithm distinctive features include:

(1) parameter set of the optimization problem is coded as a finite-length string;
(2) parallel solution space search;
(3) no auxiliary information required except for the objective function information;
(4) probabilistic transition rules are used;
(5) partial bits exchange (PBE) operator are used;
(6) random bits alteration (RBA) operator are used.

5.2. Fundamentals of Genetic Algorithms

The mechanics of simple genetic algorithm is surprisingly simple, merely involving copying of strings and swapping partial strings. Parts of the strings of integers, characters, etc. are simply swapped. Though genetic algorithm is so simple in its operation, it is one of the most powerful and effective optimization techniques. Simplicity of operation and power of effect are two of the main attractions of the genetic algorithms approach (Goldberg, 1989).

Genetic algorithms operators, process solution spaces of strings that contain certain information coded in them. A solution space of strings of job sequences may look like the following:

2	3	5	4	1
4	1	3	5	2
1	2	4	3	5
5	4	1	2	3

Such solution spaces are usually generated at random but they do represent a design or decision variable. Each string in the solution space is evaluated in turn to find its score, known as string score. The solution space is normally operated through three main operators listed as follows:

(1) Selection: selection of good strings to form an intermediate solution set of strings. This operator is sometimes called the selection operator.
(2) Partial bits exchange: partial exchange of bits in the intermediate solution set.
(3) Random bit alteration: random alteration of bits in solution strings.

From the partial bits exchange and random bit alteration operators, a new set of solution space is obtained and is further evaluated and tested for better performance. These operators are now discussed in more details.

5.2.1. *Selection*

Selection is the random selection of copies of solutions from the solution space, according to their score value, to generate one or more new structures (roulette wheel method). The selection operator is the first operator applied to a solution space of strings. It involves the selection of individual strings according to their objective function, f, which is referred to as score. The function f, is a measure of goodness, profit, validity, that we want to measure. In this way, the high quality strings will be selected into the next generation. Since strings are copied according to their objective function, or score, f, it means strings with higher values have a higher probability of contributing one or more subsequent string in the next generation.

There exists a number of selection operators in genetic algorithm literature, but in each case the essential idea is that the above-average strings are selected and the copies are inserted into the intermediate solution set in a probabilistic manner. The commonly used selection operator is the proportionate selection operator where a string is selected with a probability proportional to its score. In general, the ith string is selected with a probability p_i proportional to f_i, thus,

$$p_i = \frac{f_i}{\sum\limits_{j=1}^{n} f_j} \tag{1}$$

where n is solution space size. Another selection criterion similar to the one just described is the use of the expected count e_i that is a count of the expected number of copies in the next generation. The expected count can be calculated as,

$$e_i = \frac{f_i}{\sum\limits_{j=1}^{n} \frac{f_j}{n}} = \frac{f_i}{f_{\text{ave}}} \tag{2}$$

where f_{ave} is average score. Since e_i is usually a real number, it has a fractional part. The decimal part is then used as a probability of having extra copies of that particular string. Having selected strings for the intermediate solution set, the next operator is partial bits exchange.

5.2.2. *Partial bits exchange*

Partial bits exchange operation is a random modification of a randomly selected string. Its function is to guarantee the possibility of exploring the space of solutions for any initial population so as to permit a zone of local minimum to be abandoned during the search. In selection, good strings are probabilistically assigned a larger number of copies and an intermediate solution set is formed. It is an important point to note here that no new strings are formed in the selection phase. In the partial bits exchange operator, unlike in selection, new strings are generated by exchange of information among strings of the intermediate solution set. There are a number of partial bits exchange operators applied to genetic algorithms.

The single-point partial bits exchange operator

This type of partial bits exchange operator is the commonly used operator in the published genetic algorithm applications. Single partial bits exchange operation takes place through three steps:

Step 1: Two strings are picked at random from the intermediate solution set obtained from the selection phase;
Step 2: A common exchanging site, or a point of exchanging, is randomly chosen along the string;
Step 3: Exchange all bits on the right hand side of the exchanging site.

Two strings participating partial bits exchange are known as *current strings* and the resulting strings are the *subsequent strings* as illustrated as follows:

Current strings					*Subsequent strings*			
1	3 ----- 5	2	4		1	3 ----- 4	5	2
				$\Longrightarrow$				
1	3 ----- 4	5	2		1	3 ----- 5	2	4
	Exchange site							

As can be observed, the single-point partial bits exchange operator exchanges mainly on the right of the strings. Thus, the string-bit to the left are not usually exchanged. The worst case is that the first string bit is never used in the partial bits exchange operation. Consider the following solution space.

1	3	3	2	2
1	2	2	3	3
1	3	3	2	3
1	3	1	2	1
3	1	1	2	3

The main limitation is that the single-point operation cannot exchange the first bit, so this may hinder genetic algorithm's effectiveness of operation. Therefore, the multi-point partial bits exchange operation to be described is preferred due to its greater improvement.

The multi-point partial bits exchange operator

The operation of a multi-point, partial bits exchange operator is shown as follows:

Step 1: Two strings are picked at random from the intermediate solution set obtained from the selection phase;
Step 2: Two common exchanging sites, or points of exchanging, are randomly chosen along the string;
Step 3: Swap all bits that are inside the exchanging sites as shown in the following illustration.

	Current strings							
				Subsequent strings				
3	2 ---- 1	4 ---- 5		3	2	4	1	5
			⟹					
5	3 ---- 4	1 ---- 2		5	3	1	4	2
	Exchange site 1	Exchange site 2						

The extreme case of the multi-point operator is called *the uniform partial bits exchange operator* where a bit at any location is chosen from either of the current string with a probability. This operator appears to have great search power, and should be incorporated in implementation.

Partial bits exchange may, or may not yield good strings. Selection operator will select against those strings that have lower score-values in subsequent generations. Therefore the effect of partial bits exchange may be detrimental or beneficial. In order to preserve some good strings in the present solution space, not all strings in the intermediate solution set are used in the partial bits exchange operation. When a partial bits exchange probability, *pexchange*, is used, only 100* *pexchange* % in the solution space are used in partial bits exchange, and 100*(1-*pexchange*) % of the solution space remains intact.

A partial bits exchange operator is mainly used for the search of new strings, and random bit alteration operator is also used for the same purpose sparingly in accordance with a probability, *pmut*.

5.2.3. *Random bit alteration*

Random bit alteration operator is also used for searching new strings. For a binary problem, the operator changes a "1" to "0" and vice versa, that is in a binary representation or code, with a probability *pmut*. If the probability comes true then the bit is altered, otherwise it remains unchanged. Random bit alteration is essential in the genetic algorithms operation for the following reasons:

(1) There is need for random bit alteration to generate a point in the neighborhood of the current point, and in so doing, it achieves a local search around each string in the solution space.
(2) There is need to maintain diversity in a given solution space.

For example, in the following solution space of 5-bit string, all the 4 strings having a "0" are on the left.

0	1	1	0	1
0	0	1	1	1
0	1	0	1	0
0	0	0	1	0

Now if the optimum solution has a "1" on the left, then neither single-point partial bits exchange nor multiple-point partial bits exchange can result in a "1" in that solution space. Only random bit alteration can alter the zero's on the left so as to get a "1". Therefore, random bit alteration will be very necessary for such cases.

Convergence before the optimum solution is another potential problem that may occur as generations proceed. This may lead to loss of potential solution. This problem is easily solved by random bit alteration of the whole solution space to increase diversity. In other words, random bit alteration is a form of insurance against loss of potential solutions.

The genetic algorithms operators discussed so far are the fundamental operators. There are some more advanced operators that include multi-point exchanging, uniform exchanging, and score scaling. The selection operator selects good strings, the partial bits exchange operator recombines good substrings from good strings together to hopefully generate a better sub-string.

The random bit alteration operator then alters a string locally to hopefully generate a better string. If bad strings are generated, they are eliminated by the selection-operator in the next generation. However if good strings are generated they will be increasingly emphasized.

5.3. Manual Simulation of Genetic Algorithms

The terminology and various design issues of genetic algorithms are considered in this section. These include the following: representation or coding, initialization, objective function, selection, partial bits exchange, and random bit alteration. These issues are essential for the development of an algorithm for any combinatorial optimization problem.

As already discussed, genetic algorithms candidate-solution, which is a point in the search space is represented by a sequence of numbers (*bits*) and is known as a *string*. The potential of a string being a solution is determined by its *score function* which evaluates the string with respect to the objective function of the optimization problem under consideration. A randomly selected set of strings is called a *solution space* and the solution space at a given time is called a *generation* or *iteration*. The size of the solution space remains the same from iteration to iteration and has been found to have a significant impact on the performance of the genetic algorithm.

In order to have a good grasp of the fundamental features of genetic algorithms discussed so far, we work through an example. Including the aggregate production planning-problem at this early stage will put off new comers to genetic algorithms. As the goal at this stage is to understand the fundamental principles of genetic algorithms, we first use scheduling problem as an example since it is easier to follow. Then problem on aggregate production planning is deferred toward the end of the chapter.

Example 1

The scheduling problem we are considering is the one that seeks to minimize makespan for the flowshop problem, the specific example in Table 5.1 is taken from Nawaz, Enscore, and Ham (1983). Use genetic algorithms fundamental features for the scheduling problem

Table 5.1. Processing times.

	Jobs			
Machines	1	2	3	4
1	5	9	9	4
2	9	3	4	8
3	8	10	5	8
4	10	1	8	7
5	1	8	6	2

Solution

Let us apply our genetic algorithm to solve this problem of minimizing the flowshop makespan. Since there are four jobs, there is need to generate sequence of a job. For example, {1, 2, 3, 4} is a sequence. So a sequence occupies a solution string.

To start off, we select an initial solution space. For the problem, we choose a solution space size of 10, tossing a fair coin 40 times. Let us make sure at this stage that we understand how the *objective function values*, *score functions* and *expected counts* are calculated from the string representation. Let us use the first solution string for illustration purposes. The genetic algorithm scheme tossed fair coins 5 times to obtain sequence {4, 3, 1, 2}. The objective function is obtained from Table 5.2. We are only doing this so that readers can follow the steps involved, otherwise, genetic algorithm will use an objective value formulation included as a subprogram to calculate the makespan for each sequence automatically. The solution space shown in Table 5.3 consists of ten sequences, their objective junction values, score functions and expected counts.

Table 5.2. Makespan for sequence {4, 3, 1, 2}.

	Machines				
Jobs	1	2	3	4	5
4	4/4	8/12	8/20	7/27	2/29
3	9/13	4/17	5/25	8/35	6/41
1	5/18	9/27	8/35	10/45	1/46
2	9/27	3/30	10/45	1/46	8/54

Table 5.3. Solution space report from genetic algorithm.

No.	String				objvalue	score	expecc
(1)	4	3	1	2	54.0000	8.0000	1.5094
(2)	4	3	2	1	55.0000	7.0000	1.3208
(3)	1	3	2	4	56.0000	6.0000	1.1321
(4)	1	3	2	4	56.0000	6.0000	1.1321
(5)	1	3	2	4	56.0000	6.0000	1.1321
(6)	1	3	2	4	56.0000	6.0000	1.1321
(7)	1	3	2	4	56.0000	6.0000	1.1321
(8)	3	4	2	1	58.0000	4.0000	0.7547
(9)	3	1	2	4	58.0000	4.0000	0.7547
(10)	2	1	4	3	62.0000	0.0000	0.0000

The makespan (objective function) value is obtained from the last job on the last machine, which is 54 in this case. This explains how the objective function value for the first solution string is obtained. Other objective function values may be obtained similarly.

Then, how do we calculate the score function values? Genetic algorithms make life easy! All that is required to compute the score function values is to look for one thing: the minimum objective function-value (if your objective is to maximize profit) or the maximum objective function value (if your objective is to minimize profit). Since we are minimizing makespan, we look for the maximum objective function value (see sixth column in Table 5.3). The value in the example is 62. Then we subtract all other objective function values from this maximum. This operation explains why we have *score function* of 8 (see seventh column in Table 5.3) for the first or topmost solution string {4, 3, 1, 2}. What information does this give us? It simply tells us that solution string {4, 3, 1, 2} is the best solution amongst the 10 for that particular state. The least attractive solution is string {2, 1, 4, 3} with score function value of 0. Let us prove this further. The expected count is now calculated. How do we calculate it? We sum up all the score function values. This total is 53 (8 + 7 + 6 + 6 + 6 + 6 + 6 + 4 + 4 + 0). Then we divide this total by the size of the solution space, which in this case is 10. This gives us a mean score function value of 5.3. Then one by one, each score function value is divided by this mean value of 5.3. The first or topmost string {4, 3, 1, 2} then has an expected count of $8/5.3 = 1.5094$. Other expected count values may be calculated similarly (see last column in Table 5.3). It is that easy!

The fundamental underlying mechanisms of genetic algorithms operates on a generation and, generally consists of three main operations enumerated as follows:

(1) Selection: the selection of copies of strings according to their score values as calculated from a given objective function;
(2) partial bits exchange: this means the partial exchange of the portion of two strings chosen from one or two solution spaces;
(3) random bit alteration: this is the random modification of the string, or strings in a solution space.

Let us explain each of these three operations: selection, partial bits exchange, and random bits exchange. A generation of the genetic algorithm commences with selection. We select the temporary solution space of the next generation by spinning the weighted roulette wheel. The number of times of spinning the roulette wheel is equal to the solution space size. Figure 5.2 shows the roulette wheel for the problem in Table 5.3. Simple selection allocates subsequent strings using a roulette wheel with slots sized according to expected count values. Each slot is obtained by multiplying the expected count by the solution space (10 in this case). For example, from Table 5.3, we multiply the last column by the solution space of 10 to give the slots in the roulette wheel. We notice that string 1 has the highest slot value of 15.09%, followed by string 2 (13.21%), followed by 3, 4, 5, 6, and 7 (11.32% each), strings and finally followed by 8 and 9 (7.55%), while string 10 is dumped. The actual simulation of this process gives two copies of string 1, one copy each for strings 2, 3, 4, 5, 6, 7, 8 and 9 and no copy of string 10, which agrees with the expected counts (see last column of Table 5.3).

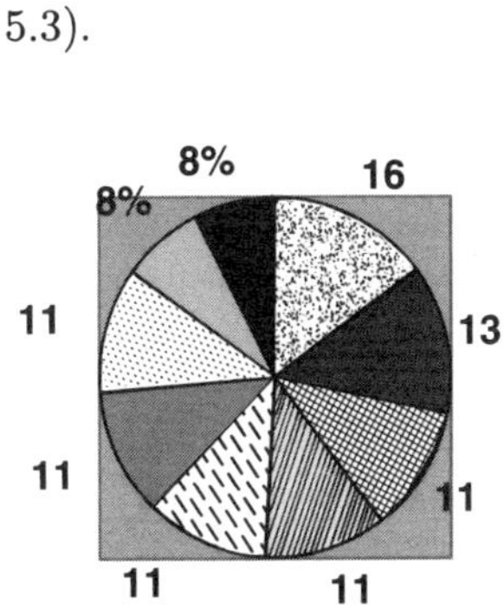

Fig. 5.2. Single roulette wheel sized for Table 5.3.

Table 5.4. Partial bits exchange simulation.

Current string Exchange site (–)	Select	Exchange site	Subsequent string	Objvalue
4 3–1 2	1	2	4 3 2 1	55
3 4–2 1	8	2	3 4 1 2	57

Table 5.5. Random bits exchange simulation.

Current string Exchange site (–)	Select	Exchange site	RBA site	Valid solution
4–3 1–2	2 and 3	4 1 2 2*	4	4 1 2 3
3–1 2–4	2 and 3	3* 3 1 4	1	2 3 1 4

Simple single-point, partial bits exchange proceeds by choosing randomly two strings and choosing randomly an exchange site. Table 5.4 shows a random choice of strings 1 and 8 from Table 5.3, with an exchange site of 2, the two strings {4 3 1 2} and {3 4 2 1} exchange and yield two new strings {4 3 2 1} and {3 4 1 2}. The objective function values are 55 and 57 respectively.

Random bits alteration is illustrated by first, applying a 2-point partial bits exchange to strings 2 and 9 of Table 5.3 as shown in Table 5.5. These two solutions are not valid. Randomly generating two numbers 4 and 1 and swapping the bits in these positions (marked asterisk) for strings 1 and 2 respectively gives two valid solutions {4 1 2 3} and {1 3 2 4}.

The strings that result from the above-mentioned fundamental operations are called *subsequent strings* and they form the next generation's solution space. The process of selection, partial bits exchange, and random bit alteration are repeated for a desired number of iterations, generally found to be the point when the system ceases to improve or when the solution space has converged to a few well-performing sequences.

5.4. Aggregate Production Planning

Aggregate production plan specifies planned rates of production, inventory level, and employee staffing rates and policies to fulfill the charter of a plant.

The charter of a plant is the assignment of product families to a specific plant or sub-process within a plant. The environment in which production planning exists comprises of: overall business strategy, human resources policy, financial profile, short-term marketing plan, and product/plant charters. The focus of production planning is on one product family rather than separate plans for individual combinations of features. For example, production planning might direct the process to make 2000 Toyota Corolla saloon cars rather than 1500 Toyota Corolla 1.3 model and 500 Toyota Corolla 1.5 model. The planning horizon for production planning is usually from 6 months to one and a half years ahead. Production planning horizon must be sufficiently long to allow decisions, such as the hiring and firing of personnel, to be optimal in the long term and not merely on a short-term basis. The goal of production planning is to define the combination of production rates, employee staffing levels, and inventory patterns that satisfies the needs of the firm, and in addition, reduces the overall costs of manufacturing. Therefore, it is not necessary for planner to waste time searching for optimal arrangements for every resource in an environment subject to change. Emphasis should be placed on searching for a satisfactory plan rather than an ideal one.

The six basic steps involved in the development of a aggregate production plan are as follows:

(i) determine relevant company objectives and planning policies;
(ii) determine demand in each planning period;
(iii) determine production capacities for each planning period;
(iv) determine unit costs for regular labour, hiring and firing, over-times, subcontracting, and inventory;
(v) develop alternative plans and choose the minimum;
(vi) select the optimum plan that satisfies the objectives.

Development of the aggregate production plan initially commences with the identification of the long-term objectives of the organization and the constraints under which the functional departments within the organization must operate. There are two important issues here: objectives and constraints. Leaving out any of these in planning, defeats the whole purpose of the planning and therefore limits the performance of the organization. Both long-term objectives and constraints have a major effect on the feasibility of a production plan. For example, a production plan may require organizations to ensure that customer's orders are always delivered on time, but that inventory levels remain low. Delivery of customer's orders on time is a strategic aim, while keeping low inventory levels is an operational constraint. Often such aim and constraint conflict. For example, in many organizations an overriding constraint when developing production plans is the provision of stable and secure employment for the work force. A stable work-force size would enable stock to be built up during low sales demand periods that may be available during high sales demand periods. However, during the low sales demand periods, there exists a conflict between a low stock level strategy and stable work-force policy.

For each planning period it is necessary to forecast the sales demand for the aggregated product group and determine the normal production capacity that is available for achieving his demand. The production plan forms the constraints under which detailed scheduling of facilities and personnel then proceeds. Decisions must be made concerning how to correct variances existing between forecast and available manufacturing capacity.

Sales demand forecasts for each period within the planning horizon for each product-group is determined. Since forecasts of the sales demand for each product-group are inputs to production planning, accurate methods for forecasting should be used by planners. Market globalization, increase of quality standards, stronger competition on service and price are driving companies to revise their demand forecasting process. A good sales forecasting system leads to a reduction in production planning disturbances and to an increase in service levels.

For each planning period and each product group it is then necessary to estimate the level of manufacturing resources required to produce the forecast volume of finished goods. In production planning process, individual products are aggregated into groups according to similarities using such criteria as the types of processing operations required, production times required, types of labour skills required or the fact that each product in the group makes use of specific limited resources. The objective is to allow the capacity requirements of a product group to be measured in a common unit, such as man-hours of production time on a specific time of processing equipment. This common resource units is used to estimate the level of manufacturing resources required to produce the forecast volume of finished goods earlier mentioned. When differences exist between required and available capacity then one or more of the following capacity management techniques should be used to achieve a balance:

(1) recruit/fire staff/redundancy;
(2) work overtime/shorter hours;
(3) vary levels of inventory held;
(4) vary lot sizes processed;
(5) subcontract work;
(6) transfer multi-skilled labour between work areas.

The essential problems involved in generating an optimal and feasible production plan include choosing which technique to use, when to commence using the technique, the number of planning periods over which it will be used and the amount of production capacity to add or remove.

Aggregate production planning is essentially a stochastic process since the aim is to plan for events that will take place in the near future. Between when the plan is made and when it is carried out, several events would have taken plane. For example, worker morale and motivation, level of operator experience and the processing reliability of equipment, which affect the initial estimates of the available resources would have changed. Moreover, the accuracy of sales forecasts would have deteriorated the further into the future. The majority of these factors are themselves affected by production planning. For example, equipment availability is greatly influenced by productive maintenance policies, and operator morale is greatly influenced by hiring and firing policies. The stochastic (unpredictable) nature of these variables results in a plant's manager having to consider the trade-off between costs of holding excess inventories and the cost of lost sales. Holding excess inventories occur when demand is lower than forecast, and lost sales occur when demand is greater than forecast. The costs for each of these situations have some consequences, for example obsolescence costs, interest charges, warehousing costs are attributed to holding excess inventories, while lack of revenue and lost customers are attributed to lost sales. The plant manager must therefore, consider these relative costs when deciding to carry stock from one planning period to the next.

The costs must be determined for each alternative method that could be used to vary capacity in order to satisfy the customer. In practice the options available for varying capacity will be limited by the need to meet management policies. Some management policies might be: fulfilling customer's orders within a specific time period, allowing only an occasional stock-out, or never running out of stock. Costs involved in production planning are divided into two categories: (i) quantifiable and (ii) non-quantifiable. Quantifiable costs include: inventory, overtime, subcontracting, and shift premium, recruitment, redundancy, training, and purchasing. Non-quantifiable costs include: stock-outs, loss of orders, low workers' morale, inexperience personnel, handling complaints, and loss of efficiency

From the several alternative plans, planners should choose the optimum that satisfies the firm's objectives. The search for a satisfactory plan rather than an ideal plan should be emphasized.

In a comprehensive survey, Nam and Logendran (1992) classified at least eight different techniques for aggregate planning. These techniques include:

(1) graphical technique;
(2) integer linear programming;
(3) linear decision rule;
(4) management coefficient approach;
(5) search decision rule;
(6) hybrid fuzzy logic and multi-decision rules;
(7) heuristics;
(8) spreadsheet.

Graphical techniques (Stevenson, 1993) consist of developing graphs that aid planners to visually compare forecast demand with future estimate of production capacity. By comparing the forecast demands and anticipated production, alternative plans can be developed which can then be analyzed on cost-basis for determining their relative merits and demerits. The drawback of graphical technique is that it does not normally result in optimal plans being developed.

Linear programming (LP) models have been formulated for solving the aggregate planning problem, for example as a transportation-type LP (Bowman, 1993). Feiring and Sastri (1990) quantified the uncertainty in demand by a probability distribution and then formulated it as a LP problem. Here, the expected demand is substituted by a stochastic demand modeled as a probability distribution. The consequence of this approach is that the probabilistic nature of the problem is converted into an equivalent deterministic nature. Taha (1971) studies the inaccuracies in forecast demand requirements using sensitivity analysis. For situations in which significant amount of uncertainty is involved, Love and Turner (1993) used stochastic optimal control to address the aggregate planning problem. Gen (1992) employed fuzzy logic to generate a crisp multiple-objective linear programming model. The drawback of the LP technique is that although the stochastic nature of the data can be taken into consideration, costs are linearly related, leading to non-optimal plans been developed.

Linear decision rule (LDR) which uses linear equations as decision rules for specifying optimal work-force size and production levels has been applied to the aggregate planning problem (Holt *et al.*, 1960). In LDR cost curves arising through hiring and firing employees, the use of overtime, holding inventory and cost arising through stock shortages are initially developed. These cost functions are differentiated and combined with the labour costs to realize two rules that can help the planner to obtain the production rate and work-force size for the same planning period. Kamien and Li (1990) extended the LDR to include a third component, of subcontracting. The drawback of the LDR technique is that multiple rules are tedious to develop and do not perform well in situations where the assumed cost relationship differs from the actual company cost.

Management coefficient approach (Bowman, 1963, Riggs, 1987) which involves applying regression analysis to past aggregate planning decisions in order to generate formulae that can be used to compute future decision costs has been applied to the aggregate planning problem. The drawback of the technique is that undesirable bias may be set up from the formulae which has the effect of weakening the decisions previously made.

Search decision rule (SDR) which is a guided search through alternative aggregate plans until the minimum cost is found (Taubert 1968) has been applied to the aggregate-planning problem.

Hybrid fuzzy logic and multi-criteria decision rule is another technique applied to the has been applied to the aggregate planning problem (Satyadous and Chen, 1992). The method enables approximate reasoning to take place, consequently, enabling planning decision to be made with incomplete and imprecise information.

Gilbert and Madan (1991) applied heuristic technique to the aggregate-planning problem. Starr and Mahmoodi (1994) and Onwubolu and Mhlanga (1997) applied spreadsheet-based simulation for aggregate production planning.

Several shortfalls have been identified with most of the above models (Buxey, 1993, Bowman, 1990, Vollman *et al.*, 1992). Models require specific types of information that are difficult if not impossible to gather and quantify.

Examples of such information include cost of hiring and training of new employee, consequence of reduction in productivity due to long hour worked during overtime, the time it takes a new employee to become productive and effects of redundancy in employee morale. Each technique discussed so far is situation-specific, that is, it is not directly transferable to other planning situations. This means that models are limited and are not very useful in an organization when operations conditions (e.g. increased competitive actions in the marketplace and increases in labour and material costs) change. Moreover, none of the existing aggregate production planning techniques can identify optimal or near-optimal plans for practical, real-life problems involving a range of complex planning variables.

Consequently, genetic algorithm (GA), which is an emerging optimization technique is now employed in solving aggregate production planning techniques to identify optimal or near-optimal plans for practical, real-life problems involving a range of complex planning variables. We now formulate the aggregate production planning-problem, and apply genetic algorithm to seek an optimum solution.

5.4.1. *Mathematical model for aggregate production planning*

Notation

x_{OU}	overtime upper limit
x_{SU}	subcontract upper limit
x_{LU}	late deliveries upper limit
x_{FU}	finish stock upper limit
x_O	overtime units
x_S	subcontract units
x_{OC}	overtime cost
x_{SC}	subcontract cost
x_1	$x_{OC} + x_{SC}$
y_1	random $(1, x_{OU})$
y_2	random $(1, x_{SU})$
p_{XO}	probability of x_O
p_{XS}	probability of x_S
x_{DEM}	forecast demand

x_{PROD}	normal production rate
x_{LD}	late deliveries
x_{FS}	finish stock
x_{OS}	opening stock
z	stock
p	period
$fact$	$x_{DEM} - x_{PROD} - x_{FS(P-1)}$

Randomly generated parameters

Overtime units and subcontract units are generated randomly according to following logic:

$$p_{XO} = 1 - \left[\frac{\text{random}(1, x_{OU})}{(fact^* x_{OU})}\right] \bullet \left(\frac{x_{SC}}{x_1}\right) \tag{3}$$

$$p_{XS} = 1 - \left[\frac{\text{random}(1, x_{SU})}{(fact^* x_{SU})}\right] \bullet \left(\frac{x_{OC}}{x_1}\right) \tag{4}$$

$$x_O = \begin{cases} \text{random}(0, y_1) & \text{if } p_{XO} > 1 \\ 0 & \text{if otherwise} \end{cases} \tag{5}$$

$$x_S = \begin{cases} \text{random}(0, y_2) & \text{if } p_{XS} > 1 \\ 0 & \text{if otherwise} \end{cases} \tag{6}$$

Deduced parameters

Late deliveries, finishing stock, and opening stock are deduced from the overtime units and subcontract units respectively using the following logic:

From the notation already given, we define the stock

$$z = x_{OS} + x_{PROD} + x_O + x_S + x_{DEM} \tag{7}$$

$$\left.\begin{aligned} x_{LD} &= -z \\ x_{FS} &= 0 \\ x_{OS+1} &= z \end{aligned}\right\} \quad \text{if } z < 0 \tag{8}$$

$$\left.\begin{aligned} x_{LD} &= 0 \\ x_{FS} &= z \\ x_{OS+1} &= z \end{aligned}\right\} \quad \text{if } z \geq 0 \tag{9}$$

5.4.2. *Aggregate production planning fundamentals*

Company policy

A company-policy is normally put in place by management, rather than the production planner. Several constraints that constitute company policy are listed as follows:

- Units gained through the use of overtime incur a penalty in terms of cost per additional charge;
- The maximum units that can be gained through the use of overtime has
- to be set to a maximum; Units gained through the use of
- subcontracting incur a penalty in terms of cost per additional
- charge; The maximum units that can be gained through the use of
- subcontracting has to be set to a maximum; Stock-outs or late
- deliveries incur a penalty in terms of cost per unit in order for
- the organization providing a high level of customer service; Stocks
- remaining at the end of the last planning period incur a penalty
- cost per unit in order to discourage an organization carrying
- unnecessary expensive inventories.

As a guide for formulating company policy we provide the upper limits for overtime, subcontract, late deliveries and finish stock as follows:

- Overtime upper limit = (Maximum demand − Maximum production rate)/α
- Subcontract upper limit = (Maximum demand − Maximum production rate)/α
- Late deliveries upper limit = Overtime upper limit + Subcontract upper limit
- Finish stock upper limit = Beginning inventory + Late deliveries upper limit

where $\alpha \in [2, 4]$.

Steps involved in generating production plan

Three steps are involved in generating production plan; these are given as follows:

Step 1: Deduce overtime and subcontract parameters upper limits (x_{OU}, x_{SU}) from the given forecast demand and normal production rate;

Step 2: Generate randomly, the number of overtime and subcontract units respectively using Eqs. (3)–(6);

Step 3: Deduce opening stock, late deliveries and finishing stock using Eqs. (7)–(9).

Example 2

Table 5.6 shows the forecast demand and normal production rate of aggregated products in a firm, for which an aggregate production plan is to be developed.

Solution

Step 1: Using $\alpha = 2$ gives

$$x_{OU}, x_{SU} = (\text{Maximum demand} - \text{Maximum production rate})/\alpha$$
$$= (2000 - 1000)/2 = 500$$

Step 2: Random values for overtime units and subcontract units are generated within the bounds established to give values in columns 5 and 6 respectively of Table 5.7.

Table 5.6. Forecast demand and normal production-rate (usually known).

Forecast demand	Normal production
1400	1000
900	1000
800	1000
1500	1000
1500	1000
800	800
500	800
1200	800
600	1000
2000	1000
500	1000
1500	900

Table 5.7. Aggregate plan.

Period	Opening stock	Forecast demand	Normal production	Additional units Overtime ($10)	Subcontract ($50)	Late deliveries ($200)	Finishing stock ($100)
Jan	0	1400	1000	36	16	348	0
Feb	−348	900	1000	21	283	0	56
March	56	800	1000	258	27	0	541
April	541	1500	1000	35	198	0	274
May	274	1500	1000	262	84	0	120
June	120	800	800	198	16	0	334
July	334	500	800	150	244	0	1028
Aug	1028	1200	800	329	86	0	1043
Sep	1043	600	1000	71	159	0	1673
Oct	1673	2000	1000	238	57	0	968
Nov	986	500	1000	43	294	0	1805
Dec	1805	1500	900	20	94	0	1319
				1661	1558	348	
				16610	77900	69600	131900

Total cost = $16,610 + $77,900 + $69,600 + $131,900 = $296,010

Step 3: Opening stock, late deliveries and finishing stock values are deduced to give values in columns 2, 7, and 8 respectively of Table 5.7. For example using the first row of Table 5.6, and first row, columns 5 and 6 of Table 5.7,

$$z_1 = 0 + (1000 + 36 + 16) - (1400) = -348$$

When Eq. (8) is applied, values for the first row, columns 7 and 8 and second row, column 1 of Table 5.7 are obtained.

Let us continue with row 2 of Table 5.7. For example using the second row of Table 5.6, and second row, columns 5 and 6 of Table 5.7,

$$z_2 = -348 + (1000 + 21 + 283) - (900) = 56$$

When Eq. (9) is applied, values for the second row, columns 7 and 8 and third row, column 1 of Table 5.7 are obtained. Continuing in this manner, Table 5.7 is completely filled up. The values for the last column from March to December are given as follows:

$$
\begin{aligned}
z_3 &= 56 + (1000 + 258 + 27) - 800 = 541 \\
z_4 &= 541 + (1000 + 35 + 198) - 1500 = 274 \\
z_5 &= 274 + (1000 + 262 + 84) - 1500 = 120 \\
z_6 &= 120 + (800 + 198 + 16) - 800 = 334 \\
z_7 &= 334 + (800 + 150 + 244) - 500 = 1028 \\
z_8 &= 1028 + (800 + 329 + 86) - 1200 = 1043 \\
z_9 &= 1043 + (1000 + 71 + 159) - 600 = 1673 \\
z_{10} &= 1673 + (1000 + 238 + 57) - 2000 = 968 \\
z_{11} &= 968 + (1000 + 43 + 294) - 1805 = 1805 \\
z_{12} &= 1805 + (900 + 20 + 94) - 1500 = 1319
\end{aligned}
$$

Summing up the fifth, sixth and seventh columns gives values of 1661, 1558, and 348 respectively. The units gained through the use of overtime and subcontracting incur penalty of $10/unit and $50/unit each. Late deliveries incur a penalty of $200/unit. Stocks remaining at the end of the last planning period incur a penalty of $100/unit. The total cost is then obtained by summing up the products of the total units for overtime, subcontract, late deliveries and finishing stocks and their respective penalty cost. This total cost is shown at the bottom of Table 5.7.

The typical production plan normally consisting of a number of monthly planning periods is generated as shown in Table 5.7. For each planning period it is necessary to forecast the sales demand for the aggregated product group and determine the normal production capacity that is available for matching the demand. Consequently, the goal of such a plan is to seek the best combination of manufacturing resources and finished goods inventory that meets management and sales objectives, with overall reduction in manufacturing costs.

5.5. Genetic Algorithms Design Issues

In this section, we discuss in details, a number of design issues: representation or coding, initialization, score function, selection, partial bits exchange, and random bits alteration.

5.5.1. *Representation: genetic code*

An important aspect of genetic algorithm is coding or representation. Representation is the process of transforming a problem at hand into a code that can be easily interpreted, and can be used in the evaluation of the information it represents using the score function. Once a problem is coded into a string of bits it can be solved easily by genetic algorithm. Bits may be represented in the form of binary digits, e.g. 1 0 1 0 1, or even real numbers. It may also be in form of characters such as "efgacffdd", depending on the nature of the problem. Quite a number of representations have been suggested, and these include adjacency measure, ordinal representation, and precedence matrix representation. Representations are problem specific. For example, in the aggregate production planning-problem, each bit represents overtime units and subcontract units. The length of the string therefore represents twice the planning period for a problem. In this section, we show details of how genetic algorithms are used to determine an aggregate plan of the type shown in Table 5.7. The table indicates the additional units that must be achieved using overtime or by subcontracting work so as to meet forecast demand requirements. Additionally, when all demand cannot be satisfied due to operations constraints the plan should indicate when demand will be provided for, in following planning periods, that is late deliveries. Specific cost is attached to each method of gaining these additional units. Normally, premium costs are paid for outsourcing jobs through subcontracting, and overtime premiums are paid to employees for engaging them in extra hours during overtime schedules. Consequently, any unit manufactured using subcontracting or overtime will cost more than normal production. Traditionally, the objective in developing production plan is to ensure that these quantities minimize the overall costs involved in implementing the aggregate production plan.

The plan shown in Table 5.7 is one of the many alternatives that can be identified. Hence it is necessary to represent this and other alternative plans in the form of genetic code. Since we have subdivided our data as randomly generated and deduced types, only randomly generated data are represented in

Table 5.8. Genetic code for aggregate plan: 1 solution.

36 16	21 283	258 27	35 98	262 84	198 16	150 244	329 86	71 159	238 57	43 294	20 94
Jan	Feb	Mar	April	May	June	July	Aug	Sept	Oct	Nov	Dec

the genetic code. Table 5.8 shows the genetic code for one aggregate production planning solution. It can be seen that each number in the string represents a specific aspect of the problem solution. Such a string represents a single solution to the aggregate production-planning problem.

5.5.2. *Initialization*

Initialization is the second most important step in the genetic algorithms that entail the generation of an initial solution space. The usual method of generating an initial solution space of the desired size is done randomly. For example, a typical randomly generated solution space of size 15 is shown in Table 5.9. It can be seen that each number in the string represents a specific aspect of the problem solution. Such a string represents a single solution to the aggregate production-planning problem. In genetic algorithm terms each such string represents a solution and a group of solutions represents a solution space. Table 5.9 shows a solution space of fifteen solutions that are processed simultaneously in order to move towards an optimum or near optimal solution. In fact, this is the major advantage of genetic algorithms over other search techniques that deal with single solutions at a time.

As can be seen the overtime units and subcontract units of columns 5 and 6 in Table 5.7 are coded in the last column of Table 5.9 as a particular solution of the 15 solutions. It is not necessary to code the late deliveries and finishing stock because these can always be deduced from the known forecast demand, normal production rate, and randomly generated overtime units and subcontract units. By so doing, enormous computer memory space is conserved. Table 5.9 requires further explanation. The labels 1 and 2 in the second column represent overtime units and subcontract units respectively for each of the planning periods (months). The numbers in the first row represent the number of solutions, which in this case run from 1 to 15. The number of solutions is user-specified.

5.5.3. *Evaluation or score function values*

After the random generation of a solution space, the next stage is to evaluate each string using a suitably formulated objective function, and then use a score function for the evaluation of the score of each string in relation to a calculated average. First, we find the string with the maximum objective function

Table 5.9. Complete genetic code for aggregate plan: 15 solutions.

		1	2	3	4	5	6	7	8	9	10	11	12	13	14	15
Jan	1	265	265	265	228	117	265	10	144	4	289	192	297	345	88	36
	2	154	154	154	173	327	154	303	78	198	234	49	14	44	309	16
Feb	1	191	191	191	29	191	4	191	191	191	256	55	134	7	40	21
	2	73	73	73	415	73	140	73	73	73	14	122	240	74	65	283
Mar	1	91	91	91	39	88	91	203	91	92	168	25	69	45	8	258
	2	25	25	25	75	34	25	81	25	155	211	117	8	422	104	27
Apr	1	96	96	96	194	237	96	96	91	96	145	164	269	199	23	35
	2	76	76	76	9	110	76	76	43	76	2	8	135	469	134	198
May	1	148	148	148	22	115	114	98	148	148	328	51	91	5	1	262
	2	19	19	19	162	221	123	9	19	19	134	97	122	108	5	84
June	1	373	373	373	3	373	69	16	4	373	15	396	143	50	67	198
	2	37	37	37	187	37	207	49	178	37	243	1	180	140	59	16
July	1	66	66	66	59	66	66	198	34	66	0	8	139	87	27	150
	2	0	0	0	48	0	0	32	204	0	135	15	365	174	4	244
Aug	1	146	146	146	3	146	17	130	146	146	287	232	17	97	3	329
	2	8	8	8	29	8	25	59	8	8	250	12	135	79	205	86
Sept	1	0	0	0	0	0	0	0	70	2	9	222	3	30	152	71
	2	1	1	1	1	1	1	1	388	143	42	222	60	47	15	159
Oct	1	48	48	48	4	28	83	147	48	87	109	74	91	96	5	238
	2	21	21	21	20	58	84	371	21	16	85	108	181	15	303	57
Nov	1	7	7	7	12	7	9	2	173	122	84	23	53	27	88	43
	2	0	0	0	58	0	62	123	99	337	87	245	167	471	242	294
Dec	1	7	7	7	19	7	7	171	7	8	18	263	43	97	82	20
	2	44	44	44	103	44	44	188	44	173	45	48	161	65	150	94
e_i		1.97	1.97	1.97	1.56	1.51	1.37	0.96	0.80	0.80	0.58	0.50	0.42	0.37	0.23	0.00

value. In most practical problems, e.g. the aggregate production planning-problem, the objective function usually minimizes some cost function, such as the minimization of the cost of the alternate plans.

The underlying objective function is mapped to the score function through one or more mappings. In the genetic algorithm minimization process, the following objective function-to-score function transformation is commonly used:

$$f(p) = \begin{cases} f_{\max} - g(p) & \text{when } g(p) < f_{\max} \\ 0 & \text{when otherwise} \end{cases} \tag{10}$$

where $g(p)$ is the objective function value of a string and $f_{\max}$ is the largest objective function value in the current generation or may be taken as a large input coefficient.

In the genetic algorithm maximization process, the following objective function-to-score function transformation is commonly used:

$$f(p) = \begin{cases} g(p) - f_{\min} & \text{when } g(p) > f_{\min} \\ 0 & \text{when otherwise} \end{cases} \tag{11}$$

where $g(p)$ is the objective function value of a string and $f_{\min}$ is the least objective function value in the current generation or may be taken as a small input coefficient.

The score for each string (solution) is determined (see second to the last row of Table 5.9). Since we are minimizing the production plan, the score for each string is given as

$$score = maximum\ objective\ function\ value - objective\ function\ value$$

The individual solutions within a solution space are compared using score values. Those plans that resulted in low implementation costs would have correspondingly high score values. In the implementation, the string having the best score value is retained and passed on to the next regeneration or state. This is what is referred to as the *isbest* strategy, and it ensures that the best solution in any state is retained and not lost.

Let us work through the first solution of the initial solution of Table 5.9 in order to understand how to obtain the evaluation or score functions. Defining z_1 = stock for January, z_2 = stock for February,.., z_{12} = stock for December, we obtain:

$z_1 = 0 + (1000 + 265 + 154) - 1400 = 0$ late deliveries; 19 finish stock.
$z_2 = 19 + (1000 + 191 + 73) - 900 = 0$ late deliveries; 373 finish stock.
$z_3 = 373 + (1000 + 91 + 25) - 800 = 0$ late deliveries; 699 finish stock.
$z_4 = 699 + (1000 + 96 + 76) - 1500 = 0$ late deliveries; 371 finish stock.
$z_5 = 371 + (1000 + 148 + 19) - 1400 = 0$ late deliveries; 38 finish stock.
$z_6 = 38 + (800 + 373 + 37) - 800 = 0$ late deliveries; 448 finish stock.
$z_7 = 488 + (800 + 66 + 0) - 500 = 0$ late deliveries; 814 finish stock.
$z_8 = 814 + (800 + 146 + 8) - 1200 = 0$ late deliveries; 568 finish stock.
$z_9 = 568 + (1000 + 0 + 1) - 600 = 0$ late deliveries; 969 finish stock.
$z_{10} = 969 + (1000 + 48 + 21) - 2000 = 0$ late deliveries; 38 finish stock.
$z_{11} = 38 + (1000 + 7 + 0) - 500 = 0$ late deliveries; 545 finish stock.
$z_{12} = 545 + (900 + 7 + 44) - 1500 = 4$ late deliveries; 0 finish stock.

Total for overtime = 215 + 191 + 91 + 96 + 148 + 373 +66 + 146 + 0 + 48 + 7 + 7 = 1438
Total for subcontract = 154 + 73 + 25 + 76 + 19 + 37 + 0 + 8 + 1 + 21 + 0 + 44 = 458
Total for late deliveries = 4.
Last finishing stock = 0.
Associated costs for overtime, subcontract, late deliveries and finishing stock are \$10, \$50, \$200 and \$100 respectively. Hence,
Total cost or objective value (*objvalue*) for solution-1 = 1438*\$10 + 458*\$50 + 4*\$200 + 0*\$100 = \$38080.

Other values are calculated and shown as follows:

Objective value for solution-2 = \$38,080;
Objective value for solution-3 = \$38,080;
Objective value for solution-4 = \$91,720;
Objective value for solution-5 = \$98,200;
Objective value for solution-6 = \$117,260;
Objective value for solution-7 = \$170,970;
Objective value for solution-8 = \$191,170;
Objective value for solution-9 = \$191,300;
Objective value for solution-10 = \$220,180;
Objective value for solution-11 = \$229,950;
Objective value for solution-12 = \$241,390;
Objective value for solution-13 = \$241,390;

Objective value for solution-14 = \$247,750;
Objective value for solution-15 = \$296,010.

Since we are minimizing the objective function, the score for each solution is obtained by subtracting the objective function value from the maximum function value (equation 12). The maximum objective function value is 296010 for solution 15. Denoting score-1 as score for solution 1, ..., score-2 as score for solution 2, score-15 as score for solution 15, the scores are obtained as follows:

$$
\begin{aligned}
&\text{Score-1} = 296010 - 38080 = 257930\\
&\text{Score-2} = 296010 - 38080 = 257930\\
&\text{Score-3} = 296010 - 38080 = 257930\\
&\text{Score-4} = 296010 - 91720 = 204290\\
&\text{Score-5} = 296010 - 98200 = 197810\\
&\text{Score-6} = 296010 - 117260 = 178750\\
&\text{Score-7} = 296010 - 170970 = 125040\\
&\text{Score-8} = 296010 - 191170 = 104710\\
&\text{Score-9} = 296010 - 191300 = 104710\\
&\text{Score-10} = 296010 - 220180 = 75830\\
&\text{Score-11} = 296010 - 229950 = 66060\\
&\text{Score-12} = 296010 - 241390 = 54620\\
&\text{Score-13} = 296010 - 247750 = 48260\\
&\text{Score-14} = 296010 - 226290 = 29720\\
&\text{Score-15} = 296010 - 241390 = 0\\
&\sum_{i=1}^{15} \text{Score-}i = (257930 + \cdots + 29720 + 0) = 1963720.
\end{aligned}
$$

5.5.4. *Selection and selection*

Selection of the best strings is carried out by the use of the score function values. Thus the strings with the highest score values are selected with higher priority using stochastic sampling without replacement (Goldberg, 1989). The method of stochastic sampling without replacement demonstrates to be highly superior to other selection methods like deterministic sampling, stochastic sampling with replacement, remainder stochastic sampling with replacement, and stochastic tournament Booker (1987). In the experimentation of the genetic algorithms for the several classes of problems, the *stochastic sampling without replacement* yield better results as compared to *stochastic sampling with replacement.*

During the selection process, a comparative parameter known as expected count is used. The expected count is calculated as follows,

$$\text{expected count} = \frac{\text{score value of the } i\text{th string}}{\text{average score of all the strings in the selection}} \tag{12}$$

$$= \frac{f_i}{\sum \bar{f}} \tag{13}$$

This evaluation parameter gives a measure of the expected number of the duplicates of a string in the next generation. Each string in the generation is allocated samples according to the integer part of the expected count values, and the fractional part is treated as a probability of getting the same string in the next generation. Thus, one by one, the fractional parts are used as success probabilities.

Using the scores the expected count, e_i for each string is now computed as

$$e_i = \frac{\text{score} \bullet \text{popsize}}{\sum^{\text{population}} \text{score}} \tag{14}$$

For example, in Table 5.9, the sum of all the scores equal to 1963720, while the score for the first column-string is 257930. Since there are 15 strings, the expected count for the solution in the first column = 257930*15/1963720 = 1.97. The values for each string is accordingly computed (see last row of Table 5.9). These values determine how many copies of the current solution are passed on to the next generation. For example in the illustration given here, a copy of the string in the first column is automatically copied, and a coin is tossed with a probability corresponding to the fractional part (0.97) for possibly considering it as a duplicate solution. This way, the worst solutions are dumped and have no chance of passing on to the next generation.

5.5.5. *Partial bits exchange: mechanism to obtain new solutions*

The selection process using the principle of expected count produce a *temporary solution space* from where random selection of strings to be exchanged and exchanging point(s) is accomplished. The partial bits exchange technique

		X1									
36 16	21 283	258 27	35 98	262 84	198 16	150 244	329 86	71 159	238 57	43 294	20 94
192 49	55 122	25 117	164 8	51 97	396 1	8 15	232 12	222 222	74 108	23 245	263 48

36 16	21 283	258 27	164 8	51 97	396 1	8 15	232 12	222 222	74 108	23 245	263 48
192 49	55 122	25 117	35 98	262 84	198 16	150 244	329 86	71 159	238 57	43 294	20 94

Fig. 5.3. Example of 1-point exchange operation.

adopted in the genetic algorithm for aggregate production planning is the two-point partial bits exchange technique.

The partial bits exchange is done using a probability called the partial bits exchange probability. Let us first consider how the single-point partial bits exchange technique works. The genetic algorithm operator, e.g. the single-point partial bits exchange operator is applied to produce improved solutions. This operation simply involves randomly identifying section of one solution string and swapping this with another section randomly identified from another solution string. The process is illustrated in Fig. 5.3, where the partial bits exchange point is designated as $x1$. From this point the numbers in the first and second strings for solutions 11 and 15 of Table 5.7 are swapped to form two new solutions. The partial bits exchange point $x1$ is generated randomly. Combining two solutions using partial bits exchange operation may result in poor quality or better quality solutions. However using genetic algorithms there is always the possibility that two poor-quality solutions may combine to form better solutions. By calculating the scores and expected counts, inferior solutions are abandoned, when the algorithm progresses to the next generation.

The partial bits exchange and random bit alteration operations are the most important operations which drive the genetic algorithms towards the right solution. In the partial bits exchange process for instance, the first-level subsequent strings and second-level subsequent strings that result are stored in a solution space known as the *selection pool.* Basically it is a pool of strings resulting from random exchanging the strings in the temporary solution space already described. By comparing the score of the strings in the *pool solution space* with those in the temporary solution space, those that fare better are retained whilst those that perform badly are replaced by those that perform better. It is the repeated exchanging of the strings that drives the solution space towards the desired solution.

5.5.6. *Improving partial bits exchange operation*

Although the single-point partial bits exchange technique was found to be effective, a better technique is preferred for improvement of solutions. Computational-tests prove that the multi-partial bits exchange point technique is better than the single-point partial bits exchange technique. In the two-point partial bits exchange technique:

- Two exchange-sites are chosen instead of one as in single point partial bits exchange;
- Exchanging or interchange of string bits is done for those bits in between the two exchange sites.

The genetic algorithm operator, e.g. the two-point partial bits exchange operator is applied to produce improved solutions. This operation simply involves randomly identifying section of one solution string and swapping this with another section randomly identified from another solution string. The process is illustrated in Fig. 5.4, where the partial bits exchange points are designated as $x1$ and $x2$. Between these points the numbers in the first and second strings for solutions 15 and 11 of Table 5.9 are swapped to form two new solutions. The partial bits exchange points, $x1$ and $x2$ are generated randomly. Combining two solutions using partial bits exchange operation may result in poor quality or better quality solutions. However using genetic algorithms there is always the possibility that two poor-quality solutions may combine to form better solutions. By calculating the scores and expected counts, inferior solutions are abandoned, when the algorithm progresses to the next generation.

After each partial bits exchange operation, the random bits exchange operation is carried out on each current solution, at a randomly chosen

			X1					x2			
36 16	21 283	258 27	35 98	262 84	198 16	150 244	329 86	71 159	238 57	43 294	21 94
192 49	55 122	25 117	164 8	51 97	396 1	8 15	232 12	222 222	74 108	23 245	264 48

36 16	21 283	258 27	164 8	51 97	396 1	8 15	232 12	71 159	238 57	43 294	21 94
192 49	55 122	25 117	35 98	262 84	198 16	150 244	329 86	222 222	74 108	23 245	264 48

Fig. 5.4. Example of 2-point exchange operation.

random-bits-exchange-operation point. This involves replacing a digit (bit) in the string with a randomly chosen number between certain specified limits.

5.5.7. *Random bit alteration: mechanism to obtain new solutions*

Another effective operator for realizing improved solutions is random bits exchange operation. The random bit alteration operation is a means of getting new and better performing strings. The operator drives the solution space towards the desired solution. The random bit alteration is implemented in the following cases:

(1) When the diversity of a particular solution space falls below a certain level: As the generation proceeds, the solution space usually converges to a certain solution, but there is a possibility that the solution space may converge prematurely such that the strings in a solution space will be almost similar. At that point, there is a need to diversify the solution space by the random bit alteration operation so that new different strings are generated.
(2) After every partial bits exchange operation: This can conveniently be termed the *normal random bit alteration* operation. It is the operation that is applied to the solution space regardless of the above constraint, thus it is carried out after each and every partial bits exchange operation.

In random bits exchange operation, two non-equal numbers are generated randomly for two strings chosen at random, and the string bits corresponding to these numbers are exchanged. The process is illustrated in Fig. 5.5 (taken from new solution in Fig. 5.4), where the mutated bits are underlined.

Experimentation with the algorithm proves that random bit alteration under the above-mentioned conditions drastically improves the operation of genetic algorithms. On the long run, better performing strings are obtained

36 16	21 283	258 27	164 8	51 97	<u>222</u> 1	8 15	232 12	71 159	238 57	43 294	22 94
192 49	55 122	25 117	35 98	262 84	198 16	150 244	329 86	222 <u>396</u>	74 108	23 245	265 48

Fig. 5.5. Example of random bits exchange operation.

and they do replace the worst strings. This way, the right solution is obtained with fewer iterations or generations.

5.5.8. *Replacement strategy*

In every partial bits exchange and random bit alteration operation, new strings are generated. The new strings may be better or worse. This means that performing strings are to be replaced using a replacement strategy. Several strategies have been suggested in the past (Liepins and Hilliard, 1989). These include the following:

(1) Probabilistic replacement: a strategy that uses the method of probabilistically replacing poorly performing strings the previous generation;
(2) Crowding strategy: a strategy that probabilistically replaces the most similar strings in the previous generation;
(3) Elitist Strategy: this strategy appends the best performing strings of the previous generation to the present generation. This was designed so as to ensure that the strings with better performance will proceed to the next generation.

Combination of these methods has been formulated and is outlined as follows:

(1) After every partial bits exchange and random bit alteration, a pool of newly generated strings are generated to form a *selection pool.* The selection pool size is calculated as *pool size* = *probability of exchange* * *solution space size.* Therefore the pool size is lower than the solution space size.
(2) The strings are evaluated using the objective function. The objective function values are then used as a performance measure;
(3) A close comparison is made between the new strings and the old strings, and those performing better are advanced into the next generation;
(4) Since pool size is less than solution space size, the remaining strings to fill up the solution space is selected randomly using a certain predetermined probability of acceptance called *beta.*

5.5.9. *Convergence and divergence policies*

In every genetic algorithm application, it is observed that as generations proceed, the solution space is gradually filled with better strings. The solution space will be converging to a certain solution. A possible outcome is that the convergence may take place before the optimum solution is obtained, this is known as *premature convergence.* A large number of subsequent strings for a specific string-type may deprive the rest of the strings from producing new subsequent string. This is a phenomenon called *genetic drift.* This results in the loss of the best solution and should be avoided. To avoid the problem of premature convergence, a number of proposals have been suggested, e.g. increasing the solution space size, or solution space. Grefenstette (1987) suggested an entropy measure H_i in a solution space of strings.

5.5.10. *The entropy measure*

For each overtime or subcontract unit, i, the entropy measure is found as follows:

$$H = \frac{\sum_{j=1}^{c} (n_{ij}/p) \bullet \log(n_{ij}/p)}{\log(c)} \tag{15}$$

where, n_{ij} is the number of strings in which overtime or subcontract unit i is assigned to a planning period j in the current solution space. Other parameters are p, the solution space size, and c, the solution space size. The divergence, H, then is calculated as

$$H = \sum_{i=1}^{m} H_{ij}/m\,. \tag{16}$$

As the solution space proceeds the divergence parameter, H, approaches zero. Therefore, this parameter (preset at 0.491) is used to monitor the diversity of each and every generation.

5.5.11. *Diversification*

When diversity falls below the predetermined value the whole solution space is diversified using the random bit alteration operator, until the diversity is above the preset value. The diversification results in generation of new strings, which may be better or worse and this may result in the loss of better strings. This is a great disadvantage inherent in the convergence policies previously.

5.5.12. *Preventing loss of solutions*

The best strings in old solution space should be maintained. The diversified solution space is evaluated using the objective function and the best performing strings are compared with those of the previous undiversified solution space. The best between the diversified and the undiversified is advanced into the next generation. This prevents the loss of the best solution. The technique used here produces better results.

5.6. Genetic Algorithm Implementation

In this section, the different important issues for the implementation of genetic algorithm are discussed. Genetic algorithms process much data. These include solution space strings, their objective function values, scores, as well as expected counts are some of the data arrays that need careful handling in the implementation of genetic algorithms.

5.6.1. *Data structures*

The data structure for the genetic algorithm developed for the aggregate production planning-problem consists of a record of *stg*, *objvalue*, *score*, and *expecc*. Referring to Fig. 5.6, constant parameters need to be defined. The bit-string stg is an array of integer-type that records string bits of integer number that represents overtime units and subcontract units respectively for each planning period. The record-position, *objvalue* is of real-type for storing the objective function values, *score* stores the goodness of an objective function value, and *expecc* stores the expected count. The *population* of solution is an array of the record described. Global populations (oldpop, newpop, temppop, spoolpop)

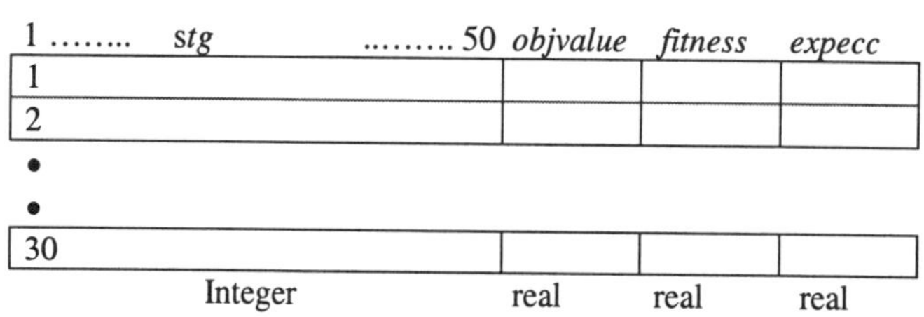

Fig. 5.6. Data structure for genetic algorithm for aggregate production planning.

are defined as type of population through the following declaration

$$oldpop, newpop, temppop, spoolpop : population\,.$$

The two populations, *oldpop* and *newpop* are non-overlapping, while *temppop* and *spoolpop* are temporary and selection pool populations respectively. Other useful declarations include *f_demand*, *norm_prod*, *op_stock*, *ld*, and fs representing the forecast demand, normal production rate, operating stock, late deliveries and finish stock. All these are arrays of integer-type with length equal to the total planning period.

5.6.2. *The genetic algorithm procedures*

Procedures for all the models are described in this section. The steps involved in the genetic algorithm are summarized in an algorithmic form and the flow chart is shown in Fig. 5.7.

Step 1: Initialization
Select the initial input parameters and randomly generate an initial diversified solution space.

- Test for upper and lower bounds of overtime units, subcontract units, late deliveries, finishing stock, and operating stock–test for constraints.
- Set the values for solution space size (*popsize*), number of generations (maxgen), probabilities of partial bits exchange (*pexchange*) and random bit exchange (*pmut*).
- Generate randomly, an initial population, *oldpop*.

Step 2: Selection/Selection
Select strings using stochastic sampling without replacement.

- Evaluate strings by objective function, score, and expected count calculation.
- Generate a temporary or intermediate solution set (temppop) using the integer parts of expected count and use the fractional parts as success probabilities.

Step 3: Partial bits exchange/Recombination
Apply the multi-partial bits exchange operator to the temporary population, temppop to generate a selection pool of size: $popsize \bullet pexchange = poolsize$.

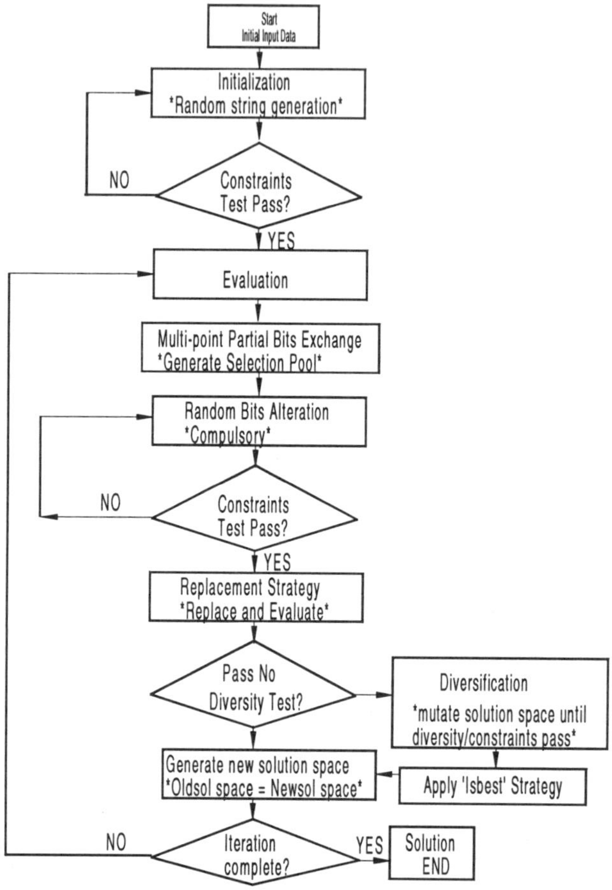

Fig. 5.7. Flow chart for the genetic algorithm procedure.

- Candidates for partial bits exchange are selected randomly by remainder selection without replacement.
- The multi-partial bits exchange operator is applied to two chosen strings, with a probability, *pexchange*.
- Apply random bit alteration to subsequent strings–normal random bit alteration regardless of constrains.
- Test for constrains. If violated, apply random bit alteration until constraints are satisfied.

- Evaluate the strings by calculating their objective function values.
- Sort out the selection pool in increasing order of objective function values.

Step 4: Replacement strategy
Compare each of the strings of sorted selection pool with corresponding old solution space strings.

- Compare corresponding strings in selection pool and old solution space, and take the best in turn.
- Choose the one that fares better than the other in each turn. For other subsequent string, a random selection is made with a probability, 0.6491.

Step 5: Diversification
Diversify the solution space if diversity falls below a predetermined minimum.

- Calculate the diversity, H, of the solution space.
- Compare the calculated diversity with the given minimum, the acceptable diversity. If less, then apply random bit alteration operator to the whole population, otherwise proceed. Repeat this process until diversity is satisfied.
- Calculate the objective function values of the diversified solution space.
- Sort out the new pool of strings in increasing order of objective function values.

Step 6: New Generation
The value of the current generation number determines the next step.

- If generation is less than current solution space, then current solution space becomes old solution space and the process is repeated.
- Otherwise, stop. The solution is string with lowest objective function value in the current solution space. Print out the solution.

5.7. Qualitative Innovations and Improvements

A number of observations were made, and quality innovations were done to improve the genetic algorithm operations during implementation. Tests and experimentation proves the validity of these changes. The improvements that are discussed in this section are:

(a) Parameter setting;
(b) Genetic-algorithms operations.

5.7.1. *Parameter setting*

Observation 1: Number of Iterations or Generations

Solutions quality improves significantly between 0–200 generations for the class of aggregate production planning problems solved. It can be seen that the solution settles at generation 250, and then drops down for a better solution at generation of about 500 as show in Fig. 5.8.

Setting: The number of generations is set to an average of 500.

Observation 2: Solution Space Size

Most good solutions are obtained with solution space size within the interval [15, 30]. Increasing the size would not improve anything since this will be mere duplication of strings.

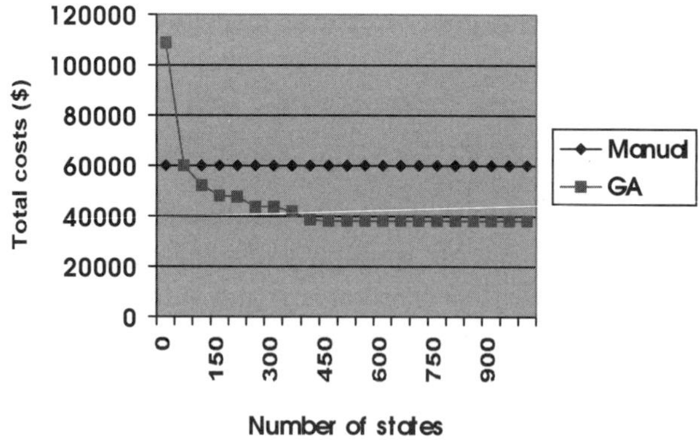

Fig. 5.8. Variation of number of states with costs for GA.

Setting: The solution space size is set to be within the interval [15, 30] on average, but should be increased for large problems.

Observation 3: Partial Bits Exchange Probability

It is also observed that low probabilities of partial bits exchange prevent creation of good new strings.

Setting: Through experimentation and tests the optimum probability is found to be within the interval [0.67, 0.77] on average.

Observation 4: Random Bit Alteration Probability

Again it is observed, that low probabilities of random bit alteration prevent creation of good new strings.

Setting: By careful experimentation and tests the optimum probability is found to be within the interval [0.45, 0.66] on average. All the average parameter-setting values are shown in Table 5.10.

5.7.2. *Genetic algorithm operation*

Observation 1: Selection

Most selection methods reported in the literature are deterministic sampling, stochastic sampling with replacement, and probabilistic sampling. These methods proved to be less effective since they result in duplication of the best string leading to premature convergence (Booker, 1989).

Table 5.10. Summary of parameter settings.

Parameter	Setting
Solution space size	15–30
Iterations or generations	400–500
Partial bits exchange bit probability	0.67–0.77
Random bits exchange bit probability	0.45–0.66

Improvement: A superior method known as stochastic remainder sampling without replacement is used as demonstrated in Fig. 5.9.

Observation 2: Multi-Point Partial Bits Exchange

In the previous applications of genetic algorithms, the partial bits exchange operator mostly used is the single-point operator that again is found to have a low search power.

Improvement: A new and better approach, the multi-point partial bits exchange technique that was described earlier is used. This technique is found to be more effective.

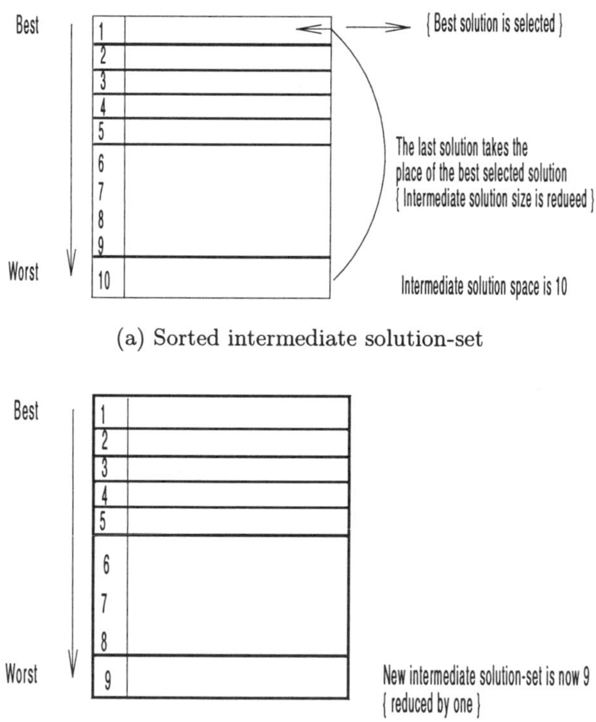

(a) Sorted intermediate solution-set

(b) New intermediate solution-set

Fig. 5.9. Remainder selection without replacement.

Observation 3: Compulsory Random Bit Alteration

Most of the previous applications of genetic algorithm did not implement random bit alteration operator after each and every partial bits-exchange and this is found to have a low search power since less new strings will be generated.

Improvement: After each and every partial bits exchange operation, random bit alteration was applied to each subsequent string. This is found to have better search power.

Observation 4: Replacement strategy

Replacement involves the selection of good strings to form the new solution space for the next generation. Methods that have been used include:

(1) comparison of selection pool and old solution space and taking the best;
(2) probabilistic replacement — which is done by a probability *Beta*, found as $Beta = \exp\{-(f_1 - f_0)\}$ where f_0 = objective function value of the worst string in the old solution space, and f_1 = objective function value of the subsequent string in the selection pool.

However, a problem arises with this method when $f_1 < f_0$, which means that $Beta > 0$ cannot be a probability.

Improvement: A simple method is used by comparing corresponding strings of selection and old solution space, and the winner is the better one in each case.

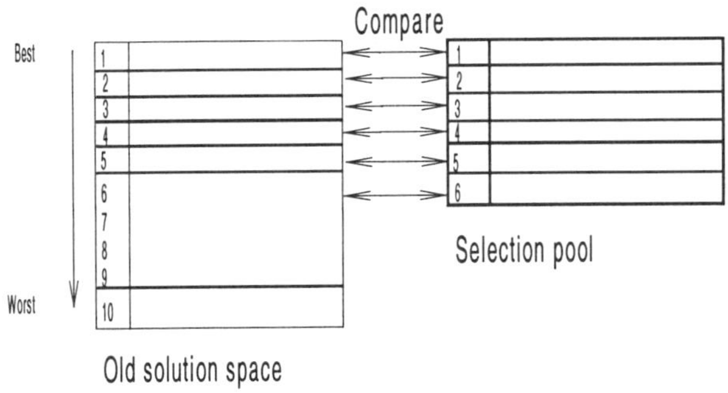

Fig. 5.10. Replacement strategy.

This process is illustrated in Fig. 5.10. The steps involved are:

(1) compare corresponding between the old solution space and the selection pool space, e.g. 1 in selection pool and 1 in old solution space;
(2) Take the better;
(3) Repeat up to 6 strings, which is the *poolsize*;
(4) The last 4 are selected probabilistically with a chosen probability, which is 0.641.

Observation 5: Diversification and Loss of Solution

After diversification by random bit alteration of the whole solution space new strings are generated and this may result in generation of worse strings losing the previous best solutions. Thus the genetic algorithm procedure jumps out of the good, local optimum solution, which may be the current solution.

Improvement: "Isbest" strategy
The best solution in the current is retained before random bit alteration is applied to a solution space. After random bit alteration, the best is compared with the previous best and a better one is taken into the next generation as shown by the following example in Fig. 5.11 with three *isbest* solutions. The worst strings are pushed out of the new solution space. This greatly improves the genetic algorithm operation.

For more information on the qualitative innovations and improvements on genetic algorithms, see Onwublu and Mutingi (2000).

5.8. Computational Tests and Results

The data-input to the genetic algorithm procedure developed for the aggregate production- planning problem consists of:

Planned period: for example, a six-month planned period is represented by a value of 6 Starting inventory: if there is no starting inventory, the value of zero in input

For $i = 1$ to planned period
Demand : normally known
Production : normally known
End *For*

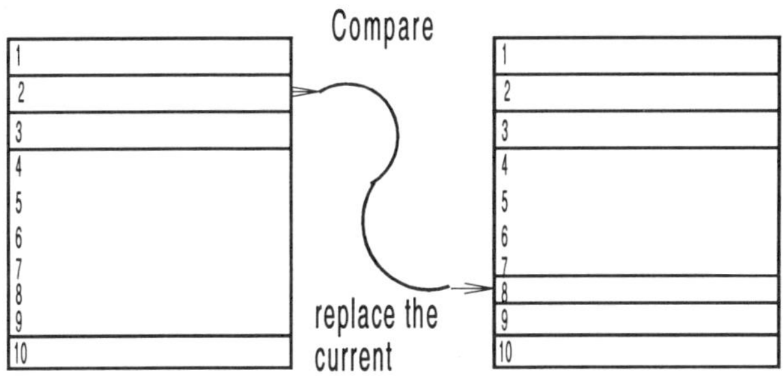

(a) Current solution space and Random bits exchange solution space

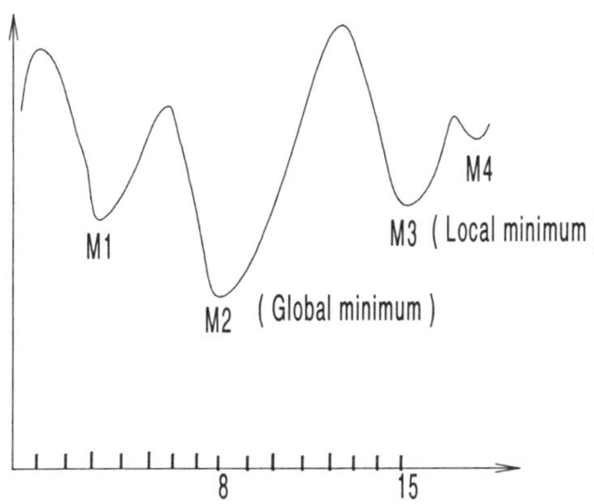

(b) Best minimum, M2 at iteration 8 should not be lost at iteration 15

Fig. 5.11. "Isbest" strategy.

Overtime cost : this cost is from company policy
Subcontract cost : this cost is from company policy
Late deliveries cost : this cost is from company policy
Finishing stock cost : this cost is from company policy

5.8.1. *Comparing genetic algorithms with other methods*

In order to validate the genetic algorithms functions presented here for generating aggregate production plan, trial problems from literature are used for comparison. Problem 1 is taken from Stockton and Quinn (1995), while

Problems 2, 3, and 4 are modified versions from Melnyk and Denzler (1996). In fact, the problem solved and shown in Tables 5.6 to 5.9 are from Stockton and Quinn (1995). By relaxing the company policy, the genetic algorithm procedure presented in this chapter outperforms (with a total cost of $38,080) the solutions of Stockton and Quinn (1995) (with a total cost of $58,900), who also applied genetic algorithm based on Genesis software (Grefenstette and Schraudolph, 1990). This shows the effect of company policy on the solution.

The optimum solution obtained by the genetic algorithms presented in this chapter for generating aggregate production plan is shown at the bottom, column 3 (marked with asterisk) of Table 5.9. The company policy of Stockton and Quinn (1995) are too rigid, hence they obtained inferior solution. They specified that, the maximum units that can be gained through the use of overtime is set at 90 and that overtime units could only be added in units of 30. They also specified that the maximum units that can be gained through the use of subcontracting was set at 300, and subcontracted units could only be added in units of 100. Figure 5.12 shows the comparison of the genetic algorithms presented here for generating aggregate production plan and manually derived aggregate production plans. It can be seen that after 50 generations, GA finds a solution equal to that obtained using normal method. GA continues to search for better solutions up to 450 generations and settles for optimum solution from

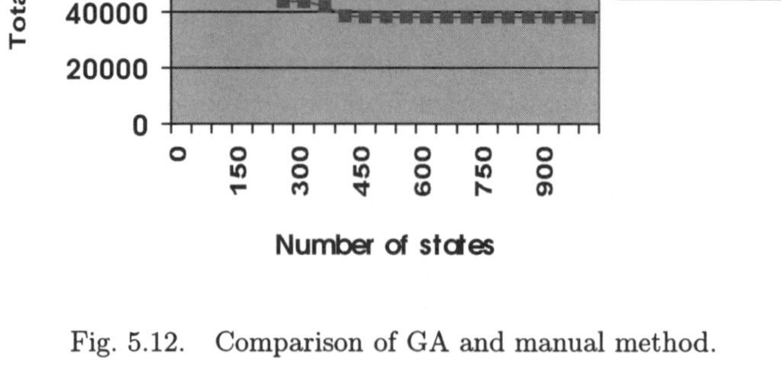

Fig. 5.12. Comparison of GA and manual method.

500 generations. After extensive experimentation, the parameter settings for the GA procedure for solving wide range of planning problems are found to be population size, *popsize* = 15, number of states or generations, *nstate* = 500, exchange probability, *pexchange* = 0.67, and mutation probability, *pmutate* = 0.45.

Stockton and Quinn (1995) neglected labour cost in their formulation, and only considered additional costs. In other to compare with most conventional methods that include labour cost, labour cost is considered when comparing solutions for Problems 2 (Chase Strategy) 3 (Level-Production Strategy), and 4 (Flexible Work Hour Strategy) which as earlier mentioned, are modified versions from Melnyk and Denzler (1996). Cost attributed to labour is given as

$$\text{Labour cost} = \sum^{\text{period}} [\text{Regular workers required} \bullet \text{daily rate}] \bullet \text{workday} \bullet \text{hours/day}\,.$$

The data for problems 2, 3, and 4 are shown in Table 5.11.

Further labour-intensive study has yielded the following costs estimates:

Labour hrs/unit	5 hours
Labour cost	$10 per hour
Overtime premium	25%
Hiring Cost	$400 per hour
Firing costs	$250 per fire
Beginning inventory	1000 units
Beginning employment level	75 employees
Inventory holding costs	$8 per period
Material cost per unit	$95

The labour costs for the Chase strategy, Level-Production strategy, and Flexible work hour strategy are $873,200, $876,960, and $613,200 respectively. The additional costs (overtime and subcontracting) from the genetic algorithms procedure presented here are $10,842.50, $9024.5, and $44,007.50 for the Chase strategy, Level-Production strategy, and Flexible work hour strategy respectively. This results in genetic algorithm finding total costs for Chase strategy, Level-Production strategy, and Flexible work hour strategy as being $884,042.5 ($10,842.50 + $10,842.50), $885,984.5 ($876,960 + $9024.5), and $657,207.5

Table 5.11. Demand forecast for half-year period.

Period	Forecast demand	Number of working days
January	3000	21
February	3200	19
March	3000	22
April	2000	21
May	2800	21
June	3600	22

Table 5.12. Comparison of GA and manual method for problems 2, 3, and 4.

	GA Total cost ($)	Manual Total cost ($)	GA improvement Over Manual (%)
Chase Strategy	$884,042.5	$961,250	8.032%
Level-Production Strategy	$885,984.5	$938,310	5.580%
Flexible hour Strategy	$657,207.5	$732,860	10.32%

($613,200 + $44,007.50) respectively. These results are obtained at 500 generations or states. Table 5.12 shows the comparison between the GA-based solution and manual solution. It can be seen that GA out-performs the manual method for all three strategies normally used in manual procedure. This means that GA produced a better solution than the best solution produced by manual method.

5.8.2. *Comparing genetic algorithms with integer linear programming*

The integer linear programming model for the aggregate production planning-problem is given as

$$\text{Minimize } Z = \sum_{t=1}^{T}(c_x X_t + c_w W_t + c_o O_t + c_u U_t + c_i I_t^+ + c_s I_t^- + c_h H_t + c_f F_t)$$

subject to:

$$X_t + I_{t-1}^+ - I_{t-1}^- - I_t^+ + I_t^- = d_t \quad \text{for } t = 1, 2, 3, \ldots, T$$

$$W_{t-1} + H_t - F_t = W_t$$

$$O_t - U_t = kX_t - W_t$$

$$X_t, W_t, O_t, U_t, I_t^+, I_t^-, H_t, F_t \geq 0$$

Z	total system cost
X_t	production units scheduled in period t
W_t	regular workforce level in period t
O_t	overtime scheduled in period t
U_t	under-time allowed in period t
H_t	new labour force added in hours in period t
F_t	layoffs in hours in period t
I_t^+	inventory in stock at the end of period
I_t^-	shortage in stock at the end of period
d_t	aggregate demand t
c_x	variable cost of production excluding labour per hour
c_w	cost of regular time labour per hour
c_o	overtime labour cost per hour
c_u	opportunity cost of not using the equipment
c_i	cost of unit inventory holding per hour
c_s	cost of a shortage per hour
c_h	cost of hiring regular time labour per hour
c_f	cost of layoffs per hour
k	labour hours required d to produce a unit
T	number of periods in the planning period

Example

Data on expected aggregated sales of products A, B, and C, over a planning period of six 4-week periods are given in Table 10.13. Products A and B require 2 cell-hours per units while product C requires 1 cell-hour per unit. The company has developed machining cell-hour as common basis for aggregating products. Overtime is permitted up to 60 units per period. The maximum working capacity of the company is 300 units per period, which can be varied up to 300 units per period. Other data are: \$100/hour, \$14/hour, \$14/hour, \$30/hour, \$3/unit/hour, \$400/unit/hour, \$50/hour, and the initial work-force level is 240 labour-hours per hour. Two planning strategies are available in the company from which to choose:

(1) Develop an aggregate production plan if there is a constant production of 400 units per period for the entire period.
(2) Develop an aggregate production plan if production rate of 500 units is allowed for first four periods and then at the rate of 350 for the remaining periods.

Table 5.13. Data for demand forecasts for three products.

Period	Product A		Product B		Product C	
	Units	Equivalent cell-hours	Units	Equivalent cell-hours	Units	Equivalent cell-hours
1	60	120	40	80	100	100
2	70	140	50	100	160	160
3	50	100	70	140	210	210
4	55	110	65	130	170	170
5	40	80	55	110	100	100
6	40	80	40	80	80	80

Table 5.14. ILP solution for aggregate production planning problem: Option 1.

Period	d_t	X_t	W_t	O_t	H_t	F_t	U_t	I_t^+	I_t^-
1	300	300	240	60	0	0	0	0	0
2	400	400	240	160	0	0	0	0	0
3	450	450	240	210	0	0	0	0	0
4	410	410	240	170	0	0	0	0	0
5	300	300	240	60	0	0	0	0	0
6	240	240	240	0	0	0	0	0	0

Table 5.15. ILP solution for aggregate production planning problem: Option 2.

Period	d_t	X_t	W_t	O_t	H_t	F_t	U_t	I_t^+	I_t^-
1	300	290	240	50	0	0	0	0	10
2	400	430	380	50	140	0	0	20	0
3	450	430	380	50	00	0	0	0	0
4	410	410	380	30	00	0	0	0	0
5	300	300	300	0	00	80	0	0	0
6	240	240	240	0	00	60	0	0	0

Solution

The solutions of the ILP formulation are given in Tables 5.14 and 5.15. The objective function and problem-constraints are listed in the appendix at the end of this chapter.

Using integer linear programming (ILP) gives an optimum solution but the major problem with this technique is that too many variables are required for a moderate problem. Hence it becomes difficult if not impossible to solve large problems (long planning periods), taking the additional costs into consideration.

5.9. Summary

In this chapter, various terminology have been described, design issues explained and the process of genetic algorithms discussed in brief. The underlying mechanics of the genetic algorithm have also been dealt with, namely *selection* (selection of copies of strings or strings according to their score values), the *partial bits exchange* (exchange of the portion of a pair of strings), and *random bit alteration* (the random modification of the strings).

A variety of aggregate production-planning techniques are available, with each yielding acceptable results only under a limited range of specific situation. The setback of this is that planners have to be careful in choosing from the variety of techniques that meet their company requirements. For example several strategies are used such as Chase strategy, Level-Production strategy and Flexible work hour strategy. In Chase strategy, the market demand is set equal to the normal production rate and inventory remains constant. This is of course, not the case usually encountered in practice. In Level strategy, the normal production rate is set to be equal for all planning periods and the strategy features a level work force, and emphasis is placed on hiring and firing employees. Again, this is not true in practice, as there are restrictions caused by labour-relations law with respect to downsizing. In Flexible hour strategy, it is assumed that workers would be willing to accept less number of workdays during a period of slow demand, although without contemplating hiring or firing. It all depends whether or not some employees would want their workdays cut merely because there is periodic slow demand. Again, this is not the case in practice. These are just few of the restrictions of the variety of aggregate production planning techniques that are currently available. The diverse planning conditions and the possible changes that occur require that planning procedures should be able to adapt to these changes. Genetic algorithm is well suited to accommodate these planning complexities because it uses a single planning technique for all planning situations.

Finding optimal or near-optimal production planning schedules are currently difficult because this requires the use of complex procedures that are often difficult for managers and production planner to understand. The result of this is that there is absence of use of many aggregate production-planning techniques in the industry. Genetic algorithm approach is easily understood by manager and production planners because it employs simple rules and is versatile because it can be easily applied to similar planning requirements. In this regard, the production planner would have more time to seek appropriate objective functions that give alternative plans from which to find the one that is most cost-effective. A genetic algorithm-based procedure developed for finding optimal or near-optimal production plan is presented and compared with other results form literature. The results show that GA procedure identifies low-cost aggregate production planning solutions and is therefore recommended for use by planner in the industry.

Exercises

1. What do you understand by aggregate production planning?
2. Describe the environment in which aggregate production planning exists.

3. What is the relationship between strategic business plan and aggregate production planning?
4. What is the relationship between aggregate production planning and master production scheduling?
5. Define late deliveries, opening stock, and finishing stock in aggregate production planning.
6. For the market demand and normal production shown in Table 5.14, carry out an aggregate production plan. Take initial inventory level equal to zero, overtime cost = \$10/unit, subcontracting cost = \$45/unit, late delivery penalty = \$250/unit, penalty for finish stock for the last planning period = \$120.
7. For the market demand and normal production shown in Table 5.17, carry out an aggregate production plan. Take initial inventory level equal to zero, overtime cost = \$15/unit, subcontracting cost = \$55/unit, late delivery penalty = \$200/unit, penalty for finish stock for the last planning period = \$100.
8. What difference does it make in problem 6 if the normal production is kept constant at 1000 units per planning period?
9. What difference does it make in problem 7 if the normal production is kept constant at 950 units per planning period?

Table 5.16. Forecast demand and normal production-rate (usually known).

Forecast demand	Normal production
1000	1200
900	1000
900	1000
1500	1100
1500	1000
700	900
600	800
1200	800
800	1000
2000	1200
600	1000
1500	1000

Table 5.17. Forecast demand and normal production-rate (usually known).

Forecast demand	Normal production
700	1200
800	1000
1500	1000
1500	1100
1500	1000
900	600
600	800
1200	900
1000	1000
2200	1000
900	1000
1500	1000

Table 5.18. Data for demand forecasts for three products.

	Product A		Product B		Product C	
Period	Units	Equivalent cell-hours	Units	Equivalent cell-hours	Units	Equivalent cell-hours
1	80	160	50	100	120	120
2	70	140	60	120	150	150
3	60	120	60	120	200	200
4	55	110	70	140	150	150
5	50	100	50	100	110	110
6	50	100	45	90	100	100

10. Data on expected aggregated sales of products A, B, and C, over a planning period of six 4-week periods are given in Table 5.18. Products A and B require 2 cell-hour per units while product C requires 1 cell-hour per unit. The company has developed machining cell-hour as common basis for aggregating products. Overtime is permitted up to 60 units per period. The maximum working capacity of the company is 300 units per period, which can be varied up to 300 units per period. Other data are: \$100/hour,

$14/hour, $14/hour, $30/hour, $3/unit/hour, $400/unit/hour, $50/hour, and the initial work-force level is 240 labour-hours per hour. Develop an aggregate production plan if there is a constant production of 400 units per period for the entire period.

11. Develop an aggregate production plan for problem 10 if production rate of 500 units is allowed for first four periods and then at the rate of 350 for the remaining periods.
12. Outline the differences between genetic algorithm and other optimization methods.
13. What do you understand by: (i) representation, (ii) selection, (iii) genetic operators?
14. For the aggregate production planning, illustrate the features listed in problem 13.
15. Compare partial bits exchange and diversification mechanism in tabu search.
16. Compare random bits exchange and intensification mechanism in tabu search.
17. What are the advantages of separating randomly generated data and deduced data in the aggregate production planning methodology?
18. For problem 6, how would you go about constructing the randomly generated data and deduced data for the aggregate production planning methodology?
19. Suggest what other objective function formulation would be useful for the aggregate production planning-problem (see Eqs. 3–9).
20. What modifications to the genetic algorithm procedure presented in this chapter may be lead to its performance improvement?

Appendix

$$
\begin{aligned}
\text{MIN}\quad & 100\ X1 + 14\ W1 + 20\ O1 + 50\ U1 + 3\ I1 + 400\ B1 + 14\ H1 + 30\ F1 \\
& + 100\ X2 + 14\ W2 + 20\ O2 + 50\ U2 + 3\ I2 + 400\ B2 + 14\ H2 + 30\ F2 \\
& + 100\ X3 + 14\ W3 + 20\ O3 + 50\ U3 + 3\ I3 + 400\ B3 + 14\ H3 + 30\ F3 \\
& + 100\ X4 + 14\ W4 + 20\ O4 + 50\ U4 + 3\ I4 + 400\ B4 + 14\ H4 + 30\ F4 \\
& + 100\ X5 + 14\ W5 + 20\ O5 + 50\ U5 + 3\ I5 + 400\ B5 + 14\ H5 + 30\ F5 \\
& + 100\ X6 + 14\ W6 + 20\ O6 + 50\ U6 + 3\ I6 + 400\ B6 + 14\ H6 + 30\ F6
\end{aligned}
$$

SUBJECT TO:

$$
\begin{aligned}
&X1 - I1 + B1 = 300\\
&X2 - I1 + B1 - I2 + B2 = 400\\
&X3 - I2 + B2 - I3 + B3 = 450\\
&X4 - I3 + B3 - I4 + B4 = 410\\
&X5 - I4 + B4 - I5 + B5 = 300\\
&X6 - I5 + B5 - I6 + B6 = 240
\end{aligned}
$$

$$
\begin{aligned}
&- W1 + W0 + H1 - F1 = 0\\
&- W2 + W1 + H2 - F2 = 0\\
&- W3 + W2 + H3 - F3 = 0\\
&- W4 + W3 + H4 - F4 = 0\\
&- W5 + W4 + H5 - F5 = 0\\
&- W6 + W5 + H6 - F6 = 0
\end{aligned}
$$

$$
\begin{aligned}
&O1 - U1 - X1 + W1 = 0\\
&O2 - U1 - X2 + W2 = 0\\
&O3 - U2 - X3 + W3 = 0\\
&O4 - U3 - X4 + W4 = 0\\
&O5 - U4 - X5 + W5 = 0\\
&O6 - U5 - X6 + W6 = 0
\end{aligned}
$$

References

Production Planning

Bowman, E. H., 1956, Production planning by the transportation method of linear programming, *Journal of Operations Research Society*, **4** 100–103.

Bowman, E. H., 1963, Consistency and optimality in managerial decision making, *Management Science*, **9**(4) 310–321.

Buxey, G., 1993, Production planning and scheduling for seasonal demand, *International Journal of Operations and Production Management*, **13**(9) 4–21.

Feiring, B. R. and Sastri, T., 1990, Improving production planning by utilizing stochastic programming, *Computers and Industrial Engineering*, **19**(1–4) 53–56.

Gen, M., 1992, Method for solving multi-objective aggregate production planning problems with fuzzy parameters, *Computers and Industrial Engineering*, **23**(1–4) 117–120.

Gilgeous, V., 1989, Modeling realism in aggregate planning: A goal-search approach, *International Journal of Production Research*, **27**(9) 1179–1193.

Gilbert, K. C. and Madan, M. S., 1991, A heuristic for a class of production planning and scheduling problems, *IIE Transactions*, **23**(5) 282–289.
Holt, C. C., 1960, *Planning, Production, Inventories and Work Force* (Prentice-Hall).
Kamien, M. I. and Li, L., 1990, Subcontracting, coordination, flexibility and production smoothing in aggregate planning, *Management Science*, **36**(11) 1352–1363.
Love, C. E. and Turner, M., 1993, Note on utilizing stochastic optimal control in aggregate production planning, *European Journal of Operational Research*, **65**(4) 199–206.
Melnyk, S. A. and Denzler, D. R., 1996, *Operations Management: A Value-Driven Approach* (Irwin).
Nam, S.-J. and Logendran, R., 1992, Aggregate production planning — a survey of models and methodologies, *European Journal of Operational Research*, **61**(5) 255–272.
Nawaz, M., Enscore, E. E., and Ham, I., 1983, A heuristic algorithm for the m-machine, n-job flow-shop sequencing problem, *OMEGA, The International Journal of Management Science*, **11**(1) 91–95.
Onwubolu, G. C. and Mhlanga, S., 1997, Development of a computer-aided production management course in industrial engineering curriculum, *Production Planning and Control*, **8**(3) 297–306.
Riggs, J. L., 1987, *Production Systems: Planning, Analysis, and Control*, 4th edition (John Wiley).
Satyadas, A. and Chen, H. C., 1992, Multi-criteria multi-goal decision making — the fuzzy paradigm, *Computers and Industrial Engineering*, **23**(1–4) 393–396.
Starr, P. J. and Mahmoodi, F., Development and teaching a computer-assisted course in production planning, *Production Planning and Control*, **5**(4) 199–205.
Stevenson, W. J., 1993, *Production/Operations Management*, 4th edition (Irwin).
Taha, H. A., 1971, *Operations Research and Introduction*, 2nd edition (Macmillan Publishing).
Taubert, W. H., 1968, A search decision rule for the aggregate scheduling problem, *Management Science*, **14**(8) 343–359.
Vollman, T. E., Berry, W. L., and Whybark, D. C., 1992, *Manufacturing Planning and Control Systems*, 3rd edition (Irwin).

Genetic Algorithms

Booker, L. B., 1987, Improving search in genetic algorithms. In *Genetic Algorithms and Simulated Annealing*, Davis, L. (ed.) (Los Angeles: Morgan Kaufmann), 61–73.
Goldberg, D. E., 1989, *Genetic Algorithms in Search, Optimization and Machine Learning* (Addison-Wesley Publishing Company, Inc., Reading MA).
Grefenstette, J. J., 1986, Optimization of control parameters for genetic algorithms, *IEEE Transactions on Systems, Man, and Cybernetics*, **16**(3) 122–128.
Grefenstette, J. J. and Schraudolph, N. N., 1990, *A User Guide to Genesis 1.2USCD*, Navy Center for Applied Research in Artificial Intelligence, Naval Research Laboratory, Washington.

Liepins, G. E. and Hilliard, M. R., 1989, Genetic algorithms: Foundations and applications, *Annals of Operations Research*, **21** 31–58.

Stockton, D. J. and Quinn, L., 1995, Aggregate production planning using genetic algorithms, *Proceedings of the Institution of Mechanical Engineers*, B07593, **209** 201–09.

Chapter 6

Tabu Search

6.1. Background to Tabu Search

Artificial intelligence (AI) and operations research (OR) are fields that grew up together, and originally shared many common notions for improving problem solving. Both fields started out with the same objective of developing methods to solve real life problems encountered in every day practical life situations. In the late 1960s however, the fields of AI and OR began to diverge. Researchers in AI began to emphasize symbolic and qualitative analyses, while researchers in OR began to concentrate on quantitative study of optima and convergence of optimization problems.

During this era of divergence between AI and OR, researchers directed their effort toward introducing probabilistic and integrated design concepts into heuristic procedures. Simultaneously, some researchers in OR began to emphasize discrete search spaces instead of continuous differentiable search spaces. The idea behind this shift is that differentiable search spaces emphasize monotonic convergence, which in practice is not the case. Real life problems are often characterized by discrete search spaces. Another useful development during the same era, was the re-thinking in terms of rigid memory structures such as in branch-and-bound technique. The need for flexible memory structure was awakened amongst researchers. These developments were supposed to bring together the divergent AI and OR communities, but failed over the years. These two corner stones: discrete search spaces or combinatorial optimization and flexible memory, provided the foundations on which tabu search strategies were built in the late 1980s.

Tabu search is therefore, based on selected concepts that unite the fields of artificial intelligence and optimization (an early emphasis of operations research). Four main early developments were instrumental to the success of tabu search: (1) flexible memory (already discussed in Chapter 1) based on recency and frequency, (2) strategic oscillation, (3) surrogate constraint, and (4) logical restructuring and variable depth search.

Fred Glover, who proposed tabu search in its present form, pioneered these four main early developments. The first development was embodied in a method for solving integer-programming problems by referencing corner polyhedral relaxation (Glover, 1966, 1969). In this method, each variable has its own memory in order to create restrictions so as to avoid the generation of unrealizable solutions. The restrictions carried partial attributes of the solutions, instead of the full solutions. Strategies were devised for selectively combining the solutions, to provide subsequent solutions that contributed toward optimality. The methodology for combining these solutions became the premises for the scatter search approach that eventually emerged. This idea of flexible memory structure paved the way to one of the primary components of tabu search, know as adaptive memory.

The second development also involves an exact approach for integer programming problems. Here, the simplex method of linear programming was augmented using cutting plane processes to introduce new constraints implied by the integer requirement for the variables. The cutting plane approach requires that the linear programming commence with feasible starting point in a *primal* or *dual* sense and then to remain feasible toward an optimal solution while advancing monotonically. However, since monotonicity has to be discarded when dealing with combinatorial optimization or discrete search spaces, a pseudo primal-dual method was created (Glover, 1968) to "cut across" various regions of the solution space by periodically violating and then restoring feasibility conditions. This strategy of purposely visiting both feasible and unrealizable regions periodically paved the way to one of the primary components of tabu search, known as strategic oscillation.

The third development also involves the introduction of surrogate constraint methods for integer programming (Glover, 1965). Surrogate constraint methods were based on the premises of combining constraints to produce new ones, with the objective of providing information not separately available in the parent constraints. This led to the uses of recency-based and frequency-based memory for extracting the information and exploiting it within heuristic processes. The strategy of using recency-based memory for short term and frequency-based memory for long term is the hallmark of tabu search approach.

The fourth development involved the introduction of contrasting strategy (Glover, 1963) for exploiting a collection of decision rules by establishing a way of combining them to create new rules. This led to the concept of logical restructuring which is one of the important components of tabu search.

Although it can be seen that the primary components of tabu search emanated from Glover's work on integer programming, tabu search has found favour with many researchers, and it is now applied to virtually any combinatorial optimization problem encountered in practice.

Does tabu search qualify as an intelligent optimization technique? Of course, yes, because memory-based strategies are the hallmark of tabu search approach. In fact, tabu search has all the features of intelligent optimization techniques shown in Fig. 1.4 and discussed in Chapter 1. Remember, the "order-qualifiers" for an approach to be considered as an intelligent optimization technique are: adaptive memory and responsive exploration at strategic level, neighbourhood structure, intensification and diversification strategies at tactical level, move mechanism and constructive seed solution method at operational level. Tabu search passes these tests and therefore, qualifies as one of the intelligent optimization techniques.

Traditionally, the seed solution to tabu search is obtained from a simple descent or hill climbing method. This solution is usually one of the local optima, which can be improved upon by tabu search using the features at the strategic and tactical levels. This accounts, for why the simple decent or hill climbing method is usually first discussed before tabu search approach is discussed. We do the same here, by first presenting the simple descent method in the next section before presenting tabu search thereafter.

6.2. The Dilemma of Hill Climbing

Combinatorial optimization problems have discrete search space, made up of individual points. For example, these may relate to the information concerning jobs visiting machines or production parts to be assigned to a manufacturing cell, or tools to be assigned to certain jobs being produced by a machine, or cities to be visited in the traveling salesman problem, etc. The search space consists of all possible points in the solution space with nothing in between. Searching for the solution requires "hopping" from one point to the other unlike traditional mathematical models that require search space to be continuously differentiable. In combinatorial optimization problems, there is no such thing as continuous differentiability that is, all simplistic and unrealistic in practice.

There has to be a measure for a point x in space. Let that measure be $f(x)$. If the current position in space is x_k and a neighbourhood position is designated as x_{k+1}, then we can approximate the gradient by $[f(x_{k+1}) - f(x_k)]$. A mechanism is needed for the search process to hop from the currently examined point to another. For example, the points in space could be arbitrarily laid out in one-dimensional solution space (as shown in Fig. 6.1(a)). Looking at the numbers, it is easy to identify three optima (minimum) each for the solution space types of one-dimensional and two-dimensional. These numbers are boxed.

For the one-dimensional solution space, the local minima are 2, 4, and 3, and of these, the global minimum is 2. For the two-dimensional solution space, the local minima are 2, 3, and 4 respectively, and of these, the global minimum is 2.

In order to find the global minimum, you have to examine all the neighbours to see if any were better. This illustrates hill climbing or simple descent method, which simply ensures that the value selected at the current iteration is better (less than or greater than) than the value selected at the previous iteration.

Let the best solution so far be at the point $x = x_k$. The associated objective function is $f(x_k)$, for which the neighbouring point is x_{k+i}. The associated

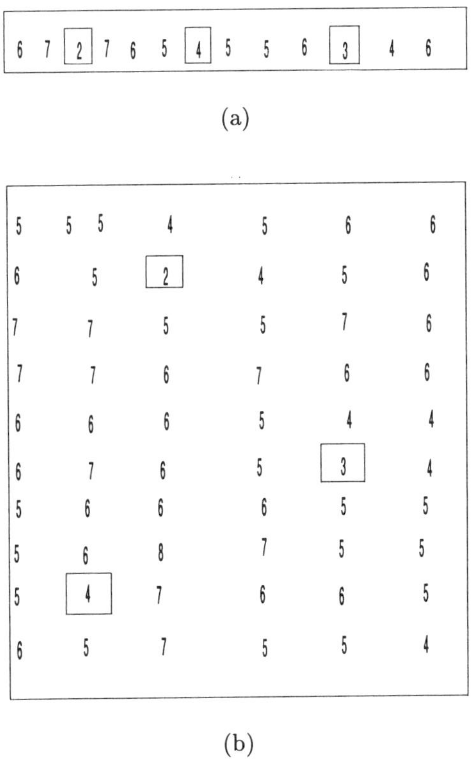

Fig. 6.1. Local optima for 1-D and 2-D solution spaces.

objective function for the neighbouring point is $f(x_{k+i})$, $i = 1, \ldots, n$; $k = 1, \ldots, N$. Here, N is the total number of points in the space. But n has a different meaning since it related to the neighbourhood size. For the one-dimensional search space, n can take a value up to 2, i.e. points to either side of the point x. For the two-dimensional space (see Fig. 6.1(b)), n can a value up to 8, i.e. all the points surrounding the point x within its immediate vicinity. In both cases, when $n = N - 1$, then we are examining the whole search space for each move, and this is steepest descent method. Mathematically stated, the search moves from x_k to x_{k+1} according to the following rule:

$$x = \begin{cases} x_{k+1} & \text{if } f(x_{k+1}) \leq f(x_k) \\ x_{k+1} & \text{if } f(x_{k+1}) \geq f(x_k)\,. \end{cases}$$

Consequently, a solution is finally found which can not be improved by any further moves. Hill climbing moves down (or up) the slope until it reaches a minimum (or maximum). But what if this movement takes the search procedure to a local optimum that is far in quality from the global optimum? This is where hill climbing runs into problem because it simply gets stuck! Herein, lies the dilemma of hill climbing. It simply gets stuck like a ball that soars high over a school fence and lands in a deep pit. It is simply stuck. The players have to seek a way of getting the ball out of the pit. The dilemma of hill climbing are three-fold. Firstly, we are not sure how we can avoid ending up in a local optimum (maximum or minimum). Secondly, we are not sure whether the optimum is simply a local optimum or the global optimum. Thirdly, we are not sure how far we are from the global optimum. These are major problems. It is not merely getting a solution! What is important is getting *good* solution. A good solution may not be the global optimum, but it should be close to it. We need an approach that will consistently move the solution search toward the global optimum, irrespective of what point in the solution space the search commences from. Hill climbing is helpless in this circumstance and therefore, must be disregarded as a dependable search approach for global optimum. Since the neighbourhood examination involves all the points in the solution space, this approach is not feasible for large problem instances encountered in practice. This search dilemma is the motivation for intelligent optimization techniques. And indeed, this is the motivation for tabu search, which qualifies as one of the intelligent optimization techniques.

6.3. Tabu Search Fundamentals

A general combinatorial optimization problem (P) is formulated as

$$\min f(s)$$
$$s.t.\ s \in S\,,$$

where $f(s)$ is any function of a discrete variable s, and S is a discrete set of feasible solutions as has been earlier discussed. Local search approaches for solving problem (P) are based on search procedures in the solution space, S. These procedures commence from an initial solution $s_0 \in S$. At each iteration, a heuristic is invoked to obtain a new solution s' in the neighbourhood $N(s)$ of the current solution s. New solutions are obtained through atomic changes in s. Every feasible solution $\tilde{s} \in N(s)$ is evaluated according to the objective function $f(\cdot)$, which is eventually, locally optimized. The current solution propagates toward better neighbour solutions, thereby improving the best obtained solution s^*. The basic local search approach corresponds to the hill climbing heuristic (algorithm) that has been sufficiently discussed in the preceding section.

Tabu search is an adaptive search procedure that may be used for solving combinatorial optimization problems, which guides a hill climbing heuristic to continue exploration for better solutions without being confounded in a local optimum. At every iteration, tabu search applies an admissible move to the current solution, transforming it into its neighbour with the least cost. Moves toward a new solution that increases the cost function is permitted. This is where tabu search departs from a hill climbing approach. Tabu search prohibits or taboos a reverse move along some iterations, in order to avoid cycling which leads to getting stuck. These tabu restrictions are based on maintenance of a short term memory (STM) function that determines how long a tabu restriction will be enforced or which moves are admissible at each iteration. We now discuss the basic features of an STM in the next section.

6.4. Short Term Memory in Tabu Search

Tabu search uses memory to forbid certain moves that have been labeled tabu-active in order to avoid cycling. This notion can be formalized in general by conceptualizing the solution neighbourhood as depending on the time stream, and consequently on the iteration number k. This leads to neighbourhood $N(s,k)$ that is time-dependent, instead of the usual $N(s)$. A form of memory embodied in a tabu list T that records the $|T|$ solutions most recently visited is given as:

$$N(s,k) = N(s) \backslash T\,.$$

This means that $N(s,k)$ consists of the solution of $N(s)$ not included on the tabu list. Characterized in this way, tabu search may be viewed as a *variable neighbourhood method.* This means that the neighbourhood of s is not static but dynamic, changing with respect to the search history over some time interval. Consequently, short-term memory (STM) of tabu search is recency-based. Such a recency-based memory will prevent cycles of length less than or equal to $|T|$ from occurring in the search trajectory. Again, a recency-based memory will allow a solution s to be visited more than once, although $N(s,k)$ would have changed.

In this section, we present tabu search that is based on STM by specifying its four components: the recency-based memory, tabu tenure, aspiration criteria, and candidate list strategies. Tabu search commences in the same way as ordinary local or neighbourhood search, proceeding iteratively from one solution to another until a chosen termination criteria is satisfied. Each $s \in S$ has an associated neighbourhood $N(s) \subset S$, and each solution $s' \in N(s)$ is reached from x by an operation called a *move*. A function $m \in M(s)$ transforms s into s' which is a new feasible solution ($s' = m(s)$). The neighbourhood $N(s)$ is defined as $\{s' : s' = m(s); s, s'(X; m \in M(s)\}$.

A crucial aspect of tabu search is the mechanism to avoid as much as possible cycling. As a starting point, consider a form of memory embodied in a *tabu list* T that explicitly contains t different solutions, that is $T = \{s_1, s_2, \dots, s_t\}$. An element t is associated with m and s; this element defines a set of moves that are tabu or forbidden now. In particular t forbids to apply m' to s', which would transform s' back to s, thus cycling. It is possible for t to forbid other moves too. Since the elements of T define all tabu moves and therefore can grow indefinitely, the size of T is bounded by a parameter s known as the *tabu list size.* If $|T| = s$, before adding t to T, the principle of first-come-first-leave (FCFL) must be applied to remove an element. The generic scheme of tabu search based on STM is given in Fig. 6.2.

// TS generic scheme for STM //
Choose an initial solution s in χ
$s^* \leftarrow s$ and $k \leftarrow 1$
while not stopping condition **do**
$k \leftarrow k + 1$
Generate $V^* \subseteq N(s,k)$
Choose the best s' in V^*
$s \leftarrow s'$
if ($f(s') < f(s^*)$) **then** $s^* \leftarrow s'$
end while

Fig. 6.2. Generic scheme of STM-based tabu search.

6.4.1. *Recency-based memory*

Recency-based memory is a short-term memory since it keeps track of solutions attributes that have changed during the recent past. In other word, recency-based memory is time-dependent as has already been explained. In this memory structure, selected attributes that occur in solutions visited in the recent past are labeled tabu-active over a time interval defined by $|T|$, and solutions that contain tabu-active elements or their combinations, are declared tabu. Consequently some solutions from the recent past are excluded from $N(s, k)$ and therefore, are excluded from being revisited. In addition, other solutions that share in the tabu-active attributes are excluded from being visited. Similarly, any move that leads to such solutions are declared tabu.

In order to fix ideas, we illustrate the concept of recency-based memory using an graphical example of minimum k-tree problem. Because the problem is graphical, readers can easily follow the ideas of reduced neighbourhood, tabu search-active states, tabu length $|T|$, move, etc. used in describing recency-based memory.

Example 1

An undirected graph is shown in Fig. 6.3. Using recency-based tabu search memory, solve the minimum k-tree problem for $k = 4$, assuming a stopping criterion at iteration equal to ten. Edge weights are shown on arcs that join two nodes.

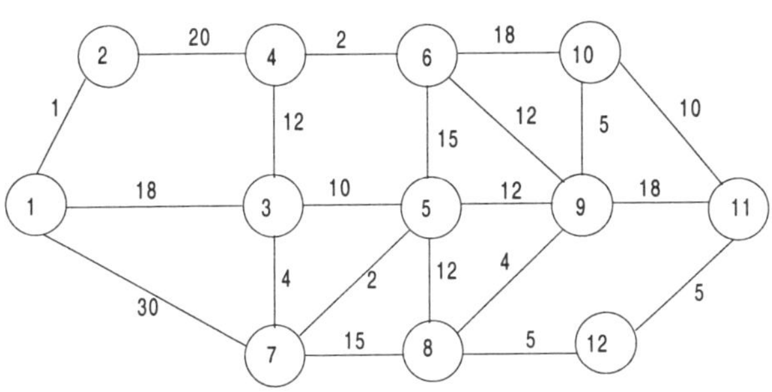

Fig. 6.3. Undirected graph for the minimum k-tree problem.

Solution

As a starting point, we employ a greedy construction approach to obtain an initial solution. The greedy construction seeks an edge having the least weight in the graph. This becomes a partial tree. In this case, edge (1,2) is chosen with weight equal to 1. The next candidates are edges (2,4), (1,3), and (1,7) each having weights 20, 18, and 30 respectively. Edge (1,3) is chosen since it has the least weight. The partial tree becomes 2-1-3 with total weight equal to 19. The next candidates are edges connected to node 3, which in this case, are edges (3,4), (3,7), and (3,5) with weights 12, 4, and 10. The next candidates are edges (7,5) and (7,8) connected to node 7 with weights 2 and 15 respectively. Edge (7,5) is chosen because it has the least weight. The initial solution corresponds to a tree 2-1-3-7-5 wit total equal to 25. The greedy construction propagation and the resulting initial solution are shown in Fig. 6.4 while the construction details are shown in Table 6.1. The initial solution has a total weight of 25.

We define a move as dropping an edge in the current 4-edge tree and adding an edge from the parent tree such that the resulting sub-graph is a 4-edge tree. The labels of the edges, which are dropped or added, are used for tabu search classification in this kind of problem.

Starting from the initial solution, we commence seeking a new solution by considering which edge to drop in the initial solution and which edge to

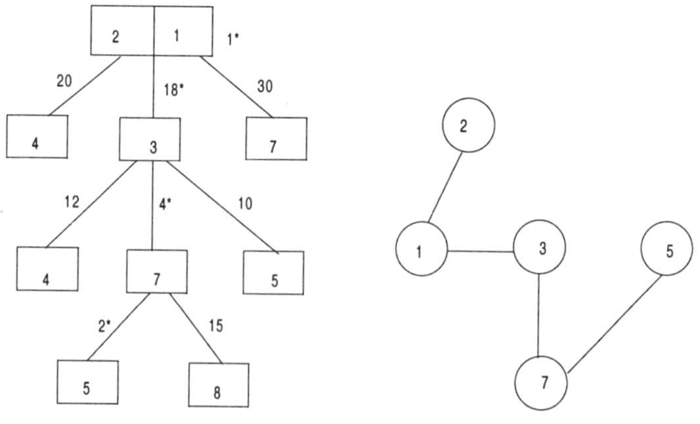

(a) construction (b) initial solution

Fig. 6.4. Initial solution generation.

Table 6.1. Initial solution construction details.

Iteration	Candidates	Chosen	Total weight
1	(1,2)	(1,2)	1
2	(2,3), (1,3), (1,7)	(1,3)	1 + 8 = 19
3	(2,3), (1,7), (3,4), (3,5), (3,7)	(3,7)	19 + 4 = 23
4	(2,3), (1,7), (3,4), (3,5), (7,5), (7,8)	(7,5)	23 + 2 = 25

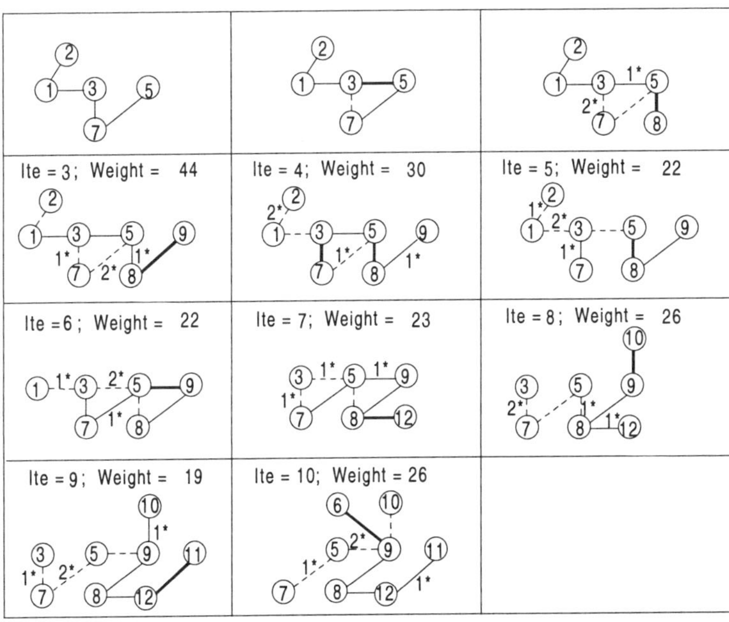

Fig. 6.5. Progression of STM solution generation.

add from the tree so that a minimum total weight can be obtained. All the possible dropping and adding of edges to the current solution constitute a neighbourhood. For the initial solution the move obtained by dropping edge (3,7) and adding edge (3,5) gives the best move, with weight equal to 31. This occurs in the first iteration. Step by step, the tabu search progresses as shown in Fig. 6.5. In this example, we use an asymmetric tabu classification in which the dropped edge has tabu tenure equal to 2 and added edge has

Table 6.2. STM tabu-active tenure, move, move value and weight.

Iteration	Tabu-active net tenure 2	1	Drop	Add	Move value	Weight
0						25
1			(3,7)	(3,5)	6	31
2	(3,7)	(3,5)	(5,7)	(5,8)	10	41
3	(5,7)	(5,8), (3,7)	(1,2)	(8,9)	3	44
4	(1,2)	(8,9), (5,7)	(1,3)	(3,7)	−14	30
5	(1,3)	(3,7), (1,2)	(3,5)	(5,7)	−8	22*
6	(3,5)	(5,7), (1,3)	(5,8)	(5,9)	0	22*
7	(5,8)	(5,9), (3,5)	(3,7)	(8,12)	1	23
8	(3,7)	(8,12), (5,8)	(5,7)	(9,10)	3	26
9	(5,7)	(9,10), (3,7)	(5,9)	(11,12)	−7	19*
10	(5,9)	(11,12), (5,7)	(9,10)	(6,9)	7	26

tabu tenure equal to 1. This means that for a dropped edge, tabu-active implies that the edge is not allowed to be added until after 2 iterations. For an added edge, tabu-active implies that the dropped edge is not allowed to be dropped until after every other iteration. This asymmetric tabu classification finds application in other problem instances where drop or add do not apply. Table 6.2 summarizes the tabu-active tenure, move, move value and weight for the undirected graph of Fig. 6.3. The added edge has a thick line. The dotted lines having numbers with asterisks show the tabu-tenure status of the edges. The recently dropped edge does not have any number. The reader can study Table 6.2 in conjunction with Fig. 6.5 for clearer understanding of the ideas discussed in this section.

6.4.2. *Tabu tenure*

Recency-based memory will prevent cycles of length less than or equal to $|T|$, the tabu tenure, from occurring in the search trajectory. The tabu tenure is an important feature of the tabu search approach that needs to be decided. It determines how restrictive is the neighbourhhood search. If it is too small, the probability of cycling increases. On the other hand, if it is too large, there is the possibility that all moves from the current solution may be visited and the algorithm may be trapped, leading to deterioration in the quality of the solution. Hence, effort should be made to find suitable tabu tenure in between these two extremes.

There are two broad classes of tabu tenure: static and dynamic. In static tabu tenure, a decision is first made on weight priorities for tabu active net tenure. For example, in the case of the minimum-k tree problem already discussed, the impact of preventing an element from being dropped is quite small. However, the impact f preventing an element from being added has more weight. In this case, we define the tabu active net tenure $\mathrm{T}_{\mathrm{net}}$, as follows

$$\mathrm{T}_{\mathrm{net}} = \alpha^*(\text{dimension of problem})^{1/2}, \text{ where } 0.5 < \alpha < 2\,.$$

Let us apply this rule to the undirected graph of Fig. 6.3 where there are 21 edges and 12 nodes. Since the number of edge attributes is greater than node attributes, we choose the dimension of the problem to be 21. Choosing $\alpha = 0.5$, the tabu active net tenure $= trunc(0.5^*\sqrt{21}) = 2$. This explains why the tabu active net tenure for dropping and adding edges equal to 2 and 1 respectively in the minimum-k tree problem. In fact, this concept can be applied to other classes of problems other than undirected graphs. Allowing the tabu tenure to vary can increase the probability of finding a global optimum. There are two methods of implementing this situation: random dynamic tenure and systematic dynamic tenure.

In random dynamic tenure, the tabu tenure t is randomly selected following a uniform distribution between the lower and upper bounds $[t_{\min}, t_{\max}]$. The values of 5 and 10 for $t_{\min}$ and $t_{\max}$ respectively have been suggested by Glover and Laguna (1993) and is commonly used and quoted in the literature. However, it is possible to define these lower and upper bounds that are problem-specific. For example, the values of 0.9^*n and 1.1^*n for $t_{\min}$ and $t_{\max}$ respectively for an n-size quadratic assignment problem (QAP) is acceptable. For a problem size of 50, $t_{\min} = trunc(0.9^*50) = 45$ and $t_{\max} = trunc(1.1^*50) = 55$. In a scheduling-problem, it is possible to have $t_{\min} = 5$ and $t_{\max} = trunc[\alpha^*(n)^{1/2}]$, where $0.5 < \alpha < 2$. Hence, for a problem size of 50, with $\alpha = 1.5$, $t_{\max} = trunc(1.5^*\sqrt{50}) = 10$. In this case the lower and upper bounds of the tabu tenure are [5,10]. There are several variations in determining tabu tenure and there is no one rule for doing this.

In systemic dynamic tenure, a sequence of tabu tenure is generated in the range defined by $t_{\min}$ and $t_{\max}$. Using the example of the tabu tenure bounds for scheduling already discussed, the following sequence can be defined {6, 10,

5, 8, 7, and 9}. This sequence is then used to assign the current tabu tenure value. This kind of definition of sequence for assigning tabu tenure is very flexible and in some cases is preferred to the uniform distribution approach between lower and upper bound values.

6.4.3. *Aspiration criteria*

Since the tabu list contains only the modifications made to a solution state, aspiration criteria are used to provide an added flexibility in overruling the tabu classification or status or a move. A basic aspiration criterion often used is to override the tabu status of a move, if it gives an objective value better than the best obtained so far.

Another aspiration criterion is to override the tabu status of a move, if it results in a better objective value than the best obtained when the move was made tabu. In this case, an aspiration criterion is used for each of the tabu moves and is updated simultaneously with the tabu list.

While the *best-so-far* criterion is global, an aspiration criterion may be regional. In such a case, the aspiration criterion is used to override the tabu status of a move, if it gives an objective value better than the best found in the region that contains the solution.

So far, we have considered aspiration criteria that depend on moves. There are aspiration criteria that depend on attributes. One such aspiration by attribute criteria is used to override the tabu status if the direction of search, whether improving or deteriorating, has not changed.

Another aspiration by move and metric is the influence-based aspiration. In this case, the aspiration criterion overrides the tabu status if a high influence move has been performed since the last time low influence move has had its tabu status established. Influence-based aspiration criteria make use of moves, or ratio of *move-value* to *move metric.* By metric, we mean distances between the elements that have moved. For example, in a 6-job schedule having sequence {3, 4, 1, 5, 6, and 2}, swapping jobs 4 and 6 incurs a move metric equal to 3. Dividing the move value of this move by this move metric gives us a clue whether the influence of the move is low or high. When the solutions in a region are good, we go for low influence moves. However, if no improvement is recorded for quite some time, high influence moves have the effect of pushing the search out of local optimum into a "green land" for further exploration.

6.4.4. *Basic ideas for implementing recency-based memory*

The implementation-strategies of recency-based memory is based on the premises of the metric depicted in Fig. 6.6. A generic representation of the memory structure has the following representations:

$$Ite \leq TabuStart(\bullet) + TabuTenure(\bullet) \qquad \text{(Premise 1)}$$
$$TabuEnd(\bullet) = TabuStart(\bullet) + TabuTenure(\bullet) \qquad \text{(Premise 2)}$$
$$NetTenure = TabuStart(\bullet) + TabuTenure(\bullet) - Ite \qquad \text{(Premise 3)}\,.$$

Glover and Laguna (1997) give four classifications based on these three premises. Premises 1 and 2 directly follow from Fig. 6.6 and Ite is the current iteration.

To see how this memory structure works, let us refer to Table 6.2. During the first iteration, edge (3,7) is dropped while edge (3,5) is added. This implies $TabuStart(3,7) = 1$. Since dropping edge (3,7) has $TabuTenure = 2$, this means that by applying Premise 1, edge (3,7) remains tabu-active so long as $Ite \leq (1+2)$ or $Ite \leq 3$.

Similarly, $TabuStart(3,7) = 1$ and edge (3,5) has $TabuTenure = 1$, (to prevent it being dropped), leading to this edge remaining tabu-active so long as $Ite \leq (1+1)$ or $Ite \leq 2$.

The reader can verify that this is the case from Table 6.2. We see that at iteration 1, edge (3,5) remains tabu-active until iteration 3, while edge (3,7) remains tabu-active until iteration 2. The concept is simple to follow.

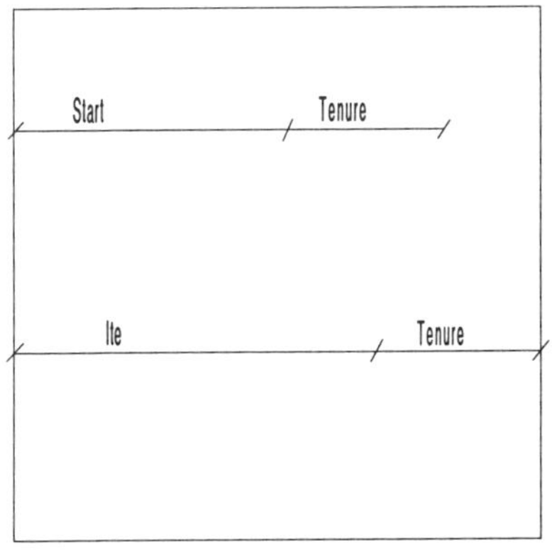

Fig. 6.6. Metric for tabu classifications.

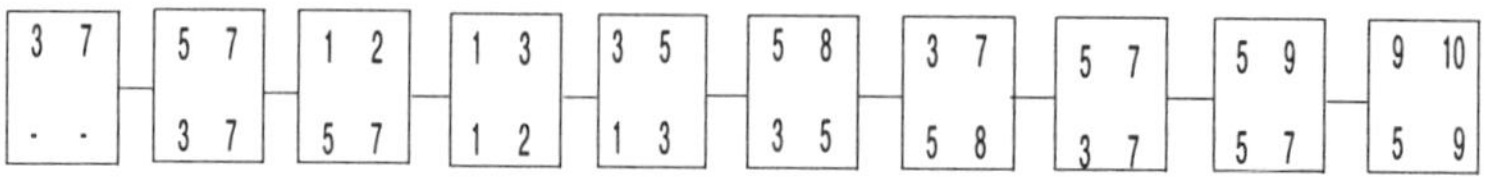

Fig. 6.7. Tabu-active status of drop moves.

But how is the short-term memory implemented in practical sense? Let us revisit Table 6.2 that contains tabu active status of each move. Two arrays $TabuDrop(*,*)$ and $TabuAdd(*,*)$ are created. We use the $TabuDrop(*,*)$ array for illustration. In iterations 1 and 2, edge (3,7) and edge (5,7) are dropped respectively and they enter the $TabuDrop(*,*)$ stack using the principle of first-come-first-served (FCFS) as shown in Fig. 6.7. In iteration 3, edge (1,2) is dropped and it enters the stack. Meanwhile, since the tabu tenure = 2, edge (3,7) is popped off and it is no longer tabu-active. In iteration 4, edge (1,3) is dropped and it is pushed into the stack while edge (5,7) pops off. Edge (3,7) re-enters the stack at iteration 7. In fact, it is free to re-enter the stack from iteration 3 if a tabu move favours the dropping of this edge.

Much information can be gleaned for the recency-based memory by considering Figs. 6.7 and 6.5 together. Let us start our observation from iteration 2. We notice that edge (3,7) is at the bottom of the stack and there is no edge at the bottom of the preceding stack. In iteration 2, edge (3,7) has tabu-active status of 2 (see Fig. 6.5). Now let us move to iteration 3. Edge (5,7) is at the bottom of the stack while edge (3,7) is at the bottom of the stack for the preceding iteration. Hence, edge (5,7) has tabu-active status of 2^* while edge (3,7) gas tabu-active status of 1^* for this iteration (again see Fig. 6.5). Let us try one move, this time iteration 10. During this iteration, Fig. 6.7 shows that edge (5,9) is at the bottom of the stack while edge (5,7) is at the bottom of the preceding stack. This mean that edge (5,9) has tabu-active status of 2^* while edge (5,7) has tabu-active status of 1^*. For each iteration the edge on top of the stack is the edge currently dropped and carries no tabu-active status value. Figure 6.5 confirms our finding. What we have not done is to consider the add moves. These are shown in Fig. 6.8, with the tabu tenure equals 1.

Let us examine iteration 2. Edge (5,8) is added and represents an arrow in Fig. 6.5. We consider the preceding edge (3,5) which should have a tabu-active status equal to 1^*. Let us consider iteration 10. Edge (6,9) is added and so has a thick line while edge (11,12) of the preceding iteration has a tabu-active status of 1^*.

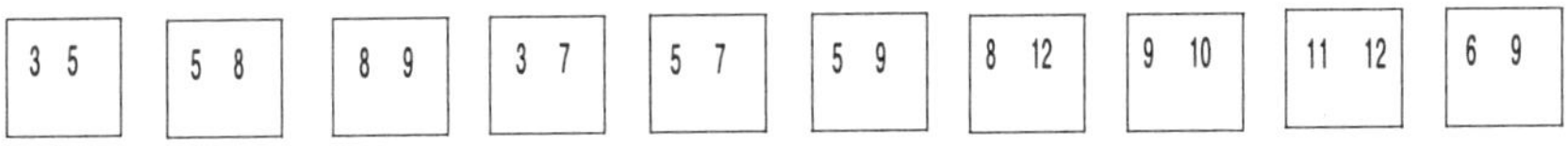

Fig. 6.8. Tabu-active status of add moves.

Hence, we can use Figs. 6.7 and 6.8 to fully describe the tabu-active status of the moves for each iteration and consequently construct Fig. 6.5. The practical implementation is based on Figs. 6.7 and 6.8 when developing software for frequency-based memory for tabu search. The data structure of Figs. 6.7 and 6.8 is commonly referred to as being circular. The principles presented in this section are crucial for the implementation of recency-based memory for tabu search. The reader should have a good grasp of this section before proceeding to the next section on long-term memory.

6.4.5. *Candidate list strategies*

The discussion so far on tabu search alludes to the fact that tabu search approach attaches much importance to selecting elements judiciously rather than randomly. The implication of this, is that efficient rules for generating and evaluating good candidates are crucially important for the success of the search. When the solution space is complex as is often the case in real practical situations, the reduced neighbourhood, $N(s,k)$ is also large and it becomes necessary to restrict the number of solutions being examined at any iteration. The goal of candidate list strategies is to isolate regions of the neighbourhood that contain moves that have favourable features and then temporarily save these moves on a candidate list so that they can be examined during the current iteration. Five well-known candidate list strategies are available: aspiration plus, elite candidate list, successive filter strategy, sequential fan, and bounded change candidate lists. The first two are often used and are therefore, described here.

Aspiration plus strategy

Based on the search history of the search pattern, the aspiration plus strategy establishes a threshold for the quality of a move. It should be emphasized that tabu search relies on the search history of the search pattern because there is no present without the past. This is a useful characteristic of intelligent optimization techniques. They remember the past history and use the information derived thereby to drive the search toward productive regions. In aspiration plus strategy, moves are examined until the procedure finds one that satisfies the threshold. The selected value "Plus" is useful for the implementation of this strategy, because additional moves are examined, equal in number to "Plus", and then the best move overall amongst those examined is selected.

To ensure that not too few or not too many moves are considered, the rule is qualified such that at least "min" moves and at most "max" moves are examined. The number of moves examined when the aspiration threshold is first satisfied is denoted by "First". The rule can be mathematically stated as follows:

$$If \begin{cases} first + \; plus < \min\,, \; then \text{ 'min' } moves\ are\ examined \\ first + \; plus > \max\,, \; then \text{ 'max' } moves\ are\ examined\,. \end{cases}$$

The values of min $= 7$ and $Plus = 5$ suggested by Glover and Laguna (1997) work well.

Elite candidate list

The elite candidate makes use of elite solutions. Elite solutions as the name suggests, are high quality local optima. Again, referring to the minimum-k tree information in Table 6.2, the weights are plotted as shown in Fig. 6.9. Clearly,

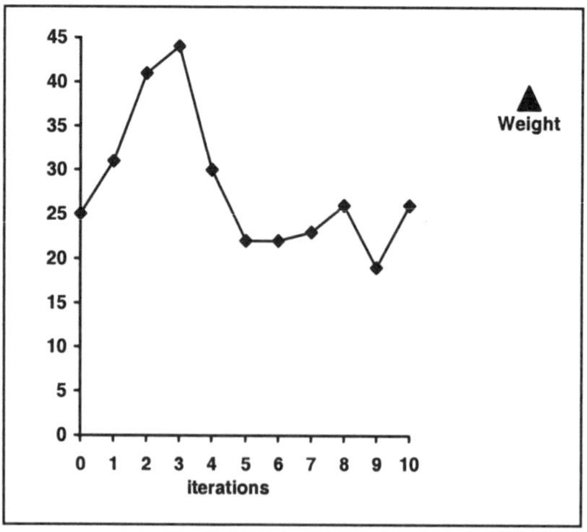

Fig. 6.9. Solution pattern for each iteration.

the solutions at iterations 5, 6, and 9 are good local optima. Elite candidate list first builds a "master list" by examining all or a relatively large number of moves, selecting the k best moves encountered, where k is a parameter of the search process. Then, at subsequent iteration, the current best move from the "master list" is chosen to be executed. This process continues until such a move falls below a given quality threshold, or until a given number of iterations have elapsed. Then, a new "master list" is constructed and the process is repeated. This way, the search process drives the search along the best search trajectory in the solution space.

6.5. Long Term Memory in Tabu Search

The short term memory (STM) incorporates a restricted portion of the tabu search design and is sufficient to produce very high quality solutions in some applications. STM tabu search approaches are frequently quite easy to implement, and consequently continue to appear very often in the literature. The motivation for its use in some applications is not only because of ease of implementation, but also due to the fact the STM tabu search approaches are some times surprisingly successful in solving combinatorial optimization problems. Because STM-based tabu search can be strongly dominated by long term memory (LTM)-based memory, it has become necessary for users to be aware of the broader framework and strategies of the latter.

In the LTM-based tabu search strategies, the modified neighbourhood produced may contain solutions that differ in various significant ways from those seen before. This memory structure retains and integrates elite solutions encountered during the search process in order to follow a more favoured trajectory. The LTM function guides the search process to regions that markedly contrast with those examined so far. While recency-based memory is the basis for STM, frequency-based memory is the foundation on which LTM is laid. The five components of LTM are: frequency-based memory, logical restructuring, intensification, diversification, and strategic oscillation. These features are now discussed in the following sub-sections.

6.5.1. *Frequency-based memory*

Frequency connotes the idea of how often an event has taken place. This event, in the context of tabu search may be *transitional* or *residential*. Consequently, in the LTM discussion on frequency-base memory, two metrics are identified: *transition measure* and *residence measure*. It is important to clarify the difference between these two measures.

Transition measure is the frequency of an event "entering" or "leaving" solutions during the search process. This can be very easily understood using the solution of Fig. 6.5 for the minimum-k tree problem. For example, how many once is edge (3,7) dropped in the solution? The answer is that it is dropped only once in iterations 1 and 7. What of edge (3,5), how many times is it added in the solution? It is added once in iteration 1.

Although these are quite useful information, it will be useful to find out how many times an event is part of the solutions. This brings us to the second metric, residence measure. The residence measure gives useful clue in terms of the frequency of an event in the solution. Of course, if an event occurs very frequently in the solutions, then that particular event is highly attractive or promising, especially if the domain consists of elite solutions. Again, let us refer to the minimum-k tree problem where the solutions are graphically shown in Fig. 6.5. Let us determine the number of times edge (3,5) is part of the solutions. The answer is four times. How do we know that it is 4 times? This can be obtained by noting that edge (3,5) is part of the solutions at iterations 1, 2, 3 and 4. How many times is edge (5,7) part of the solutions? The answer is: four times. How do we know that is 4 times? This occurs at iterations 1, 5, 6, and 7. Examining the solution very closely reveals that edge (8,9) is highly attractive. It is highly attractive because it is part of the solution 8 times. This occurs in iterations 3, 4, 5, 6, 7, 8, 9, and 10. In fact, edge (8,9) is the most attractive. Notice that edge (8,9) is introduced in iteration 3. It is important to observe that the solutions from iteration 4 to 10 contain the elite solutions. Referring to Fig. 6.9, it is observed that these elite solutions occur at iterations 5, 6, and 9. What deductions can we make concerning the most attractive event? One obvious one is that it consistently remains part of the best solutions found so far. Again, it helps drive the search process toward the best solutions. It is important to identify these so-called attractive events or elements and exploit them in the design of our search strategies.

Having treated both the transition and residence measures as mere counts, frequency-based memory would normally use ratios as frequencies. These counts are usually divided by the total number of iterations associated with the measure, or some absolute values related to the counts of the numerator, in order to translate the counts made in transition or residence measure into ratio type or frequency.

Frequency-based memory is usually implemented by defining penalty and incentive values to modify the evaluation of moves. Following an approach suggested by Taillard (1992) and Gendreau *et al.* (1993), if a move from s to s' reduces the value of the objective function such that $(f(s') < f(s))$, then the move is not penalized. However, if the cost of the move is non negative (i.e. $f(s') \geq (f(s))$, then the cost of the move is increased by the product of the following four factors:

(i) $\Delta_{\max}$, the maximum absolute difference in the objective function between two successive iterations;
(ii) N, which is approximately proportional to the square root of the size of the neighbourhood;
(iii) *Div-factor*, a scaling factor which is judiciously chosen, as too low a value results in cycling while too high a value leads to deteriorating solution;
(iv) *freq*, which is defined as $freq = \max(n_a, n_b)/n$, where n is the iteration number, n_a is the number of times, an element a has been moved, n_b is the number of times an element b has been moved. Here, a refers to the first element in the section being evaluated for a move, and b refers to the final element in the section being evaluated for a move.

Although the LTM function is problem-dependent and can be implemented in a variety of ways, two different implementations are commonly used: the penalized frequency method and the constrained frequency method. Both methods use an integral array *freq* of size n as has been discussed, which keeps a count of the number of times an element has been "added to" and "removed from" the solutions over the history of the search. The definition of *freq* of course, is the transition measure. When defined as a residence measure, *freq* keeps a count of the number of times an element has being part of the solutions over the history of the search.

The penalized frequency method is designed to create a bias against moves that incorporate high transition measures. This is done by calculating an adjusted weight coefficient, P_i, for each element by adding a penalty to its weights, so that

$$P_i = w_i + \mu(freq_i)\,,$$

where μ is a constant penalty parameter set at value based on the initial trial runs. The best non-improving move based on the adjusted weight P_i, is chosen. The constrained frequency method restricts consideration to only those elements $i \in F$ to be removed from the solution, such that

$$F = \{i : freq_i/n\_iterations < critical\_freq\}\,.$$

It must be indicated that the use of LTM does not imply long solution runs. After all, the whole idea of LTM is to guide the search process beyond regions where solutions are not improving within some finite time. This helps in driving the search very quickly toward the global solution. In fact, some of the known fastest methods of solving some combinatorial optimization problems are based on LTM tabu search approach. Having said that, it is not always true that LTM requires shorter time than STM during the search process. Where very high quality solution is required, of course, the price in terms of time has to be paid.

6.5.2. *Logical restructuring*

Logical restructuring combines quality and influence to drive the search process toward improved solutions. It anticipates an improved solution and seeks a move that creates this condition. The idea of anticipation is important in intelligent optimization techniques. As the search progresses, a history of promising solutions are kept in memory so that they can be exploited to anticipate improved solution. To illustrate the concept of logical restructuring, consider undirected graph shown in Fig. 6.10. Again, we use a greedy method to provide an initial solution as shown in Fig. 6.11. The moves are obtained by considering the total neighbourhood moves of the initial solution and then

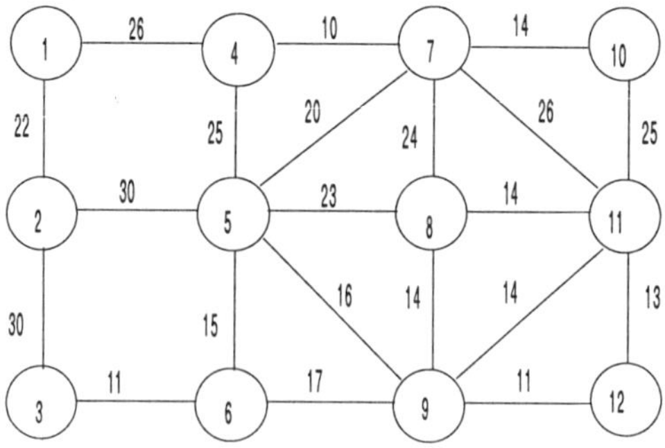

Fig. 6.10. Undirected graph.

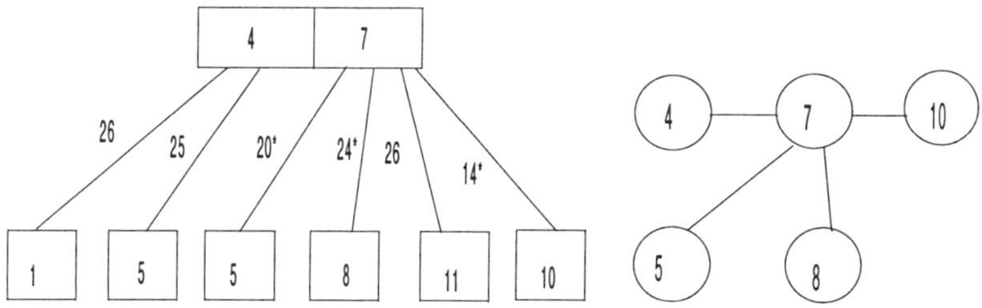

(a) (b) Cost =68

Fig. 6.11. Initial solution for undirected graph of Fig. 6.10.

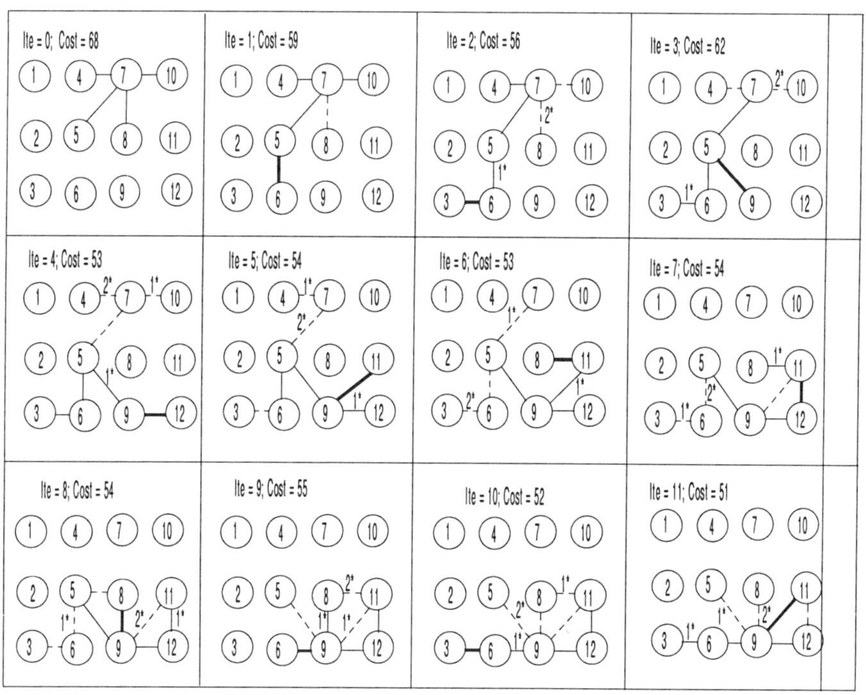

Fig. 6.12. Complete iterations for Fig. 6.10.

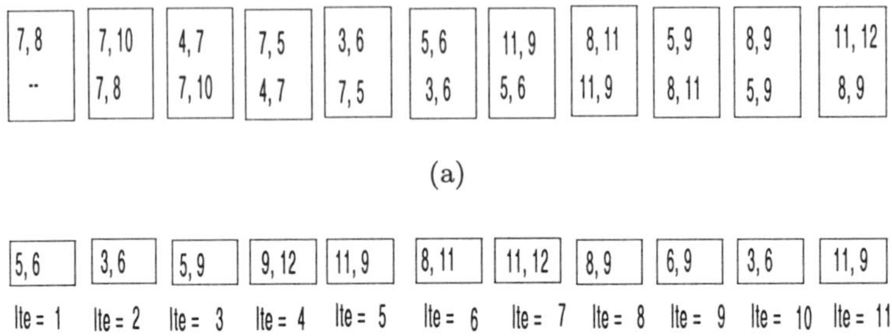

Fig. 6.13. Tabu-active status of (a) drop move (b) add moves.

Table 6.3. Tabu-active tenure, move, move value and weight.

Iteration	Tabu-active net tenure 2	1	Drop	Add	Move value	Weight
0						68
1			(7,8)	(5,6)	−9	59
2	(7,8)	(5,6)	(7,10)	(3,6)	−3	56*
3	(7,10)	(3,6), (7,8)	(4,7)	(5,9)	6	62
4	(4,7)	(5,9), (7,10)	(7,5)	(9,12)	−9	53*
5	(7,5)	(9,12), (4,7)	(3,6)	(11,9)	1	54
6	(3,6)	(11,9), (7,5)	(5,6)	(8,11)	−1	53*
7	(5,6)	(8,11), (3,6)	(11,9)	(11,12)	1	54
8	(11,9)	(11,12), (5,6)	(8,11)	(8,9)	0	54
9	(8,11)	(8,9), (11,9)	(5,9)	(6,9)	1	55
10	(5,9)	(6,9), (8,11)	(8,9)	(3,6)	−3	52
11	(8,9)	(3,6), (5,9)	(11,12)	(11,9)	−1	51*

choosing the best (which gives the least cost). This move corresponds to "add" edge (5,6) and "drop" edge (7,8) with a cost of 59. The complete iterations are shown in Fig. 6.12. We have used asterisks to denote the tabu-active status. The tabu-active status progressions are shown in Fig. 6.13. Then, as we did previously, the tabu-active status progression can be used to build the tabu-active net tenure shown in Table 6.3.

Let us glean some useful information from Fig. 6.12 required to understand the logical restructuring strategy. The information so far in Fig. 6.12 and Table 6.3 are for STM tabu search. We now want to apply the logical restructuring strategy in order to move toward the LTM tabu search. We start by identifying edges that have high frequency of occurrence in the solutions. Edge (3,6) occurs 5 times, being part of the solutions in iterations 2, 3, 4, 10, and 11. Edge (5,6) occurs 5 times and is part of the solutions in iterations 1, 2, 3, 4, 5. Edge (5,9) occurs 6 times, being part of the solutions in iterations 3, 4, 5, 6, 7, and 8. Edge (9,12) occurs 8 times, being part of the solutions in iterations 4, 5, 6, 7, 8, 9, 10, and 11. These are useful information. The first observation is that edge (9,12) and edge (3,6) are members of the best solution at iteration 11. In fact, there are other elite-solutions at iterations 2, 4, and 6, marked with asterisks in Table 6.3. In iteration 2, edge (3,6) and edge (5,6) are members of the solution at this elite solution. In iteration 4, edges (3,6) and (9,12) are members of the elite solution. In iteration 6, edges (5,9) and (9,12) are members of the elite solution. There is much to gain by gleaning these particular attractive edges: (3,6), (5,6), (5,9), and (9,12) and exploiting their goodness in anticipating productive moves during the search process. There are two rules to apply to these particularly attractive edges:

*Rule*1: identify the two best edges to add in a dynamic swap, subject to these edges meeting different nodes of the tree.

*Rule*2: create a modified anticipatory move value for each candidate static swap that results in a terminal edge; subtract the weight of this terminal edge from the move value.

Applying these rules to the solution in iteration 5, two edges can be added from node 8 to nodes 9 and 11, with weights of 14 each. Any of these can be added, in favour of dropping the terminal edge (5,6) which has weight of 15. This is a dynamic swap and corresponds to Rule 2. The resulting solution is now that of iteration 6. By using the search history to glean attractive edges, it has become possible to explore the current solution without the need for the usually exhaustive neighbourhood search. This simple example illustrates the logical restructuring strategy, which is very useful in LTM tabu search implementation. The next two sections consider the intensification and diversification strategies that add strength to LTM.

6.5.3. *Intensification strategies*

Intensification is one of the important components of tabu search. In the framework of intelligent optimization techniques discussed in Chapter 1, intensification and diversification are positioned at the tactical level. Intensification strategies are based on modifying choice rules in order to favour move combinations and elite solutions. In our previous discussions in this chapter, we have noted that elite solutions are solution features that have been historically found good. Intensification strategies may also return to attractive or green regions to search them more thoroughly. This concept agrees with natural phenomenon. Take for example, a company seeking or exploring the existence of natural oil in an area. The geology department commences work by carrying out a seismic survey of the area. The results may show traces of mineral deposits here and there in the geographical region. From the initial results, attractive regions are identified and explored further for more deposits. This is essentially what intensification strategies do. They initiate a return to attractive regions to search them more thoroughly. Keeping track of elite solutions, intensification strategies require that explicit memory structure be put in place. Explicit memory records complete solutions, typically consisting of elite solutions visited during the search. These memorized elite solutions are used to expand the local search to *greener* regions. During intensification, the neighbours of elite solutions are examined more thoroughly. Two things happen when the neighbours of elite solutions are examined: (1) neighbours to elite solutions are considered using standard move mechanism, (2) neighbours are generated by using modified evaluation or choice strategies that favour the introduction of such components into a current solution. Two approaches for implementing intensification exist (Glover and Laguna, 1997) that are commonly used by researchers.

The first approach makes use of diversification measure to ensure that the solutions recorded are quite distinct from each other by a desired degree. Once this is done, all STM are erased before commencing from the best of the recorded solutions. Since the recorded solutions are the elite solutions, and since the best of these is chosen for re-commencing search after all STM are erased, it becomes evident that this strategy gleans or focuses on examining neighbours of the winner amongst the elite solutions. This is a very powerful strategy in fine tuning the search around the best solutions.

The second approach is dependent on keeping a bounded length sequential list that is updated with a new solution at the end of the list if the new solution is the best so far. As already explained, explicit memory structure is used to keep track of the elite solution. In this strategy, the last member of the list, which is the best so far is always the candidate for resuming search. The explicit memory then enforces the tabu active net tenure, which then forbids the move of the best solution so far. Consequently, a new solution trajectory is triggered. Since the new solution trajectory is found which is in the neighbourhood of the best solution so far, the strategy achieves the goal of focusing around the best elite solution.

6.5.4. *Diversification strategies*

Diversification is another one of the important components of tabu search. In the framework of intelligent optimization techniques discussed in Chapter 1, intensification and diversification are positioned at the tactical level, thus, showing their importance. Diversification favours the search process to examine unvisited regions in the solution space and generates solutions that differ in various significant ways from those already obtained. There are several approaches for implementing diversification in tabu search. Many of the tabu search implementations in the literature incorporate simple and sometime naïve designs of diversification strategies in an attempt to find quick solutions. In this section, we present two well-defined approaches that tabu search researchers and practitioners are encouraged to include in their diversification strategies. These include restarting, and modifying choice rules.

Restarting

The simplest restarting method is one that is done randomly. This is done either after some number of iterations or when the solution is not getting better. This is not the kind of restarting that is proposed in this section.

The restarting mechanism that is useful in tabu search design depends on frequency information derived from the frequency-based memory. We devoted quite a substantial amount of time this concept in Sec. 4.5.1 of this chapter and need not revisit the details here. By now, the reader should be quite familiar with the frequency information needed for our restarting mechanism based on frequency-based memory.

Let us use the NEH constructive method discussed in Chapter 3 for sequencing flow shop as an example. The first and second steps of NEH are all that we require for designing a restarting mechanism. For completeness, let us recapitulate the operation time matrix to avoid flipping back to Chapter 3.

Operation time matrix					
	Machines (m)				
Jobs (n)	1	2	3	4	5
1	5	9	8	10	1
2	9	3	10	1	8
3	9	4	5	8	6
4	4	8	8	7	2

The first and second steps of NEH assign the highest priority to the job with the highest total processing time:

$$T_1 = 5 + 9 + 8 + 10 + 1 = 33$$

$$T_2 = 9 + 3 + 10 + 1 + 8 = 31$$

$$T_3 = 9 + 4 + 5 + 8 + 6 = 32$$

$$T_4 = 4 + 8 + 8 + 7 + 2 = 29\,.$$

Since the total processing times are {33, 31, 32, 29}, then the initial sequence is {1, 3, 2, 4}. A restarting sequence can be obtained by using frequency information from the frequency-based memory to modify the priority index of each job. Here, the priority indices are simply given by the total processing times, which are then biased using an appropriate frequency measure represented as:

$$priority\_index' = priority\_index + \alpha^* frequency\,.$$

As an example, let the frequency measures for jobs {1, 3, 2, 4} be {0.02, 0.30, 0.15, 0.50} and $\alpha = 15$. Then the priority indices, $priority\_index'$ are, for:

$$\text{job } 1, = 33 + 15^*0.02 = 33.30$$

$$\text{job } 2, = 31 + 15^*0.30 = 35.50$$

$$\text{job } 3, = 32 + 15^{*}0.15 = 34.25$$

$$\text{job } 4, = 29 + 15^{*}0.50 = 36.50\,.$$

The current priority indices for the jobs are now {33.30, 34.25, 35.50, 36.50}. From these, the restarting sequence is {4, 2, 3, 1}.

Inclusion of this restarting mechanism in a tabu search design is very straightforward and is useful in LTM structure. Now, priority indices can always be found as long as information on frequency is available. Since the frequency information is dependent on history, this mechanism of restarting described here is therefore, not random. It depends on the history of the search process.

Modifying choice rules

Similar to the priority index method for restarting, modified choice rules can be incorporated in tabu search design to bring attributes into the solution that are infrequently used. This is achieved through a penalty value that is a function of frequency measures. For example, in arc routing problems, arcs which have been moved frequently are discouraged from being accepted to move again by adding to the evaluation of the move, a factor proportional to the frequency of movement. A generic penalty function is given as follows:

$$Move\_value' = Move\_value + \beta^{*} Penalty\,.$$

The value β, is an adjustable diversification factor that penalizes or induces moves. For example, large positive β values will penalize an element (arc, job, etc.) from making a move, and hence corresponds to a desire for more diversification. On the other hand, negative β values will induce an element (arc, job, etc.) of low frequency to make a move, and hence corresponds to a less desire for diversification.

6.5.5. *Strategic oscillation*

Strategic oscillation in tabu search provides a means to drive the search process toward and away from selected boundaries of feasibility. This is achieved either by manipulating the objective function through use of penalties and incentives or by compelling the choice of moves that direct the search trajectory to specified directions (Glover and Laguna, 1997a).

Let f be the objective to be optimized. For a given s, feasible or not, let $f(s)$ be the value for solution s of the objective. The problem to be solved is

$$(P)\begin{cases} f^* = \text{min}imum f(s) \leq \bar{f} \\ s.t.\ s \in S \end{cases}$$

where $\bar{f}$ is a threshold on the value of the objective f, computed for example as the maximum accepted deterioration from the optimal value f^* of problem (P). The threshold constraint may be taken as soft constraint by means of penalty incorporated into the objective function. Let $\mu \geq 0$ be the penalty coefficient associated with the objective. The associated problem then becomes

$$(P_{penalty})\begin{cases} f^* = minimize\ \mu \bullet \max\{0, f(s) - \bar{f}\} \\ s.t.\ s \in S\,. \end{cases}$$

It is an easy matter to observe that if the penalty coefficient μ is taken sufficiently large, constraint $f(s) < \bar{f}$ will automatically be enforced. However, if the penalty coefficient μ is small, the optimal solution of $(P_{penalty})$ may violate the constraint associated with the corresponding term. The strategic oscillation scheme basically consists in periodically adjusting the penalty coefficients. Therefore, in applying the strategic oscillation strategy in the LTM tabu search design, the following penalty function is used:

$$f_{penalty}(s) = \mu \bullet \max\{0, f(s) - \bar{f}\}\,.$$

In every iteration, a new solution, $s' \in N(s)$ is obtained, where s, is the current solution. If $f_{penalty}(s')$ is less than or equal to the best value already found for the penalty function, then the best which may be eventually unrealizable or not feasible solution s^* is updated. This means we set $s^* = s'$. If the new solution s' is feasible, that is, if it satisfies the threshold constraint $f(s) < \bar{f}$, and it improves the value of the best feasible solution $s^*_{feasible}$ found so far, then the best feasible solution is updated. This means that we set $s^*_{feasible} = s'$. When all solutions found during a series of a certain number of iterations (say, 10 iterations) of the tabu search violate constraint $f(s) < \bar{f}$, the penalty coefficient μ is doubled. On the other hand, if the associated constraint is not violated by any solution during this series of iterations, the penalty coefficient μ is halved. Research has shown that, with this type of rule, the tabu search process may quickly reach appropriate values for the penalty coefficient μ that produce a mix of feasible and unrealizable solutions. This wave of movement

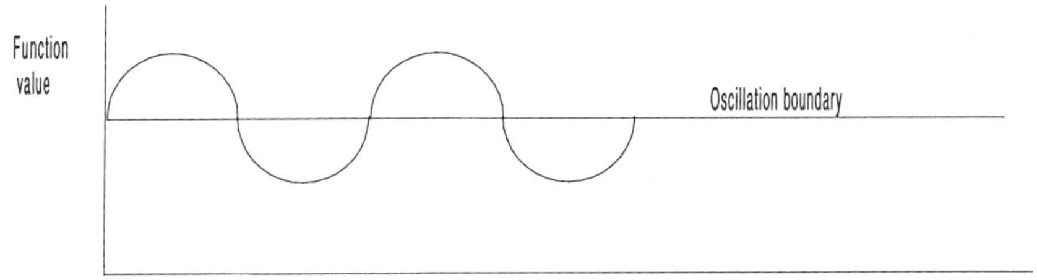

Fig. 6.14. Strategic oscillation pattern.

to a specified depth beyond the oscillation boundary, and turning around, gives an oscillatory pattern shown in Fig. 6.14.

6.5.6. *Basic ideas for implementing long term memory*

The basic ingredients required for implementing LTM tabu search have been discussed in this chapter: frequency-based memory, logical restructuring, intensification, diversification, and strategic oscillation. It is important that tabu search researchers and practitioners incorporate these strategies in their applications in order to benefit from the great potentials inherent in long term memory design. It has already been mentioned that including these strategies into a tabu search design, and realizing a long term memory does not take much more time than short term memory that is often hastily adopted by many users. As you can see, the strategies discussed in this chapter enable tabu search to qualify without any doubt as an intelligent optimization technique.

To wrap up the discussions on LTM features, we now discuss the basic implementation issues that tabu search researchers and practitioners should be familiar with. The brain behind LTM is frequency-based memory. Once this is in place, then other strategies that depend on frequency follow as presented in this chapter. Recall that there are two aspects of frequency-based memory: transition and residence.

Transition frequency-based memory is easy to implement. We simply record the number of times an element changes its *move status*. For the minimum-k tree problem, or any other tree problem, the move status may refer to keeping track of when an edge moves from being dropped to when it is being added. A transition memory, $Transition(i)$ can then be used to record increments at each move where such a change occurs. The idea is easily applied to other areas. For example, in job sequencing problem, the move status refers to the iteration where a job that "enters" a particular position changes to "be moved" from that particular position. In order word, at what iteration does a job change its "in" status to "out" status?

Residence frequency-based memory is more involving than transition memory, because it takes into account the information of the frequency-based memory stored in $TabuStart(i)$. The *move start* for transition memory applies to residence memory but a different rule is adopted:

$$Residence(i) = Residence(i) + Iter - TabuStart(i)\,.$$

We start by setting $Residence(i) = 0$ for all values of i. Whenever there is change in move status as earlier described, after updating the number of iterations $Iter$, but before updating $TabuStart(i)$, we then apply the residence rule.

To clarify issues, let us consider Fig. 6.12 for the minimum-k tree problem. We want to trace the residence information for edge (5,9). It is first added to the solutions in iteration 3 so the $TabuStart = 3$. The edge is cut in iteration 9. Applying the residence rule, residence $= 0 + 9 - 3 = 6$. As edge (5,9) is no further added in the solution after iteration 6, we stop here. Checking the solutions, edge (5,9) is part of the solution in iterations 3, 4, 5, 6, 7, and 8. Counting these iterations shows that the total is 6, meaning that edge (5,9) is part of the solution in 6 iterations.

6.6. The Theory of Constraints Product Mix Problem

There are four approaches so far in the literature that have been used in solving the product mix problem: (1) labour-based management accounting, (2) linear programming, (3) theory of constraints (TOC) heuristic, and (4) revised theory of constraints algorithm. To this list, we add the fifth method, which is addressed in this chapter, tabu search. These five approaches are discussed and compared, leading to some useful conclusions in terms of the newcomer, tabu search approach.

6.6.1. *Labour-based management accounting*

Labour-based management accounting suggests that the use of margin analysis is the solution to the product mix problem. Margin analysis is based on marginal income, which is the difference between sales revenue and all variable costs including material, labour, and manufacturing overhead. The five steps involved in margin analysis are as follows:

(1) carry out a marginal analysis;
(2) determine revenue potential;
(3) carry our resource load;
(4) order products using product margin;
(5) determine the net profit or loss per period.

To illustrate the concept of labour-based management accounting procedure, let us use the information contained in Table 6.4, taken from Patterson (1992). Table 6.4 contains the processing times of five product types A, B, C, D, E in rows 1, 2, 3, 4, 5, respectively, and resources (machines) 1, 2, 3, 4 in columns 2, 3, 4, 5 respectively. In addition, the demand, sales, and material costs for the five products are shown in columns 6, 7, and 8 respectively. The available times or capacity for each resource, the required time are shown in rows 6 and 7 respectively. The labour and overhead cost is $50 per hour.

Table 6.4. Labour requirements, and availability, unit price, material cost and demand.

	[1]	[2]	[3]	[4]	[5]	[6]	[7]	[8]
			Resources					
	Product	1	2	3	4	Demand	Sales ($)	Materials ($)
1	A	15	0	20	6	120	90	30
2	B	10	7	10	15	100	91	28
3	C	15	7	13	10	100	101	35
4	D	0	23	15	6	50	146	80
5	E	0	60	3	4	30	224	84
6	Capacity	4800	2400	7200	4800			
7	Required	4300	4350	5540	3640			

Step 1: Marginal Analysis

The labour requirements represent the processing times of product types produced on resources. Product type A has a total labour requirement of

Table 6.5. Margin analysis.

[1] Product	[2] Sales price ($)	[3] Materials cost ($)	[4] Direct labour/unit $\sum$(processing times) (min.)	[5] Labour overhead/unit ($50* min./60)	[6] Product margin/unit [2] − ([3] + [5])
A	90	30	41	34.16	25.84
B	91	28	42	35.00	28.00
C	101	35	45	37.50	28.50
D	146	80	44	36.67	29.33
E	224	84	67	55.83	84.17

41 minutes representing the total processing times on all the resources. Similarly, product type B has a total labour requirement of 42 minutes. Product types C, D, and E have labour requirement values equal to 45, 44, and 67 minutes respectively as shown in Table 6.5. Since the labour and overhead cost is $50 per hour, the direct-labour (in minutes) are converted to labour/unit in column 4. For example, the labour/unit for product type A is 41*$50/60 or $34.16 per hour. The total cost to produce product type A equal to the material cost plus the labour and overhead costs. This total manufacturing cost = $30 + $34.16 = $64.16. The unit sales price for product type A = $90. The product margin for A = $90 − $64.16 = $25.84. Similarly, the product margin for B, C, D, and E are $28, $28.5, $29.33, and $84.17 respectively as shown in Fig. 6.5.

Using the product margin/unit of Table 6.5 results in the following product rankings:

(1) Product E
(2) Product D
(3) Product C
(4) Product B
(5) Product A.

The production constraints will limit the number of each item that the plant will be able to produce.

Step 2: Revenue Potential

Market constraints do exist since there is a maximum number of each product that can be sold. The revenue potential of the this plant under consideration, if all of the market demand could be met, is shown in Table 6.6. However, potential production constraints do exist since the number of available time (minutes) in each work centre is limited. Consequently, it might not be possible to realize the revenue potential shown in Table 6.6.

Step 3: Resource Load Analysis

The resource load analysis involves summing the product of the market demand and the processing times for all the product types produced by a resource. For

Table 6.6. Revenue potential.

Product	Market demand	Sales price ($)	Product revenue
A	120	90	10800
B	100	91	9100
C	100	101	10100
D	50	146	7300
E	30	224	6720
			44020

Table 6.7. Resource load analysis.

		Resources				
	Product	1	2	3	4	Demand
1	A	15	0	20	6	120
2	B	10	7	10	15	100
3	C	15	7	13	10	100
4	D	0	23	15	6	50
5	E	0	60	3	4	30
6	Capacity	4800	2400	7200	4800	
7	Required	4300	4350	5540	3640	
8	% Overload	89.58	181.25	76.94	75.83	

example, the required production time for resource−1 = 15*120 + 10*100 + 15*100 + 0*50 + 0*30 = 4300 minutes. Similarly the required production time for resources 2, 3, and 4 are 4350 minutes, 5540 minutes, and 3640 minutes respectively. These values are shown in Table 6.7. The percentage overload is given as:

$$\%\,\text{overload} = (\text{required production time available/available time})^{*}100\,.$$

The percentage overload for resource 1 = (4300/4800)*100 = 89.58%. The percentage overloads for resources 2, 3, and 4 are 181.25%, 76.94%, and 75.83% respectively. Therefore, it appears from the analysis that work centre 2 is the only production constraint or bottleneck.

Step 4: Ordering Products Using Product Margin

We consider products using the product margin results of sequence {E, D, C, B, A}. This sequence is combined with the time available in the bottleneck, work centre 2 to determine the product mix. This approach seems quite reasonable since our more profitable products are also our highest priced ones and commissions are usually paid on dollar sales.

The constraint resource at work centre 2 makes us to produce 30 of E, 25 of D, 120 of A, 3 of C, and none or 0 of B as shown in Table 6.8. There are 2400 minutes available for resource 2. Producing 30E will leave 2400 − 30*60 = 600 minutes. Producing 25D will leave 600 − 25*23 = 25 minutes. Producing 3C will leave 25 − 3*7 = 4 minutes.

Table 6.8. Ordering products using product margin and bottleneck data.

Product	Market demand (D_i)	Product quantity (Q_i)	Product margin (PM_i)	Resources 2 t_2	t_{used}	t_{left}
E	30	30	84.17	60	1800	600
D	50	25	29.33	23	575	25
C	100	3	28.50	7	14	21
B	100	0	28.00	7	0	4
A	120	120	25.84	0	0	4

Table 6.9. Net profit (loss) per week for management accounting.

Product	Product quantity (Q_i)	Sales price ($)	Material cost ($)	Operating cost ($)
E	30	224	84	
D	25	146	80	
C	3	101	33	
B	0	91	28	
A	120	90	30	
		$21473	$8225	$14000

Step 5: Net profit (loss) per week

The profit potential for the product mix {3E, 25D, 3C, 0B, 120A} is shown in Table 6.9. The net loss per week is defined as:

$$\begin{aligned}[\text{sales} - (\text{material cost} + \text{operating cost})] &= 21473 - (8225 + 14000)\\ &= -\$752\,.\end{aligned}$$

The result of this analysis is that the plant incurs a net loss of $752 per week by producing 3E, 25D, 3C, 0B, and 120A.

Let us approach the same problem using other methods: linear programming, theory of constraints heuristic, revised theory of constraints heuristic, and then find out if the plant will incur profit or loss. If the plant incurs loss of the same magnitude as obtained using the management accounting method, then it means that management accounting principle is quite valid in the industrial setting for the product mix problem and indeed for related problems. If on the other hand, the plant incurs loss using these other methods, then it means something is wrong with management accounting principle. So let us now continue our search for the preferred approach or approaches.

6.6.2. *Linear programming for the product mix problem*

The formulation of the product mix decision problem is given as follows.

$$\text{Maximise} \quad z = \sum_{j=1}^{n} t_j x_j$$

$$\text{Subject to} \quad \sum_{j=1}^{n} p_{ij} x_j \leq c_i \quad i = 1, 2, \ldots, m$$

$$0 \leq x_j \leq b_j \qquad j = 1, 2, \ldots, n$$

where t_j is given as the throughput of the product type j; $_{pij}$ denotes the processing time required for resource i to produce a type j product; x_j is the decision variable representing the quantity of the product type j; c_i is the capacity limit of resource i; b_j is the bound for x; m is the number of resources; n is the number of product mix types.

For the product mix problem at hand, the linear program (LP) formulation is:

$$\begin{aligned}
\text{maximize} \quad & 60A + 63B + 66C + 66D + 140E & \\
\text{s.t.} \quad & 15A + 10B + 15C + 0D + 0E & \leq 4800 \\
& 0A + 7B + 7C + 23D + 60E & \leq 2400 \\
& 20A + 10B + 13C + 15D + 3E & \leq 7200 \\
& 6A + 15B + 10C + 6D + 4E & \leq 4800\,.
\end{aligned}$$

Solving this problem using standard LP software such as *STORM* gives $120A$, $100B$, $100C$, $16E$. The profit (loss) per week is shown in Table 6.10. The net profit $=$ \$33584 – (\$11244 + \$14000) $=$ +\$8340 (gain). There is problem here! Management accounting approach gave a loss of \$752, but linear programming approach gives a gain of \$8340. What is the matter? Which approach is correct

Table 6.10. Net profit (loss) per week for LP.

Product	Product quantity (Q_i)	Sales price (\$)	Material cost (\$)	Operating cost (\$)
A	120	90	30	
B	100	91	28	
C	100	101	35	
D	0	146	80	
E	16	224	84	
		\$33584	\$11244	\$14000

and which one is incorrect? Let us not rush into some conclusion at this point. Let us be patient and try out another approach. We now follow the theory of constraints heuristic.

6.6.3. *Theory of constraints heuristic for the product mix problem*

Theory of constraints (Goldratt, 1984) is a management philosophy for production control built on the foundation that a system's outputs are determined by its constraint(s). Another foundation of TOC is the money-flow concept (Preiss *et al.*, 1996). The money-flow concept has new definitions for throughput, operating expenses, inventory and return on investment. A firm is in existence to make money. It either makes money or folds up. Throughput is the difference between sales and variable costs incurred in production. Inventory is the money spent in holding parts or products not yet sold. Operating expenses is defined as all the money the system spends in turning inventory into throughput. Return-on-investment is defined as throughput per inventory. Increasing throughput and decreasing inventory simultaneously contribute to the goal of making money because return-on-investment is increased.

One key idea of TOC is that the system's constraints, known as capacity constraint resources (CCRs) determine the system's throughput and should be the focus of management attention. Constraints are defined as the most limited resources that make a system to not achieve its goal. The TOC procedure that provides on-going improvement to the throughput of a system consists of the following steps:

(1) identify the system's constraints;
(2) decide how to exploit the system's constraints;
(3) subordinate everything else to the above decision;
(4) elevate the system's constraints;
(5) if in the previous steps, a constraint has been broken, go back to Step 1. Do not let inertia become the constraint.

The product mix decision problem is one important application of the TOC five steps that has appeared quite often in the literature. A small but growing literature exists on TOC and a good description of the philosophy is found in Fry *et al.* (1992). Miltenburg (1997) shows that the material requirements planning (MRP) system for planning and controlling production systems is being replaced more and more by just-in-time (JIT) and TOC, and gives a five-step technical procedure for embedding TOC with MRP. The product mix decision problem solved using TOC can be formulated as an LP model. See for example, Ronen and Starr (1990), and Patterson (1992). Luebbe and Finch (1992) compared TOC and LP and clarifies the differences surrounding the TOC philosophy and the LP technique. In their investigation, they concluded that TOC is an overall management philosophy, which utilizes a number of specific techniques, whereas LP is an optimization technique. Plenert (1993) found that when multiple constrained resources exist, LP is a much better planning tool than TOC. In their work, Hsu and Chung (1998) propose a dominance rule-based algorithm, which applies row operations using Gauessian elimination method for solving the load calculation equation for resolving the incomplete or implicit nature of the TOC approach. They claim that their method gives the same result as a dual-simplex method with bounded variables.

The above papers seem to focus on the fact that TOC approach has the shortcoming of not optimally solving the product mix problem. What they have not done in an attempt to make the TOC approach explicit, is to justify their approaches by considering practical real-life problems which are combinatorial in nature. Although optimal solution can be obtained for small problems such as illustrated in all the above papers, their methods, which tend to find optimal solution can not solve practical real-life problems typically encountered in manufacturing firms. This chapter addresses the problem by employing a tabu search approach for solving the TOC problem. We view TOC as an overall management philosophy, which utilizes a number of specific techniques. One such technique presented in this chapter, is tabu search. More on this, later. This approach ensures that no infeasible solution exists when a plant has multiple resource constraints and it is very robust in TOC problem solving.

Step 1: Identify the system's constraint(s)

The same approach as already presented is followed. The internal constraint is a capacity constraint in work centre 2 as well as market constraints for each of the products. However, TOC calculates the product mix margin, using throughput (according to Goldratt). The throughput at the bottleneck resource = (sales price − raw material price), which for product types A, B, C, and D are 60, 63, 66, and 140 respectively. This information is contained in Table 6.11.

Table 6.11. TOC data.

	[1]	[2]	[3]	[4]	[5]	[6]	[7]	[8]	[9]
			Resources						
	Product	1	2	3	4	Demand	Sales ($)	Material ($)	Contribution margin
1	A	15	0	20	6	120	90	30	60
2	B	10	7	10	15	100	91	28	63
3	C	15	7	13	10	100	101	35	66
4	D	0	23	15	6	50	146	80	66
5	E	0	60	3	4	30	224	84	140
6	Capacity	4800	2400	7200	4800				
7	Required	4300	4350	5540	3640				
8	Difference	500	−1950	1660	1160				
9	BNs	BN2	BN1	BN4	BN3				

Table 6.12. TOC analysis ($R_i = CM_i/t_{i,BN1}$).

Product	Demand (D)	Quantity product (Q)	R_i	Resources							
				2		1		4		3	
				Used	Left	Used	Left	Used	Left	Used	Left
A	120	120	∞	0	2400	1800	3000	720	4080	2400	4800
C	100	100	9.40	700	1700	1500	1500	1080	3080	1300	3500
B	100	100	9.00	700	1000	1000	500	1500	1580	1000	2500
D	50	43	2.87	989	11	0	500	258	1322	645	1855
E	30	0	2.33	–	11	0	500	–	1322	–	1855

Clearly, resource 2 is the bottleneck because it is most overloaded. So we use the processing times at the bottleneck together with contribution margins (CM_i) to order the products to be produced in the order of the non-increasing $CM_i/t_{i,2}$. This analysis is shown in Table 6.12.

The value of R for product type $A = (90 - 30)/0 = \infty$. The R-values of product types B, C, D, and E are $63/7 = 9$, $66/7 = 9.4$, $66/23 = 2.87$, 140/60=2.33 respectively. For resource 2, the available time is 2400 minutes. To produce product type A, resource 2 requires 120*0 $= 0$ minutes $2400 - 0 = 2400$ minutes are left. To produce product type C requires 100*7 $=$

Table 6.13. Net profit per week for TOC.

Product	Product quantity (Q)	Contribution margin (\$)	Operating cost (\$)
A	120	60	
B	100	63	
C	100	66	
D	43	66	
E	0	140	
		\$22938	\$14000

700 minutes, so the time left = 2400 − 700 = 1700 minutes. To produce product type B requires 100*7 = 700 minutes, so the time left = 1700 − 700 = 1000 minutes. To produce product type D requires 43*23 = 989 minutes, so the time left = 1000 − 989 = 11 minutes. To produce product type E requires 0*60 = 0 minutes, so the time left = 11 − 0 = 11 minutes. The information for resources 1, 4, and 3 are similarly calculated and shown in Table 6.12. The net profit = \$22938 − \$14000 = \$8938. The relevant information for calculating the net profit is shown in Table 6.13.

Now, we can confirm that the labour management accounting principle for the product mix problem should be discarded in favour of linear programming and the theory of constraint heuristic.

6.6.4. *More on theory of constraints*

Example 2 TOC where there is no multiple constraint

The process chart shown in Fig. 6.15 is used to further illustrate the TOC product mix problem. It involves two products R and S. Product R is manufactured by processing raw material 1 through Department A for 15 minutes, then through Department B in 20 minutes, resulting in Part #1. In addition, raw material 2 is processed 20 minutes in Department B, and 10 minutes in Department C, resulting in Part #2. Then both Part #1 and Part #2 are assembled in Department D for 10 minutes, to produce the final product R.

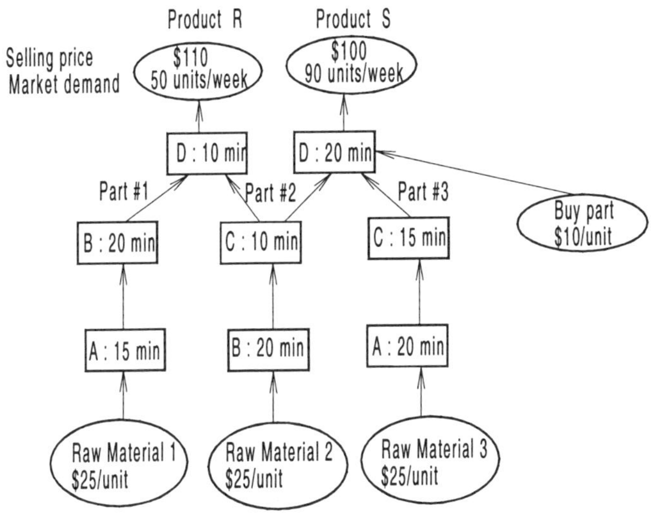

Fig. 6.15. Product mix information of a Theorey of Constraint problem.

Table 6.14. Resource loads for producing product types.

		Products
Resources	R	S
A	15	20
B	40	20
C	10	25
D	10	20
Demand	50	90
Selling price ($)	110	100
Material price ($)	50	60

The raw materials used to produce product R are raw materials 1 for $25 and raw materials 2 for $25. Similarly, product S is assembled from raw material 2 which produces Part #2, raw material 3 for $25, and a purchased part for $10, which produce Part #3. There are four work centres in this operation: A, B, C and D. The operating expenses for this operation are $4000. Each work centre has one machine, which can operate for 2400 minutes per week. The load on each resource for producing one unit of products R and S are shown in Table 6.14.

The linear program (LP) formulation for the product mix problem is:

$$\begin{aligned} \text{maximize} \quad & 60\text{R} + 40\text{S} \\ \text{s.t.} \quad & 15\text{R} + 20\text{S} \le 2400 \\ & 40\text{S} + 20\text{S} \le 2400 \\ & 10\text{R} + 25\text{S} \le 2400 \\ & 10\text{R} + 20\text{S} \le 2400 \\ & 0 \le \text{R} \le 50 \\ & 0 \le \text{S} \le 90\,. \end{aligned}$$

Solution

The first equation maximizes the profit that is the difference between the selling price and the materials cost for each product. The second equation up to the fifth equation, are the load calculation formula for each resource A–D, each of which has 2400 minutes capacity available. The sixth and seventh equations ensure that the number of resources, are within the market demand.

The following TOC steps are used to analyze the product mix problem for optimum profits.

Step 1: Identify the system's constraint(s)

The loads of the different resources are calculated. For resource A (see Table 1), the load generated by product $R = 50 \times 15 = 750$ minutes, and the load generated by product $S = 90 \times 20 = 1800$ minutes. For resource B, the load generated by product $R = 50 \times 4 = 2000$ minutes, and the load generated by product $S = 90 \times 20 = 1800$ minutes. For resource C, the load generated by product $R = 50 \times 10 = 500$ minutes, and the load generated by product $S = 90 \times 25 = 2250$ minutes. For resource D, the load generated by product $R = 50 \times 10 = 500$ minutes, and the load generated by product $S = 90 \times 20 =$ 1800 minutes. The load calculations for other resources are similarly computed. The loads and capacity-overloads for all the resources are shown in Table 6.15.

From Table 6.15, resources A, B and C are overloaded while resource D is underutilized. The most constrained resource (CCR) is B because it has the greatest overload.

Table 6.15. Load calculations and overloads for resources.

Resources	Load R	Contribution S	Total load (min)	Capacity limit (min)	Overload (min)
A	750	1800	2550	2400	150
B	2000	1800	3800	2400	1400
C	500	2250	2750	2400	350
D	500	1800	2300	2400	−100

Step 2: Decide how to exploit the system's constraint(s)

In order to decide how to make the best use of the constraint, we first calculate the throughput. It is defined as, throughput = selling price − materials costs.

For product R (see Table 6.14), the throughput = \$110 − (\$25 + \$25) = \$60. For product S, the throughput = \$100 − (\$25 + \$25 + \$10) = \$40.

Secondly, the throughput per unit of each product earned for the most critically constrained resource (CCR) which in this case is resource B is calculated. For product R, the throughput/unit earned = \$60/40 = \$1.5/minute. For product S, throughput/unit earned = \$40/20 = \$2.0/minute. Since product S offers the highest throughput per unit of the CCR (resource B), we then produce as much as possible of S, and then use the remaining available capacity to produce product R. Producing the whole of S requires capacity in resource B equal to $90 \times 20 = 1800$ minutes. The remaining capacity for producing R at resource $B = (2400 - 1800) = 600$ minutes.

The number of product R that can be produced using the available capacity at resource B is $600/40 = 15$ units. Therefore, the result of the product mix problem is that we have to produce 15 of R and 90 of S.

The net profit is obtained by subtracting the plant's operating expenses from the total throughput. Throughput for product R = 15 × \$60 = \$900. The throughput for product S = 90 × \$40 = \$3600. The total throughput for the plant = \$900 + \$3600 = \$4500. The plant's operating cost = \$4000. Therefore the total net profit = \$4500 − \$4000 = \$500.

Step 3: Subordinate everything else to the decisions made in Step 2.

Multiple constrained resources

Let us consider the TOC-problem for multiple constrained resources. An example is presented in order to deduce whether or not TOC is realizable.

Example 3 TOC-problem for multiple constrained resources.

The process chart shown in Fig. 6.16 is taken from Plenert (1993) to illustrate the TOC problem for multiple constrained resources. The load on each resource for producing one unit of products R, S, T and U are shown in Table 6.16. Carry out the TOC analysis.

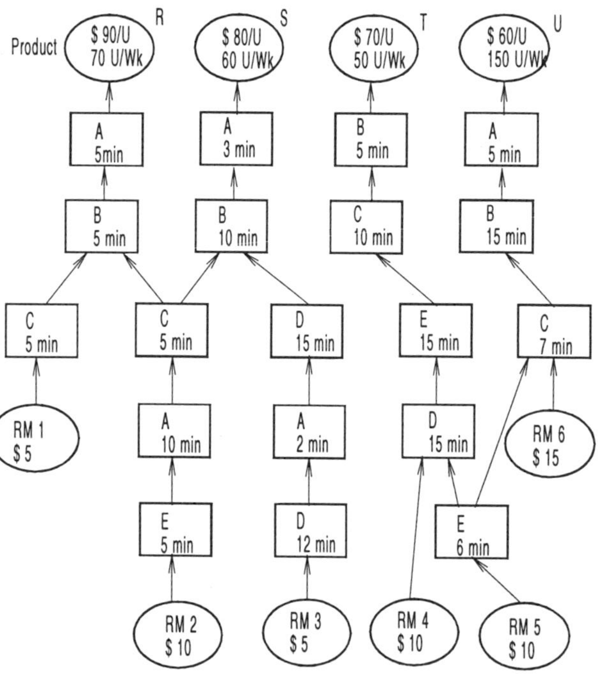

Fig. 6.16. Process chart data for product mix problem (Plenert 1993).

Solution

The TOC approach for solving the problem is as follows:

Step 1: Identify the system's constraint(s)

The loads generated by all products on resource $A = 70{*}15 + 60{*}16 + 50{*}0 + 150{*}5 = 2760$. The load generated by all products on resource $B = 70{*}5 +$

Table 6.16. Total load computations and overloads.

Resources	Products				Local Calculation				Total Load	Capacity	Overload
	R	S	T	U	R	S	T	U	(min)	limit (min)	(min)
A	15	16	0	5	1050	960	0	750	2760	2400	360
B	5	10	5	15	350	600	250	225	3450	2400	1050
C	10	5	10	7	700	300	500	0	2550	2400	150
D	0	28	15	0	0	1680	750	105	2430	2400	30
E	5	5	21	6	350	300	1050	0	2600	2400	200
							0				
							900				
Demand	70	60	50	150							
Selling price ($)	90	80	70	60							
Material price ($)	15	15	20	25							

60*10 + 50*5 + 150*15 = 3450. Table 3 shows the complete load computation values and overloads.

Resource A is overloaded by 360 minutes, resource B is overloaded by 1050 minutes, resource C is overloaded by 150 minutes, resource C is overloaded by 30 minutes, and resource D is overloaded by 200 minutes. Since resource B exceeds its constrained limit by the most, it is therefore the CCR.

Step 2: Decide how to exploit the system's constraint(s)

The throughput for product R (see Table 3) $= 90 - 15 = \$75$, for product $S = 80 - 15 = \$65$, for product $T = 70 - 20 = \$50$, and for product $U = 60 - 25 = \$35$.

The throughput per unit of production of production of the CCR (resource B) for product $R = \$75/5 = \$15/\text{min}$, for product $B = \$65/10 = \$6.5/\text{min}$, for product $T = \$50/5 = \$10/\text{min}$, and for $U = \$35/15 = \$2.3/\text{min}$. The order of production, using throughput per unit of production of the CCR is product R, T, S, and U.

The result of the product mix problem is obtained by assignment method using the knapsack heuristic:

Produce 70 of R, with resource capacity at CCR $= 350$ hours

Produce 50 of T, with resource capacity at CCR $= 350 + 250 = 600$ hours

Produce 60 of S, with resource capacity at CCR $= 600 + 600 = 1200$ hours

Produce $(2400 - 1200)/15 = 80$ of U, with resource capacity at

CCR $= 1200 + 80 \times 15 = 2400$ hours.

The total throughput $= 70^*75 + 50^*50 + 60^*65 + 80^*35 = \$14450$. The plant's operating expenses $= \$6000$. Therefore, the net profit $= \$14450 - \$6000 = \$8450$.

We now check if the TOC solution is feasible or not:

$$\text{Capacity for resource } A = 15^*70 + 16^*60 + 5^*80 = 2410 \text{ hours}$$

$$\text{Capacity for resource } B = 5^*70 + 10^*60 + 5^*50 + 15^*80 = 2400 \text{ hours}$$

$$\text{Capacity for resource } C = 10^*70 + 5^*60 + 10^*50 + 7^*80 = 2060 \text{ hours}$$

$$\text{Capacity for resource } D = 0^*70 + 28^*60 + 15^*50 + 0^*80 = 2430 \text{ hours}$$

$$\text{Capacity for resource } E = 5^*70 + 5^*60 + 21^*50 + 6^*80 = 2180 \text{ hours}.$$

The analysis shows that resources A and D have not been satisfied by the TOC solution. This means that TOC solution is unrealizable.

From the above examples and publications to date, it appears that although TOC approach is a useful management philosophy, the approach seems implicit when solving product mix decision problem when multiple constrained resources exist. Plenert's LP (1993) and Hsu and Chung's dominance-rule based method (1998) find feasible optimum solution for the product mix TOC problem. These methods are acceptable when small-scale problems are considered. But in real life, large-scale problems are encountered for which it is difficult if not impossible to find optimum solutions in acceptable computation time. This chapter addresses the problem by presenting a tabu search approach for solving large, practical problem instances, with very high quality results, which are optimum or very close to optimum in most cases. Onwubolu (2001) proposed the tabu search heuristic in making the TOC approach explicit for the product mix problem where multiple resource constraints exist.

6.6.5. *Revised theory of constraints algorithm for the product mix problem*

Fredendall and Lea (1997) proposed the revised theory of constraints algorithm for the product mix problem described in this section. The capacity resource planning heuristic schedules n products on m resources. Each product has market demand $D_i(i = 1, 2, \ldots, n)$ and a contribution margin (CM_i) calculated as the product's sales price (S_i) minus its raw material (RM_i) costs. Each resource has capacity $CP_j(j = 1, 2, \ldots, m)$. The heuristic is given here without proof of optimality:

Step 1: Identify the system's constraint(s);

(a) Calculate the difference (d_j) between a resource's capacity and demand on it;

$$d_j = CP_j - \sum_{i=1}^{n} t_{ij} D_i\,, \quad i = 1, 2, \ldots, n \quad \text{and } j = 1, 2, \ldots, m\,,$$

where t_{ij} is the processing time of product i on resource j.

(b) Determine $CR = \{BN_1, BN_2, \ldots, BN_q\}$, $q \le m$ where CR is the set of constrained resources with $d_q \le 0$ and $d_1 \le d_2 \le \cdots \le d_q$.

Step 2: Decide how to exploit the system's constraint(s):

(a) Assume that BN_1 is the dominant bottleneck resource and calculate the ratio of the CM to the product's processing time as $R_i = CM_i/t_{i,BN_1}$.

(b) Check that BN_1 is the dominant bottleneck:

(i) If the number of resources (q) in CR equals 1, then the dominant bottleneck constraint is BN_1. Go to (c).

(ii) if $q > 1$ and every product uses every resource then the bottleneck constraint is BN_1 and the ranked constrained resources are $BN_2, BN_3, \ldots, BN_q$. Go to (c).

(iii) Otherwise, determine the dominant bottleneck as follows:

(1) Schedule the products in non-ascending order of their R_i ratio until either one of the following two conditions is met. (Note, if two products to be scheduled have the same R_i, then schedule the product with the higher CM_i first):

   (a) The market demand is met (i.e. $Q_i = D_i$);
   (b) At least one of the constrained resources does not have sufficient capacity to schedule more of the product.

(2) If the first resource exhausted was not BN_1, make this resource BN_1. The current BN_1 increases rank by 1 to be BN_2. The rank for other bottlenecks is changed in a similar manner. The R_i for the new dominant bottleneck is recalculated. Go to (c).

(c) Schedule all the products requiring the dominant bottleneck resource in non-ascending order of their R_i ratio. For each product P_i schedule the largest quantity possible (i.e. market demand is not exceeded and there is adequate capacity on all constrained resources). If two products to be scheduled have the same R_i, then schedule the product with the largest CM_i first.
(d) When the dominant bottleneck's capacity is exhausted or there is insufficient capacity remaining to produce another unit of product, then treat BN_2 as the new bottleneck. Repeat steps (2c) and (2d) until every resource in CR is exhausted, or there are no products remaining that need these constrained resources.
(e) Schedule the "free" products (defined here as a product that does not use any of the constrained resources) until their market demand is met. The initial schedule is $P_i, P_2 \ldots, P_i \ldots, P_n$ where P_i is the ith product in the schedule.
(f) Calculate the time left ($t_{left,q}$) on the bottleneck BN_q as follows:

$$t_{left,q} = CP_{BN_q} - \sum_{i=1}^{n} t_{i,BN_{q_i}} Q_i$$

where Q_i is the quantity of product i scheduled.

(g) Examine whether reducing a Q_k and increasing some Q_i (where $i > k$) increases throughput (i.e. examine whether trading-off production of some products for the production of other products increases throughput). *A* neighbourhood search is conducted to determine which product will be candidates for this interchange. In this search, the neighbourhood of product (k) is defined as the nearest downstream product (i.e. $k+1$).

Let X be the set of candidate products where reducing some Q_k and increasing some $Q_i (i > k)$ may increase throughput. The set X is found as follows:

(i) For $j = 1$ to q
For $i = 1$ to $(n-1)$
if > 0
then if $R_{i+1}(t_{left,j} + t_{i,BN_j})/CM_i \geq 1$, **then** go to (ii)
else $i = i + 1$
else $j = j + 1$

(ii) if $Q_i < D_i$ OR $Q_{i+1} < D_{i+1}$ **then** set $k = i$ go to (iii)
else $i = i + 1$ go to (i)

(iii) $X = \{P_k, P_{k+1} \dots, P_n\}$. If $X = \phi$, the current master production plan is optimal, so stop.

Otherwise, go to Step 2(h).

(h) Reduce the current quantity Q_k of product P_k one unit at a time. This creates $(t_{k,BN_q} + t_{left,q})$ of available time on the dominant bottleneck resource. Schedule this available time to process P_{k+1}, $P_{k+2} \dots$, and P_n based on their R_i ratios $(i = k+1, k+2, \dots, n)$ in non-ascending order using the procedure in Step 2(c). Calculate the throughput gain of this change as:

$$\text{Gain} = \Delta Q_k CM_k + \Delta Q_{k+1} CM_{k+1} + \dots + \Delta Q_n CM_n\,,$$

where $\Delta Q_k$is the change in units for P_k.

Stop if the gain is less than zero, OR if the solution becomes infeasible (i.e. $Q_k > D_k$).

Example 4

Table 6.17 shows the processing times for machines A, B, C, D, E to produce product types R, S, T, U. The master production plan, sales price and material costs are shown on the last three columns. Machine capacities are shown on the last row. Determine if the master production is realistic. What is the maximum throughput of the system?

Table 6.17. Processing times for resources A, B, C, D, and E.

(1) Products	(2) A	(3) B	(4) C	(5) D	(6) E	(7) Market demand	(8) Sales price ($)	(9) Raw material cost ($)
R	15	5	10	0	5	70	90	15
S	16	10	5	28	5	60	80	15
T	0	5	10	15	21	50	70	20
U	5	15	7	0	6	150	60	25
Capacity (min)	2400	2400	2400	2400	2400			

Table 6.18. Determining bottlenecks.

(1) Product	(2) A	(3) B	(4) C	(5) D	(6) E	(7) Market demand	(8) Sales price ($)	(9) Raw material cost ($)	(10) CM
[1] R	15	5	10	0	5	70	90	15	75
[2] S	16	10	5	28	5	60	80	15	65
[3] T	0	5	10	15	21	50	70	20	50
[4] U	5	15	7	0	6	150	60	25	35
[5] Capacity	2400	2400	2400	2400	2400				
[6] Required	2760	3450	2550	2430	2600				
[7] Difference (D_j)	−360	−1050	−150	−30	−200				
Order of BNs	BN2	BN1	BN4	BN5	BN3				

Solution

Step 1: (a) This is shown in Table 6.18.
(b) $CR = \{A, B, C, D, E\}$ and $q = 5$.

Step 2: (a) The ratio R_i of each products' CM to its processing times on Resource $B(BN_1)$ are shown in column 4 of Table 6.19.

Table 6.19. Used and left times for resources B, A, E, C, and D.

1	2 D_i	3 Q_i	4 R_i	5 U	6 L	7 U	8 L	9 U	10 L	11 U	12 L	13 U	14 L
R	70	70	15	350	2050	1050	1350	350	2050	700	1700	0	2400
T	50	50	10	250	1800	0	1350	1050	1000	500	1200	750	1650
S	60	60	6.5	600	1200	960	390	300	700	300	900	168 0	−30
U	150	80	2.33	1200	0	400	−10	480	220	560	340	0	−30
				$BN3$		$BN2$		$BN4$		$BN5$		$BN1$	

where $U = Used$, and $L = Left$.

(b) Since $q = 5$ and U does not use every constrained resource in CR, go to step 2b (iii):

(iii) Schedule the products in non-ascending order of R_i to meet their market demand until the resource is exhausted as shown in columns 5 to 14 of Table 6.19. Since resource D is out of capacity before other resources, resource D is the dominant bottleneck (i.e. BN_1) and others follow as shown. Recalculate R_i for resource D.

Table 6.20. Ranking the products by their R_i.

(1)	(2)	(3) R_{iBN1}	(4) R_{iBN2}	(5) R_{iBN3}	(6) R_{iBN4}	(7) R_{iBN5}	(8) CM	(9) D_i	(10) Q_i	(11) $t_{i,D}$	(12) t_{left}
P_1	T	3.33	—	—	—	—	50	50	50	15	1650
P_2	S	2.32	—	—	—	—	65	60	58	28	26
P_3	R	—	5.0	15.0	15.0	7.5	75	70		0	26
P_4	U	—	7.0	2.33	5.83	5.0	35	80		0	26

(c) Rank the products by their R_i for resource D (column) 3 of Table 6.20). Plan to produce T and S to meet market demand (column 9). Rank product R before U since its CM is larger than U's (column 8) and plan to produce R and U until resources are exhausted. We can produce $50T$ and $57S$.

(d) The remaining 26 minutes on resource D (column 11) is not adequate to produce another product, so resource $A(BN_2)$ is the dominant bottleneck. Plan to produce R.
Since T and S have been previously scheduled, these are dealt with first. Moreover, the values of Q_i for R and U cannot be greater than values already obtained in the previous iteration ($R = 70, U = 80$, respectively). Hence, the solution at this step is $70R$, $58S$, $50T$, $80U$ as shown in Table 6.21:

Table 6.21. Used and left times for products T, S, R, and U.

Product	Q_i	$t_{i,A}$	*Used*	*Left*
T	50	0	0	2400
S	58	16	928	1472
R	70	15	1050	422
U	80	5	400	22

(e) There are no "free" products to schedule.

Table 6.22. Throughput calculations for products T, S, R, and U.

(1)	(2)	(3) R_{iB}	(4) R_{iB} $N1$	(5) R_{iB} $N2$	(6) R_{iB} $N3$	(7) C $N4$	(8) D $N5$	(9) D_i M	(10) Q_i	11 t_1	12 t_2	13 *Obj*	14 *Factor*
P_1	T	3.3	—	—	—	—	50	50	50	15	1650	2500	2.3(26 +15)/ 50 > 1
P_2	S	2.3	—	—	—	—	65	60	58	28	26	6270	—
P_3	R	—	5.0	15.0	15.0	7.5	75	70	0	26	11520	—	
P_4	U	—	7.0	2.3	5.8	5.0	35	80	0	26	14320	—	

where $t = t_{i,D}, t_2 = t_{left}, obj = throughput, factor = R_{i+1}(t_{left,j} + t_{i,BN_j})/CM_1$

(f) The time left on resource D is 26 (column 12) and the throughput is \$14 320 (column 13).

(g) Determine if decreased production of some products will increase throughput. As shown in column 14, product T is the first product where the ratio $R_{i+1}(t_{left,j} + t_{i,BN_j})/CM_1$ is greater than or equal to 1. Since the master production plan for S is less than the market demand (column 10), T is the first product in set $X = \{T, S, R, U\}$ considered for interchange to increase throughput.

(h) The impact of increasing or decreasing the order quantity of each number of candidate set X is evaluated in Table 6.23. By not processing one unit of product T, 41 minutes are available on resource D. this is obtained from I1 and column 14 where $(t_{left,j} + t_{i,BN_j}) = 26 + 15 = 41$. This information is used to obtain the gain. We decrement T by one in each iteration and determine how many units of S, R, and U that have to be produced. Therefore, $41/15 = 2.7$. Rounding down gives 2 units that can be produced. This is shared between S, R and U. Assign 2 to S, 0 to R and 0 to U since any further increase for R and U will violate the constraint of demand. The gain is calculated as:

$$\text{Gain} = -1^*50 + 1^*65 + 0^*75 + 0^*35 = 15\,.$$

The remaining time on D becomes $2400 - [(50 - 1)^*15 + (58 + 1)^*28] = 13$.
If R and U were incremented then there will not be enough time left in resource D.

Table 6.23. Impact of increasing or decreasing the order quantities.

	T	S	R	U	Gain	Q<D	D_{remain}	Throughput
(0)								14320
(1)	−1	(2)1	0	0	15	Yes	13	14335
(2)	−2	(4)2	0	0	30	Yes	0	14350
(3)	−3	(6)2	0	0	−20	Yes	15	14300

During the second iteration, T is decremented by 2 units. The possible number of products that can be produced = 2*41/15 = 4 approximately. Again, we cannot add to R and U otherwise they will exceed their demand of 70 and 80 respectively. But if 2 units are added to S, the gain can be calculated as $-2*50 + 2*65 = 30$. The remaining time on resource D $= 2400 - [(50 - 2)*15 + (58 + 2)*28] = 0$.

During the third iteration, T is decremented by 3 units and the possible number of products that can be produced equal to $3*41/15 = 6$ approximately. Only S can be considered. Only 3 units can be added to S. Remaining time on resource D $= 2400 - [(50 - 3)*15 + (58 + 2)*28] = 15$. Gain $= -3*50 + 2*65 = -20$. The heuristic stops once the gain is negative. The largest throughput of 14335 gives the optimum solution.

6.7. Application of Tabu Search to the Product Mix Problem

The product mix problem algorithm of TOC is the decision process that develops the master production schedule (MPS). The MPS is the detailed statement of top level planned production over the short to medium term. It seeks to balance incoming customer order and forecast requirements with available material and capacity. In this section, we discuss the process of applying a tabu search heuristic to the product mix decision problem. The discussion of the application of a tabu search to the TOC product mix decision problem in this section follows the proposal by Onwubolu (2001).

According to previous research (Taillard, 1990), the performance of TS is influenced by the following factors: approach used to generate the initial solution, type of move, size of the neighbourhood examined, tabu list size, and the termination criterion. For this problem, the selection of the parameters is documented as follows.

6.7.1. *Initial solution*

There are two ways to generate an initial solution, randomly or using a heuristic. *A* good initial solution may lead to a better final solution. For the product mix problem, a random initial solution is generated randomly. The decision variable, x_j, representing the quantity of the product type j is generated randomly to satisfy the product mix-equation.

Table 6.24. Total load computations and overloads.

Products	Resource number 10	20	30	40	50	60	Market demand	Sales price $	Material price $	Through-put $
A	2.5	9.5	6.5	12.0	4.0	30.0	20	30	10	20
B	5.5	3.5	1.5	16.0	1.0	10.0	30	50	42	8
C	3.5	8.5	9.5	25.0	2.0	9.0	40	50	25	25
D	2.0	8.0	10.0	30.0	1.0	10.0	30	40	25	15
E	10.0	20.0	15.0	0.0	10.0	2.0	60	20	15	5
Total load (min)	1015	2075	1755	2620	820	1680				
Capacity (min)	2400	1825	2400	2400	2400	2400				
Overload (min)	–	250	–	220	–	–				

For example, a good initial solution for the problem in Table 6.24 is {20, 20, 40, 25, and 50}. This solution is equivalent to 20A, 20B, 40C, 25D and 50E. This is a feasible solution because it satisfies all the conditions of Eqs. (2) and (3) and using Eq. (1), a total throughput value of \$2185 (20*20 + 8*20 + 25*40 + 15*25 + 5*50) is obtained.

6.7.2. *Move*

The move is a process that transforms the search from the current solution to its neighbourhood solution. There are two kinds of moves that are commonly used: *swap* and *insert* moves. In the *swap* move, a product type in the ith position is interchanged with the product type in the jth position. In the *insert* move, a product type removed from the ith position is inserted in the jth position. The insert move is preferred to the swap move by many researchers because it is known to advance the search to near regions within the neighbourhood. However, experiments should be carried out to conclude which move type is more appropriate. From the problem, the *insert* moves provide better solutions than the *swap* moves (Onwubolu, 2001). These two moves are illustrated as follows:

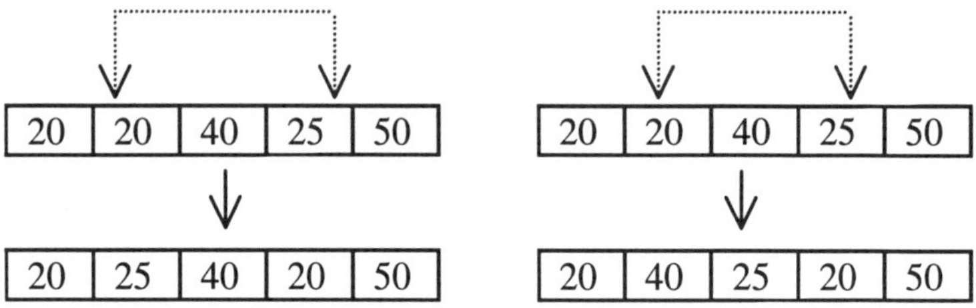

swap move *insert* move

6.7.3. *Neighbourhood size*

The neighbourhood size is the number of candidate solutions to be evaluated in each iteration of the search process. TS use two types of neighbourhood size. The first type is the whole size neighbourhood in which all possible neighbours are evaluated and the best non-tabued solution in each iteration, is selected. The second type is the random size neighbourhood in which the neighbours are evaluated and the first non-tabued solution that improves the solution in each iteration, is selected.

The exhaustive or whole size neighbourhood is used. The *move* described in the previous sub-section illustrates the concept of neighbourhood as follows:

Neighbourhood 1	20	20	40	25	50
Neighbourhood 2	20	40	25	20	50

The first neighbourhood is made up of product types {20, 20, 40, 25, 50}. By inserting the second product type in the fourth position, a new neighbourhood is obtained which is made up of product types {20, 40, 25, 20, 50}. The move mechanism is therefore, instrumental to generating new neighbourhood structures. Each neighbourhood is checked for feasibility by subjecting it to the constraint conditions and if feasible, it is evaluated using the objective function equation.

6.7.4. *Tabu list size*

The tabu list contains a list of moves that have been made, and are therefore prohibited or tabued. In designing the TS scheme, too small a tabu list size may cause cycling of the tabu search, while too large a tabu list size may prevent the tabu search from reaching certain good solution regions. Glover (1989) has suggested tabu list sizes between 5 and 12. A tabu list size of 7 is chosen.

6.7.5. *Aspiration criterion*

Several aspiration criteria are employed to avoid missing good solutions. At each step, the largest weighted non-tabu move is selected from those available, and an aspiration criterion is used to allow a *move* to be considered admissible in spite of leading to a tabu solution. It is possible to implement the tabu list in such a wary that the aspiration criterion becomes unnecessary. For a neighbourhood, all the feasible solutions are declared tabu and the solution with the largest objective function value is declared *currentbest.* The elite candidates having the *currentbest* values for each neighbourhood are examined for the one having the largest objective function value. The *currentbest* values for each neighbourhood represent the local optima. The largest of all these solutions is found, and the best, known as *isbest*, results in global minimum solution. It is possible to implement the tabu search heuristic for the product mix problem in such a way that allows no better solution to be found than the elite solution for each neighbourhood.

6.7.6. *Intensification*

The mechanism for intensification enhances the search to focus on examining elite solutions in a neighbourhood. It tends to move the search to a neighbouring position in the search space, and so could be considered a local search. The *intensification* length is taken to be equal to the number of the product mix types.

6.7.7. *Diversification*

The mechanism for diversification allows large jump to be made in the solution space. This ensures that large areas of the space are searched and solutions do not get stuck in local minima. This mechanism is also referred to as the *restarting* procedure. The *diversification length* is equal to 100. For each diversification process, a different initial schedule is randomly generated. This way, the search is able to explore a large solution space, thereby enhancing the possibility of finding the optimum solution in a very short time.

6.7.8. *Stopping criteria*

The most accepted stopping criterion relies on the search being terminated if the objective function value has not improved within a certain number of iterations, which is usually specified at the start of the run. Another criterion relies on the search being terminated when a maximum number of iterations have been reached to avoid an extremely long run. The problem with this other criterion is that it is difficult to determine the maximum number of iterations because the value may either lead to Premature-termination, or expensive termination. Onwubolu (2001) used the intensification and diversification lengths to terminate the solution search. This stopping rule is faster than the ones described above. We terminate search after 15^*n^2 total iterations, where n is the number of, product mix types.

By applying the different steps of the TS-based heuristic to the product mix problem shown in Table 6.24, the results of the TS-based heuristic are compared to the original TOC heuristic solution, integer linear programming solution and the revised TOC heuristic solution of Fredendall and Lea (1997). TS-based heuristic results in product mix of 20A, 27B, 40C, 24D and 49E with a total throughput of $2221. The optimal solution for this problem using ILP and the revised TOC heuristic is $2230. The original TOC product mix heuristic identifies a throughput of $2325. But this solution is infeasible, since one resource is over-utilised. This means that the TS-based heuristic is 99.6% as good as the ILP technique for this example problem. The iteration history of the tabu search-based heuristic for this example is shown in Fig. 6.17.

The tabu search-based heuristic is tested with the problems previously cited in the literature by comparing its solutions to the original TOC, integer liner programming, revised-TOC heuristic of Fredendall and Lea (1997). For comparison, these heuristics are referred to as original TOC (TOC abbreviated), integer liner programming (ILP abbreviated), and revised-TOC heuristic by Fredendall and Lea (1997) (R-TOC, abbreviated). Tables 6.25(a) and 6.25(b) show comparison of results obtained between the TS-based heuristic and these other methods. The results show that compared with other methods, the

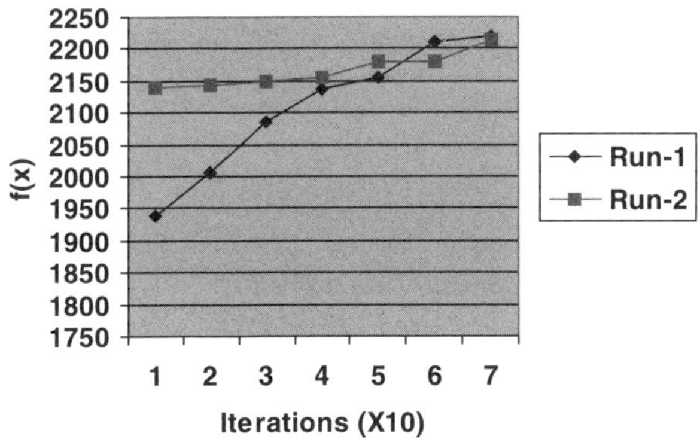

Fig. 6.17. Iteration history of the TS-based heuristic for Frendendall and Lea (1997) problem.

TS-based heuristic performs slightly worse than ILP in four (problems 4, 7, 8, 9) out of nine cases (44%). It performs equally to ILP in two cases (problems 3, 6) or 22%, and competes with ILP in two cases (problems 1, 2) or 22%. ILP clearly performs better than TS in one case (problem 5). The TS-based heuristic equal to R-TOC in two cases (problems 3, 6). R-TOC clearly outperforms the tabu search-based heuristic in two cases (problems 5, 7).

The number of factorial experiments for the tabu search-based heuristic for the product mix problem consists of $8(2^3)$ combinations, which include two types of moves (*swap* and *insert*), two diversifications (25 and 50), and two tabu list sizes (7 and 12). The experiments carried out show that the *insert* moves are superior to the *swap* moves, 50 diversifications produce better results than 25 diversifications for insert moves, and tabu list size of 7 is adequate. From the experiments, it is concluded that for the tabu search-based heuristic the best combination is insert moves, 50 diversification and tabu list size of 7.

The comparison of the proposed tabu search-based heuristic results and other methods in the literature show that proposed approach is appropriate for solving the product mix problem. However, all the problems reported to date are of small sizes (see Table 6.25(a)) and are considered too simplistic compared to typical manufacturing situations. Therefore, a second data set (Table 6.26), typical of manufacturing environments are generated randomly and solved using the proposed tabu search-based heuristic. For this class of problems, other techniques listed in Table 6.25 cannot perform due to

Table 6.25(a). Description of known problems.

Problem no.	Source	Problem size $m \times n$
1	Lee and Plenert (1993)	4×3
2	Lee and Plenert (1993)	4×3
3	Luebbe and Finch (1992)	4×4
4	Luebbe and Finch (1992)	4×4
5	Patterson (1992)	4×5
6	Plenert (1993)	4×2
7	Plenert (1993)	5×4
8	Posnack (1994)	4×3
9	Posnack (1994)	4×3
10	Mayday (1994)	—
11	Fredendall and Lea (1997)	6×5

Table 6.25(b). Performance/comparison of four heuristics.

No.	Original TOC solution $	Revised TOC algorithm solution	Integer linear programming solution $	Tabu search heuristic solution $	% improvement of ILP to TS-heuristic
1	6310	6444	6444	6420	0.374%
2	6460	6775	6775	6715	0.893%
3	6600	6600	6600	6660	—
4	9110	9110	9110	8955	1.73%
5	8872	8872	9050	8392	7.84%
6	4875	4875	4875	4875	—
7	Unrealizable	14 370	14 370	14 165	1.447%
8	6460	6444	6444	6378	1.035%
9	6800	6775	6775	6760	0.222%
10	—	—	—	—	—
11	2325	2230	2230	2221	0.405%

computational time constraints. Table 6.26 shows total throughput and CPU times for randomly generated data for the problem mix TOC problem instances. The resources processing times (in minutes) for producing problem mix were generated within the lower and upper bound ranges of [0, 10]. The problems do not have a routing specified and consequently, it will not matter whether it is flow shop or a job shop. Only the bottleneck resources are loaded over 100%. For the product mixes, demands (units/period), selling prices (in $) and raw material prices (in $) were generated within lower and upper bound limits of [20, 100], [20, 60], and [5, 20] respectively. Ten data sets were randomly generated for resources ranging from 10 to 30 in number, and

Table 6.26. Total throughput and CPU times for 2nd problem type.

No.	Resources	Product mix	Total throughput	CPU time (sec.)
1	10	5	7300	0.05
	10	10	12 987	1.76
2	15	10	12 633	2.31
	15	25	31 971	32.68
3	20	25	36 837	41.25
	20	50	70 335	66.13
4	25	50	72 044	83.65
	25	75	104 147	425.30
5	30	75	109 823	329.66

Table 6.27. Comparing TS and ILP for the 2nd problem type.

No.	Resources	Product mix	*TS-based heuristic* Total throughput	*ILP* Total throughput
1	10	5	7300	7340
	10	10	12 987	16 384
2	15	10	12 633	13 063
	15	25	31 971	**

** Too large for GINO/PC

for product mixes ranging from 5 to 100 in number. These randomly generated data sets are typical of problem instances encountered in manufacturing firms. Table 6.27 shows comparative results of the TS-based heuristic and for the two smaller problems (10×5 and 10×10 and 15×10 and 15×25). The integer linear programming formulation for the other three larger problems (20×25 and 20×50, 25×50 and 25×75, and 30×75) are extremely cumbersome and are therefore not included. The TS-based heuristic competes with ILP for two of the problems (10×5 and 15×10) but is inferior for the 10×10 problem. The results for these small size problems show that GINO/PC cannot solve the 15×25 problem due to memory space problem. This confirms the point made earlier in this book (Chapter 1) that optimization algorithms are only best suited for small size problems.

6.8. Summary

The tabu search fundamentals are first presented in this chapter. The basic features for short term-memory (STM) tabu search were examined, i.e. recency-based memory, tabu search, aspiration criteria, and candidate list strategies. Then the basic features for long term-memory (STM) tabu search were examined, i.e. frequency-based memory, logical restructuring, diversification, intensification, and strategic oscillation. Next, the chapter follows the TOC five steps and discusses four approaches to the product mix problem: (1) labour-based management, (2) linear programming, (3) theory of constraints (TOC) heuristic, and (4) revised theory of constraints algorithm. The chapter discusses the tabu search-based approach for the theory of constraints product mix problem, proposed by Onwubolu (2001). Published methods are only applied to simple problems for which optimal solutions can be found using linear programming technique. When dealing with the problem mix problem with sizes typical of practical real life situations, these published techniques become less effective because of prohibitive computational time associated with combinatorial optimization problems. Onwubolu's proposal therefore, is a contribution to the product mix problem in the theory of constraints because tabu search is a heuristic that solves combinatorial optimization problems with high quality solutions and reasonable computation times. The tabu search-based heuristic is applied to solve all the problems published in Frendendall and Lea (1997) which are considered to be of small sizes. The results obtained are very encouraging and therefore confirm that tabu search is appropriate for solving the theory of constraint product mix problem. Additionally, large problem instances typically encountered in the manufacturing industry are generated randomly and solved. The results show that the total throughput values and computation times increase with the problem size. In all the problems solved, the tabu search-based heuristic found feasible solutions.

Exercises

1. Clearly distinguish between the short-term memory (STM) and the long-term memory (LTM).
2. How does tabu search short-term memory (STM) differ from, hill-climbing search strategy?
3. What is the effect of a good seed solution as an input to tabu search?
4. Discuss the advantages and disadvantages of randomized diversification strategy as opposed to memory-based diversification strategy.
5. Discuss the advantages and disadvantages of problem-size dependent tabu tenure.
6. Discuss the components of tabu search that constitutes its memory.
7. How can partial neighbourhood search be achieved and what are its advantages over exhaustive neighbourhood search?
8. Table 6.28 shows the relevant data for 6-resource, 5-product TOC problem. Use the integer linear programming (ILP) method to solve the product mix problem. Is the solution feasible?

Table 6.28. Relevant data for Problem 8.

	Resource number						Market demand	Sales price $	Material price $
Products	1	2	3	4	5	6			
A	2.5	9.5	6.5	12.0	4.0	30.0	20	30	10
B	5.5	3.5	1.5	16.0	1.0	10.0	30	50	42
C	3.5	8.5	9.5	25.0	2.0	9.0	40	50	25
D	2.0	8.0	10.0	30.0	1.0	10.0	30	40	25
E	10.0	20.0	15.0	0.0	10.0	2.0	60	20	15
Capacity (min)	2400	1825	2400	2400	2400	2400			

9. For problem 8, use the labour-based management accounting method to solve the product mix problem.
10. For problem 8, use the original theory of constraints (TOC) method to solve the product mix problem. Is the solution feasible?
11. For problem 8, use the revised-original theory of constraints (R-TOC) method to solve the product mix problem.
12. For problem 8, use the tabu search (TS) method to solve the product mix problem.
13. What is the disadvantage of ILP for large problem sizes typically found in industry?
14. Discuss the results obtained from problems 8–12. Which methods are preferred and why?
15. What improvement strategies can be incorporated into the tabu search technique for solving the product mix problem?

References

Tabu Search

Gendreau, M. P., Soriano, P., and Salvail, L., 1993, Solving he maximum clique problem using a tabu search approach, *Annals of Operations Research*, **41** 385–403.

Glover, F., 1963, *Parameter combinations of local job shop rules*, Chapter IV, ONR Research Memorandum No. 117, GSIA Carnegie–Mellon University, Pittsburgh, PA.

Glover, F., 1965, A multiphase-dual algorithm for the zero–one integer programs, *Cahiers de Recherche Operationelle*, **6**(3) 131–168.

Glover, F., 1966, *An algorithm for solving the linear integer programming problem over a finite additive group, with extension to solving general non-linear integer programs*, CRC. 66-29, University of California at Berkeley.

Glover, F., 1968, Surrogate constraints, *Operations Research*, **16** 741–749.

Glover, F., 1969, Integer programming over a finite additive group, *SIAM Journal on Control*, **7** 213–231.

Glover, F., 1986, Future paths for integer programming and links with artificial intelligence, *Computers and Operations Research*, **13** 533–549.

Glover, F., 1989, Tabu Search — Part I, *ORSA Journal on Computing*, **1** 190–206.

Glover, F., 1990, Tabu search — Part II, *ORSA Journal on Computing*, **2** 4–32.

Glover, F. and Laguna, M., 1997, General purpose heuristics for integer programming — Part I, *Journal of Heuristics*, **2**(4) 343–358.

Glover, F. and Laguna, M., 1997a, *Tabu Search* (Kluwer Academic Publishers, Norwell, MA).

Onwubolu, G. C., 2001, Tabu search-based algorithm for the TOC product–mix decision, *International Journal of Production Research*, **39** 2065–2076.

Taillard, E., 1990, Some efficient heuristic methods for the flow shop sequencing proble, *European Journal of Operational Research*, **47** 65–67.

Applications

Buxey, G., 1989, Production scheduling: practice and theory, *European Journal of Operational Research*, **39** 17–31.

Bylinsky, G., 1983, An Israeli shakes up US factories, *Fortune*, September 5th, 120–132.

Fredendall, L. D. and Lea, B. R., 1997, Improving the product mix heuristic in the theory of constraints, *International Journal of Production Research*, **35**(6) 1535–1544.

Fry, T. D., Cox, J. F. and Blackstone, J. H., 1992, An analysis and discussion of the optimized production technology software and its use, *Production and Operations Management*, **1**(2) 229–242.

Goldratt, E. M. and Cox, J., 1984, *The Goal* (Croton-on-Hudson, NY: North River Press).

Goldratt, E. M., 1990a, *The Haystack Syndrome* (Croton-on-Hudson, NY: North River press).

Goldratt, E. M., 1990b, *What Is This Thing Called Theory of Constraints*? (Croton-on-Hudson, NY: Worth River Press).

Goldratt, E. M., and Fox, R. M., 1986, *The Race*, 1st ed. (Croton-on-Hudson, NY: North River Press).

Hsu, T-C. and Chung, S-H., 1998, The TOC-based algorithm for solving product mix problems, *Production Planning and Control*, **9**(1) 36–46.

Lea, T. N. and Plenert, G., 1993, Optimized theory of constraints when new product alternatives exist, *Production and Inventory Management Journal*, **34**(3) 51–57.

Luebbe, R. and Finch, B., 1992, Theory of constraints and linear programming: a comparison, *International Journal of production Research*, **30**(6) 1471–1478.

Mayday, C. J., 1994, Proper use of constraint management, *Production and Inventory Management Journal*, **35**(1) 84.

Miltenburg, J., 1997, Comparing JIT, MRP and TOC and embedding TOC into MRP, *International Journal of Production Research*, **35**(4) 1147–1169.

Patterson, M. C., 1992, The product-mix decision: A comparison of theory of constraints and labor-based management accounting, *Production and Inventory Management Journal*, **33**(3) 80–85.

Plenert, G., 1993, Optimized theory of constraints when multiple constrained resources exist, *European Journal of Operational Research*, **70**(1) 126–133.

Posnack, A. J., 1994, Theory of constraints: Improper applications yield improper conclusions, *Production and Inventory Management Journal*, **35**(1) 85–86.

Preiss, K., Goldman, S. L., and Nagel, R. N., 1996, *Cooperate to compete: Building agile business relationships* (Van Nostrand Reinhold, New York).

Ronen, B. and Starr, M. K., 1990, Synchronized manufacturing as in OPT: From practice to theory, *Computer Industrial Engineering*, **18** 585–600.

Schragenheim, E. and Ronen, B., 1990, Drum-buffer-rope shop floor control, *Production and inventory Management Journal*, **31**(3) 18–23.

Chapter 7
Ant Systems

7.1. The Ant System Paradigm

One of the most promising innovations in the field of general heuristics is certainly the development of adaptive methods. These methods may be non-population-based approach or population-based approach. The non-population-based approach moves from one current solution to the next after each iteration and uses the properties of the solution for searching intelligently in the solution space. On the other hand, population-based methods use the collective properties of a group of distinguishable solutions, called *population*, for searching intelligently in the solution space. The central idea in the population-based approach is to manipulate a set of solutions located in different promising areas of the solution space. The population goes through an adaptive process that is made up of two distinct phases. In the *self-adaptation phase*, solutions change their internal structure without any interaction with the other members of the population. In the *co-operation phase*, solutions of the current population exchange information with the aim of producing new solutions that inherit good attributes. Hopefully such an adaptive behaviour will make good solutions of the problem under study to emerge.

The collective performance of social insects, such as ants, bees, wasps or termites has intrigued entomologists for several years. Their main concern is about the mechanisms that allow the individuals of the same colony to co-ordinate their activities and to favour the survival of the species. Apparently everything works out because of an underlying factor which regulates the activities of each individual. Studies have shown that this global adaptive behaviour arises from a multitude of very simple local interactions. The nature of these interactions, the treatment of the information, the differences between the solitary behaviour, and the social behaviour have remained unclear for a long time. The realisation of a specific task by a colony of insects has shown that the co-ordination of the work does not depend on the insects but rather on the advancing state of the task. Co-ordination emerges from an auto-catalytic chain retroaction between stimuli and responses. An insect does not control directly its work, the whole process progresses as if each insect were guided by its work. While working an insect modifies the form of the stimulation which triggers its behaviour. This induces the emergence of a new stimulation that will trigger new reactions in the colony.

In order to illustrate the emergence of collective structures in an insect society, let us cite the example of an ant colony in search of a nearby feeding source. Initially, ants leave the nest and move randomly. When an ant discovers a feeding source, it informs its other ants belonging to the same colony by laying a temporary trail on the ground on its way back to the nest. The trail is nothing else than a chemical substance called pheromone, which guides the other ants towards the same feeding source. On their way back, the latter also lay pheromone on the ground and thus reinforce the marking of the path that leads from the nest to the discovered feeding source. The reinforcement of the marking by pheromone optimises the collection of food. All trails laid on the ground evaporate progressively as time goes by. Because of the larger elapsed time between two passages of an ant on the paths leading to remote feeding sources, trails will get undetectable faster on these paths. In the long term, ants will thus, all prefer the closest feeding source. This example shows that an ant colony converges towards an optimal solution whereas each single ant is unable to solve the problem by itself within a reasonable amount of time. In this case, as pointed out (Théraulaz *et al.*, 1991) 'the environment plays the role of spatio-temporal memory keeping track of the swarm past actions while selecting its own dynamic regime'.

For many years, engineers have been interested in the behaviour of social insects in order to define new models of collective problem solving. The Ant System, developed recently in 1992 by Colorni, Dorigo, and Maniezzo, is an adaptive algorithm which takes its roots in the collective behaviour of an ant colony. In the co-operation phase of an ant algorithm each solution of the population is examined with the aim of updating a global memory keeping track of important structures of the set of all feasible solutions which have been successfully exploited in the past. The self-adaptation phase uses a problem-specific constructive method to create a new population of solution on the basis of the global memory. Therefore, the ants are able to optimise their paths by this process.

An ant system, as we use the term here, is a computational paradigm inspired by ants collective contributions in solving a common problem. A similar process can be transposed to combinatorial optimisation: solutions of the problem are built using a statistics on solutions previously generated. This statistics play the role of the pheromone traces and it gives a higher weight to the best solutions. After a while, it is observed that such a procedure is able to build solutions of better quality than a procedure guided by partial objective function evaluations only. Examples of adaptive methods are Genetic Algorithms, Scatter Search and Ant Systems.

In this chapter we discuss the ant systems fundamentals, and then deal in details with the fast ant variant using the intelligent optimisation techniques framework of Sec. 1.3. This is followed by hybrid systems. The fast ant system simulator is then discussed in details in terms of its data structures and algorithm. Then flow shop application is then discussed. The fast ant simulator output for the flow shop scheduling is then given so that readers can follow how the simulator works.

7.2. Ant Systems Fundamentals

In this section, we present the different components of fast ant system. These are the memory or pheromone trail, solutions manipulation, intensification, and diversification.

7.2.1. *Pheromone trail or memory*

The most important component of the ant system is the management of pheromone trails. In a standard ant system, pheromone trails are used in conjunction with the objective function for constructing a new or provisory solution. Informally, pheromone levels give a measure of how desirable it is to insert a given element into a solution. For example, in Dorigo and Gambardella's TSP application, links get a higher amount of pheromone if they are used by ants that make shorter tours. Pheromone trails in this case are maintained in a pheromone matrix $T = [\tau_{ij}]$ where τ_{ij} measures how desirable the edge between cities i and j is. For the quadratic assignment problem, the set of the pheromone trails is maintained in a matrix $T = [\tau_{ij}]$ of size $n \times n$, where the entry τ_{ij} measures the desirability of setting $\tau_{ij} = j$ in a solution.

There are two different uses of pheromone: *exploration* and *exploitation*. Exploration is a stochastic process in which the choice of the component used to construct a solution to the problem is made in a probabilistic way. Exploitation chooses the component that maximises a blend of pheromone trail values and partial objective function evaluations.

After having built a new solution, a standard ant system updates the pheromone trails as follows: first, all the pheromone trails are decreased to simulate the evaporation of pheromone. Then, the pheromone trails corresponding to components that were chosen for constructing the solution are reinforced, taking into consideration the quality of the solution.

Since the first implementation of an ant system by Dorigo, Maniezzo and Colorni in 1992, it has been changed in many ways to improve its efficiency. In the best ant system to date (Dorigo and Gambardella, 1997) update of pheromone trails is very loosely coupled with what happens with real ants. Indeed, pheromone trails are not only modified directly and locally by the artificial agents during or just after the construction of a new solution, but also globally, considering the best solution generated by all the agents at a given iteration or even the best solution ever constructed. In the hybrid ant system, local updates of the pheromone trails are not performed, but only a global update. This makes the search more aggressive and requires less time to reach good solutions. Moreover, this has been strengthened by an *intensification* mechanism. An inconvenience of such a process is the risk of an early convergence of the algorithm. Therefore, a *diversification* mechanism that periodically erases all the pheromone trails has been added to exploit other areas in the solution space.

7.2.2. *Solutions manipulation*

The first ant colony optimisation system which is a computational intelligence approach to combinatorial optimisation was called ant system (AS). More recently, Dorigo and Gambardella have been working on various extended versions of the AS paradigm. Ant-Q is a hybridisation of AS with Q-learning, a well-known reinforcement learning algorithm. Ant colony system (ACS) is a further extension of Ant-Q. Dorigo and Gambardella have also proposed new hybrid versions of ant colony optimisation with local search. In the early applications of ant systems, pheromone trails were exploited to build a completely new solution. In hybrid ant system, pheromone trails are used to *modify* an existing solution, in the spirit of a neighbourhood search. After an artificial agent has modified a solution, taking into account only the information contained in the pheromone trails, an improvement phase that consists in performing a fast local search that takes into consideration only the objective function is applied. From a memory point-of-view, hybrid ant system is between genetic algorithms that use a population of solutions only and fast ant systems that use a statistics-matrix only.

Other research workers have identified adding local search to ant systems as very promising. For the travelling salesman problem (TSP), Dorigo and Gambardella succeeded in designing an efficient ant system by adding a simple 3-opt phase after the construction phase. For the sequential ordering problem (SOP), Gambardella and Dorigo alternate an ant system with a local search specifically designed for the SOP, obtaining very good results and new upper bounds for a large set of test problems. For both of these applications, the hybrid ant system is reported to outperform all known algorithms on vast classes of benchmark problems.

In fact, many of the most efficient heuristic methods for combinatorial optimisation problems are based on a neighbourhood search, either greedy local search or a more elaborate intelligent optimisation technique.

7.2.3. *Intensification and diversification*

As mentioned, in hybrid ant system each ant is associated with a problem solution that is first modified using pheromone trail and later is improved using a local search mechanism. This approach is completely different from that of previous ant systems where solutions were not explicitly associated with ants but were created in each iteration using a constructive mechanism.

Therefore, in hybrid ant system, there is the problem of choosing, during each iteration, the starting solution associated to each ant. To solve this problem two mechanisms, called intensification and diversification, have been identified. Intensification is used to explore the neighbourhood of good solutions more completely. When intensification is active, the ant comes back to the solution it had at the beginning of the iteration if this solution was better than the solution it had at the end of the iteration. In all other cases, the ant simply continues working on its current solution. Diversification implements a partial restart of the algorithm when the solutions seem not to be improving any more. It consists in a re-initialisation of both the pheromone trail matrix and the solutions associated to the ants.

7.3. FANT: Fast Ant System

It has already been mentioned that since the first implementation of an ant colony optimisation system, it has been refined in may ways to improve its efficiency. One such change is the implementation of fast ant system by Taillard and Gambardella (1997). Now, let us see how an ant system can be implemented in practice. The FANT method presents however a number of differences regarding to basic ant systems as described by Dorigo, Maniezzo and Colorni (1992). First, a local search is added for improving the solution built by the ant. This has been identified as very promising for other problems such as the travelling salesman problem (Dorigo and Gambardella, 1997) or the sequential ordering problem (Gambardella and Dorigo, 1997).

Initialise the memory or pheromone trail.
For I^{max} iterations
Repeat, until a stop criterion is satisfied:
    Construct a *new provisory solution*, using the information contained in the memory
    Improve the provisory solution with a *local search.*
    Update the memory.
End For

Fig. 7.1. General scheme of a fast ant system.

Another difference is the statistics that is computed. Indeed, this statistics gives more and more weight to the best solution found so far as the search progresses. This makes the search to converge faster to good solutions. However, this may also cause the search to be trapped in local optima. In order to avoid this situation, a diversification mechanism that resets the memory has been developed. Finally, FANT does not use a population of artificial agents.

In this section we present the FANT algorithm i.e. by specifying its four components: the memory structure, the constructing procedure and the way the memory is updated. The general scheme of the fast ant system is given in Fig. 7.1.

7.3.1. *Memory implementation*

The basic idea of our method is to evaluate *a priori* the desirability or interest of setting $\tau_i = j$ in a solution (equal to permutation) τ. Therefore, our memory is principally constituted by a matrix T of size $n \times n$ whose entry τ_{ij} measures the preference of setting $\tau_i = j$.

From an ant system point-of-view, this matrix represents the pheromone trail left by the ants. The entries of this matrix consist in a statistics derived from the solution previously generated by the search process.

The memory structures in ant systems are primarily *influenced-based* and *quality-based.* The notion of *influence* integrates the two key aspects of *intensification* and *diversification* in ant systems by seeking change that holds

$I = \varnothing$, $J = \varnothing$
While $I| < n$ repeat:
Choose i, randomly, uniformly, $1 \leq i \leq n$, $i \notin I$.
Choose j, randomly, $1 \leq j \leq n$, $j \notin J$, with probabilit $\tau_{ij} / \sum_{1 \leq k \leq n, k \notin J} \tau_{ik}$ and set $\mu_i = j$.
$J = J \cup \{j\}$, $I = I \cup \{i\}$

Fig. 7.2. Construction of a provisory solution.

indication of promise. Influence measures the degree of change induced in solution structure. The memory of an ant system makes use of the solution structure for its update. The memory pattern is cyclic in the sense that every time a better solution is found the memory is initialized. In between, the memory takes a pattern that reflects the structure of the current solution. This influenced-based memory structure also depicts the *quality* of the solution found. We shall examine the issue of *influence-based* and *quality-based* ant memory structures in more details in Sec. 7.3.4.

7.3.2. *Provisory solution generation*

For generating a provisory solution μ, we use a constructive method that chooses the elements of μ successively, in a random order and with a probability proportional to the values contained in the T matrix. More formally, the constructive method is presented in Fig. 7.2:

An initial pheromone trail consists of a statistics initialized as $\tau_{ij} = 1$. Let us now see how fast ant system builds a provisory solution.

Example 1

Describe how FANT generates a provisory solution for scheduling a 5-job problem.

Solution

For the 5-job problem, the set $i = \{4\,5\,1\,2\,3\}$ is generated. This set may be obtained by generating a number randomly between 1 and n. Let this number be $\alpha = 2$, for example. Let us explain how the set i is obtained. Each number in the set is obtained as $(\alpha + I)$ $modulus(n) + 1$, where $I \in [1, n]$. When $I = 1$, the first term in the set becomes $(2 + 1)$ $modulus(5) + 1 = 4$. When $I = 2$, the second number in the set is $(2+2)$ $modulus(5)+1 = 5$. When $I = 3$, the third number in the set is $(2 + 3)$ $modulus(5) + 1 = 1$. The fourth number in the set is $(2 + 4)$ $modulus(5) + 1 = 2$, while the last number $(I = 5)$ in the set is $(2 + 5)$ $modulus(5) + 1 = 3$. Hence the sequence is $i = \{4\,5\,1\,2\,3\}$. Since the statistics matrix is of size $n \times n$, for the initial pheromone trail, we can define $\alpha_i = n$. This is equivalent to summing the unit entries of each row of the phenomena trail. This means for our illustration that $\alpha_i = 5$. A number $u \in [1, m\alpha_i]$ is generated randomly. Assuming this number is $u = 4$. These initial values of i and u are used in conjunction with randomly generated j to commence the constructive method for generating a provisory solution that is permutative.

We need to generate $j \in [1, n]$ randomly so that an entry in the memory can be accessed. An entry in the memory is defined by (i, j). Assuming $j = 3$ is generated randomly, then $\tau_{43} = 1$ because $\tau_{ij} = 1$. The indices of τ are obtained from $i = 4$ and $j = 3$. A new set of j is obtained from the current value of j using the expression (j) $modulus(n)+1$. For example, since $j = 3$, we generate a set of j as (3) $modulus(5) + 1 = 4$, (4) $modulus(5) + 1 = 5$, (5) $modulus(5) + 1 = 1$, (1) $modulus(5) + 1 = 2$. We stop here to avoid cycling. This leads to a set of $j = \{4, 5, 1, 2, 3\}$. The pair of (i, j) now available correspond to $i = 4$, $j = \{4, 5, 1, 2, 3\}$. It is now easy to evaluate $\Sigma\tau_{ij}$ using the pairs $(i, j) = \{(4, 4), (4, 5), (4, 1), (4, 2)\}$ as $\tau_{43} + \tau_{44} = 1 + 1 = 2$; $\tau_{43} + \tau_{44} + \tau_{45} = 3$; $\tau_{43} + \tau_{44} + \tau_{45} + \tau_{41} = 4$; $\tau_{43} + \tau_{44} + \tau_{45} + \tau_{41} + \tau_{42} = 5$. But there is a constraint that this summation is only possible when $\Sigma\tau_{ij} < u$. Since $u = 4$, the second summation is accepted because its summation value is equal to 3. The value of j^* is 5 corresponding to the second summation that met the constraint condition. This value of j^* is used to assign the partial solution $\mu_i = j^*$ or $\mu_4 = 5$. The current partial solution is $\mu = \{0, 0, 0, 5, 0\}$. The process just described is used to construct an initial solution. To avoid duplicating a partial solution, the expression j $modulus(n) + 1$ is invoked. In order to simulate the evaporation of the pheromone trail, the value of α_i is decremented by τ^*_{ij}, so that $\alpha_i = 4$.

Let us now see how the provisory solution is built progressively in a constructive manner. Assuming a new set $u \in [1,4]$ generated with a value 4, and a $j = 4$ is generated randomly. In the second iteration $i = 5$ so that $\tau_{54} = 1$. A new set of j is obtained from the expression $(j)\ modulus(n)+1$ as {5, 1, 2, 3, 4}. Again, the set is obtained as $(4)\ modulus(5) + 1 = 5$, $(5)\ modulus(5) + 1 = 1$, $(1)\ modulus(5)+1 = 2$, $(2)\ modulus(5)+1 = 3$. We stop here to avoid cycling. This leads to $\Sigma\tau_{ij}$ being evaluated as $\tau_{54} + \tau_{55} + \tau_{52} + \tau_{53} = 4$. The summation that yields a value of 3 corresponds to $j^* = 2$. Hence the partial provisory solution becomes $\mu_5 = 2$. During this second iteration, $\mu = \{0, 0, 0, 5, 2\}$. Meanwhile, the pheromone trail would have evaporated, so α_i is decremented by τ_{ij^*} or τ_{52}. Since $\tau_{52} = 1$, $\alpha_i = 3$. A new value of u is generated randomly between 1 and α_i. Suppose this new value is $u = 3$. For $j = 3$, generated randomly, this leads to a set of $j = \{4, 5, 1, 2, 3\}$. The next value of $i = 1$ is used. During this third iteration, $\tau_{13} = 1$ and $\tau_{13}+\tau_{14} = 1+1 = 2$; $\tau_{13}+\tau_{14}+\tau_{15} = 3$. There is no need to proceed because the upper bound for the summations that can be accommodated is 3. Therefore, the first summation is acceptable, with $i = 1$ and $j^* = 4$. Since the first position in the partial solution μ_i is not already visited, we accept that $\mu_i = 4$. This leads to the current partial solution being $\mu = \{4, 0, 0, 5, 2\}$. Meanwhile, the pheromone trail would have evaporated, hence $i = 2$. The value of u is generated randomly within the range $[1, \alpha_i]$ or [1, 2]. Assuming $u = 1$ is generated randomly. During this fourth iteration $i = 2$. We generate a j value within the range $[1, n]$ or [1, 5]. Let this number be 2. Then $\tau_{22} = 1$. The entries of the set j are obtained as $(2)\ modulus(5) + 1 = 3$, $(3)\ modulus(5) + 1 = 4$, $(4)\ modulus(5) + 1 = 5$. Therefore, the set of $j = \{3, 4, 5\}$ will now be considered. The summations of the pheromone trail matrix entries $\tau_{22} + \tau_{23} = 2$; $\tau_{22} + \tau_{23} + \tau_{24} = 3$ are obtained. As the upper bound is 2 there is no need to continue. Therefore, only $\tau_{22} = 1$ qualifies. But $j^* = 2$ is not feasible because it is already in the fourth position. This means $i = 2$ and $j^* = 3$, leading to $\mu_2 = 2$ and $u = \{4, 3, 0, 5, 2\}$. Meanwhile, the pheromone trail would have evaporated so that $\alpha_i = 2 - \tau_{22} = 1$. A new u is generated within the range $[1, 1] = 1$. A new j is generated randomly with a value of 4, for example. Then $i = 3$ and $j = 4$. The entries of the set if j are obtained as $(4)\ modulus(5) + 1 = 5$, $(5)\ modulus(5) + 1 = 1$, $(1)\ modulus(5) + 1 = 2$, $(2)\ modulus(5) + 1 = 3$, $(3)\ modulus(5) + 1 = 4$. Therefore, the set j = {5, 1, 2, 3, 4} are possible.

The first summation becomes $\tau_{34} + \tau_{35} = 2$. There is no need to continue because the upper bound is 1. Therefore, τ_{34} is considered leading to $\mu_3 = 4$.

Although the third position is empty, the value of 4 has already been assigned to the first position. To resolve this problem the expression (j) *modulus*$(n)+1$ is again invoked so that $j^* = 5$. This is not feasible since it already occupies the fourth position in the partial solution. Repeating the process yields (5) *modulus*$(5) + 1 = 1$. Hence, $j^* = 1$. This is the only feasible candidate. Hence $\mu_3 = 1$ and $\mu = \{4, 3, 1, 5, 2\}$ is a complete provisory solution. As can be observed, this process of generating a provisory solution is constructive and depends on the previous information contained in the memory statistics. □

7.3.3. *Improvement procedure*

The provisory solution μ generated at the previous step is not so good, generally; at the first iteration, μ is just a random permutation. So we apply to μ the first steps of a first improving neighbour procedure. More formally, we repeat the procedure of Fig. 7.3 twice, where $\Delta(\pi, i, j)$ is the difference in the objective function value when exchanging the elements π_i and π_j in π.

```
I = ∅
While  I| < n repeat:
        Choose i, randomly, uniformly, 1 ≤ i ≤ n, i ∉ I.
        J = {i}
        While  J| < n repeat:
                Choose j, randomly, 1 ≤ j ≤ n,  j ∉ J.
                If Δ (π, i, j) < 0, exchange πi and πj in π
                J = J ∪ {j}
        I = I ∪ {i}
```

Fig. 7.3. Fast descent procedure (to be repeated twice).

The evaluation of $\Delta(\pi, i, j)$ can be performed in $O(n)$ using the formula:

$$\Delta(\pi, i, j) = (c_{\pi_j \pi_j} - c_{\pi_i \pi_i}) + (c_{\pi_j \pi_i} - c_{\pi_i \pi_j}) + \sum_{k \neq i,j} (c_{\pi_k \pi_j} - c_{\pi_k \pi_i}) + (c_{\pi_j \pi_k} - c_{\pi_i \pi_k})$$

The first improving neighbour procedure is simple to understand. Consider a *move* from position i to another position j in the search space. The search procedure can move from I to i. For example, in scheduling problem, this situation can arise during swapping of jobs and corresponds to when the same job is considered. Moreover, the search procedure can move from position j to position j. another possibility is for a move from position i to position j, or from position j to position j. Applying the law of triangular inequalities and considering a point $k \neq i, j$, leads to some other possibilities. It is possible for a move from the position k to position i, and from position k to position j. Moreover, other possibilities are moves from position j to k and from position i to k. this analysis can be easily illustrated using the following triangle:

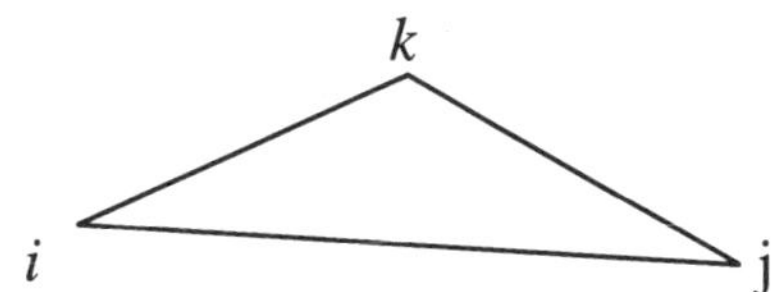

Applying the inequality rule leads to $ij \geq ik + kj$. This leads to $0 \geq ik + kj - ij$. Expanding the terms, while accepting duplication for i or j, but not for k, yields the following expressions:

$$0 \geq ik - ki + kj - jk - (ii - jj + ij - ji)$$

$$0 \geq ((jj - ii) + (ji - ij) + (kj - ki) + (ik - jk) = \Delta(i, j)$$

We simply accept a move when $\Delta(i, j) < 0$ for an objective function. The reason for applying this neighbourhood examination twice is the possibility of moving from position i to j or from position j to i. Therefore this procedure can be executed in $O(n^3)$. It does not necessarily return a local optimum, but it is fast and may produce different solutions when starting with the same initial, not locally optimal solution. From now on, we are going to call π the improved solution.

Local search consists in applying a complete neighbourhood examination twice with first improving strategy to a solution $\hat{\pi}_k$ to produce a solution $\tilde{\pi}_k$. The neighbourhood of a solution is defined as the set of all solutions that can be arrived at by a *move*. The word 'solution' should be explicitly defined to reflect the problem being solved. To define a specific problem being solved at this juncture seems premature, and this is deferred to Sec. 7.7 in this chapter where applications are treated. A *move* is defined as a feasible change of state of a current solution that produces a subsequent solution.

Since a large number of moves is evaluated during the improvement procedure, it is essential for overall computational efficiency to make the evaluation efficiently. When a move is made, the effect of the move need be reconsidered. Let $\Delta(\pi, i, j)$ be the change in the objective function due to the possible exchanging of the elements π_i and π_j in π. The worth of this move is calculated very efficiently by repeating the procedure of Fig. 7.3 twice. The reason for repeating the procedure is simply to reconsider the *move* made. Fore example if a move is from i to j, it is necessary to consider the other possibility of j to i. To account for these two options, the procedure is repeated.

Let us see how the improvement procedure works. First generate a solution π_j. For a problem size equal to five, let an initial solution will be {1, 2, 3, 4, 5}. The elements from the second position to the last, third to the last, fourth to the last are considered respectively. Any two numbers within these brackets are randomly chosen and their positions swapped. Assuming for the first operation elements 1 and 3 are swapped, for the second operation elements 2 and 3 are swapped, and for the third operation elements 3 and 5 are swapped, and for the fourth operation elements 4 and 5 are swapped. This leads to the following sets {3, 2, 1, 4, 5}, {3, 1, 2, 4, 5}, {3, 1, 5, 4, 2} and {3, 1, 5, 2, 4}. The last term represents a randomly generated permutative solution π_i. The same procedure is followed to generate another random generated permutative solution π_j. This could be a set {4, 3, 1, 5, 2}.

Then each element of π_i is considered with each element of π_j for possible moves that will result in improving the objective function. For example, consider the two solutions $\pi_i = \{3,1,5,2,4\}$ and $\pi_j = \{4,3,1,5,2\}$. Each i in π_i and each j in π_j are paired up and considered. For example, pairing these two solutions gives $\Delta(i,j) = \{\{3,4\},\{3,3\},\{3,1\},\{3,5\},\{3,2\},\{1,4\},\{1,3\},\{1,1\},\{1,5\},\{1,2\},\{5,4\},\{5,3\},\{5,1\},\{5,5\},\{5,2\},\{2,4\},\{2,3\},\{2,1\},\{2,5\},\{2,2\},\{4,4\},\{4,3\},\{4,1\},4,5\},\{4,2\}\}$. This operation is of the order $O(n^2)$. Since the evaluation of $\Delta(\pi, i, j)$ can be performed in $O(n)$, the overall procedure can therefore be performed in $O(n^3)$. The change in $\Delta(\pi, i, j)$ is evaluated. If this value is negative the objective function value for the problem at hand is decremented by the $\Delta(\pi, i, j)$ This way, a provisory solution is improved towards the optimal solution.

7.3.4. *Memory updates*

The statistics the memory stores uses two parameters, r and r^*, that represent the re-enforcement of the matrix entries corresponding to the solution produced during update of the memory and, respectively, π^*, the best solution produced by the algorithm. During the entire process, we set $r^* = 4$ while r may vary. At the beginning $r = 1$ and we set $\tau_{ij} = r$, $1 \leq i, j \leq n$, meaning that the memory does not contain any information initially. There are three ways the memory is updated. Usually, the entries of matrix T are updated as follows:

(1) For $i = 1$ to n do:
$\tau_{i\pi_i} = \tau_{i\pi_i} + r$
$\tau_{i\pi_i^*} = \tau_{i\pi_i^*} + r^*$
End For

Where π is the solution produced at the current iteration and π^* the best solution produced so far.

(2) If τ^* has been improved, then r is reset to 1 and all the entries of T are set to 1. The aim of this resetting is to *intensify* the search around π^* by giving less importance to the past of the search.
(3) If the provisory solution μ generated is equal to π^*, then r is increased by one unit and all the entries of T are set to r. This situation occurs when π^* has not been improved for a large number of iterations, meaning that the re-enforcement of the entries corresponding to π^* is too high. The aim of this is to *diversify* the search when the information contained in T is not very different from π^*.

The memory structure and its update are important for enabling an ant system to achieve its best performance level. From an ant system perspective, the memory is a matrix of size $n \times n$ that represents the pheromone trail left by the ants during their search optimization. Initially the memory is empty. To simulate this, we initialize the matrix with a 1 for all its entries. Subsequent entries of the matrix depend on the previous solution generated by the search process.

The memory updates of ant algorithm is *influenced-based* and *quality-based* as already mentioned. The three basic ways the ant memory is updated are now discussed to clarify the concepts of *influence* and *quality.* We illustrate these points with the following example.

Example 2

An initial sequence of $\{2\ 3\ 4\ 1\}$ is obtained by FANT and the makespan is 58. How does FANT update its memory in order to find better solutions?

Solution

Since this is an initial solution, the memory is initially empty and has the value 1 everywhere in the matrice T. The memory is updated using two parameters r and r^*, that represent the re-enforcement of the matrix entries corresponding to the solution produced. There are two solutions involved here, the current solution and the best solution so far. Both are equal at this juncture: $\pi_i = \pi_i^* = \{2, 4, 3, 1\}$. We set r^* to any positive value, but here we set it to 6. This means that the memory will have $6 + 1 + 1 = 8$ in positions $j = \{2, 4, 3, 1\}$ for $i = \{1, 2, 3, 4\}$ respectively. The memory has therefore memorized the structures of both the best solution and the current solution. Now let us advance.

Assuming the provisory solution found by the algorithm, *build_new_solution* (to be described in Sec. 7.5), is equal to the best solution so far. Then the memory updates all the entries of the matrice T by one and diversifies the search because π^* has not been improved for a large number of iterations. This corresponds to π_i^*, a solution where the re-enforcement of the entries corresponding to is too high. This means that the memory now has the value of 2 every where in the matrix T. The memory is now updated using the previous sequence $\{2\ 3\ 4\ 1\}$ and the best sequence so far $\{2\ 3\ 4\ 1\}$.

This means the memory will have the value, 10 $(6 + 2 + 2)$ in positions $j = \{2, 3, 4, 1\}$ for $i = \{1, 2, 3, 4\}$ respectively. Ant continues to memorize the structure of both the previous solutions and the best solution so far. When the procedure *build_new_solution* is activated, it constructs a solution $\{3, 1, 4, 2\}$. Let us now see how FANT will deal with this in its memory. Here the two memory update strategies are invoked:

$$\tau_{i\pi_i} = \pi_{i\pi_i} + r \text{ and } \pi_{i\pi_i} = \tau_{i\pi_i} + r^* .$$

In one step, FANT will use the value 16 $(10 + 6)$ to update positions $j = \{2, 3, 4, 1\}$ for $i = \{1, 2, 3, 4\}$ respectively. Then in another step, FANT will use the value, 2 to update positions $j = \{3, 1, 4, 2\}$. Since the position $j = 4$

is common to both steps, it is effectively updated to 18 $(16+2)$. Let us stop here and see the contents of the memory during these particular updates. The memory propagates like this:

$$\begin{bmatrix} 1&1&1&1 \\ 1&1&1&1 \\ 1&1&1&1 \\ 1&1&1&1 \end{bmatrix} \rightarrow \begin{bmatrix} 1&8&1&1 \\ 1&1&8&1 \\ 1&1&1&8 \\ 8&1&1&1 \end{bmatrix} \rightarrow \begin{bmatrix} 2&2&2&2 \\ 2&2&2&2 \\ 2&2&2&2 \\ 2&2&2&2 \end{bmatrix} \rightarrow \begin{bmatrix} 2&10&2&2 \\ 2&2&10&2 \\ 2&2&2&10 \\ 10&2&2&2 \end{bmatrix} \rightarrow \begin{bmatrix} 2&16&4&2 \\ 4&2&16&2 \\ 2&2&2&18 \\ 16&4&2&2 \end{bmatrix}$$

What FANT is doing in its memory update is to memorize both the immediate previous solution and the best solution so far. These are used to update the memory. We can observe similar patterns in the memory structures. The first and third matrices reflect when either a solution that is best so far or a provisory solution that is equal to the best so far is found. In both cases all entries of the matrix equal to a constant number. The third matrix is different from the first matrix by an increment of one for each entry and represents *diversification.* The second and fourth matrices are also similar in pattern. They represent memory updates. The fifth matrix represents an update but a new solution is found that is different from the best solution so far. A careful examination of the matrix shows that the structure of the memory is now *influenced* by the new solution. The fifth matrix is of lower *quality* than all the previous matrices because no better solution has been found so far. Let us dig deeper to explore how far ant systems can remember. Accessing the *build_new_solution* procedure, the provisory solution found is equal to the best solution so far. FANT simply levels the memory to entries equal to 3. When the memory is updated the memory will have the value 12 $(6+3+3)$ in positions $j = \{2,3,4,1\}$ for $i = \{1,2,3,4\}$. The two propagations are as follows:

$$\begin{bmatrix} 3&3&3&3 \\ 3&3&3&3 \\ 3&3&3&3 \\ 3&3&3&3 \end{bmatrix} \rightarrow \begin{bmatrix} 3&12&3&3 \\ 3&3&12&3 \\ 3&3&3&12 \\ 12&3&3&3 \end{bmatrix} \qquad \square$$

What has FANT done? It has simply remembered that the best solution so far is still $\{2\ 3\ 4\ 1\}$, so it ignores an inferior solution $\{3\ 1\ 4\ 2\}$. FANT is now *influenced* by the structure of the best solution so far. The term, *quality* is reflected by the best solution so far.

We can go on and on with this illustration. It may be more fruitful to stop here, now that the principle of memory update is clear. The concept of memory update will be revisited in the Sec. 7.7.1.4. on FANT Simulator outputs when results will be listed.

7.3.5. *Intensification strategies*

The memory update strategy accomplishes intensification in ant systems. When the best solution so far has improved, the parameter r is set to 1 and all the entries of the memory matrix T are set to 1. However, the positions of j that reflect the best solution so far have entries of $r^* + 2$. The aim of this resetting is to intensify the search around the best solution so far, π^* by giving less importance to the past of the search. Intensification strategies are based on *intensifying* rules to favour solution features that are historically good. During intensification stage, the ant search concentrates on examining neighbours of good solutions. It is a time to glean the neighbour of good solutions in order to drive the search to possibly better solutions.

7.3.6. *Diversification strategies*

When the best solution so far has not improved for a long time, it becomes necessary to *diversify* the search. Diversification strategies involve erasing all short term memory before resuming from the best of the recorded solution. Operationally, when the provisory solution μ generated in the *build_new_solution* procedure is equal to the best solution so far, then the parameter r is increased by one unit and all the entries of T are set to r. This simple process described favours search away from the best solution so far. It is therefore a mechanism to *explore* new regions of solutions. Another way of simulating diversification is to include a *restarting procedure*. Diversification is achieved in FANT through the following procedure.

```
repeat
{
build_new_solution;
:
}
until {last iteration}
```

Every time the *build_new_solution* is accessed, a new solution is constructed. This helps the ant algorithm to diversify. On the other hand, too many iterations may lead to excessive computation cost with no effect of solution quality. Therefore a trade-off is required in specifying the number iterations necessary to actuate diversification.

7.4. HAS: Hybrid Ant System

Hybrid ant system (HAS) is a completely different way of implementing an ant system that has been proposed by Gamberdella, Taillard and Dorigo (1997). Indeed, traditional ant systems build a solution with the memory from scratch. In hybrid ant system, the memory is used to modify existing solution, in the spirit of local search. This ant variant has been applied to the quadratic assignment problem (QAP). Hybrid ant system is based on the schematic algorithm of Fig. 7.4 and its salient aspects are presented in details in Fig. 7.5. The algorithm uses the fast ant procedure of Fig. 7.3 as in FANT. In this section, we describe the hybrid ant system in some detail. We consider the memory or pheromone trail, solutions manipulation, intensification, and diversification.

```
Generate m initial solutions, each one associated to one ant
Initialise the pheromone trail
  For I^max iterations repeat
        For each ant k = 1,...., m do
              Modify the solution associated to ant k using the
                    pheromone trail
              Apply a local search to the modified solution
              Associate a new starting solution to ant k using an
                    intensification mechanism
        End For
        Update the pheromone trail
        Apply a diversification mechanism
  End For
```

Fig. 7.4. The HAS algorithm structure.

{Initialisation}
Generate m random initial permutation $\pi^1(1), \ldots, \pi^m(1)$, each one associated to an ant
Improve $\pi^1(1), \ldots, \pi^m(1)$ with the local search procedure
Let π^* be the best solution
Initialise the pheromone trail matrix T
Activate intensification
{main loop}
For $i = 1$ to $\mathrm{I}^{\max}$ repeat
    {solution manipulation}
    For each permutation $(1 \leq k \leq m)$ do
        Apply R pheromone trail swaps to $\pi^k(i)$ to obtain $\hat{\pi}^k(i)$
        Apply the local search procedure to $\hat{\pi}^k(i)$ to obtain $\tilde{\pi}^k(i)$
    End For
    {intensification}
    For each ant k do
        If intensification is active
          Then
                Best permutation between $\pi^k(i)$, and $\tilde{\pi}^k(i)$
          Else
                $\pi^k(i+1) \leftarrow \tilde{\pi}^k(i)$
    End For
    If $\forall\,\mathrm{k}\pi^k(i+1) = \pi^k(i)$ then deactivate intensification
    If $\exists\,\mathrm{k}$ such that $f(\pi(\tilde{i}^*)) < f(\pi^*)$
        Then
        Update π^*, the best solution found so far
        Activate intensification
        {pheromone trail updating}
        Update the pheromone trail matrix
        {diversification}
        If S iterations have been performed without
        improving π^* then perform a diversification
End For

Fig. 7.5. The HAS algorithm.

7.4.1. *Memory implementation*

There are two different memories in hybrid ant system. The first one is a matrix that records a statistics on the best solutions produced by the search. The second one is a population of m (equal to 10) solutions, in addition to the best solution produced by the search. Initially, no information is contained in the pheromone trail matrix. This means that all pheromone trails τ_{ij} are set to the same value τ_0. Since pheromone trails are updated by taking into account the absolute valued of the solution obtained, τ_0 must take a value that depends on the valued of the solutions that will be visited. For the hybrid ant system for the quadratic assignment problem, Gambardella, Taillard and Dorigo chose to set $\tau_0 = 1/(Q.f(\tau^*))$, where τ^* is the best solution found so far and Q a parameter. The re-initialisations of the pheromone trails are done in the same way, but π^* might have changed. From a memory point-of-view, hybrid ant system is between genetic algorithms that use a population of solutions only and fast ant systems that use a statistics-matrix only.

7.4.2. *Provisory solution generation*

Instead of building a completely new solution, hybrid ant system modifies all the solutions in memory by swapping a number of times two elements of the solutions. Pheromone trail based modification is applied by each ant to its own permutation π^k.

It starts with π^* and produces a permutation $\hat{\pi}^k$. It consists in repeating R of the following swaps. First, an index r is chosen, randomly between 1 and n. Then, a second index $s \neq r$ is chosen and the elements are swapped in the current solution π_k. The second index is chosen according to one of two different randomly chosen policies. With a probability given by a parameter q, s is chosen in such a way that $\pi^k_{r\pi_s} + \pi^k_{s\pi_r}$ is a maximum. This policy consists in exploiting the pheromone trail. The second policy, chosen with probability $(1-q)$, consists in exploring the solution space by choosing the second index s with a probability proportional to the values contained in the pheromone trail. More precisely, s is chosen with probability:

$$\frac{\pi^k_{r_{\pi s}} + \pi^k_{sr_s}}{\sum_{j\neq r}(\pi^k_{r\pi_j} + \pi^k_{j\pi_r})}$$

7.4.3. *Improvement procedure*

Local search consists in applying a complete neighbourhood examination twice with first improving strategy to a solution to produce a solution. The local search procedure is shown in Fig. 7.3, where is the difference in the objective function value when exchanging the elements π_i and π_j in π. The evaluation of $\Delta(\pi, i, j)$ can be performed in $O(n)$ using the formula:

$$\Delta(\pi, i, j) = (c_{\pi_j \pi_j} - c_{\pi_i \pi_i}) + (c_{\pi_j \pi_i} - c_{\pi_i \pi_j}) \\ + \sum_{k \neq i,j} (c_{\pi_k \pi_j} - c_{\pi_k \pi_i}) + (c_{\pi_j \pi_k} - c_{\pi_i \pi_k})$$

This procedure systematically examines all the possible swaps and immediately performs an improving swap, if one is found. The order in which the swaps are examined is randomly chosen. Therefore, the local search procedure examines $O(n^2)$ swaps and can be executed in $O(n^3)$. it does not necessarily reach a local optimum, but is fast and may produce different solutions when starting with the same initial, not locally optimal solution. To speed-up the local search, it is clear that it is not necessary to compute $\Delta(\pi, i, j)$ if $i = j$ and the second neighbourhood examination is useless if the first one did not find an improving move.

7.4.4. *Memory update*

The update of the pheromone trails is done differently from the standard ant system where all the ants update the pheromone trails with the result of their computations. Indeed, this manner of updating the pheromone trails determines a very slow convergence of the algorithm. To speed-up the convergence, the pheromone trails are updated by taking into account only the best solution produced by the search to date. First, all the pheromone trails are weakened by setting $\tau_{ij} = (1 - \alpha_1).\ \tau_{ij}$, $(1 \leq i, j \leq n)$ where $(0 < \alpha_1 < 1)$ is a parameter that controls the evaporation of the pheromone trail. A value of close to 0 implies that the pheromone trails remain inactive a long time, while a value close to 1 implies a high degree of evaporation and a shorter memory of the system. Then, the pheromone trails are reinforced by considering only τ^*, that is, the best solution generated by the system so far and setting for all i.

7.4.5. *Intensification*

The function of intensification is to explore the neighbourhood of good solutions more completely. The intensification mechanism is activated when the best solution produced by the search so far has been improved. When intensification is active each ant starts its next iteration with the best permutation between τ^k and $\tilde{\pi}^k$. This is different from what happens when intensification is not active, in which case the permutation maintained is $\tilde{\pi}^k$. The intensification flag remains active at least one ant succeeds in improving its solution during an iteration. The rationale for intensification is that it favours search in the neighbourhood of the new best solution found. In fact. since pheromone trail updating is governed by the value of the solution τ^* at a given iteration, the pheromone trail distribution is determined by previous τ^*'s. When a new solution τ^{k*} is found such that $f(\tau^{k*}) < f(\tau^*)$, τ^{k*} becomes the new τ^*. In general it will take some iterations before the influence of the new τ^* on pheromone trail distribution will have all its effect. Intensification focuses the search around the new τ^* while information about the new τ^* grows into the pheromone trail matrix T.

7.4.6. *Diversification*

The aim of diversification is to encourage the search process to examine solutions with different structure. The diversification mechanism is activated if during the last S iterations no improvement to the best solution generated is detected. Diversification consists in erasing all the information contained in the pheromone trails by a re-initialisation of the pheromone trail matrix and in generating randomly a new current solution for all the ants but for one ant that preserves the best solution produced by the search so far.

7.5. The FANT Simulator

In this section, we shall present the design for the fast ant simulator. We first construct the data structures and then present the algorithms necessary to implement the fast ant system described already. Specifically, we write a Pascal computer code called fast ant (FANT) which contains all the ingredients required for solving combinatorial optimization problems. Specifically, FANT is applied to solving flow shop problems in Sec. 7.7.

7.5.1. *FANT data structures*

Fast ant system processes a string of information. Therefore we simply construct a record containing the pheromone trail or memory, the solution produced at the current iteration, the best solution produced so far, and auxiliary information. The Pascal code for the structure is given as follows:

```
        const n_max = 20; {problem size}
type
        type_matrice   =  array [1..n_max, 1..n_max] of integer;
        type_vector    =  array [1..n_max] of integer;
        data-matrice   =  array [1..n_max, 1..n_max] of real;
        data-vector    =  array [1..n_max] of real;
        solutionData = record
                memory: type_matrice;   {memory}
                p       : type_vector;  {current solution}
                mp      : type_vector;  {best solution so far}
                ptemp  : type_vector;   (for house keeping processing}
                processt:data-matrice;  {input data}
                complete: data-matrice; {an objective function used for most
                                        problems}
                zigmaT: real;           {objective function value}
end;
var solution: solutionData;
         Cost, minimum_cost: real;
         nb_iterations, R, no_iteration, m, n, increment: integer;
```

There is one constraint that is declared: the maximum problem size, $n_$max. This sets the upper bound on the problem size. Although there are other ways of implementing FANT so that the maximum size is not fixed *a priori*, we have tried to keep things simple. The *solutionData* is defined in the *type* block as a record composed of a type *type_matrice* called memory, a type *type_vector* called *p*, a type *type_vector* called *mp*, a type *type_vector* called *ptemp*, a type *data-matrice* called *processt*, a type *data-matrice* called *complete*, and a *real* variable called *zigmaT*. These represent the pheromone trail memory, the current solution, the best solution so far, a temporary house keeping vector used in the local search, the input data, an objective function that is used by a number of other objective functions, and the objective function value for the problem being solved respectively. The type *type_matrice* is itself a square array of *integer* type of size *n*_max by *n*_max. The type *type_vector* is itself an array of *integer* type, indexed between 1 and *n*_max. The type *data-matrice* is a square array of type *real* of size *n*_max by *n*_max. The type *data-vector* is an array of type *real*, indexed between 1 and *n*_max.

The declaration of solution together with the declaration of a number of other global program variables set the scene for the FANT simulator. The *real* variables *cost*, *initial_cost* correspond to the best cost of an objective function so far and the initial cost respectively. The *integer* variables R, ant *increment* are used together with r^*, r for the memory updates respectively. The *integer* variable *nb_iterations* defines the number of iterations for FANT. The problem size being solved is defined by the integer variables m and n. We shall now proceed to the development of the algorithms necessary to simulate the FANT system.

7.5.2. *FANT algorithms*

In Sec. 7.3, we saw the three main components of FANT as being: provisory solution generation, improvement procedure, and memory update. These three were discussed in detail and now we present the algorithms. It is important to understand how these three components work because they constitute the power-house of FANT. When you understand the underlying principles of a power-house then it becomes simple to apply the power generated to meet different needs. This is what we aim to achieve in this chapter.

The more you understand how a problem-solver works, the more you can apply it to solve different problems. By the time you complete reading through this section, you would have understood all about fast ant systems, at least the fundamental principles involved and how it can be simulated to solve real life problems. Now let us plunge into the details of the three algorithms that drive the ant system. We commence with *build_new_solution* for building a new solution, then move on to *local_search* for the neighbourhood local search, and conclude with *update_memory* for updating the pheromone trail or memory.

```
  : Final Version: 6 October 1999 : GC Onwubolu }
{**************************************************}
procedure build_new_solution;
{  build a new solution probabilistically }
 var i, j, ii, k,uu,u,s,flag :integer;
  choice : type_vecteur_boolean;
  nb : type_vecteur;
  identify : boolean;
 BEGIN

  s := 0; flag := 0;
  for i := 1 to n do nb[i] := 0;

  for i := 1 to n do begin
   for j := 1 to n do begin
    nb[i] := nb[i] + solution.memory[i][j];   {sum rows of memory}
   end;
  end;

  for j := 1 to  n do choice[j] := false;    {initialize flag}
   uu := unif(init_sol_seed,1, n);
  for ii := 1 to n do begin                  {I}
    i := ((ii + uu) mod n) + 1;              {i <> I}
    u := unif(init_sol_seed,1, nb[i]);       {u}
    j := unif(init_sol_seed,1, n);           {J}
    while (choice[j] = true) do  j := (j mod n) + 1;  {j already considered }
    s := solution.memory[i][j];              {s}
    flag := flag + 1; if ( flag > 50 ) then exit; {avoid getting stock}
    while (s < u) do begin                   {probability}
     j := (j mod n)+1;                       {different j, not existing }
      while (choice[j] = true) do j := (j mod n) + 1; {j already considered }
      s := s + solution.memory[i][j];        {sum s}
     end; {while s<u}
    solution.p[i] := j; choice[j] := true;   {partial provisory solution}

    for k := 1 to n do nb[k] := nb[k] - solution.memory[k][j]; {evaporation}
  end; {ii}
   identify := true;
  for k := 1 to n do begin
   if (identify = true) then begin   {ensure only new solution is considered}
    if (solution.p[k] <> solution.mp[k]) then identify := false;
   end; {if}
  end; {k}
  if (identify = true) then begin
   increment := increment + 1;       {counter to strengthen memory}
   initialise_memory(n, increment, solution);   { strengthen memory}

  if  ( another = 1)   then begin   {print memory in file}
  writeln(f2[g],'Provisory = Best solution: diversify ');
  print_memory_content;
  end; {if print}

   end;  {if}

 END;
{**************************************************}
```

The algorithm to build a new solution, *build_new_solution*, commences by assuming that the pheromone trail or memory is already initialized. The algorithm sums up each row of the memory. The sum per row of the memory is used to simulate the probability from which a variable j is chosen. This variable j generated randomly with a probability that is dependent on the statistics of the memory together with another variable i generated randomly, constitute the mechanism for constructing a provisory solution.

If the provisory solution generated is equal to the best solution produced so far, then the memory is incremented by one and all the entries of the memory are set to the incremented value. The aim of this is to *diversify* the search when the information contained in the memory is not very different from the best solution produced so far. Once a provisory solution is generated, on objective function value is then computed.

The provisory solution generated in the procedure *build_new_solution* is not so good. Therefore, we apply to this solution the first steps of a first improving neighbour procedure. We achieve this first improvement through the algorithm *local_search* which we now define as follows:

```
  : Final Version: 6 October 1999 : GC Onwubolu }
{*************************************************}

procedure local_search(var m : integer; n : integer;
               Cout : real; solution: solutionData );
{ local search }
 var a_tester : type_matrice_boolean;
     r, s, i, ii, rr, ss, j : integer ;
     delta  :  real;
     ps, pr : type_vecteur ;
     flag : boolean;

 BEGIN

   flag := true;

  for i := 1 to n do begin
   for j := 1 to n do begin
    a_tester[i][j] := true;
   end;
  end;
  for i := 1 to n do a_tester[i][i] := false;
```

```
  for ii := 1 to 2  do begin
   if ( flag = true) then begin
    flag := false;
    permutative_solution(n, solution);
    for k := 1 to n do pr[k] := solution.ptemp[k];
    for rr := 1 to n do begin
     r := pr[rr];
      permutative_solution(n, solution);
      for k := 1 to n do ps[k] := solution.ptemp[k];
      for ss := 1 to n do begin
       s := ps[ss];
      if (a_tester[r][s] = true) then begin
       delta := calc_delta(m, n, r, s, solution{,b});
        if (delta < 0) then begin
         Cout := Cout + delta;
           swap(solution.p[r], solution.p[s]);
           flag := true;
          for i := 1 to n do begin
            for j := 1 to n do begin
               a_tester[i][j] := true;
              end;
            end;
          for i := 1 to  n do  a_tester[i][i] := false;
         end;
        a_tester[r][s] := false; a_tester[s][r] := false;
       end;

       end;
     end;
    end;
  end;
 END;
{*************************************************}
```

The algorithm, *local_search* commences by generating two permutative solutions randomly. The generation of these two permutative solutions is simulated in the function *permutative_solution*. These are represented as arrays *pr* and *ps* each of the problem size. An exhaustive enumeration is carried out between these permutative solutions using the indices r and s to calculate the change in cost due to a move in function *calculate_delta*. If *calculate_delta* is less than zero (negative) it means that the move is favourable and so the total cost obtained from the provisory solution is decremented by this value. When this condition holds, then we swap *pr* and *ps* at positions r and s.

We have already explained that the whole process is repeated when considering a move. The simple reason is that if we move from r to s, it is possible to move from s to r. We note that this algorithm is very fast and can lead to very quick convergence of FANT. We want a fast algorithm that can easily solve difficult real life problems in reasonable computation time. But we do not want the algorithm to converge too soon.

This leads to it being stuck in a local optimum. To alleviate this problem, we *diversify* the search. We have already discussed how FANT diversifies it search. It is simply by updating the statistics of the memory matrix as we shall now discuss further.

There are two tactical issues that are involved in improving FANT: intensification and diversification. We intensify search around an improved solution to glean the neighbour for possible better solutions. But when carrying out a local search, the tendency is for the algorithm to converge to a local optimum. Therefore, we update the memory to diversify or jump out from a possible local optimum. These two key issues are important and help us to move towards better solutions. Remember that the goal of intelligent optimization techniques is to improve solutions toward the optimum solution by means of an adaptive memory. This is what we are doing when we consider the different attributes of FANT necessary for optimization.

7.6. HAS Simulator

As we discussed earlier in this chapter, the FANT model varies from the HAS model primarily in the implementation of the pheromone trail based modification. With the HAS model, two probability policies exist in the choice of the indices r and s as already discussed. These two probability policies differ from the FANT model. In addition, while a population of artificial ant agents are considered in the HAS model, one artificial ant agent is considered in the FANT model. Both models utilise local search for advancing towards the global optimum solution. In both the HAS and FANT models, the pheromone trails are used to modify an existing solution, in the spirit of neighbourhood search. Local updates of the pheromone are not performed, but only a global update.

This makes the search more aggressive and requires less time to reach good solutions. These attributes in both the HAS and FANT models are a remarkable departure from the traditional ant systems. Both models accommodate intensification and diversification mechanisms.

Therefore, we note that the major differences between the HAS and FANT models is the probability policies for solution modifications and the number of artificial agents involved. Other than these two minor variations, the implementation of the HAS simulator should be straightforward. Using the HAS and FANT discussions as a guide, we leave it as an exercise for you to develop the data structures and algorithms needed to implement a HAS simulator.

7.7. Application of FANT to Flow Shop Scheduling: 1-Criterion

Table 7.1 lists some current applications of ant systems so that readers become familiar with current applications, and they can then decide on possible area of extension of ant systems applications.

My work has largely centred on engineering applications of ant systems. One such application is scheduling in flow shops. Therefore, in this section, we present flow shop scheduling and the important performance measures that are used for solving scheduling problems. Then we present results of some flow shop scheduling problems that I have solved using ant system. This way, the computational power of ant can be appreciated as well as its advantages in applications to practical, real life problems.

7.7.1. *Flow shop scheduling*

A flow-shop schedule is one in which all jobs must visit machines or work centres in the same sequence. Processing of a job must be completed on a current machine before processing of the job is started on a succeeding machine. This means that initially all jobs are available and that each machine is restricted to processing only one job at any particular time. Since the first machine in the facility arrangement is the first to be visited by each job, the other machines are idle and the other jobs are queued. Although queuing of jobs is prohibited in just-in-time (JIT) manufacturing environments, flow-shop manufacturing

Table 7.1. Ant system applications in search and optimisation.

Year	Investigators	Description
	SYMMETRIC AND ASYMMETRIC TSP	
1992	Dorigo	Seminal work in which Ant System (AS) was first introduced.
1995	Dorigo and Gambardella	Hybridisation of ant system and Q-learning, a well known learning algorithm.
1996	Dorigo, Maniezzo, and Colorni	Fundamental application of Ant System to travelling salesman problem (TSP).
1996	Dorigo and Gambardella	A study of some properties of Ant-Q
1997a	Dorigo and Gambardella	Quick, short overview of what ant colony optimisation (ACO) means: applied to TSP.
1997b	Dorigo and Gambardella	Description of the latest and most performing version of ant colony optimisation (applied to TSP).
1997	Stutzle and Hoos	Improvement of the Ant System through MAX-MIN.
	SEQUENTIAL ORDERING PROBLEM	
1997	Gambardella and Dorigo	Hybrid ant system applied to SOP
	QUADRATIC ASSIGNMENT PROBLEM	
1997	Taillard and Gambardella	Application of ant system to structured QAPs.
1999	Gambardella, Taillard, and Dorigo	Hybrid ant system (HAS) applied to QAP.
1999	Stutzle and Dorigo	Ant colony optimisation (ACO) application to QAP.

Table 7.1 (*Continued*)

Year	Investigators	Description
	VEHICLE ROUTING PROBLEM	
1999	Bullnheimer	Seminal work on application of ACO to the vehicle routing problem.
1999	Bullnheimer, Hartl, and Strauss	Describes ACO as improvement on earlier ant systems.
	SCHEDULING	
1994	Colorni, Dorigo, Maniezzo, and Trubian	Ant system application to job shop scheduling.
1997	Forsyth and Wren	Ant system application to bus driver scheduling.
	GRAPH COLOURING	
1997	Costa and Hertz	Co-operative artificial ant agents applied to graph colouring.
	PARTITIONING PROBLEMS	
1997	Kuritz, Layzell, and Snyers	
	TELECOMMUNICATIONS NETWORKS	
1997	Schoonderwoerd, Holland, Bruten, and Rothkrantz	Developed an ant colony algorithm called ABC for routing and load balancing in circuit switched telecommunication networks.
1998	Caro and Dorigo	Application of ant colony to routing in packet switched Telecommunication networks.

Table 7.1 (*Continued*)

Year	Investigators	Description
	PARALLEL IMPLEMENTATION	
1998	Bullnheimer, Kotsis, and Strauss	Parallel ant system implementation.
	OPTIMISATION OF CONTINUOUS FUNCTION	
1995	Bilchev and Parmee	Ant colony applied to continuous design spaces.

continues to find applications in electronics manufacturing, and space shuttle processing, and has attracted much research work. Scheduling an n-job m-machine with the constraint that all machines process jobs in the same order leads to the permutation situation in which there are $n!$ possible sequences to be considered. For large numbers of jobs this easily leads to combinatorial explosion. One-machine and two-machine flow-shop scheduling have been studied extensively using techniques such as branch and bound (Hariri, 1981; McMahon and Burton, 1967) and others have resorted to heuristic techniques (for example, Kochar and Morris, 1987; Park, Pegden, and Enscore, 1984) for seeking optimal solutions. An important aspect of scheduling is sequencing. *Sequencing* is the process of defining the order in which jobs visit a machine. Scheduling is the process of adding start and finish time information to the job order dictated by the sequence. Johnson's algorithm is well suited for two-machine problems and can be extended to three-machine case by splitting machines into two pseudo machines which have processing times equal to the sum of the processing times on the actual machines. A generalization of Johnson's algorithm is that proposed by Campbell, Dudek and Smith (1970) for solving general n-jobs m-machine problems in which $m - 1$ two-machines problems are solved and the sequence having the least makespan is selected.

There are basically two ways in which the scheduling problem can be solved. The first method is interactive, in which the user can interfere with the scheduling system. In this case, the user can input some sequences derived from some heuristics to the scheduling system while several performance measures can then be evaluated. The second method is non-interactive, in which the user does not interfere with the scheduling system. When heuristics that give near optimal solutions are used then there is no need to interfere with scheduling system. Graphical method of scheduling is discussed in Sec. 7.7.1.1 while non-graphical method that relies entirely on heuristic solutions is discussed in Sec. 7.7.1.2.

7.7.1.1. *Graphical method for flow shop scheduling*

Scheduling outputs are usually graphically represented by Gantt charts drawn using machine processing times for each job and ensuring that delay times are taken into consideration. This process is repeated for different sequences until a minimum makespan is found which represents the minimum time taken to complete all the jobs. The corresponding sequence is considered to be the best solution. It becomes immediately obvious that manual method for scheduling is tedious and prone to error. Clearly, an interactive graphical system that displays the work centres and their loading is very useful to a scheduler. The advantages are numerous: the processing of the load pattern for each machine is sped up, the scheduler can easily visualize the output. Moreover, the makespan computed by the system can be compared with the lower bound for each sequence and the scheduler is then easily guided to consider other sequences until the minimization of the makespan is virtually optimal. This section presents an interactive graphical system for scheduling flow-shop environment for any number of jobs and machines. There are two modes in which the scheduler inputs data into the scheduling system. Sequences of operations are either input by the user or generated by the system. The processing times of jobs on machines are input interactively and using this information, the system computes the maximum value of the lower bounds so that the scheduler can be guided during the interactive session to seek for the optimal schedule. The sequences and the processing time information are used by the system to display the Gantt charts for each set of sequence. The ability of such a system to accept user suggestions as well as utilize system-generated sequences makes it flexible for scheduling environment. First, schedulers may not like to depend entirely on sequences suggested by the system. Secondly, even when sequences are suggested by the system, schedulers may want to decide which sequences to use as input in order to proceed with the scheduling.

Data-entry function

There are three types of data input that can be input into a graphical scheduling system: the *primary*, *static* and *system* data. *Primary* data input from the scheduler is made up of the number of machines, number of jobs, and the processing times with inclusion of set-up and transport times. The *static data* are the sequences input to the scheduling system by the scheduler, if the scheduler chooses this option. If there are m-machines and the user chooses to utilize sequences generated by the scheduling system, the system uses the *primary* data to generate $m-1$ sequences, based on the algorithm of Campbell, Dudek and Smith (1970). This is a generalized form of Johnson's algorithm (1954). The collection of the $m-1$ sequences is referred to as *system data.* The *system* or *static* data is used to drive the processor of a graphical scheduling system.

Editor

When the primary data are input in the data-entry mode, it is possible for the scheduler to make mistake. An editor function is available so that primary data are verified before any processing starts. If the primary data is wrong, the system data will also be wrong as well as other results which the processor outputs.

Scheduling processor

The processor performs the scheduling functions. It determines the delay times for each job on the various machines. Computation of the delay time is very important because it determines the loading pattern on the Gantt chart. Delay time is either zero or positive. The delay time determines whether a succeeding machine will have to idle or commence as soon as the subsequent machine finishes operation. The processor also computes the makespan and the maximum lower bound (MLB). The MLB gives the scheduler an indication of the near-optimal makespan, which in turn gives the near-optimum sequence. The makespan is the total time required to complete all jobs. This information is important to the scheduler because it helps to indicate whether or not meeting due-date is possible.

Graphic output function

The processor utilizes the input data and the delay times generated by graphical system to output the Gantt chart. The Gantt chart is made up of time scale on one axis of a graph and the machine centres on the other axis. Blocks having lengths equivalent to their processing times are displayed on the correct machine axis for the particular job being scheduled. The correct sequencing is maintained and placed in the blocks for proper identification. The Gantt chart is a powerful tool when used by a supervisor on the factory shop-floor because it graphically displays the work status of the various machines at any time interval within the makespan. It also shows the makespan, which in this case is the time the last job in the sequence is completed on the last machine. The advantages of a graphical scheduling system are:

(1) shortens scheduling time because only limited data are initially input and processing is fast;
(2) offers flexibility because by using the initial input data various sequences can be used by the system to offer alternative scheduling solutions from which the optimum can be obtained;
(3) conveys scheduling in graphical forms to scheduler and shop-floor operators. Some examples are now given in order to clarify issues concerning scheduling.

Example 3

The 4-machine, 4-job processing times shown in Table 7.2 are input data. Determine the lower bounds for the 4-stage schedule.

Table 7.2. Processing times (minutes).

Job	Machines (m)			
(n)	1	2	3	4
1	2	3.5	1.5	2
2	4.5	3	2.5	1
3	1.5	1.5	5	0.5
4	4	1	2.5	0.5

Solution

Based on the input processing times, a graphical scheduling system computes the lower bounds on schedule makespan for each machine. The lower bound is the summation of the minimum of processing time before the machine can begin processing, the total processing time on the machine, and the minimum of the remaining processing time for the last job after it leaves the machine. From Table 7.2, the minimum processing time for *machine* 1 before the machine can begin processing is zero, for *machine* 2 it is the minimum of {2, 4.5, 1.5, 4} = 1.5 minutes, for *machine* 3 it is the minimum of {(2 + 3.5), (4.5 + 3), (1.5 + 1.5), (4 + 1)} = 3.0 minutes, and for *machine* 4 it is the minimum of {(2 + 3.5 + 1.5), (4.5 + 3 + 2.5), (1.5 + 1.5 + 5), (4 + 1 + 2.5)} = 7.0 minutes. Considering all jobs, the total processing times for *machine* 1 is 12 minutes, for *machine* 2 is 9 minutes, for *machine* 3 is 11.5 minutes, and for *machine* 4 is 4 minutes. The remaining processing times for the last job after it leaves *machine* 1 is the minimum of {(3.5 + 1.5 + 2), (3 + 2.5 + 1), (1.5 + 5 + 0.5), (1 + 2.5 + 0.5)} = 4 minutes, for *machine* 2 is the minimum of {(1.5 + 2), (2.5 + 1), (5 + 0.5), (2.5 + 0.5)} = 3 minutes, for *machine* 3 is the minimum of {2, 1, 0.5, 0.5} = 0.5 minutes, and for *machine* 1 is zero. From these computations, the lower bound for *machine* 1 is the sum of {0, 12, 4} = 16.0 minutes, for *machine* 2 is the sum of {1.5, 9,3} = 13.5 minutes, for *machine* 3 is the sum of {3, 11. 5, 0.5} = 15.0 minutes, and for *machine* 4 is the sum of {7, 4, 0} = 11.0 minutes. The maximum value of 16.0 minutes becomes the lower bound for the 4-stage scheduling system. This maximum lower bound helps the scheduler to judge if a schedule is near optimal solution. □

Example 4

Using the CDS method, find the schedule sequences for Example 3.
Graphical scheduling system now uses the processing times input by the scheduler to generate three possible sequences based on Campbell, Dudek and Smith (CDS). The sequences are {3, 1, 2, 4}, {1, 3, 2, 4,}, and {1, 2, 4, 3}. These form the initial options available to the scheduler as *system data* for input. □

So far, the lower bound and the schedule system sequences generation have been system-driven. This means that the scheduler is not in control. We prefer a situation in which the scheduler can direct the system towards an optimal solution through the utilization of *system* and *static* data. We refer to this as schedule optimization. First, the scheduler inputs the system sequences for use by the processor. The sequences {3, 1, 2, 4}, {1, 3, 2, 4,}, and {1, 2, 4, 3} obtained using the CDS method are input by the user. Their makespans are also computed as {17.5 minutes, 17.5 minutes and 20 minutes}. Since the lower bound is 16 minutes, sequence {1, 2, 4, 3} need no further consideration by the scheduler because it has a deviation of 25%. Sequences {3, 1, 2, 4} and {1, 3, 2, 4,} each has deviation of 9.38%.

For sequence {3, 1, 2, 4}, we start by assigning *job* 3 to *machine* 1 at time zero. The job lasts for 1.5 minutes on *machine* 1. Since all jobs must visit *machine* 1 first, while *job* 3 is being processed by *machine* 1, jobs 1, 2, and 4 are queued and machines 2, 3, and 4 are idle. At the end of 1.5 minutes, *job* 3 visits *machine* 2 while *job* 1 goes on to *machine* 1. Jobs 2 and 4 are queued for *machine* 1 while machines 3 and 4 idle. Meanwhile, job 3 is completed on *machine* 2 at 1.5 minutes while *job* 1 is still being processed on *machine* 1. This means that *machine* 2 idles for 0.5 minutes, waiting for *machine* 1 to complete *job* 1 before it commences. *Job* 1 then commences on *machine* 2 and completes 3.5 minutes later while *job* 1 is still being processed on *machine* 1. *Machine* 2 idles for 1 minute, waiting for *machine* 1 to complete to complete *job* 2. *Job* 2 takes 3 minutes on *machine* 2 and is completed while *job* 4 is still being processed on *machine* 1, so that *machine* 2 idles for 1 minute before *job* 4 commences on *machine* 2. *Job* 3 takes 5 minutes on *machine* 3 while *job* 1 on *machine* 2 is already completed 1 minute earlier on *machine* 2. *Job* 1 is completed on *machine* 3 while *job* 21 is still being processed on *machine* 2. *Machine* 3 therefore idles for 2.5 minutes before *job* 2 commences. *Job* 4 takes only 1 minute on *machine* 2 while *job* 2 takes 2.5 minutes to be completed on *machine* 3. This means that *job* 4 starts as soon as *job* 1 is completed on *machine* 3 and therefore there is no idling. On *machine* 4, *job* 3, *job 1*, *job* 1 and *job* 4 commence when they are completed on *machine* 3. This means that *machine* 4 idles for 1 minute before *job* 1 starts, idles for 3 minutes before *job* 2 commences, and idles for 1.5 minutes before *job* 4 starts. The processor unit of the scheduling system keeps track of the idle times and the output module uses the idle times and sequenced processing times to draw the Gantt chart. The makespan is computed and displayed so that the scheduler can decide whether to accept the schedule or not. The deviation of the makespan from the lower bound helps the scheduler to take an appropriate decision.

The scheduler may decide to try a few combinations of the system sequences that gave the least makespan. By swapping the 1, and 2 in sequence {3, 1, 2, 4}, it is observed in that the sequence {3, 2, 1, 4} gives the least makespan of 17 minutes, with a deviation of 6.25% form the lower bound. The worst case is with sequence {4, 2, 3, 1} resulting in a deviation of 40.63%.

Complete enumeration

The methodical approach in which lower bound, system sequence and schedule optimization are used as scheduling tools, cuts down scheduling time by at least 80%. For the 4-*machine* 4-job problem considered, if the methodical approach discussed in this paper was not adopted, the scheduler would have to generate 4! (24) sequences and then search for the schedule that yields the minimum makespan. Since only 4 sequences are considered in order to obtain the optimum solution and there are 24 possible sequences, scheduling time is therefore cut down by $(1 - 4/24) \times 100\%$, giving about 80% saving. The Gantt charts solutions to the corresponding complete enumeration solution for this 4-*machine* 4-job problem are not shown but their makespans and the deviations from the lower bound of 16.0 minutes of the 24 complete enumeration scheduling solution are given in Table 7.3. The results show that sequences starting with *job* 3 have the least deviations, followed by sequences starting with *job* 1, then followed by those starting with *job* 1, with the worst case being sequences starting with *job* 4.

Machine and labour utilization

The various sequences for the 4-machine, 4-job show that machine utilization is not very high. We consider as an example, sequence {3, 2, 1, 4} which gave the optimal makespan value of 17 minutes. *Machine* 1 is only in operation for a total of 12 minutes after which it idles for the remaining 5 minutes before the makespan time. This means that the non-utilization time for *machine* 1 is 5 minutes. *Machine* 2 does not start operation till 1.5 minutes when *job* 3 is completed on *machine* 1; there is a further idle time of 3 minutes after *job* 3 is complete on *machine* 2 before *job* 2 starts. *Machine* 2 is operational for total of 9 minutes after which it idles for the remaining 3.5 minutes before the

Table 7.3. Deviations for complete enumeration solution.

Sequence	Makespan	Deviation (%)
{4, 2, 3, 1}	22.5	40.63
{4, 2, 1, 3}	22	37.50
{4, 1, 2, 3}	21.5	34.38
{1, 4, 2, 3}	21.5	34.38
{2, 4, 1, 3}	21	31.25
{2, 3, 4, 1}	21	31.25
{2, 4, 3, 1}	21	31.25
{4, 1, 3, 2}	20.5	28.13
{2, 1, 3, 4}	20.5	28.13
{2, 1, 4, 3}	20.5	28.13
{1, 4, 3, 2}	20.5	28.13
{4, 3, 2, 1}	20	25
{1, 2, 3, 4}	20	25
{1, 2, 4, 3}	20	25
{3, 4, 2, 1}	20	25
{2, 3, 1, 4}	19.5	21.88
{3, 2, 4, 1}	19	18.75
{4, 3, 1, 2}	18.5	17.63
{3, 4, 1, 2}	18.5	17.63
{3, 1, 4, 2}	18.5	17.63
{1, 3, 4, 2}	18.5	17.63
{1, 3, 2, 4}	17.5	9.38
{3, 1, 2, 4}	17.5	9.38
{3, 2, 1, 4}	17	6.25

makespan time is reached. The non-utilization time for *machine* 2 is 8 minutes. *Machine* 3 does not start operation until 3 minutes from the starting time of *machine* 1. The machine is operational for 10.5 minutes and *job* 1 idles for 2 minutes after *job* 2 is completed. The machine idles for 0.5 minutes before the makespan time; so that there is a total non-utilization time of 7.5 minutes. *Machine* 4 wastes 8 minutes from the starting time of *machine* 1 before commencing operation. The sum of the idle times for *job* 3, *job* 2 and *job* 1 is 5 minutes, so the total non-utilization time for *machine* 4 is 13 minutes.

If machine utilization is the priority of the operations/production manager, the next schedule of jobs may be assigned to *machine* 1 after 12 minutes from start until the makespan of 17 minutes; the next jobs may also be assigned to *machine* 2 after 13.5 minutes from when *machine* 1 commenced first operation. *Machine* 3 may utilize 3 minutes to complete an earlier schedule before *job* 3 is started on it. *Machine* 4 could be busy finishing a previous schedule for 8 minutes before *job* 3 commences. The difficult times to utilize are idle times between machines since all jobs follow the same sequence in flow-shop environment. This is one of the major disadvantages.

If the priority of the operations/production manager is manpower utilization, then workers are only engaged on machines only from the time job start on a machine till when it finishes on the machine. For *machine* 2, workers should be assigned to it for 12 minutes; this time includes the 3 minutes idle time for *job* 2. The machine is operational for only 9 minutes and reassigning the worker on that machine for 3 minutes may not be necessary. For *machine* 3, workers should be assigned for 12.5 minutes, again the machine being actually busy for 10.5 minutes and is idle for 2 minutes just before *job 1* starts on the machine. For *machine* 4 workers may be assigned for 9 minutes with the machine being busy for only 4 hours. The only problem with *machine* 4 is that it has the highest *in-process idle time* i.e. idle times from when a machine starts the first job until the last job is completed. For example, the *in-process idle time* for *machine* 2 is 3 minutes, for *machine* 3 is 2 minutes, and for *machine* 4 is 5 minutes. If *machine* 4 has no previous scheduled jobs to complete be *job* 3 is put on it, the workers for *machine* 4 can be assigned elsewhere for 8 minutes before *job* 3 starts on it.

The decision to schedule for machine or labour utilization depends on how far the work-order target is met at a particular time. Re-deployment of workers when they are not usefully engaged on machine is known as *flexible labour* in Just-in-time operation. Companies operating JIT encourage operators to idle rather than produce unnecessary stock. They may be re-deployed help other workcentres that require attention.

Although graphical method of scheduling is flexible, it can is only applicable to very simplistic problems. Real life problems are not simplistic. How do we know, for example, what sequences to suggest to a scheduling system? Since graphical systems can be applicable to trivial scheduling problems, we abandon this method since we are interested in solving combinatorial optimization problems.

7.7.1.2. *Heuristic methods for flow shop scheduling*

In this section, well-known scheduling heuristics of Palmer (1965), Campbell, Dudek and Smith known as CDS (1970), Gupta (1971), Dannenbring (1977), Hundal and Rajagopal (1988) and Ho and Chang (1991) are discussed. These are heuristics often referred to in the literature. The reason for discussing them is to ascertain how why compare with FANT. The following notations apply.

m: number of machines in series
n: number of different jobs processed by the m machines in the same order
i: machine index ($i = 1, \dots, m$)
j: job index ($j = 1, \dots, n$)
k: index for a sub-problem when using the CDS heuristic ($k = 1, \dots, m-1$)
t_{ij}: processing time (setup included) for the operation of job j on machine i
s_j: slope index when using Palmer, and Hundal and Rajagopal methods ($j = 1, \dots, n$)
$f(j)$: slope index for Gupta's method ($j = 1, \dots, n$)
g_{ij}^k: gap values when using Ho and Chang's method
X, Y: largest and least values of gaps for associated positions in Ho and Chang's method

Palmer. A job priority function known as the slope index is specified. Jobs with processing times that tend to decrease from machine to machine will have higher priority. Conversely, jobs with processing times that tend to increase from machine to machine will have lower priority. The jobs are then sorted in ascending order given by the priority indices. The following description is given for Palmer:

(a) Compute for each job j

$$s_j = \sum_{i=1}^{m}(m - 2i + 1)t_{ij} \quad V_j \,,$$

(b) Rank jobs in ascending order of the s_j values.

CDS. The method generates $M - 1$ artificial two-machine problems from the M machines. Using Johnson's algorithm (1954), an optimal sequence for each artificial sub-problem is determined. For each of the sequences, the actual performance measure is computed and the schedule with the least performance measure is chosen. The following description is given for CDS:

(1) Set $k = 1$ and value° $= \infty$
(2) If $k \geq m - 1$, stop, else go to (3)
(3) Compute for each job j:

$$d_{1j}^k = \sum_{j=1}^{k} t_{ij} \quad \text{and} \quad d_{2j}^k = \sum_{j=m-k+1}^{k} t_{ij}\,,$$

where d_{1j}^k and d_{2j}^k are the artificial processing times for the first and second stage, respectively, for the kth problem.
(4) Using the artificial processing times for the jobs, compute a job sequence using Johnson's two-machine algorithm and compute the performance measure (PM^k) for this sequence. If value$^{k-1} > PM^k$, then set value$^k = PM^k$ and record the current job sequence as the "best" sequence so far. Else value^k = value^{k-1}. Set $k = k + 1$ and go to (2).

Gupta. A job index is specified as the minimum of the reciprocal of the sums of the processing times of a job on adjacent machines. The form of the priority function is chosen so that jobs with processing times from machine to machine that are greater than the processing time on the first machine are penalised. The jobs are then sorted in descending order by the priority indices. The following description is given for Gupta:

(1) Compute for all jobs j:

$$t_j = \min(t_{i,j} + t_{i+1,j})\,, \quad i \in [1, m-1]$$

$$f(j) = \begin{cases} \dfrac{1}{t_j} & \text{if } t_{ij} \leq t_{ij} \\ -\dfrac{1}{t_j} & \text{if otherwise} \end{cases}$$

(2) Rank jobs in descending order of the $f(j)$ values.

Dannenbring. Known as the Random Access Procedure (RAP), the method attempts to combine the advantages of Palmer's slope index and the CDS's method. Two artificial machine problems are generated with processing times reflecting the same behaviour as Palmer's slope index. Johnson's algorithm (1954) is then applied to obtain the schedule. The following description is given for Dannenbring or rapid access procedure:

(1) Compute for each job j:

$$d_{1j} = \sum_{i=1}^{k}(m - i + 1)t_{ij} \quad \text{and} \quad d_{2j} = \sum_{i=1}^{k}(i)t_{ij}\ V_j\,,$$

where d_{1j} and d_{2j} are the artificial process
where d_{1j} and d_{2j} are the processing times for the first and second jobs.

(2) Using the artificial processing times for the jobs, compute a job sequence using Johnson's two-machine algorithm and compute the performance measure.

Hundal and Rajagopal. In addition to the Palmer job priority function, two more slope indices are constructed to account for odd number of machines. The best schedule is then selected from Palmer's slope index and the two additional slope indices. The following description is given for Hundal and Rajagopal:

(1) Compute for each job j the Palmer index and two additional indices:

$$s_j = \sum_{i=1}^{m}(m - 2i + 1)t_{ij} \quad \forall\ j$$

$$s_j = \sum_{i=1}^{m}(m - 2i + 2)t_{ij} \quad \forall\ j$$

$$s_j = \sum_{i=1}^{m}(m - 2i)t_{ij} \quad \forall\ j\,.$$

(2) Choose best from the three schedules.

Ho and Chang. Any of the above heuristics is first applied to get an initial schedule, called the incumbent solution. The method utilises 'gaps' which are the times between the end of the last job in a machine and the start of the last job in a subsequent machine, to construct revised values of gaps. Using the incumbent solution and the revised values of gaps constructed, a complete sequence for the job is computed. The following description is given for Ho and Chang:

(1) Apply any heuristic method to get an initial solution, called the incumbent solution.

(2) (a) Construct the gap values:

$$g_{ij}^k = t_{i,k+1} - t_{j,k}, \quad V_{i,j}\ (i,j = 1,2,\ldots,n),\ k=12,\ldots,m-1 \text{ and } i \neq j\,.$$

(b) Construct the discount factors:

(c) Construct the revised gap values

$$g_{ij}^R = \sum_{k=1}^{m-1} g_{ij}^k \delta_{ij}^k \quad V_{i,j}\ (i,j = 1,2,\ldots,n)\,.$$

(3) Let $l(l = 1,\ldots,n)$ represent the positions in the incumbent solution and P_l represent the job in the position l. Set a to 1, and b to n

(4) Search for the largest value (called X) of $(g_{pa\,pl}^r)$, where l is between a and b. Let v represent this l associated with Y.

(5) Search for the least value (called Y) of $(g_{pl\,pa}^r)$, where l is between a and b. Let v represent this l associated with Y.

(6) If $(X < 0)$, $(y > 0)$, and $(|X| \leq |Y|)$, then go to (9)

(7) If $(X < 0)$, $(y > 0)$, and $(|X| > |Y|)$, then go to (10)

(8) If $(|X| > |Y|)$, then go to 9. Else go to (10)

(9) Let $a = a + 1$, and swap the jobs in the positions a and u. Go to (11)

(10) Let $b = b - 1$, and swap the jobs in the positions b and v.

(11) If the new schedule is better than the incumbent in terms of the performance measure, it becomes the new incumbent solution. Else, swap back to the original incumbent solution.

(12) If $b = a + 2$, then stop. Else, go to (4)

In recent years, a growing body of literature suggests the use of heuristic search procedures for combinatorial optimisation problems. Glover and Greenberg (1989) reviewed five search procedures: simulated annealing, genetic algorithms, tabu search, target analysis and neural networks, and concluded that these methods have great potential to address practical optimisation problems. Consequently, over the past few years, several researchers have demonstrated the applicability of these methods, i.e. tabu search to combinatorial optimisation problems such as the flow-shop scheduling (see for example, Taillard, 1990, 1993 and Widmer and Hertz, 1989).

7.7.1.3. *Performance measures for flow shop scheduling*

There are several performance measures used for flow shop scheduling problems described in the literature. This section briefly discusses these performance measures. The next section then presents some solutions for the FANT simulator discussed previously.

Basically, performance measures are divided into two classes: single objective functions and multiple objective functions. Single objective functions consider only one objective. In practice, it is usual to come across real life problems that necessitate multiple criteria.

The following are the common single objective functions in use in flow shop scheduling problems:

(1) Makespan ($C_{\max}$) $= \text{Max}(C_{i-1,j}, C_{i,j-1}) + P_{ij}$

(2) Flow time (FT) $= \frac{1}{mn}\sum_{i=1}^{m}\sum_{j=1}^{n} C_{ij}$

(3) Weighted flow time (WFT) $= \frac{1}{mn}\sum_{i=1}^{m}\sum_{j=1}^{n} w_j C_{ij}$

(4) Waiting time (WT) $= \frac{1}{mn}\sum_{i=1}^{m}\sum_{j=1}^{n} W_{ij}$

(5) Weighted waiting time (WWT) $= \frac{1}{mn}\sum_{i=1}^{m}\sum_{j=1}^{n} w_j W_{ij}$

The following are single, non-regular performance measures:

(6) Flow time variance (FTV) $= \frac{1}{mn}\sum_{i=1}^{m}\sum_{j=1}^{n}(C_{ij} - \bar{C})^2$

(7) Weighted flow time variance (WFTV) $= \frac{1}{mn}\sum_{i=1}^{m}\sum_{j=1}^{n} w_j(C_{ij} - \bar{C})^2$

where $C_{ij} = \text{Max}(C_{i-1,j}, C_{i,j-1}) + P_{ij}$ and $\bar{C} = \frac{1}{mn}\sum_{i=1}^{m}\sum_{j=1}^{n} C_{ij}$

(8) Waiting time variance (WTV) $= \frac{1}{mn}\sum_{i=1}^{m}\sum_{j=1}^{n}(W_{ij} - \bar{W})^2$

(9) Weighted waiting time variance (WWTV) $= \frac{1}{mn}\sum_{i=1}^{m}\sum_{j=1}^{n} w_j(W_{ij} - \bar{W})^2$

where $W_{ij} = [C_{i-1,j} - C_{i,j-1}]^+$ and $\bar{W} = \frac{1}{mn}\sum_{i=1}^{m}\sum_{j=1}^{n} W_{ij}$

(10) Tardiness (T) $= \frac{1}{mn}\sum_{i=1}^{m}[C_{m,j} - d_j]^*$

(11) Weighted tardiness (WT) $= \frac{1}{mn}\sum_{i=1}^{m} w_j[C_{m,j} - d_j]^*$

(12) Number of tardy job (NT) $= \sum_{j=1}^{n} Y_j$

where $Y_j = 1$ if $C_{m,j} > d_j$
0 if otherwise.

These performance measures listed in this section are typical in flow shop scheduling problem. The FANT simulator described in this chapter solves all the performance measures that we have listed in addition, FANT solves bi-criterion problems that are basically combinations of single performance measures.

Now, what are the meanings of these performance measures? It is important to understand them. Makespan is the time from when the first job visits the first machine for processing until when the last job leaves the last machine for processing. The operations manager minimizes the makespan when maximizing utilization is to be achieved. Remember, we discussed utilization in the section on graphical scheduling. Flow time is the total time that all the jobs spend in the manufacturing system. It is the sum of all their completion times. Tardiness is positive lateness. This performance measure is important when we want to produce goods or deliver services just-in-time. Number of tardy jobs is obvious. Total waiting time, as the name implies, is the time that the machines are idle. Again, minimizing the total waiting time encourages high machine utilization. Weights or priorities can be attached to jobs, hence we have weighted flow time, weighted waiting time, waited tardiness. All these performance measures are regular.

It may be useful to measure fluctuations in the performance measures. To achieve this, the variances of the regular performance measures are computed. This is the square of the deviation of measures from their mean values. Variances magnify positive or negative fluctuations and may be useful performance measures if fluctuations are to be avoided.

7.7.1.4. *FANT simulator output for flow shop scheduling*

All the ingredients required for simulating fast ant system have been presented in this chapter. We started by presenting the ant system paradigm, then the ant system fundamentals. The fundamental components of memory, provisory solution generation, improvement of solutions using local search technique, memory updates, intensification and diversification, were discussed in some details. It was indicated that the traditional ant system does not use the memory to modify its solution. As a result of developments in improving artificial ant's solutions, two variants have been implemented quite recently. These include the hybrid ant system and the fast ant system, both of which share common characteristics. The FANT simulator was also presented to include data structures, and algorithms. The HAS simulator was left to readers as an exercise since enough information was presented for the FANT model.

Since the FANT simulator presented in this book is specifically applied to several aspects of flow shop scheduling, we presented flow shop scheduling in some details. First, a flexible method in which the scheduler is in control of directing the scheduling system was presented. The method was found to have several disadvantages. Then several well-known heuristics were presented. These include Palmer, Campbell, Dudek and Smith known as CDS, Gupta, Dannenbring or rapid access, Hundal and Rajagopal and Ho and Chang. Out of all the Heuristics, Ho and Chang's method is the most recent and best amongst them for the makespan performance measure. We do not discuss other heuristics such the ones by Newaz, Enscore and Ham (NEH) or the hybrid tabu search-insertion heuristic of Widmer and Hertz. The latter is reported to be the best-known method to date. We shall soon discover in this section that our FANT competes with the best known methods and indeed, we have solved a wide range of flow shop problems. We now present an output for the FANT simulator. We use the problem taken from the published paper by Newaz, Enscore and Ham (1983) as input to FANT so that its solution quality can be validated. The entire output of the FANT memory is deliberately listed so that the reader can fully follow the discussions on memory management discussed in Sec. 7.3 in this chapter. By the time you examine each output, you will appreciate the fundamental ideas of FANT as one of the *intelligent optimization techniques*, which is the title of this book! You will notice that it takes ant half of a second to solve the problem. This is because FANT is both of very high quality in terms of solutions and it is very fast because of the local search improvement, diversification and intensification mechanisms in-built. For the problem now presented, FANT actually finds the optimum solution, as does NEH.

Numerical Illustration on FANT for Flow Shop Scheduling

To illustrate the FANT procedure outlined in this chapter, the four job, five machine flow-shop problem in Table 7.4 taken from Newaz, Enscore and Ham (1983) is solved.

Let us now see the solution for FANT listed below. FANT commences by finding an initial makespan equal to 58 at the first iteration. This is the initial

Table 7.4. Operation time matrix.

		Machines (m)				
		1	2	3	4	5
Jobs	1	5	9	8	10	1
(n)	2	9	3	10	1	8
	3	9	4	5	8	6
	4	4	8	8	7	2

cost. And the solution sequence is {2 3 4 1}. The memory is initialized with all entries equal to one. Intensification is carried out because FANT has found a solution. Remember, every time a new solution is found, FANT intensifies search for a better solution. Then the memory is updated according the initial sequence. Another provisory solution is found and the memory is updated. FANT continues with a cycle of memory update, diversification when provisory solution equal to best solution, intensification when new best solution is found until it finds the optimum solution. We went through all these operations in Sec. 7.3.4 of this chapter and it is time consuming repeating all that again. Simply recall our discussion in Sec. 7.3.4 on memory updates and then follow FANT memory output listed below:
New best solution value, cost: 58.00 Found at iteration: 1

2 3 4 1

Improved solution: intensify

1	1	1	1
1	1	1	1
1	1	1	1
1	1	1	1

Memory update

1	8	1	1
1	1	8	1
1	1	1	8
8	1	1	1

Provisory = Best solution: diversify

2	2	2	2
2	2	2	2
2	2	2	2
2	2	2	2

Memory update

2	10	2	2
2	2	10	2
2	2	2	10
10	2	2	2

Memory update

2	16	4	2
4	2	16	2
2	2	2	18
16	4	2	2

Provisory = Best solution: diversify

3	3	3	3
3	3	3	3
3	3	3	3
3	3	3	3

Memory update

3	12	3	3
3	3	12	3
3	3	3	12
12	3	3	3

Provisory = Best solution: diversify

4	4	4	4
4	4	4	4
4	4	4	4
4	4	4	4

Memory update

4	14	4	4
4	4	14	4
4	4	4	14
14	4	4	4

Provisory = Best solution: diversify

5	5	5	5
5	5	5	5
5	5	5	5
5	5	5	5

Memory update

5	16	5	5
5	5	16	5
5	5	5	16
16	5	5	5

New best solution value, cost: 57.00 Found at iteration: 7

1	3	4	2

Improved solution: intensify

1	1	1	1
1	1	1	1
1	1	1	1
1	1	1	1

Memory update

8	1	1	1
1	1	8	1
1	1	1	8
1	8	1	1

Memory update

14	2	1	1
1	1	15	1
2	1	1	14
1	14	1	2

Provisory = Best solution: diversify

2	2	2	2
2	2	2	2
2	2	2	2
2	2	2	2

Memory update

10	2	2	2
2	2	10	2
2	2	2	10
2	10	2	2

Provisory = Best solution: diversify

3	3	3	3
3	3	3	3
3	3	3	3
3	3	3	3

Memory update

12	3	3	3
3	3	12	3
3	3	3	12
3	12	3	3

Provisory = Best solution: diversify

4	4	4	4
4	4	4	4
4	4	4	4
4	4	4	4

Memory update

14	4	4	4
4	4	14	4
4	4	4	14
4	14	4	4

Memory update

20	4	8	4
8	4	20	4
4	4	4	24
4	24	4	4

New best solution value, cost: 54.00 Found at iteration: 13

4	3	1	2

Improved solution: intensify

1	1	1	1
1	1	1	1
1	1	1	1
1	1	1	1

Memory update

1	1	1	8
1	1	8	1
8	1	1	1
1	8	1	1

Memory update

1	2	1	14
1	1	15	1
15	1	1	1
1	14	1	2

Provisory = Best solution: diversify

2	2	2	2
2	2	2	2
2	2	2	2
2	2	2	2

Memory update

2	2	2	10
2	2	10	2
10	2	2	2
2	10	2	2

Memory update

2	4	2	16
2	2	18	2
18	2	2	2
2	16	2	4

Provisory = Best solution: diversify

3	3	3	3
3	3	3	3
3	3	3	3
3	3	3	3

Memory update

3	3	3	12
3	3	12	3
12	3	3	3
3	12	3	3

Memory update

3	3	3	21
6	3	18	3
18	6	3	3
3	18	6	3

Provisory = Best solution: diversify

4	4	4	4
4	4	4	4
4	4	4	4
4	4	4	4

Memory update

4	4	4	14
4	4	14	4
14	4	4	4
4	14	4	4

Memory update

4	4	8	20
8	4	20	4
20	4	4	8
4	24	4	4

Memory update

4	4	8	30
12	4	26	4
26	4	8	8
4	34	4	4

Provisory = Best solution: diversify

5	5	5	5
5	5	5	5
5	5	5	5
5	5	5	5

Memory update

5	5	5	16
5	5	16	5
16	5	5	5
5	16	5	5

Memory update

5	10	5	22
5	5	27	5
27	5	5	5
5	22	5	10

Provisory = Best solution: diversify

6	6	6	6
6	6	6	6
6	6	6	6
6	6	6	6

Memory update

6	6	6	18
6	6	18	6
18	6	6	6
6	18	6	6

Memory update

6	12	6	24
6	6	30	6
30	6	6	6
6	24	6	12

Memory update

6	12	6	36
12	6	36	6
36	6	12	6
6	36	6	12

Memory update

6	12	6	48
18	6	42	6
42	6	18	6
6	48	6	12

Provisory = Best solution: diversify

7	7	7	7
7	7	7	7
7	7	7	7
7	7	7	7

Memory update

7	7	7	20
7	7	20	7
20	7	7	7
7	20	7	7

Memory update

7	7	14	26
7	7	26	14
33	7	7	7
7	33	7	7

Provisory = Best solution: diversify

8	8	8	8
8	8	8	8
8	8	8	8
8	8	8	8

Memory update

8	8	8	22
8	8	22	8
22	8	8	8
8	22	8	8

Provisory = Best solution: diversify

9	9	9	9
9	9	9	9
9	9	9	9
9	9	9	9

Memory update

9	9	9	24
9	9	24	9
24	9	9	9
9	24	9	9

Memory update

9	9	18	30
9	18	30	9
39	9	9	9
9	30	9	18

Memory update

9	9	18	45
9	27	36	9
54	9	9	9
9	36	18	18

Memory update

9	9	27	51
18	27	42	9
60	9	9	18
9	51	18	18

Memory update

18	9	27	57
18	27	57	9
66	9	9	27
9	66	18	18

Provisory = Best solution: diversify

10	10	10	10
10	10	10	10
10	10	10	10
10	10	10	10

Memory update

10	10	10	26
10	10	26	10
26	10	10	10
10	26	10	10

Memory update

10	10	20	32
10	20	32	10
42	10	10	10
10	32	10	20

Memory update

20	10	20	38
10	20	48	10
48	10	10	20
10	48	10	20

Memory update

20	10	20	54
10	30	54	10
64	10	10	20
10	54	20	20

Memory update

20	10	20	70
10	40	60	10
80	10	10	20
10	60	30	20

Memory update

20	10	20	86
10	50	66	10
96	10	10	20
10	66	40	20

Provisory = Best solution: diversify

11	11	11	11
11	11	11	11
11	11	11	11
11	11	11	11

Memory update

11	11	11	28
11	11	28	11
28	11	11	11
11	28	11	11

Memory update

11	22	11	34
11	11	45	11
34	11	11	22
22	34	11	11

Memory update

22	22	11	40
11	11	62	11
40	11	11	33
22	51	11	11

Memory update

22	22	11	57
11	11	79	11
46	22	11	33
33	57	11	11

Memory update

22	22	11	74
11	22	85	11
52	22	22	33
44	63	11	11

Memory update

33	22	11	80
11	22	102	11
58	33	22	33
44	69	11	22

Provisory = Best solution: diversify

12	12	12	12
12	12	12	12
12	12	12	12
12	12	12	12

Memory update

12	12	12	30
12	12	30	12
30	12	12	12
12	30	12	12

Provisory = Best solution: diversify

13	13	13	13
13	13	13	13
13	13	13	13
13	13	13	13

Memory update

13	13	13	32
13	13	32	13
32	13	13	13
13	32	13	13

Memory update

26	13	13	38
13	13	51	13
38	26	13	13
13	38	13	26

New best solution value, cost: 58.00 Found at iteration: 1

2 3 4 1

New best solution value, cost: 57.00 Found at iteration: 7

1 3 4 2

New best solution value, cost: 54.00 Found at iteration: 13

4 3 1 2

Solution time : 0.0500 sec
Output File name for this problem is: neh.out

Further Numerical Illustration on FANT for Flow Shop Scheduling

We do not want to bug you with too much details any longer. But to get a feel of the performance of FANT, we present its solution for another type of performance measure, this time, the average flow time. When you compare the solution of FANT with those of other heuristics discussed in this chapter, you will notice that FANT is a again a winner! Let us see some solutions obtained during simulation in Table 7.5. All the heuristic listed in Table 7.5 were coded in the same language as FANT except for Ho and Chang's algorithm that was compiled in C++ language.

To date, the two best known constructive heuristics for the flow shop makespan performance measure are NEH and Ho and Chang's algorithm. And the best know hybrid heuristic (insertion and tabu search) is the one by Widmer and Hertz (1989). FANT competes with all these and these heuristics. FANT

Table 7.5. Relative performance of FANT ATF_{fant}/ATF_{others}.

Heuristic	Solution	ATF	ATF_{fant}/ATF_{others}
(1) FANT	{4 2 5 3 1}	117.4	100.00%
(2) Ho and Chang	{4 2 5 1 3}	156.0	73.97%
(3) Dannenbring	{5 1 2 3 4}	192.4	59.98%
(4) Gupta	{2 5 1 3 4}	197.6	58.99%
(5) CDS	{4 2 1 5 3}	170.6	67.64%
(6) Palmer	{5 2 4 1 3}	178.6	64.61%
(7) Hundal and Rajagopal	{4 5 2 1 3}	160.4	71.95%

is indeed a new-comer to the flow shop scheduling race. This is the first time FANT is applied to the flow shop scheduling problems.

7.8. Application of FANT to Flow Shop Scheduling: Bi-Criteria

Much research has focused on a single criterion for the flowshop scheduling problem. Optimisation algorithms have been applied by Johnson (1954), and Ignall and Schrage (1965) to solve the two-machine and three-machine flowshop problems for minimising makespan. Branch-and-bound algorithm has also been applied by Ignall and Schrage (1965) for minimising the total flow time of jobs in a two-machine flowshop problem. Although optimisation algorithms can give optimal solutions, this is only feasible for small-size problems that are not often encountered in practical industrial settings. A vast majority of flowshop scheduling problems encountered in practice are NP-complete (Garey *et al.*, 1976). Consequently, research in flowshop scheduling has shifted focus from optimisation techniques to development of heuristics or near-optimal procedures. Literature abound in the applications of heuristics to the flowshop scheduling problems (see e.g. Gelders and Sambadam, 1978; Kim, 1993; Nawaz, Enscore, and Ham, 1983; Widmer and Hertz, 1989). However, most of these heuristics, and the optimisation algorithms already mentioned, concentrate on minimisation of a single performance measure. This approach is simplistic because in practice, multiple objectives are encountered in production planning and control (Panwalker *et al.*, 1973).

The literature on multiple and bicriteria scheduling problems for single machines is summarised by some researchers (see e.g. Dileepan and Sen, 1988, Fry *et al.*, 1989, Hoogeveen, 1992, and Lee and Vairaktarakis, 1993). Only few attempts have been made toward addressing the multiple objective problems for multiple machines. Most research work for multiple objectives has concentrated on two-machine flowshops (see e.g. Rajendran, 1992, Neppalli, 1996, Gupta *et al.*, 1999, Chakravarthy and Rajendran, 1999) while some researchers have worked on the multiple-machine flowshop problems (Daniels and Chambers 1990, Rajendran, 1994, Nagar *et al.*, 1996).

The multiple-machine, multiple-objective flowshop problem is NP-hard in strong sense (Chen and Bulfin, 1994). This class of problems can only be realistically tackled by the relatively recent intelligent optimisation techniques

(IOT) that have emerged: ant systems (AS), tabu search (TS) and genetic algorithms (GA). We observe that although simulated annealing (SA) is often listed as a modern search heuristic, it is excluded from membership of intelligent optimisation techniques. This is because simulated annealing does not have an adaptive memory neither does it employ *responsive exploration* during the search process. Even amongst the three listed procedures that pass the tests of intelligent optimisation techniques, only ant systems and tabu search are strictly members of the classification, with genetic algorithm being implicitly classified (a weak member). The in-depth classifications of these techniques are discussed in chapter one of this book.

The standard three-field notation (Lawler *et al.*, 1995) used is that for representing a bi-objectives scheduling problem as $\alpha/\beta/F(C1, C2)$, where α describes the machine environment, β describes the deviations from standard scheduling assumptions, and $F(C1, C2)$ describes the objectives $C1$ and $C2$ being optimised. When the objectives have different weights, then the problem is denoted as $\alpha/\beta/F_w(C1, C2)$, where $F_w(C1, C2)$ represents the weighted sum of the two objectives $C1$ and $C2$. When a hierarchy of priority levels for the objectives exists, then the problem is denoted as $\alpha/\beta/F_n(C1, C2)$, where $C1$ and $C2$ denote the primary and secondary objectives, respectively, and $F_n(C1, C2)$ represents the hierarchical optimisation of objective $C2$ subject to the constraint that objective $C1$ is at optimal value. In this paper, we are solving the $n/m/F /\!/ F_w(C_{\max}, T_{\max})$ problem. In this case, $\alpha = n/m/F$ describes the multiple-machines flowshop problem, $\beta = \emptyset$, $C1 = C_{\max}$, and $C2 = T_{\max}$.

In this book, we discuss the application of an ant system heuristic to the $F \| F_w(C_{\max}, T_{\max})$. Daniels and Chambers (1990) solved this problem using a constructive approach. Chakravarthy and Rajendran (1999) solved the same problem using simulated annealing approach. As far as the author's knowledge of the literature is concerned, this is the first time an ant system is applied to the multi-criteria flowshop scheduling problem.

This section is organised as follows. First, the mathematical formulation of the problem is presented. Then the constructive approach by Daniels and Chambers is summarised. This is then followed by the detailed presentation of the ant system heuristic together with how a good initial seed sequence is generated. Finally, an extensive performance analysis of the proposed ant system heuristic and that of Daniels and Chambers are discussed and conclusions are made.

7.8.1. *Mathematical formulation of the bi-criterion flow shop problem*

The following notations are used:

m: number of machines in the flowshop
n: number of jobs in the flowshop
p_{ij}: processing time of job j on machine i
d_j: due date of job j
σ: available partial schedule
$C_{(i,\sigma)}$: completion time of partial schedule σ on machine i
$C_{(i,\sigma a)}$: completion time of partial schedule σa when job a is appended to σ, obtained by the following recursive relationship:

$$C_{(1,\sigma a)} = C_{(1,\sigma)} + p_{1a}, \text{ and}$$
$$C_{(i,\sigma a)} = \max\{C_{(i,\sigma)}; C_{(i-1,\sigma a)}\} + p_{ia}, \; i \in [2, m] . \tag{1}$$

$\hat{C}_a = C_{(m,\sigma a)}$ is completion time of time of job a. The maximum tardiness of a job in a complete schedule is therefore given as:

$$T = \max\{\max\{\hat{C}_i - d_i; 0\}\}, \; \forall i \tag{2}$$

and the makespan is given as

$$S = \hat{C}_{[n]} . \tag{3}$$

Following the formulation of Daniels and Chambers, we define the objective or preference function as:

$$Z = \alpha \cdot S + (1 - \alpha) \cdot T$$

where α represents the weights of preference indicator for the individual component of the function defining the flowshop bicriterion problem. While Daniel and Chambers solved the problem subject to $T \leq K$ in a static form (by specifying the value of K from experience), we adopt a dynamic approach whereby the value of K is automatically generated during scheduling and is problem dependent.

7.8.2. *Daniels and Chambers heuristic for the bi-criteria problem*

Daniels and Chambers (1990) proposed a constructive to solve the multiple-machine, $(C_{\max}, T_{\max})$ problem. The objective is to identify the schedule that minimises $C_{\max}$ subject to $T_{\max} \leq K$, as K is varied over the range of interest. For a given value of K, the subset of jobs that may occupy the final position in sequence is initially determined, with job i eligible for the nth position if $\sum_{j=1}^{n} p_{1j} + \sum_{l=2}^{m} p_{li} - d_i \leq K$. Since $\sum_{j=1}^{n} p_{1l} + \sum_{l=2}^{m} p_{li}$ is a lower bound on the completion time job i if placed in the last position, eligibility guarantees that only jobs that potentially satisfy the tardiness constraint are considered. Then, the eligible job with the largest completion time in the schedule that maximises the makespan is identified and placed in position n of the schedule. As no computationally efficient procedure exists for determining the minimum Cmax sequence, the heuristic of Nawaz, Enscore and Ham (1983) commonly referred to as NEH is utilised to approximate this schedule.

With the NEH schedule in place, the eligible job with the largest completion time in this schedule (now referred to as job i) is chosen for placement in the nth position in the sequence. This selection fixed, a lower bound on the completion time of job i if placed in position $(n-1)$ can be computed as $\sum_{j=1}^{n} p_{1j} - p_{1i'} + \sum_{l=2}^{m} p_{li}$. This means that for the value of K, the subset of jobs that may occupy the final but one position in sequence is determined, with job i eligible for the $(n-1)$th position if $\sum_{j=1}^{n} p_{1j} - p_{1i'} + \sum_{l=2}^{m} p_{li} - d_i \leq K$. A new set of eligible jobs is identified and the eligible job with the latest completion time in the NEH schedule chosen for the position $(n-1)$. The process is continued until an n-job schedule is generated, with an associated values of $C_{\max}$ and $T_{\max}$. The descriptions of both the heuristics of Daniels and Chambers and NEH are given as follows:

Daniels and Chambers heuristic for minimizing $(C_{\max}, T_{\max})$

Step 1: Initialization.
Determine the $T_{\max} = \text{Max}_j\{T_j\}$ in this schedule. Set $K = T_{\max} - \varepsilon$ (incremental reduction in $T_{\max}$). Let I_l represent the set of jobs ineligible to occupy position l in sequence and set $I_l = 0$, $l = 1, 2, \ldots, n$.

Step 2: Begin new iteration.
Let S represent the set of unscheduled jobs and E the set of jobs eligible to be scheduled. Set $S = \{1, 2, \ldots, n\}$, $E = 0$, and $l = n$.

Step 3: Determine job eligibility.
For each $i \in S$, $i \notin I_l$, if

$$\sum_{j\in S} p_{j1} + \sum_{l=2}^{m} p_{il} - d_i \leq k$$

then $E = E \cup \{i\}$. If $E = 0$, stop. Else, go to Sept 4.

Step 4: Select job for placement in position l.
Identify job i s.t. $i \in E$ and $C_i \geq C_j$ in the NEH schedule $\forall j \in E$. Place job i in position l in the current schedule. Set $S = S - \{i\}$ and $l = l - 1$. If $l = 0$, go to Step 5. Else, $E = 0$ and go to Step 3.

Step 5: Evaluate current schedule.
Compute $C_{\max}$, and $T_{\max}$ for the current schedule.

(i) If $T_{\max} > k$, find job i' in position l' s.t. $T_{i'} > K$ and $C_{i'} \leq C_j \, \forall j$ s.t. $T_j > K$. Set $I_{i'} = l_{i'} \cup \{i'\}$ and go to Step 3.
(ii) If $T_{\max} \leq K$, record the computed values of $C_{\max}$, and $T_{\max}$. Set $K = T_{\max}$ (from this schedule) $-\varepsilon$ and go to Step 2.

NEH heuristic for minimizing $C_{\max}$

Step 1: Determine sum of processing times for jobs. $T_j = \sum_{i=1}^{m} p_{ij}$
Step 2: Sequence jobs in non-increasing order of T_j
Step 3: **while** all jobs are not sequenced **do**

3.1 choose an unsequenced job randomly
3.2 introduce it so that it causes the minimal increase in partial makespan

end while

7.8.3. *Ant System-based Heuristic (ASH) for the bi-criteria problem*

The full description of ant systems has been covered in the first part of this chapter, and need not be repeated here. In this section, we present a variant of ant system known as fast ant (FANT) system. The general scheme of the fast ant system represents the job by $\pi_i (i = 1, \ldots, n)$. Here, there are n jobs. The seven step for the bi-criteria problem are:

- initial solution;
- memory implementation;
- provisory solution generation;
- improvement procedure;
- memory updates;
- intensification strategies;
- diversification strategies;

Only the first step will be covered here, since all other steps have been fully covered in this chapter.

Initial solution

The rationale behind starting with a good seed solution is observed by Johnson *et al.* (1989). Since then, many researchers using intelligent optimisations techniques to solve complex, real-life combinatorial problems have significantly improved their final solutions by using good seed solutions. In solving the $F||F_w(C_{\max}, T_{\max})$ problem using simulated annealing approach, Chakravarthy and Rajendran (1999) employed earliest due-date (EDD) method and least static slack (SL) method to minimise the component of maximum tardiness, and NEH for minimising the makespan component. In this paper, we avoid using EDD or SL. Instead, we use NEHedd for minimising the component of maximum tardiness, and use NEH for minimising the makespan component. Our decision is guided by the extensive study of five modified heuristics and their comparisons with seven list scheduling (LS) rules and four priority list scheduling (LSP) rules by Kim (1993) for the mean tardiness problem. In his summary, Kim says "The NEH method with the EDD rule (NEHedd) for pre-processing was better than the one with the LDD rule. Moreover, it worked better than all list scheduling algorithms including GS2 (the heuristic of Gelders and Sambandam, 1978), and after pairwise interchanges it performed almost the same as ENS (extensive neighbourhood search method)". In fact, $NEHedd$ outperformed ENS and comes next to tabu search (TS) which outperforms all the methods studied. This was the motivation for using $NEHedd$ for minimising the tardiness component and NEH for minimising the makespan component in the work reported in this book. This guaranteed a good seed solution.

The objective function of the flowshop problem then becomes

$$Z = \alpha \cdot NEH + (1 - \alpha) \cdot NEHedd \tag{4}$$

$$s.t.\ NEHedd \leq K\,. \tag{5}$$

The sequence with the least value of Z, subject to $T \leq K$, is chosen. If no sequence satisfies the condition of $NEHedd \leq K$, then the sequence with the least value of $NEHedd$ is chosen. The $NEHedd$ solution is significantly improved through adjacent interchange switches (AIS). The sequence, thus obtained, becomes the seed solution that is fed as an input into the ant system heuristic. This initial solution is known as S.

Table 7.6. Relative performance of the procedures for $\alpha = 0.25$.

		Relative % increase in Z	
n	m	DCH	ASH
4	5	0.5455	0.0000
	10	0.3534	0.0000
	15	0.1444	0.0000
	20	—	0.0085
	25	0.1111	0.0000
5	5	—	0.0080
	10	0.5381	0.0000
	15	1.5225	0.0000
	20	1.6100	0.0000
	25	0.0235	0.0000
10	5	0.9761	0.0000
	10	1.3427	0.0000
	15	0.7140	0.0000
	20	1.5010	0.0000
	25	0.4761	0.0000
15	5	1.6942	0.0000
	10	1.4580	0.0000
	15	0.7600	0.0000
	20	1.5596	0.0000
	25	0.9119	0.0000

Table 7.7. Relative performance of the procedures for $\alpha = 0.5$.

		Relative % increase in Z	
n	m	DCH	ASH
4	5	0.0000	0.0000
	10	0.1293	0.0000
	15	0.0000	0.0580
	20	0.0000	0.0085
	25	0.8582	0.0000
5	5	0.3581	0.0080
	10	0.0000	0.3490
	15	0.0048	0.0000
	20	0.0000	0.0304
	25	0.0000	0.0565
10	5	0.0000	0.1318
	10	0.0809	0.0000
	15	0.0000	0.1706
	20	0.2242	0.0000
	25	0.0000	0.0613
15	5	0.0676	0.0000
	10	0.0000	0.0156
	15	0.0000	0.1905
	20	0.0853	0.0000
	25	0.0000	0.1590

Test Results

A total of 200 problems were designed for testing, and solved for a set of jobs {4, 5, 10, 15} and a set of machines {5, 10, 15, 20, 25}. Three values of α were set {0.25, 0.50, and 0.75}, giving a total of 60 settings for experimentation. Tables 7.6, 7.7, and 7.8 show the comparison of Daniels and Chambers heuristic (DCH), and the ant system heuristic (ASH) presented in this paper for the bi-criterion problem. The processing times were generated randomly and uniformly distributed over the interval [1, 15]. The due-date of a job has been generated randomly and uniformly distributed over the interval $D_j = [\sum_{i=1}^{m} t_{ij}, \sum_{i=1}^{m} t_{ij} + 0.5(n-1)\bar{t}_{mj}]$. This due-date of a job has the tendency to guarantee approximately 50% of the jobs being tardy. This due date setting is quite tight and consequently can easily distinguish between the performances of heuristics.

Table 7.8. Relative performance of the procedures for $\alpha = 0.75$.

		Relative % increase in Z	
n	m	DCH	ASH
4	5	0.0000	0.2220
	10	0.0546	0.0000
	15	0.0000	0.1390
	20	0.0000	0.0085
	25	0.0000	0.0038
5	5	0.0000	0.7140
	10	0.0000	1.0996
	15	0.0000	1.0040
	20	0.0000	1.3537
	25	0.0000	0.0810
10	5	0.0000	0.9250
	10	0.0000	0.4876
	15	0.0000	0.9951
	20	0.0000	0.2521
	25	0.0000	0.3084
15	5	0.0000	0.9030
	10	0.0000	1.0263
	15	0.0000	0.8750
	20	0.0000	0.6837
	25	0.0000	0.9487

The limit on the maximum tardiness, K, is specified using a weak lower bound (Chakravarthy and Rajendran, 1999) on makespan and the due date as follows:

$$LBM = \max_j \left\{ \sum_{i=1}^{m} t_{ij} + (n-1)\bar{t}_{mj} \right\}$$

$$K = \max_j \{LBM - D_j\}\,.$$

Experimentation shows that another way of obtaining the value of K is to use one and a half times the strong lower bound on makespan. Let Z' and Z'' be the objective function values given by the schedules of ASH and DCH respectively for any problem under consideration. The relative percent deviation (RPD) used for the comparison is defined as follows:

$$RPD(Z' - \min\{Z'; Z''\})/\min\{Z'; Z''\}.$$

The $RPD\{Z''\}$ is also similarly obtained. The analysis shows that when the combined solution is less biased toward makespan ($\alpha = 0.25$), ASH appears to perform much better than DCH (see Table 7.6), and when the solution is less biased toward tardiness ($\alpha = 0.75$), DCH appears to perform much better than ASH (see Table 7.7), with both appearing to compete at $\alpha = 0.5$. These tables should be carefully analyzed before conclusion is made. Table 7.6 shows that there are 7 cases where DCH has RPD value > 1 and in only 3 cases is the RPD < 0.5. Table 7.8 shows that there are 4 cases where ASH has RPD value > 1 and in 7 cases, the RPD < 0.5. Then Table 7.9 presents a further true picture of the performances of the two heuristics. The best values for

Table 7.9. Overall percentage improvement of ASH over DCH.

n	m	% improvement of ant system heuristic (ASH) over DCH
4	5	0.5455
	10	0.3534
	15	0.1444
	20	—
	25	0.1111
5	5	—
	10	0.4264
	15	0.4970
	20	0.2746
	25	0.7516
10	5	1.5462
	10	1.1820
	15	1.1882
	20	0.4970
	25	1.3207
15	5	0.5765
	10	0.6127
	15	0.6000
	20	0.7818
	25	0.5395

Table 7.10. Comparison of average computation times (seconds).

n	m	DCH	ASH
4	5	0.0000	0.6430
	10	0.0000	0.7900
	15	0.0000	1.0300
	20	1.1200	1.1770
	25	1.2630	1.2670
5	5	0.0000	0.8760
	10	0.0167	3.3500
	15	0.0600	9.5000
	20	0.1870	21.053
	25	0.4400	2.0200
10	5	0.0200	3.3870
	10	0.0330	7.6400
	15	0.0600	7.9100
	20	0.0570	10.107
	25	0.0700	44.877
15	5	0.0500	9.5000
	10	0.1300	17.173
	15	0.2170	24.703
	20	0.2130	32.317
	25	0.3070	39.855

both ASH and DCH are used for each problem setting, and the improvement of ASH over DCH is calculated. It is observed that ASH does overall better than DCH in all cases but two. Table 7.10 shows the comparison for both heuristics in terms of computational times. DCH takes much less time than ASH. It should be noted that although ASH takes more time, the values are within acceptable limits of similar intelligent optimization techniques such as tabu search and genetic algorithm.

Summary

An ant system heuristic for solving the bi-criterion problem of makespan and tardiness is proposed in this section. The proposed heuristic is compared with the best know constructive method in the literature, by Daniels and Chamber, for a relatively large number of problems. The results are encouraging. The analysis shows that overall, ASH is preferred to DCH. One major problem with Daniels and Chambers heuristic is that the user suggests the initial limiting value of maximum tardiness which is then gradually reduced as the search progresses, in addition to back-tracking. These are definitely problematic. In ASH, the limiting value of maximum tardiness is automated since it is problem dependent. In fact, to the knowledge of the author this is the first time ant system is employed to the makespan/tardiness bi-criterion problem in flow shop scheduling. There are challenging research directions for ant system. It will be interesting to compare its solutions with those of its counterpart intelligent optimization techniques such as tabu search and genetic algorithm. It will also be interesting to compare ant systems heuristic with simulated annealing presented by Chakravarthy and Rajendran (1999). Then, it will be useful to include some techniques that may reduce the computation times for ants. The only parameter that need to be investigated is the number of iterations setting as this affects the run time. It will also be useful to investigate other bi-criterion flow shop problems using ant systems.

7.9. Summary

In this chapter, we presented the paradigm and fundamentals for ant systems. The hybrid ant system and fast ant system variants were discussed in some details. The memory structure, provisory solution generation, local search mechanism to speed up search, memory updates, intensification and diversification mechanisms were discussed. We presented the FANT simulator. Detailed results for the makespan problem taken from the literature was presented so that readers can have more in-depth understanding of the memory management of FANT. We notice that FANT is able to find optimum or very near optimum solution in very fast time. Apart from this performance measure, FANT was tested using another performance measure, known as weighted flow time for flow shops. Again FANT went into a race with seven other heuristics and beat them. We report here, that the FANT discussed in this chapter currently deals with eleven different classes of problems. These include: makespan, tardiness, flow time, waiting time, average tardiness, average flow time, average waiting time, flow time variance, waiting time variance, and bi-criterion performance measure that combines makespan and average flow time. In all these, FANT remains a winner. FANT competes very well with other intelligent optimization techniques such as tabu search, and genetic algorithm. To validate the performance of FANT in terms of solution quality, FANT was coded in the same language with tabu search and genetic algorithm simulators. The author is not aware of the application of FANT to flow shop scheduling problems to the extent presented in this chapter. We do not want to pursue this discussion further, because these other intelligent optimization techniques are covered in other chapters.

In this chapter, FANT was applied to scheduling problem in flow shops. This is the first time in the literature that ant system has been applied to flow shop scheduling problems that cover such a wide spectrum as discussed in this chapter. Scheduling presents many day-to-day problems and challenges to production and operations mangers due to last minute changes that may occur in customer order. Due to the complexity of most scheduling problems, there are few really satisfactory techniques which have been developed. FANT scheduling system which makes scheduling flexible and which drastically cuts time for flow-shop environment has been presented in this chapter. The FANT-Scheduling system is a very helpful tool for decision makers such as production and operation managers because it can be an easy way of communicating the scheduling situation to them.

Exercises

1. What features of fast ant system enable it to have competitive advantages over traditional optimization techniques?
2. Fast ant system swaps two intermediate solutions when the local search mode is activated. Do you think that an insertion method will give a better solution?
3. Implement a simulator for an insertion-type local search for FANT and compare your results with a swap-type.
4. Implement the hybrid ant simulator. Test it using the example test data presented in Sec. 7.6. Does the simulator behave as expected?
5. What modifications are necessary to be made in the FANT simulator provided in this chapter in order to solve the TSP problem?
6. Modify the FANT simulator provided in this chapter to solve the constraint management problem. (Hint: use FANT to solve the theory of constraint problem).

7. What further modifications may be necessary to speed up FANT?
8. What further modifications may be necessary to improve the solution quality of FANT?
9. Hybrid ant system (HAS) and fast ant system (FAST) are variants or the traditional ant system. They produce better results. Can you think of any modification to the memory management that will lead to another variant of ant system?
10. How does ant system differ from tabu search in terms of memory management?

Hands-on experimentation:

Exercises 11 to 22 are designed to give you hands-on experience using FANT. Exercises 11 to 18 are randomly generated. Exercises 19 to 22 are from the literature, so that you can verify for yourself how effective and efficient FANT is. For all these exercises, the first row represents the number of machines, the number of jobs, the memory parameter r^*, and the number of FANT iterations respectively. Each of the other rows describe the processing times of jobs on machines. The row after the last machine represents the due dates of jobs. The row after that for due dates represents the weights or priorities of jobs.

11. Table 7.11 shows the information for a 4-machine, 4-job flow shop problem. Run FANT simulator and hence determine the

    (i) makespan;
    (ii) total tardiness value;
    (iii) weighted tardiness value;
    (iv) flow time;
    (v) flow time variance;
    (vi) weighted flow time variance.

Table 7.11.

4 4 4 50			
4	10	1	4
4	9	10	3
10	4	3	9
9	4	7	1
25	19	13	26
0.03	0.14	0.172	0.658

12. Table 7.12 shows the information for a 4-machine, 5-job flow shop problem. Run FANT simulator and hence determine the
    (i) makespan;
    (ii) total tardiness value;
    (iii) weighted tardiness value;
    (iv) flow time;
    (v) flow time variance;
    (vi) weighted flow time variance.
13. Table 7.13 shows the information for a 5-machine, 8-job flow shop problem. Run FANT simulator and hence determine the
    (i) makespan;
    (ii) total tardiness value;
    (iii) weighted tardiness value;
    (iv) flow time;
    (v) flow time variance;
    (vi) weighted flow time variance.

Table 7.12.

4 5 4 50				
2	7	5	5	3
7	10	8	4	9
10	9	4	3	4
9	1	2	7	6
26	17	13	27	10
0.025	0.193	0.21	0.067	0.505

Table 7.13.

5 8 4 50							
10	1	7	4	4	5	10	6
4	8	10	4	10	6	7	10
7	3	3	2	6	2	5	6
1	7	2	2	4	8	3	8
3	2	9	6	5	7	6	3
15	29	18	21	24	30	13	20
0.017	0.234	0.234	0.186	0.069	0.146	0.036	0.089

14. Table 7.14 shows the information for a 5-machine, 10-job flow shop problem. Run FANT simulator and hence determine the

    (i) makespan;
    (ii) total tardiness value;
    (iii) weighted tardiness value;
    (iv) flow time;
    (v) flow time variance;
    (vi) weighted flow time variance.

15. Table 7.15 shows the information for a 6-machine, 6-job flow shop problem. Run FANT simulator and hence determine the

    (i) makespan;
    (ii) total tardiness value;
    (iii) weighted tardiness value;
    (iv) flow time;

Table 7.14.

4 10 4 60									
4	1	4	10	2	7	1	9	2	8
8	4	8	5	3	4	6	3	9	5
2	3	2	9	4	2	7	6	10	5
3	10	9	2	3	8	6	5	9	4
7	4	7	6	2	5	7	6	3	9
30	23	23	27	14	24	14	15	16	29
0.013	0.173	0.059	0.002	0.164	0.148	0.181	0.184	0.036	0.400

Table 7.15.

6 6 4 60					
10	1	10	6	3	3
7	2	6	5	8	5
2	9	5	5	5	1
8	6	10	9	1	3
4	1	8	10	7	7
4	10	4	8	2	10
19	22	22	11	29	28
0.013	0.174	0.132	0.104	0.186	0.391

(v) flow time variance;
(vi) weighted flow time variance.

16. Table 7.16 shows the information for a 6-machine, 8-job flow shop problem. Run FANT simulator and hence determine the

    (a) makespan;
    (b) total tardiness value;
    (c) weighted tardiness value;
    (d) flow time;
    (e) flow time variance;
    (f) weighted flow time variance.

17. Table 7.17 shows the information for a 4-machine, 6-job flow shop problem. Run FANT simulator and hence determine the

    (a) makespan;
    (b) total tardiness value;
    (c) weighted tardiness value;

Table 7.16.

6 8 4 60							
6	3	2	5	8	1	6	2
3	10	6	10	2	7	4	5
3	8	6	1	5	4	5	1
4	1	8	10	3	8	5	7
6	1	3	4	8	1	6	8
4	2	10	3	9	9	8	1
28	11	29	21	25	10	29	18
0.026	0.2050	0.12	0.099	0.125	0.095	0.028	0.302

Table 7.17.

4 6 4 60					
2	5	8	1	6	2
6	10	2	7	4	5
6	1	5	4	5	1
8	10	3	8	5	7
29	21	25	10	29	18
0.12	0.099	0.125	0.095	0.302	

(d) flow time;
(e) flow time variance;
(f) weighted flow time variance.

18. Table 7.18 shows the information for a 8-machine, 8-job flow shop problem. Run FANT simulator and hence determine the
19. Table 7.19 (Gupta) shows the information for a 3-machine, 6-job flow shop problem. Run FANT simulator and hence determine the
(i) makespan;
20. Table 7.20 (modified Ho and Chang) shows the information for a 4-machine, 5-job flow shop problem. Run FANT simulator and hence determine the

(a) makespan;
(b) total tardiness value;
(c) weighted tardiness value;
(d) flow time;
(e) flow time variance;
(f) weighted flow time variance.

Table 7.18.

8 8 4 70							
10	6	3	9	4	9	1	10
3	8	3	3	5	9	10	7
4	2	2	8	1	5	9	2
8	4	1	6	6	6	7	9
2	7	3	2	4	3	9	3
9	9	10	5	4	8	3	4
3	5	6	8	5	8	5	7
3	7	3	3	8	7	6	3
18	21	25	11	26	13	20	12
0.023	0.112	0.081	0.091	0.043	0.108	0.074	0.468

Table 7.19.

3 6 6 50					
1	8	11	3	5	12
8	5	2	9	5	7
4	10	7	2	4	7

Table 7.20.

4 5 6 200				
31	19	23	13	33
41	55	42	22	5
25	3	27	14	57
30	34	6	13	19
95	100	125	100	130
.2	.2	.2	.2	.2

Table 7.21.

2 5 6 50				
15	5	16	5	7
19	10	12	20	12

Table 7.22.

3 5 6 100				
123	57	198	154	92
300	156	201	162	99
76	200	211	122	211

21. Table 7.21 (Rajendra) shows the information for a 2-machine, 5-job flow shop problem. Run FANT simulator and hence determine the
    (i) makespan;
22. Table 7.22 (Smith and Dudek) shows the information for a 3-machine, 5-job flow shop problem. Run FANT simulator and hence determine the
    (i) makespan;

Solutions to exercises

All solutions to exercises 7.1, 7.2, 7.5, 7.7, 7.8, 7.9, 7.10, and 7.11 to 7.22 are in the Reference Manual. Lecturers and students will find the Reference Manual very useful. How? Because Lecturers can use this text book and the Reference Manual for teaching and setting tests, tutorials and examinations. Students can easily check if what they are doing is correct. The Reference Manual can be also purchased from the publishers of this book.

References

Ant System

Bilchev, G. and Parmee, I. C., 1995, The ant colony metaphor for searching continuous design spaces, *Lecture Notes on Computer Sciences.*

Bullnheimer, B., 1999, *Ant Colony Optimisation in Vehicle Routing*, Doctoral Thesis, University of Vienna.

Bullnheimer, B., Hartl, R. G., and Strauss, C., 1999, Applying the ant system to the vehicle routing problem. In Voss, S., Martello, S., Osman, I. H., Rorcairol, C. (eds.), *Meta-Heuristics: Advances and Trends in Local Search Paradigms for Optimisation* (Kluwer: Boston).

Bullnheimer, B., Kotsis, G., and Strauss, C., 1998, Parallelization strategies for the ant systems. In De Leone, R., Mulli, A., Pardalos, P., Toraldo, G. (eds.), *High Performance Algorithms and Software in Non-linear Optimisation*; *series: Applied Optimisation* (Kluwer: Dordrecht), 87–100.

Colorni, A., Dorigo, M., Maniezzo, V., and Trubian, M., 1994, Ant system for job-shop scheduling, *JORBEL-Belgian Journal of Operational Research, Statistics and Computer Science*, **34**(1) 39–53.

Costa, D. and Hertz, A., 1997, Ants can colour graphs, *Journal of Operational Research Society*, **48** 295–305.

Dannenbring, D. G., 1977, An evaluation of flow shop sequencing heuristics, *Management Science*, **23**(11) 1174–1182.

Di Caro, G. and Dorigo, M., 1998, AntNet: distributed stigmergetic control for communications networks, *Journal of Artificial Intelligence Research* (*JAIR*), **9** 317–365.

Dorigo, M., 1992, *Optimisation, Learning and Natural Algorithms* (*Ottimizzazione, aprendimento automatico, et algoritmi basati su metafora naturale*), Ph.D. Dissertation, Dipartimento Elettronica e Informazione, Politecnico di Milano, Italy.

Dorigo, M. and Gambardella, L. M., 1997, Ant colony system: A co-operative learning approach to the travelling salesman problem, *IEEE Trans. Evolutionary Computing* 1.

Dorigo, M., Maniezzo, V., and Colorni, A., 1991, *The ant system: And auto-catalytic optimising process*, Technical Report 91-016, Dipartimento Elettronica e Informazione, Politecnico di Milano, Italy.

Fleurent, C. and Ferland, J., 1994, Genetic hybrids for the quadratic assignment problem, *DIMACS Series in Math. Theoretical Computer Science*, **16** 190–206.

Forsyth, P. and Wren, A., 1997, An ant system for bus driver scheduling. Presented at the *7th International Workshop on Computer-Aided Scheduling of Public Transport*, Boston.

Gambardella, L. M. and Dorigo, M., 1997, *HAS-SOP: Ant colony optimisation for sequential ordering problems*, Technical Report IDSIA-11-97, IDSIA, Lugano, Switzerland.

Gambardella, L. M., Taillard, E., and Dorigo, M., 1999, Ant colonies for the quadratic assignment problem, *Journal of the Operational Research Society*, **50** 167–176.

Glover, F. and Greenberg, H., 1989, New approaches for heuristic search: A bilateral linkage with artificial intelligence, *European Journal of Operational Research*, **39**(2) 119–130.

Ho, J. C. and Chang, Y.-L, 1991, A new heuristic for the n-job, m-machine flow shop problem, *European Journal of Operational Research*, **52** 194–202.

Hundal, T. S. and Rajgopal, J., 1988, An extension of Palmer's heuristic for the flow shop scheduling problem, *International Journal of Production Research*, **26** 1119–1124.

Kuntz, P., Layzell, P., and Snyers, D., 1997, A colony of ant-like agents for partitioning in VLSI technology, *Proceedings of the Fourth European Conference on Artificial Life* (eds.) Husbands, P. and Harvey, I., MIT Press.

Schoonderwoerd, R., Holland, O., Bruten, J., and Rorhkrantz, L, 1997, Ant-based load balancing in telecommunications networks, *Adaptive Behaviour*, **5**(2) 169–207.

Stützle, T. L. and Dorigo, M., 1999, ACO algorithms for the quadratic assignment problem. In Corne, D., Dorigo, M., and Glover, F. (eds.), *New Ideas in Optimisation* (McGraw-Hill).

Stützle, T. L. and Hoos, H. H., 1997, The MAX–MIN ant system and local search for combinatorial optimisation problems: Towards adaptive tools for global optimisation, *Proceedings of the 2nd Meta-heuristics International Conference*, Sophia-Antipolis, France, 191–193.

Taillard, E. D. and Gambardella, M, 1997, *Adaptive Memories for the Quadratic Assignment Problem*, Technical Report IDSIA-79-97, IDSIA, Lugano, Switzerland.

Theraulaz, G., Goss, S., Gervet, J. and Deneubourg, J. L., 1991, Task differentiation in polistes wasp colonies: A model for self-organising groups of robots. *Simulation of Adaptive Behaviour: From animals to Animats*, MIT Press, Bradford Books, Cambridge, Mass, 346–355.

Scheduling

Campbell, H. G., Dudek, R. A., and Smith, M. L., 1970, A heuristic for the n job, m machine sequencing problem, *Management Science*, **16**(10) 630–637.

Brah, S. A. and Hunsucker, J. L., 1987, *Optimal scheduling in an m-stage flow shop with multiple processors*, Technical Report, University of Houston, TX.

Dannenbring, D. G., 1977, An evaluation of flow shop sequencing heuristics, *Management Science*, **23**(11).

Gupta, J. N. D., 1972, Heuristic algorithms for multistage flow shop scheduling problem, *AIIE Transaction*, **4**(1) 11–18.

Gupta, J. N. D. and Darrow, W. P., 1971, The two-machine sequence dependent flow shop scheduling problem, *AIIE Transaction*, **3** 199–205.

Hariri, A. M. A., 1981, *Scheduling using branch and bound techniques*, Ph.D. thesis, University of Keele, UK.

Johnson, S. M., Optimal two and three stage production schedules with setup times included, *Naval Research Logistics Quarterly*, **1**(1) 1954.

Kochhar, S. and Morris, R. J. T., 1987, Heuristic methods for flow lines scheduling, *Journal of Manufacturing Systems*, **6** 299–314.

Palmer, D. S., 1965, Sequencing jobs through a multi-stage process in the minimum total time — a quick method of obtaining a near optimum, *Operational Research Quarterly*, **16** 101–107.

Park, Y. B., Pegden, C. D. and Enscore, E. E., 1984, A survey and evaluation of static flow shop scheduling heuristics, *International Journal of Production Research*, **22** 127–141.

McMahon, G. B. and Burton, P. G., 1967, Flow-shop scheduling with the branch and bound method, *Operations Research*, **15** 475–482.

Onwubolu, G. C., 1996, A flow-shop manufacturing scheduling system with interactive computer graphics, *International Journal of Operations and Production Management*, **16**(9) 74–84.

Wittrock, R. J., 1985, Scheduling algorithms for flexible lines, *IBM Journal Res. Dev.*, **29** 401–412.

Bi-criterion Scheduling

Chakravarthy, K. and Rajendran, C., 1999, A heuristic for scheduling in a flowshop with the bicriteria of makespan and maximum tardiness minimisation, *Production Planning and Control*, **10** 707–714.

Chen, C. L. and Bulfin, R. L., 1994, Complexity results for multi-machine multi-criteria scheduling problems, *Proceedings of the Third Industrial Engineering Research Conference*, 662–665.

Daniels, R. L. and Chambers, R. J., 1990, Multiobjective flow shop scheduling, *Naval Research Logistic Quarterly*, **37** 981–995.

Dileepan, P. and Sen, T., 1988, Bicriterion static scheduling research for a single machine, *OMEGA*, **16** 53–59.

Fry, T. D., Armstrong, R. D., and Lewis, H., 1989, A framework for single machine multiple objective sequencing research, *OMEGA*, **17** 595–607.

Garey, M. R., Johnson, D. S., and Sethi, R., 1976, The complexity of flow shop and job shop scheduling, *Mathematics of Operations Research*, **1** 117–129.

Gelders, L. F. and Sambandam, N., 1978, Four simple heuristics for scheduling a flowshop, *International Journal of Production Research*, **16** 221–231.

Gupta, J. N. D., Palanimuthu, N., and Chen, C. L., 1999, Designing a tabu search algorithm for the two-stage flow shop problem with secondary criterion, *Production Planning and Control*, **10** 251–265.

Hoogeveen, H., 1992, *Single-Machine Bicriteria Scheduling*, Ph.D. Dissertation, Centre for Mathematics and Computer Science, Amsterdam, The Netherlands.

Ignall, E. and Schrage, L., 1965, Application of the branch-and-bound technique to some flowshop scheduling problems, *Operations Research*, **13** 400–412.

Johnson, S. M., 1954, Optimal two-and three-stage production schedules, *Naval Research Logistics Quarterly*, **1** 61–68.

Johnson, D. S., Aragon, C. R., McGeoch, L. A., and Schevon, C., 1989, Optimisation of simulated annealing: An experimental evaluation: Part 1. Graph partitioning. *Operations Research*, **37** 865–891.

Kim, Y., 1993, Heuristics for flow shop scheduling problems minimising mean tardiness, *Journal of the Operational Research Society*, **44** 19–28.

Lawler, E. L., Lenstra, J. K., Rinnooy Kan, A. H. G., and Shmoys, D. B., 1995, Sequencing and scheduling: algorithms and complexity. In *Logistics of Production and Inventory*, Graves, S. C., Rinnooy Kan, A. H. G., and Zipkin, P. H., (eds) (Amsterdam, The Netherlands: North Holland), 445–522.

Lee, C.-Y. and Vairaktarakis, G. L., 1993, *Complexity of Single Machine Hierarchical Scheduling: A Survey*, Research Report No. 93–10, Department of Industrial and Systems Engineering, University of Florida, Gainesville, Fl, USA.

Nagar, A., Haragu, S. S., and Haddock, J., 1996, A combined branch and bound and genetic algorithm based approach for a flowshop scheduling problem, *Annals of Operations Research*, **63** 397–414.

Nawaz, M., Enscore, E. E., and Ham, I., 1983, A heuristic algorithm for the m-machine, n-job flowshop sequencing problem, *OMEGA*, **11** 91–95.

Neppalli, V. R., Chen, C. -L., and Gupta, J. N. D., 1996, Genetic algorithms for the bicriteria two-stage flowshop problem, *European Journal of Operational Research*, **95** 365–373.

Panwalkar, S. S., Dudek, R. K., and Smith, M. L., 1973, Sequencing research and the industrial scheduling problem. In *Symposium on the Theory of Scheduling and its Application*, Elmaghraby, S. E. (ed.) (New York: Springer).

Rajendran, C., 1992, Two-stage flow shop scheduling problem with bicriteria, *Journal of the Operational Research Society*, **43** 871–884.

Rajendran, C., 1994, A heuristic for scheduling in flowshop and flowline-based manufacturing cell with multi-criteria, *International Journal of Production Research*, **32** 2541–2558.

Widmer, M. and Hertz, A., 1989, A new heuristic method for the flowshop sequencing problem, *European Journal of Operational Research*, **41** 186–193.

Chapter 8

Simulated Annealing

8.1. Simulated Annealing Paradigm

Physical annealing process is a heat treatment process that consists of heating the metal to a suitable temperature, holding at that temperature for a certain finite time, called soaking, and then slowly cooling. It is a heat treatment process (manufacturing process) that involves heating and cooling procedures performed to effect structural changes in a metal, which in turn affects its mechanical properties. Physical annealing is performed on a metal for any of the following reasons:

(1) to reduce hardness and brittleness;
(2) to alter microstructure in order to obtain desirable mechanical properties;
(3) to soften metals for improved machinability or formability;
(4) to re-crystallize cold worked or strain hardened metals; and
(5) to relieve residual stresses induced by prior shaping processes.

The hardness of metal is its resistance to permanent indentation. In engineering applications, including most of the tooling used in manufacturing, materials that are resistant to scratching and wear are desirable. The amount of strain that the metal can endure before failure is also a mechanical property of interest in many manufacturing processes. This is commonly measured by the ductility, which, is the ability of a material to plastically strain without failure. It is an important characteristic for metals to form while hot. Several forming processes include rolling, forging, extrusion, and drawing for cylindrical billets and bars, bending, drawing and shearing, for flat metal sheets. Cold working of metals, entails metal forming performed at room temperature or slightly above. Cold forming operation results in strain hardening of the work metal, which, in turn limits the amount of forming that can be done to the metal part. In order to deal with these problems encountered in manufacturing, physical annealing is performed. Different terms are used in annealing, depending on the details of the process and the temperature used relative to the re-crystallization temperature of the metal being treated.

Full annealing is associated with ferrous metals, and involves heating the alloy into austenite zone (about 723°C), followed by slow cooling in the furnace to produce coarse pearlite (about 620°C). *Normalizing* is another similar process that involves similar heating and soaking cycles, at faster cooling rates. Cooling is effected, by allowing the steel to cool in air to room temperature, resulting in fine pearlite and higher strength and hardness, but lower ductility than full annealing. *Process annealing* is done to allow for further cold working of parts. *Annealing* is a heat treatment process similar to process annealing, but this time it performed on the completed cold worked part in order to remove the effects of strain hardening and where no subsequent deformation will be accomplished.

In some cases, the annealing conditions permit full recovery of the cold worked metal to its original grain structure, resulting in *re-crystallization*. However, when the annealing process only permits partial return of the grain structure to its original state, the result is *recovery annealing*. In *stress-relief annealing*, the operations are performed solely to relieve residual stresses in the metal work-piece due to prior shape processing. The processes described are much the same, but different terms are used to indicate the purpose of the treatments.

In physical annealing, all particles arrange themselves in the low energy ground state of a corresponding lattice, as long as the maximum temperature is sufficiently high and cooling is carried out sufficiently slowly. The physical annealing process aims at reducing the temperature of a material to its thermal equilibrium state, which corresponds to its minimum energy state. The annealing process begins with a metal in a melted-state, with the temperature being reduced gradually. At each temperature the solid is allowed to reach thermal equilibrium as it is held at that temperature for some time. Rapid lowering of the temperature particularly at the early stages of the annealing process, results in certain defects being frozen in the material, coupled with not reaching the minimum energy state. This process is known as *quenching*.

Let us use the concepts of statistical mechanics to describe the energy of the solid as it is annealed. Statistical mechanics is an aspect of physics that deals with systems constituted of large number of particles, such that the gross properties of the system is evaluated by the average behaviour of identical systems that make up the large system. A collection of identical systems is referred to as *ensemble*. The average energy, $\langle E \rangle$, of an ensemble is the average of the energies of the constituents of the ensemble, each weighted by the probability that the system has a particular energy, given as

$$\langle E \rangle = \sum_r P_r E_r \tag{1}$$

where P_r is the probability that the system has a particular energy, E_r. The probability that the system has a particular energy, E_r, is found to be proportional to an exponential factor, given as

$$P_r \propto e^{\frac{E_r}{k_B T}} \tag{2a}$$

$$P_r = Ce^{-\frac{E_r}{k_B T}} \tag{2b}$$

where T is temperature and k_B is the Boltzmann constant.

Since $\sum_r P_r = 1$, then $C = (\sum_r e^{-\frac{E_r}{k_B T}})^{-1}$. In statistical mechanics, $Z = C^{-1}$ is known as the *partition function*. Hence,

$$P_r = \frac{e^{-\frac{E_r}{k_B T}}}{\sum_r e^{-\frac{E_r}{k_B T}}} \tag{3}$$

and the average energy of the ensemble is given as

$$\langle E \rangle = \frac{\sum_r e^{-\frac{E_r}{k_B T}} E_r}{Z}. \tag{4}$$

Starting off at the maximum temperate, we can describe the cooling phase in the following manner. At each temperature value T, sufficient time must be allowed so that the solid metal reaches thermal equilibrium. In thermal equilibrium, the solid material follows the canonical Boltzmann probability distribution:

$$\Pr\{E = E_i\} = \frac{1}{Z(T)} \bullet e^{(-\frac{E_r}{k_B T})} \sum_r e^{-\frac{E_r}{k_B T}} \tag{5}$$

where $Z(T)$ is a normalization factor as already mentioned, known in condensed matter physics as the *partition function*, depending on the temperature T and k_B is the *Boltzmann constant.* The factor $e^{(-\frac{E_r}{k_B T})}$ is known as the Boltzmann factor. As the temperature is decreased, the Boltzmann-factor reduces drastically so that the Boltzmann distribution concentrated on states that are at lowest energy level. When the temperature approaches zero, only the global, minimum energy states have a non-zero probability of occurrence.

8.2. Monte Carlo Model for Simulating Physical Annealing

Metropolis *et al.* (1953) proposed a *Monte Carlo model* for simulating the transition from a given state to the thermal equilibrium state of a solid for a fixed value of temperature, T. The Monte Carlo model generates a sequence of states $S = \{S_1, \ldots, S_n\}$. For a current state of the solid, S, described by the positions of the particles in the solid, x_i, the solid is perturbed (given some small random displacement) so that the new positions of the particles in the solid are described by x_j. These states are associated with energies, E_i and

E_j. The *Metropolis criterion* for acceptance of new states is given as

$$\Pr\{acceptance\ of\ new\ states\} = \begin{cases} 1 & \text{if } \Delta E \leq 0 \\ e^{\left(-\frac{E_r}{k_B T}\right)} & \text{if } \Delta E > 0\,. \end{cases} \tag{6}$$

The *Metropolis criterion* essentially states that if the difference in energy, ΔE, between the current state and the perturbed state is negative, the perturbed (new) state is considered, while the process is continued. However, if the difference in energy, ΔE, between the current state and the perturbed state is not negative, the perturbed (new) state is accepted with a probability, given by $e^{\left(-\frac{E_r}{k_B T}\right)}$. After a large number of perturbations, using the *Metropolis criterion*, the probability of distribution approaches the Boltzmann distribution.

In physical annealing, another concept that is useful, is entropy. What is entropy? The heat of an infinitely small reversible process is a function of the change in the entropy of the process, which is a measure of disorder in the system, given as

$$\Delta q = T \Delta s\,. \tag{7}$$

Entropy s is a state function, like internal energy, u and enthalpy, h. As we can see from Eq. (7), if entropy of a reversible process increases, this means that heat is being supplied to the material. A reversible process means that a process can occur, both in the direct and reverse sense so that the thermodynamic system returns to its initial state and there occur no changes in the surroundings. If entropy of a reversible process decreases, this means that heat is being taken off from the material. When there is no change in entropy, heat is neither supplied nor removed from the material.

The concept of entropy is useful in understanding changes in heat in a system. For example, in a reversible adiabatic process, the entropy is constant. This means that the change in heat supply to, or heat removed from a system is zero. This condition is difficult to achieve because it assumes that the process takes place without friction. An isothermal (constant temperature) process can have change in heat because its entropy changes for a particular temperature. In an isobaric process, heat is supplied to the process, and hence, its entropy increases. Entropy is a function of the Boltzmann constant, k_B, and the probability that a system has an energy, P_r, is given as

$$S = -k_B \sum_r P_r \ln P_r\,. \tag{8}$$

Let us examine two systems that have equal value of entropy $S = -k_B \sum_r P_{1r} \ln P_{1r}$ and $S = -k_B \sum_r P_{2r} \ln P_{2r}$, such that both systems have the same average energy, given as:

$$\langle E_1 \rangle = \sum_r P_{1r} E_r = \sum_r P_{2r} E_r = \langle E_2 \rangle \,. \tag{9}$$

The difference in entropy values now becomes

$$S_1 - S_2 = -k_B \sum_{i=1} [P_{1r} \ln P_{1r} - P_{2r} \ln P_{2r}] \tag{10}$$

$$S_1 - S_2 = -k_B \sum_{i=1}^{q} [P_{1r} \ln P_{1r} + P_{1r} \ln P_{2r} - P_{1r} \ln P_{2r} - P_{2r} \ln P_{2r}] \tag{11}$$

$$S_1 - S_2 = -k_B \sum_{i=1}^{q} P_{1r} \ln \frac{P_{1r}}{P_{2r}} + k_B \sum_{i=1}^{q} (P_{1r} - P_{2r}) \ln P_{2r} \,. \tag{12}$$

But $\sum_i (P_{1r} - P_{2r}) = \sum_i P_{1r} = \sum_i P_{2r} = 0$, hence, Eq. (12) becomes

$$S_1 - S_2 = -k_B \sum_{i=1}^{q} P_{1r} \ln \frac{P_{1r}}{P_{2r}} = k_B \sum_{i=1}^{q} P_{1r} \ln \frac{P_{2r}}{P_{1r}} \,. \tag{13}$$

8.3. Analogy Between Physical and Simulated Annealing

The simulated annealing approach is based on a Monte Carlo model used to study the relationship between atomic structure, entropy and temperature during the annealing process of a sample of material (Van Laarhoven and Aarts, 1987). Certain analogies exist between physical annealing and simulated annealing. The problem configuration of a combinatorial optimization problem (configuration $i \to$ configuration j, through perturbation) corresponds to the state of a solid material, the objective or cost function, f, corresponds to energy of a solid, E, and the control parameter c, corresponds to the temperature, T, of a solid. The simulated annealing algorithm can now be viewed as a sequence of Metropolis algorithms evaluated at a sequence of decreasing values of the control parameter, c, which can be described as follows.

Step 1: Control parameter, c is given a high value.

Step 2: Generation mechanism: perturb current configuration i to obtain configuration j with costs $f(i)$ and $f(j)$, so that $\Delta f_{ij} = f(j) - f(i)$.

Step 3: Acceptance criterion:

$$\Pr\{acceptance\ of\ new\ states\} = \begin{cases} 1 & \text{if } \Delta f \leq 0 \\ e^{(-\frac{\Delta f}{c})} & \text{if } \Delta f > 0 \end{cases}$$

where the second exponential-condition is a non-zero probability of continuing with a configuration with higher cost than the current configuration.

Step 4: Continue the process until equilibrium is reached, i.e. until the probability-distribution of the configurations approaches the Boltzmann distribution given as

$$\Pr\{configuration = i\} = q_i(c) = \frac{1}{Q(c)} \bullet e^{(-\frac{f(i)}{c})},$$

where, $Q(c)$ is the normalization constant.

Step 5: Lower the control parameter, c, in steps, while allowing the system to reach equilibrium through Steps (2) to (4).

Step 6: Terminate when virtually no deterioration is accepted any more, i.e. when no improvement in cost exists.

Step 7: Solution of combinatorial optimization problem is the final "frozen" configuration.

An instance of a combinatorial optimization problem is a tuple $(\mathfrak{R}, f)$, where $\mathfrak{R}$ is the finite *set of configurations or solutions* of the problem, in the $\boldsymbol{R}$-space and f is a *cost function*, f: $\mathfrak{R} \rightarrow \mathbf{R}$, which assigns a real number to each configuration. For each configuration i, $\mathrm{N}(i)$ is a subset of configurations or solutions, called the neighbourhood of i. The configurations within a neighbourhood that are consecutively visited by the simulated annealing algorithm can be seen as a Markov chain, which, is a sequence of trials, where the outcome of a current trial depends on the outcome of the previous trial, defined as:

$$P_{ij}(k-1, k) = \Pr\{X(k) = j | X(k-1) = i\}. \tag{14}$$

The meaning of Markov chain model is that the probability that the outcome of the kth trial, $X(k)$ is j, given that the outcome of the $k-1$th trial, $X(k-1)$ is i. We can also define the probability of outcome i at the kth trial $a_i(k)$ as:

$$a_i(k) = \Pr\{X(k) = i\}. \tag{15}$$

If the conditional probabilities, P_{ij}, depend on k, then the corresponding Markov chain is said to be homogenous, but if not, the corresponding Markov chain, is said to be in-homogenous. For a fixed value of control parameter, c, the configurations that are consecutively visited by the algorithm can be seen as a Markov chain with transition matrix $P = P(c)$ given by (Aarts and Van Laarhoven 1985; Lundy and Mees 1986; and Romeo and Sangiovanni-Vincentelli 1985):

$$P_{ij}(c) = \begin{cases} G_{ij}A_{ij}(c) & \text{if } j \neq i \\ 1 - \sum_{k=1}^{|\mathfrak{R}|} G_{ij}A_{ij}(c) & \text{if } j = i \end{cases} \tag{16}$$

where the generation probabilities G_{ij} are given by

$$G_{ij}(c) = \begin{cases} |N(i)|^{-1} & \text{if } j \in N(i) \\ 0 & \text{otherwise} \end{cases} \tag{17}$$

and the acceptance probabilities A_{ij} are given by

$$A_{ij}(c) = \min\left\{1, e^{\left(\frac{-(f(j)-f(i))}{c}\right)}\right\}. \tag{18}$$

As can be seen, the generation probability matrix $G(c)$ is not dependent on control parameter, c. It only depends on the inverse of the size of the neighbourhood structure. To understand this, suppose there are 100 points in a neighbourhood, then the generation probabilities, $G_{ij} = 1/100 = 0.01$, that is, the probability of moving from one point to the other within the neighbourhood is 0.01. This makes some sense. Recall that, at thermal equilibrium, the solid material follows the canonical Boltzmann probability distribution (see Eq. (5)):

$$\Pr\{configuration = i\} = q_i(c) = \frac{1}{Q(c)} \bullet e^{\left(-\frac{f(i)}{c}\right)}.$$

Consequently, for a large number of transitions, i.e. as $k \to \infty$, the stationary distribution of the Markov chain exists, and can take the form:

$$q_i(c) = \frac{e^{\left(-\frac{f(i)}{c}\right)}}{\sum_j e^{\left(-\frac{f(i)}{c}\right)}} = \frac{\varphi(f(i), c)}{\sum_j \varphi(f(j), c)}. \tag{19}$$

This stationary distribution of the Markov chain is given by (Aarts and Van Laarhoven, 1985, Lundy and Mees, 1986, and Romeo and

Sangiovanni-Vincentelli, 1985):

$$q_i(c) = \frac{|N(i)|A_{i0i}(c)}{\sum_{j \in \mathfrak{R}} |N(i)|A_{i0j}(c)} \tag{20}$$

for some $i_0 \in \mathfrak{R}_{opt}$, where $\mathfrak{R}_{opt}$ is the set of globally minimal configurations, as long as the neighbourhoods, $N(i)$, are such that for each pair of configurations (i, j), there is a finite sequence of transitions that lead from i to j. Then, these conditions are refined (the generation matrix, G is irreducible) such that for decreasing c, the stationary distribution of the Markov chain, $q(c)$ converges to a uniform distribution on the set of globally minimal configurations (Aarts and Van Laarhoven, 1985):

$$G_{ij}(c) = \lim_{c \to 0} q_i(c) = \begin{cases} |\mathfrak{R}_{opt}|^{-1} & \text{if } i \in \mathfrak{R}_{opt} \\ 0 & \text{otherwise}. \end{cases} \tag{21}$$

The consequences of Eq. (8.21) require some discussion. Since the stationary distribution of the of the Markov chain is defined as the probability distribution of the configurations after an infinite number of transitions, it follows that the simulated annealing algorithm converges with probability 1 to a globally minimal configuration if the following conditions are met:

(i) the sequence of values of the control parameter, c, converges to 0;
(ii) the Markov chain generated at each value of c are of infinite length; and
(iii) the generation matrix G is irreducible.

8.4. Cooling Schedule Classifications for Simulated Annealing Schemes

Incidentally, some of the conditions of asymptotic convergence, such as generating Markov chain at each value of the control parameter, c, of infinite length, cannot be achieved in practice. Accordingly, when implementing simulated annealing in finite-time, a choice is made with respect to each of the following parameter:

(1) an initial value of control parameter, c_0;
(2) a final value of control parameter, c_f;
(3) the decrement rule of the control parameter;

(4) the length of the Markov chain, L_k; and
(5) a termination criterion.

The choice of such parameters is referred to as cooling schedule or annealing schedule. Since the performance of simulated annealing algorithm depends strongly on the choice of the cooling schedule, the remaining part of this section is devoted to its treatment. Three classifications of cooling schedules have been identified (Osman and Christofides, 1994):

- step-wise temperature reduction scheme;
- continuous temperature reduction scheme; and
- non-monotonic temperature reduction scheme.

8.4.1. *Step-wise temperature reduction scheme*

Initial value of the control parameter, c_0: If i is the current configuration and j the new configuration, let there be t_1 number of transitions for which , and let there be t_2 (the number of moves, t, minus t_1, i.e. $t_2 = t - t_1$) number of transitions for which $f(i) < f(j)$. Let the acceptance ratio be defined as

$$\chi(c) = \left[\frac{number\ of\ accepted\ transitions}{number\ of\ proposed\ transitions}\right]_c . \tag{22}$$

Then, the acceptance ratio $\chi(c)$ can be estimated as

$$\chi(c) = \frac{t_1 + t_2 \bullet e^{\left(\frac{-\Delta\bar{f}^{(+)}}{c}\right)}}{t_1 + t_2} . \tag{23}$$

Leading to:

$$\chi(c)(t_1 + t_2) = t_1 + t_2 \bullet e^{\left(\frac{-\Delta\bar{f}^{(+)}}{c}\right)} \tag{23a}$$

$$\frac{(t_2 \bullet \chi(c) - t_1 \bullet (1 - \chi(c)))}{t_2} = e^{\left(\frac{-\Delta\bar{f}^{(+)}}{c}\right)} . \tag{23b}$$

Taking logarithm of both sides lead to:

$$\ln\left(\frac{(t_2 \bullet \chi(c) - t_1 \bullet (1 - \chi(c)))}{t_2}\right) = \frac{-\Delta\bar{f}^{(+)}}{c} \tag{23c}$$

$$c_0 = \frac{-\Delta \bar{f}^{(+)}}{\ln\left(\frac{(t_2 \bullet \chi(c) - t_1 \bullet (1-\chi(c)))}{t_2}\right)} \tag{23d}$$

$$\text{and } c_0 = \frac{-\Delta \bar{f}^{(+)}}{\ln\left(\frac{t_2}{(t_2 \bullet \chi(c) - t_1 \bullet (1-\chi(c)))}\right)}\,. \tag{24}$$

Note that given, t random moves, and the number of moves where there is cost increase, t_2, $t_1 = t - t_2$.

Decrement rule. The decrement of c_k is such that, for a small value of δ (0.1 or smaller), the stationary distributions of two succeeding Markov chains approximately satisfy

$$\frac{1}{1+\delta} < \frac{q_i(c_k)}{q_i(c_{k+1})} < 1 + \delta \tag{25}$$

$$q_i(c_k) < (1+\delta) q_i(c_{k+1}) \tag{25a}$$

$$e^{\left(\frac{-(f(i)-f_{opt}}{c_k}\right)} < (1+\delta) \bullet e^{\left(\frac{-(f(i)-f_{opt}}{c_{k+1}}\right)}\,. \tag{25b}$$

Taking logarithm of both sides of Eq. (25b), and defining $\Delta f = f(i) - f_{opt}$, leads to

$$\begin{aligned} -\frac{\Delta f}{c_k} &< \ln(1+\delta) - \frac{\Delta f}{c_{k+1}} \\ \frac{\Delta f}{c_{k+1}} &< \left(\frac{\Delta f}{c_k} + \ln(1+\delta)\right) = \left(\frac{\Delta f + \ln(1+\delta) \bullet c_k}{c_k}\right)\,. \end{aligned} \tag{25c}$$

Rearranging Eq. (25c) leads to

$$c_k \bullet \Delta f < c_{k+1}(\Delta f + \ln(1+\delta) \bullet c_k)\,. \tag{25d}$$

Solving Eq. (25d), leads to

$$\begin{aligned} c_{k+1} &> \frac{c_k \bullet \Delta f}{(\Delta f + \ln(1+\delta) \bullet c_k)} \\ c_{k+1} &= \frac{c_k}{(1 + \frac{\ln(1+\delta) \bullet c_k}{\Delta f})} \end{aligned} \tag{25e}$$

$$c_{k+1} = \frac{c_k}{(1 + \frac{\ln(1+\delta) \bullet c_k}{3\sigma_{ck}})} \quad k = 0, 1, 2, \ldots \; \forall i \in \mathfrak{R}\,. \tag{26}$$

Markov chain length. Markov chain length, L_k, is set to a constant value which, is equal to the size of the largest neighbourhood given as

$$L_k = \max_{i \in \mathfrak{R}} |N(i)|\,, \quad k = 0, 1, 2, \ldots\,. \tag{27}$$

The Markov chain length determines the length of time a search is going to be performed at a certain temperature (control parameter). Hence, the computational time depends on the Markov chain length. The use of the Markov chain length has the advantage that the nieghbourhood at a certain temperature is thoroughly searched for possible improving neighbours before the temperature can be lowered in search for better solution.

Stopping criterion. The stopping criterion depends on the ratio of the difference between the average objective function value and the optimal solution, to the initial, average objective function value, given as

$$\frac{f^{-k}(i) - f(i^*)}{f^{-1}(i_0)} \leq \varepsilon \tag{28}$$

where, $f^{-k}(i)$ is the average objective function value, $f(i^*)$ is the optimal objective function value, $f^{-1}(i)$ is the initial, average objective function value, and ε is a small positive real number.

8.4.2. *Continuous temperature reduction scheme*

While the step-wise reduction scheme specifies a finite sequence homogeneous Markov chains, the continuous reduction scheme specifies a finite sequence inhomogeneous Markov chains. In this present scheme, after each iteration k, the parameter control c_k, is reduced according to some rules.

Initial value of the control parameter, c_0: The initial temperature is taken to be much greater than the value of the maximum difference in the objective function values obtained between two consecutive solutions during the initial stage. Defining U as some upper bound on $(f(i) - f_{opt})$, the initial control parameter is defined as

$$c_0 \gg U\,. \tag{29}$$

Decrement rule. The decrement of c_k is according to Lundy and Mees (1986), and depends on a small reduction in the control parameter at each iteration using a constant decrement ratio

$$\frac{c_k - c_{k+1}}{c_k \bullet c_{k+1}} \ll \frac{\gamma}{U} = \beta \tag{30}$$

where, γ is a small, positive real number. This equation is simplified through Eqs. (30a) to (30c) to obtain the final expression in Eq. (31):

$$c_k - c_{k+1} = \beta c_k \bullet c_{k+1} \tag{30a}$$

$$\beta c_k \bullet c_{k+1} + c_{k+1} = c_k \tag{30b}$$

$$c_{k+1}(1 + \beta c_k) = c_k \tag{30c}$$

$$c_{k+1} = \frac{c_k}{1 + \beta \bullet c_k}\,. \tag{31}$$

Markov chain length. Markov chain length, L_k, is set to 1, according to the original concept of Kirkpatrick *et al.* (1983), in which small constant decrement ratio that leads to the equivalent number of iterations between the same initial and final control parameters is used:

$$L_k = 1\,. \tag{32}$$

8.4.3. *Non-monotonic temperature reduction scheme*

The non-monotonic reduction scheme reduces the control parameter (temperature value) after each attempted move or iteration with occasional resets in temperature whenever the search encounters a special cycle. When the neighbourhood of the current solution is completely searched, a cycle, is said to be completed.

Initial value of the control parameter, c_0: The initial temperature is taken to be a small value, and a seed solution from a good heuristic is advocated. The initial temperature or control parameter is given by

$$c_0 = f_{\max} - f_{opt}\,. \tag{33}$$

Decrement rule. The decrement of c_k is similar to that of Lundy and Mees, with slight modification:

$$c_{k+1} = \frac{c_k}{1 + \beta \bullet c_k} \tag{34}$$

where, the parameter is given as

$$\beta_k = \frac{c_0 - c_f}{((\alpha + \gamma^* \sqrt{k}) * c_0 * c_f)} \,. \tag{35}$$

As k increases, β_k decreases, leading to the control parameter (temperature) being decreased more slowly. Consider for example, $\alpha = 10$, $\gamma = 1$, $c_0 = 100$, and $c_f = 80$. When $k = 25$, $\beta = 0.0005$ and $c_{k+1} = 24.69$. When $k = 100$, $\beta = 0.00025$ and $c_{k+1} = 24.39$. Consequently, for a particular the control parameter is decreased more slowly with increase in k.

Diversification. Diversification is implemented in simulated annealing by resetting the control parameter (temperature) to a higher value after there is no change in elite solution within a neighbourhood for a given cycle of search. Diversification favours the search process to examine unvisited regions in the solution space and generates solutions that differ in various significant ways from those already obtained. The control parameter (temperature) is reset to a higher value, but usually not more than half of the initial value.

Stopping criterion. The algorithm is terminated after a number of control parameter resets are performed without finding any improved solution or after a pre-set number of iteration known *a priori.*

8.5. Neighbourhood Search Techniques

In order to improve the efficiency and effectiveness of the simulated annealing algorithm there is need to implement appropriate neighbourhood search technique. Two of these techniques are discussed.

8.5.1. *Random search*

Random search technique involves selecting a neighbour at random and then proceeding to evaluating it. This drawback of this technique is that there is the possibility of missing the neighbours that would improve the objective function. This is because the larger number of moves that do not necessarily guarantee an improvement to the objective function are made since there is equal probability of choosing any neighbour.

8.5.2. *Systematic search*

In the systematic search technique, as the name implies, the neighbourhood is sequentially searched to ascertain if any neighbour has any effect on the objective function. Two aspects of this search technique include the *first-admissible* and the *best-admissible* strategies.

First-admissible search technique. In this strategy, the first move that lowers the objective function is accepted. If one is not found, the search continues while maintaining the least non-improving move that can be made within the whole neighbourhood search. When an improving move is made, the control parameter (temperature) decrement rule is applied.

Best-admissible search technique. In this strategy, the search is applied to the whole neighbourhood and the best move that gives the best objective function value is maintained. If one is not found, the search continues while maintaining the least non-improving move that can be made within the whole neighbourhood search. This way, the best move within a neighbourhood is made, leading to obtaining the best solution.

8.6. Production Layout Strategies

Production layout planning is a decision concerning how best machines have to be arranged in a production facility so that the facility operates at its peak effectiveness and efficiency. A growing number of literature in the area of production layout planning (Hsu and Su, 1998; Onwubolu, 1998; Sarker and Xu, 1998; Escoto *et al.*, 1998; Taboun *et al.*, 1998; Onwubolu, 2000) shows that it is now an important aspect of production planning and control. Production layout planning specifies the arrangement of processes (such as milling, turning, welding, and painting), the related equipment, and work areas, including customer service and storage areas. An effective production layout planning also facilitates the flow of materials and people within and between areas. The goal of management is to arrange the facility so that it operates at peak effectiveness and efficiency. To achieve these layout objectives, a variety of strategies have been developed; among them are four that will be discussed in this section:

- Fixed-position layout;
- Process-based layout;

- Product-based layout;
- Group technology/cell layout.

A good production layout requires determining:

(1) *Capacity and space requirements.* Capacity decision is a primary decision in production layout planning. It is only when management knows of the machines, equipment, and personnel required, that they can proceed with the layout and provide necessary space fore each component. Management should provide necessary space requirements for actual production, aisles for transportation of part-finished or finished good, storage, maintenance, offices, and allowances should be made for safety requirements.
(2) *Material handling systems.* Decisions are also needed with respect to equipment for material handling, such as hand carts, automatic carts, trucks, cranes, conveyor systems, and automated guided vehicles.
(3) *Flows of information.* Decisions concerning the best way to facilitate communications are necessary.
(4) *Cost of moving between the various work areas.* Decisions have to be made concerning the costs involved in moving processed goods between adjacent departments.
(5) *Environment and safety.* Decision are required about the positioning of windows, air flow facilitators, and attention should be given to safety issues such as noise reduction, etc.

The following information is needed before contemplating a layout plan.

(1) *Characteristics of products and materials.* The types of products and materials, their shapes and sizes determine product variety, which is a major factor in designing facilities for economical production.
(2) *Product quantities.* It is important to know the present and future quantities of each product, since the product variety and quantity relationships dictate the type of layout to a great extent.
(3) *Process routing.* Information concerning sequences, in which products are processed, is important because it is related to the type of layout to a large extent.
(4) *Services.* Information on support services, inspection stations and locker rooms have direct bearing on production, hence are important in layout planning.

Table 8.1. Comparison of major characteristics of process-based process, repetitive process, and product-based process.

Process-based process	Repetitive process	Product-based process
1. Small volume and large variety of products are produced.	1. Long runs, normally a standardized product with options, are produced from batches.	1. Large volume and small variety of products are produced.
2. General-purpose equipment is used rather than special-purpose.	2. Some special equipment and fixtures are used.	2. Special-purpose equipment is used rather than general-purpose.
3. Raw material inventories are high compared to the value of the product.	3. Just-in-time procurement techniques are used.	3. Raw material inventories are low compared to the value of the product.
4. Work-in-progress is high compared to product output.	4. Just-in-time inventory techniques are used.	4. Work-in-progress is high compared to product output.
5. Slow movement of units through the facility.	5. Hourly or daily measurement for movement in facility is typical.	5. Slow movement of units through the facility.
6. Material handling through small flexible equipment.	6. Material handling through transfer machines, conveyor, AGVs.	6. Material handling through material guides, feeders, etc.
7. Finished goods are usually made to order and not stored.	7. Finished goods are usually made to frequent forecast.	7. Finished goods are usually made to forecast and stored.
8. Complex, order-oriented scheduling focusing on trade-off between inventory availability, capacity, and customer service.	8. Simple, periodic-oriented scheduling often overlaps operations and is designed to produce various batches and option from a variety of batches to forecasts.	8. Relatively, simple, periodic-oriented scheduling focuses o establishing a rate of output sufficient to meet sales-forecasts.

Table 8.1 (*Continued*)

Process-based process	Repetitive process	Product-based process
9. Estimated cost prior to doing the job, but known after job completion.	9. Costs are usually known due to extensive prior job experience.	9. Costs are highly dependent on utilization capacity due to high fixed cost.
10. More broadly skilled operators.	10. Modestly trained operators.	10. Less broadly skilled operators.
11. Many job instructions due to many job changes.	11. Reduced training and changes in job instructions due to repetitive operations.	11. Standardized work orders and job instructions make them few.
12. Wide aisle and pronounced storage are typical.	12. Medium or narrow aisles, with little storage space needed.	12. Equipment, machinery, and product flows dictate facility.
13. Low fixed costs and high variable costs.	13. Fixed costs depend on facility flexibility.	13. High fixed costs and low variable costs.

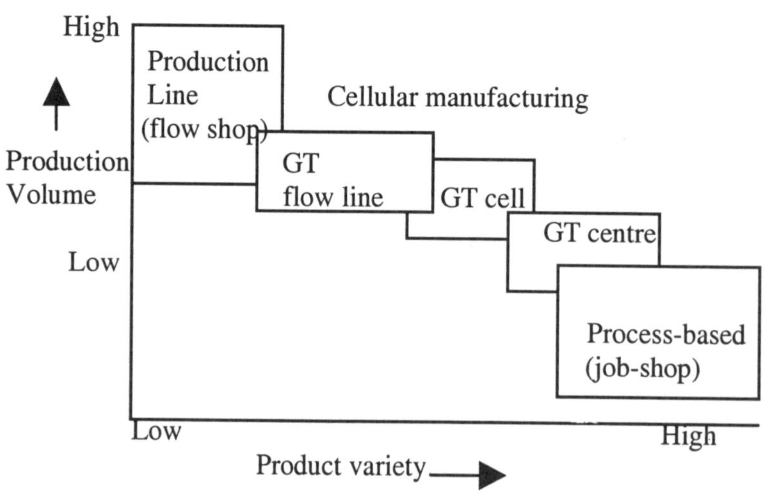

Fig. 8.1. Product variety and production volume relations.

(5) *Timing.* Information on scheduling of parts on machines is important for layout planning.

Process-based, product-based and group/cell layouts can be distinguished based on system characteristics such as product variety and production volume relations (Steudel and Desruelle, 1992) as shown in Fig. 8.1. Decision concerning which type of layout to select, is a strategic decision that has to be carefully made by management. Table 8.1 compares the major characteristics of low-volume, high-variety process (process focus), repetitive process, and high-volume, low variety process (product focus). Each of these layouts is briefly discussed in this section.

8.6.1. *Fixed-position layout*

In fixed-position layout, the project remains stationary while equipment and workers are moved to the one work area as shown in Fig. 8.2. Fixed-position layout is commonly used when the job is massive and cannot be moved to the machines required for produce it, such as in ship building, aircraft manufacturing, highway and building constructions, etc. One-of-a-kind products are most associated with fixed position layouts. Production control and coordination, space requirements and in-process inventory are complex in this type of layout. However, this type of layout is flexible and responsive to product mix, volume and design changes. Material movement is reduced considerable. On the other hand, equipment and personnel movements are higher. Overhead costs for temporary infrastructures are high since these may be pulled down

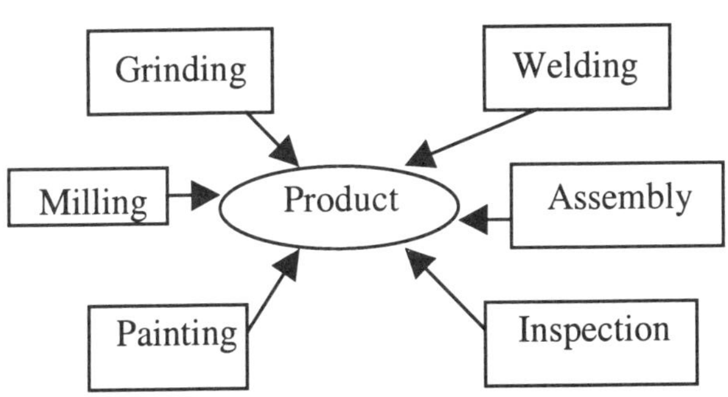

Fig. 8.2. Fixed-position layout.

after the project, as in building projects. Skill requirements are on the part of personnel, so are task identification.

8.6.2. *Process-based layout*

Process-based layouts such as are found in job-shop environments, are characterized by low-volume production and high product variety. In the process-based or job shop environment, the machines and equipment are arranged according to their functions. For instance, lathe machines are grouped together to form the turning department, milling machines grouped together to form the milling department, and welding machines to form the welding department as shown in Fig. 8.3. With the aid of the routing sheet, or the flow process chart, parts are made to move from department to department during their processing. In process-based or job shop layout, machines are more difficult to manage because of the complexity of part routing. Other disadvantages of process-based or job shop layout include:

(1) pronounced materials handling;
(2) pronounced work-in-process inventory;
(3) pronounced setups;
(4) long manufacturing lead times;
(5) production-flow is jumbled up;
(6) queue time is much;
(7) material handling is much.

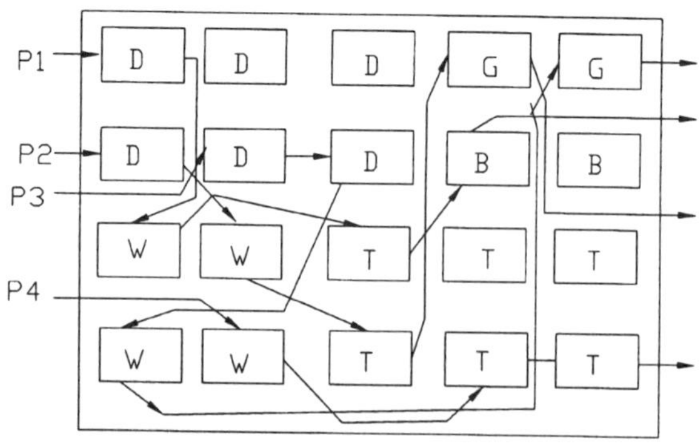

Fig. 8.3. Job-shop (functional) layout.

These disadvantages have adverse effects on operating costs. Consequently, production planners have to be aware of how planning for production layout impacts on production planning of control.

8.6.3. *Product-based layout*

Product-based layouts such as are found in flow-shop environments, are characterized by high-volume production and low product variety. In product-based layouts, using the routing sheet, machines are arranged in a line layout and all jobs visit the machines in the same sequence as shown in Fig. 8.4. This type of layout is normally used in process industries, in continuous assembly, and for mass production of discrete components. Here, there is little job variation possible. The work content at each location is balanced so that materials flow through the line in a continuous manner.

8.6.4. *Group technology layout*

Most manufacturing facilities have been built around product-based or job shop, and to transform them to cellular manufacturing system, the knowledge of group technology is so essential. Group technology is a philosophy whose central objective is to increase production efficiency by grouping various parts and products with similar design and/or production processes together. Parts may, for instance, be grouped according to geometric properties, and production processes. Geometric properties include shape and size, while production processes include the type, sequence, and the number of operations.

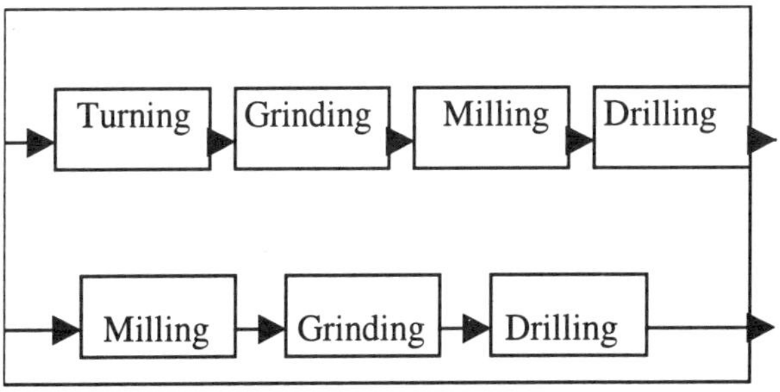

Fig. 8.4. Product-based layout.

Group technology is a manufacturing philosophy that involves grouping parts of the same family together which can be manufactured by a particular type of machine, or grouping machines together which can machine a particular part. With the philosophy of group technology, the combined benefits of mass production while dealing with multi-product, and small batch-production are enhanced.

Cellular manufacturing facility is product-based. This type of layout is also referred to as group-type or cell layout. Using the routing sheets for each product and with the concept of group technology, machines are arranged in a group, in order to cater for products having similar manufacturing characteristics.

Several benefits are derived from using group technology layout (Wemmerlov and Hyer, 1989). When compared to job shop environment, group technology layout is easier to manage. There is reduced work-in-progress inventory, material handling, set up times and manufacturing lead times. Other benefits that can be derived from successful implementation of group technology layout are:

- improved design, design reviewed improvement is systematic;
- reduced raw material and purchases;
- simplified production planning and control;.
- improved routing and machine loading;
- reduced tooling setup time, WIP and production time.

The group technology facility layout can be broadly classified into three categories (Askin and Standrigde, 1993, Singh and Rajamani, 1996) as shown in Fig. 8.5. These three categories are briefly discussed.

GT flow line layout

The GT flow line layout operates as a mixed-model assembly line, with automated transfer mechanisms sometimes used for material handling, and all jobs visit the machines in the same sequence (see Fig. 8.5(a)). In GT flow line layout, there is little job variation possible. The work content at each location is balanced so that materials flow through the line in a continuous manner.

GT cell layout

GT cell layout allows any job to move from any machine to another. Movement can be bi-directional as shown in Fig. 8.5(b), as opposed to GT flow-line where

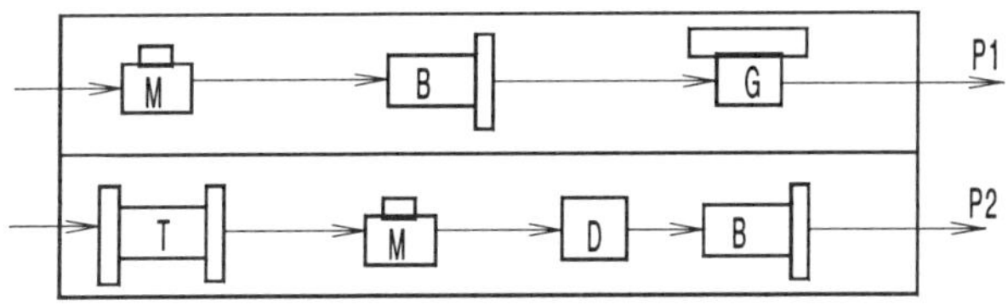

(a) GT flow-line

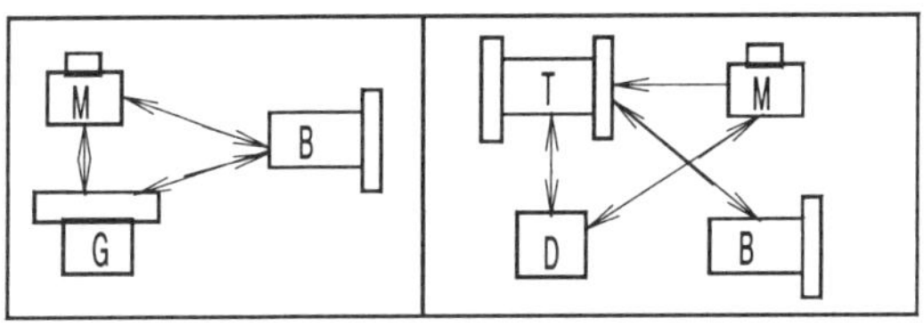

(b) GT cell

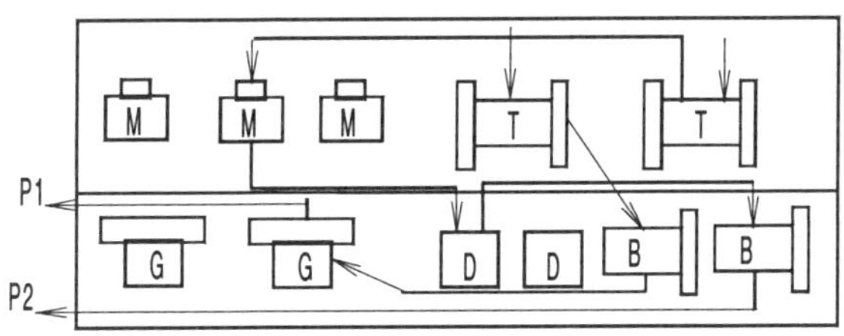

(c) GT centre

Fig. 8.5. Group technology layout.

all jobs visit the machines in the same sequence. In this layout material-handling movement is minimized since machines are close to one another. We shall return to this type of layout in our further discussion in this chapter.

GT centre layout

GT centre layout inherits the process-based layout arrangement, but each machine is dedicated to specific part families. This way, flexibility of the facility is increased and material handling is also minimized. Flexibility in this case is the ability of the facility to process a variety of product-mix more quickly. The arrangement of a GT centre layout is shown in Fig. 8.5(c).

Attributes of group technology layout

- *Product.* The group of machines in a cell produces a specialized family of parts/components, for instance castings from casting cell, and machined parts from machine cell;
- *Working Team.* The cell is a sociological unit conducive to team-work leading to, enhanced motivation for process-improvements;
- *Facility.* The groups are equipped with a set of machines and other production equipment;
- *Group layout.* The facilities are laid out in one area reserved solely for that group or cell;
- *Objective.* The team of workers in a cell share a common end-product output target at the end of every working period;
- *Minimal inter-cell relations.* The cells should be as independent from each other as possible. The normal operation and output of the cell should not depend on the other cell;
- *Cell size.* The cell or group size should be limited in number of machines per cell, or number of workers per cell. Generally, 16 to 20 cell workers, and two to ten machines are recommended maximum numbers per cell.

Benefits of group technology layout

A number of potential benefits are associated with the successful implementation of group technology layout.

- *Benefits in product design and redesign.* The personnel charged with the responsibility of design have a systematic way to review the product design via group technology codes to see if an existing component might not suffice on a new project. The use of an existing design will eliminate costs of design and development of new part. This is a potential major cost reduction. With the use of similar designs coded in GT there are few or no changes to the manufacturing system to suit the newly introduced product or part type, which is also a great potential saving. Introduction of a different design (not in the groups coded in GT) into the group technology facility will complicate parts sequencing, manufacturing management, and also requires new extra machines and equipment. This is often encountered in most of the job shop and batch production manufacturing. Applying the concepts of group technology to group technology layout encourages standardization of design features leading to design for ease of assembly (DFA). Redesign of similar parts is easier since similar designs are retrieved and modified.

- *Benefits in manufacturing.* Cost reduction is achieved through implementation of group technology layout. Production planning is modularized and therefore simplified. Manufacturing processes and tool design and layout are also simplified. In process-based or job shop and batch production layouts there is usually large product variety, and in most cases many similar parts and subassemblies exist in the products. Implementation of group technology layout will incorporate the philosophy of group technology thereby reducing the variety of components. Similar parts will now be manufactured in one cell resulting in reduced round about motions and intercellular moves, hence reduction in material handling costs.

The concept of group technology layout is product-based, where a cell operates on similar parts. In this type of manufacturing environment implementation of computerized manufacturing philosophies such as computer aided process planning (CAPP), and automated guided vehicle (AGV) systems are enhanced greatly.

Group technology facilitates group layout of manufacturing environment where a group of machines perform all processing required by one or more families of parts. Time studies and calculations for each operation and for each part are easy to carry out. This also enhances simpler capacity requirements planning (CRP) and capacity planning (CP). The system also leads to a smooth production flow.

Group technology layout simplifies production planning and control. Production scheduling is simplified due to reduction in the number of types of parts that are processed in a cell. Group tooling and fixtures common to a part family in a cell is possible. A set of tools to work on a part family can also be standardized and this will simplify the manufacturing environment since for different groups of parts a tool set is simply interchanged with the other.

- *Benefits to management sector.* Group technology layout involves grouping of machines together into cells. In each cell are assigned a group of workers to work on and complete the processing of parts in that cell. This leads to a high degree of specialization and in-depth knowledge. Higher degree of production efficiency, accuracy, and control of quality are inevitable. Since the manufacturing facility is now defined into groups (cells) where a machine or part and processes in a particular cell have their own problems, the problems can be rectified in a simplified hierarchical manner, bottom to top. It is easier to solve manufacturing problems for a small group than for the whole system at once. Management decisions, control, and supervision are enhanced.

- *Benefits to market.* Design and redesign of a product according to the customer wants are made simpler by the application of group technology with group technology layout. With this set-up it is easier to change a module only in a particular cell rather than whole system in order to meet market demands.

In this way the market demand is met within a shorter period of time, i.e. customer responsiveness is improved. Efficient production is inevitable with group technology layout leading to faster deliveries and short lead times. It's therefore evident that group technology layout is a tool for world-class manufacturing competitiveness.

- *Benefits to flexible manufacturing systems.* As already discussed, the basic manufacturing facilities are process-based or job shop, batch production, flow line and assembly line. With the knowledge of GT these basic systems are transformed to cellular manufacturing system. Group technology layout forms the primary building block for flexible manufacturing systems (FMS), and just-in-time (JIT) production facilities.

Disadvantages of group technology layout

Although group technology layout has many benefits as outlined earlier, we consider some of its disadvantages.

(1) Installation of a coding system for GT may be expensive;
(2) Realization of the importance of a GT coding system may be difficult. It is no easy task to convince management on the necessity of moving from process-based or job shop to group technology layout;
(3) High costs may be encountered in relocation of plant equipment and machines;
(4) The design of group technology layout is expensive and it requires expertise to implement.

8.7. Production Layout Planning

Layout planing for manufacturing facilities is one of the several integrated elements in production system planning, and in the context of cellular manufacturing facilities, it can be considered as a hierarchical process involving the following five principal stages (Singh and Rajamani, 1996):

(1) determining part families based on part design and process similarities;
(2) cell formation: part families assignment to groups of machine cells;
(3) part families and workloads rationalization;
(4) selection of type of cell layout;
(5) laying out machines and auxiliary facilities in cell.

Cellular manufacturing layout is an important application of group technology in which sets/groups, or families of parts are produced in cells or a group of various machines that are physically close together and can entirely process a family of parts. A part family consists of different parts having similar processing requirements. A cell can produce more than one part family, as long as each part can be entirely processed in that cell. Machines are arranged in-groups to cater for products with similar characteristics. The implementing of cellular manufacturing layout can be considered as a hierarchical process involving the following four principal stages (Wemmerlov and Hyer, 1989):

(1) Planning and justification of conversion of the present system to cellular system;
(2) Design and evaluation of cell system structure;
(3) Design of cell system infrastructure;
(4) The implementation, operation and improvement.

A combination of the hierarchical process involved in production layout planning suggested by Singh and Rajamani (1996) and Wemmerlov and Hyer (1989) should be adopted during implementation of cellular manufacturing layout. For example, it is preferred to justify conversion of the present system to cellular system. In fact, improvement of the system should be an on-going activity.

8.7.1. *Determining part families*

Determining part-families based on part design and process-similarities make use of the group technology-concept already treated in Sec. 8.6. As already discussed, group technology is a philosophy whose central objective is to increase production efficiency by grouping various parts and products with similar design and/or production processes together. Parts may, for instance, be grouped according to geometric properties, and production processes. Geometric properties include shape and size, while production processes include the type, sequence, and the number of operations.

Although, early applications of group technology used the classification and coding techniques to identify part families, Burbidge (1989, 1991) proposed production flow analysis (PFA). The PFA method finds a complete division of all parts into families and also a complete division of all existing machines into associated groups by analyzing information the process routes for parts. The identification of part families and machine groups is known as cell formation, and is one of the steps in PFA. Since the proposal of PFA, numerous approaches have been reported for cell formation. Some of these are treated in the next section.

8.7.2. *Cell formation*

Cell formation involves assigning part families to groups of machine cells. Cell formation is an important step in layout planning for cellular manufacturing. Many manufacturing firms which hitherto satisfied their customers while operating job shop production systems have recently had to rethink because of the superiority of group technology (GT) philosophy of cellular manufacturing. Cellular manufacturing systems (cellular manufacturing layout) house a large variety and combinations of processes and shorten manufacturing lead times, reduce work-in-progress (WIP), lower set-up times and operations costs, improve quality, increase workers job satisfaction and subsequent customer satisfaction (Wemmerlov and Hyer, 1989). Manufacturing firms which embrace modern manufacturing improvement philosophies such as just-in-time (JIT), total quality management (TQM), business process re-engineering (BPR) and time-based competition (TBC) use the principles of cellular manufacturing layout in their manufacturing system restructuring efforts so as to meet world-class manufacturing status (Wemmerlov and Johnson, 1997).

Solving flexible manufacturing systems (FMS) optimal production planning and scheduling problems through size reduction require grouping of parts into part families at aggregate planning level (Bitran and Hax, 1981).

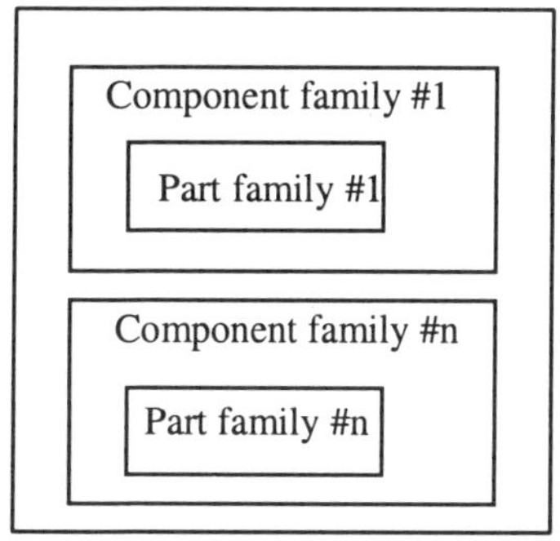

Fig. 8.6. Decomposition of FMS planning at disaggregation level.

Planning and scheduling FMS at the disaggregate-level involves machines, parts and additional components such as tools, fixtures, grippers and material handling components. The costs of these components are usually high and they may be in limited quantities. Figure 8.6 shows the decomposition of FMS planning situation at disaggregation level, with all the parts to be manufactured and the associated components being decomposed into disjoint part and component families.

This results in each of the individual part and component families being easier to manage. Individual part family is manufactured by machine-groups that utilize associated components such as the tools, fixtures, grippers and material handling components already mentioned.

The grouping problem in GT involves diagonalisation of a given machine-part incidence matrix into a cluster of machines that produce a cluster of parts. In FMS, the corresponding cluster of machines is referred to as flexible manufacturing-cell (FMC). At lower level in the classical GT physical cells, the cluster of machines is ordinarily referred to as manufacturing cell. Since manufacturing cell is the building block of FMC, the discussion in the remaining parts of this section concentrate on machine and part families grouping without loss of generality.

Machine cell formation techniques

Various approaches to cell formation have since been reported and may fall under one of the six major classifications:

- array-based clustering;
- similarity coefficient;
- mathematical programming;
- graph and network;
- heuristic; and
- combinatorial optimization techniques.

There are two types of array-based clustering techniques; rank order clustering (ROC) and bond energy analysis (BEA). In ROC, a positional based value is assigned to each "1" in the machine-part incidence matrix and the values of all the "1"s in each row. Each column is summed, the rows and columns being rearranged in decreasing order based on the values from top to bottom, and from left to right respectively (King, 1980). This process is continued until an invariant matrix is obtained resulting in a clustering of machines and parts along the diagonal. Extensions to ROC to include cost data (Askin and Subramanian, 1987) and prevention of positional based value getting extremely large for large number of machines and parts (King and Narkornchai, 1982) have been reported. Progressive slicing of identified group after one iteration of row ranking and one iteration of column ranking (Chandrasekharan and Rajagopalan, 1986a) have led to drastic improvement of ROC results. The bond energy which is defined as the product of the values of the adjoining row and column elements in the machine-part incidence matrix to determine the degree of clustering has been successfully implemented, with an optimal solution being one which maximizes the bond energy (McCormick *et al.*, 1972).

The first work to use similarity measures to group machines and parts into cells (McAuley, 1972) utilizes a similarity matrix which contains all pair-wise similarity coefficients between each machine. The similarity matrix obtained is then used by the single linkage clustering-algorithm (SLCA) to form the machine groups. These methods have been adopted by other researchers (de Witte, 1980; Waghodekar and Sahu, 1984; Seifoddini and Wolfe, 1986; Tam, 1990). The clustering algorithms based on similarity coefficient, falls under two classifications: hierarchical and non-hierarchical. Both the single linkage clustering algorithm and the average clustering algorithm are hierarchical clustering algorithms. The only known non-hierarchical clustering reported so far include the MacQueen's k-means method (MacQueen, 1967) which requires the number of clusters to be specified in advance, the ideal-seed algorithm (Chandrasekharan and Rajagopalan, 1986b) and its improved version ZODIAC (Chandrasekharan and Rajagopalan, 1987).

The *p*-median model based on an integer programming formulation which seeks to maximize the sum of similarity coefficients for a fixed number of groups with the constraints that a part will be assigned to not more than one family, has been used to solve the GT grouping problem (Kusiak, 1987). Kusiak's similarity coefficients matrix has been used as input to an assignment model (Srinivasan *et al.*, 1990) to solve disjoint and non-disjoint part family's problems. The part grouping-problem in GT has been modelled as a linear transportation problem, solved in a two-phase algorithm to approximate the optimal solution (Kumar *et al.*, 1986). The *p*-median formulation (Mulvey and Crowder, 1979) is an integer-programming model with an objective function, which seeks to minimize the total sum of clusters-to-cluster median distances.

Kumar *et al.* (1986) formulated the GT grouping problem as an optimal *k*-decomposition problem in graph theoretic terms in which decomposition of networks are considered rather than block diagonalisation of matrices. Lenstra (1974) formulated the clustering of a two-dimensional data array as equivalent to solving two travelling salesman-problem. This problem has been mapped as a *k*-decomposition of weighted networks which is NP-complete (Ravikumar *et al.*, 1986).

Several heuristics which give polynomial bounds on the required computation time, but which do not guarantee optimal solution include (Ballakur and Steudel, 1987; Tabucannon and Ojha, 1987; Kumar and Vanelli, 1987), to mention a few.

Two formulations based on similarity order clustering (SOC) for grouping of machines and parts (Onwubolu, 1998) with the inclusion of single-pass heuristic, and the similarity coefficient-distant measure (Kerr and Balakrishnan, 1996; Onwubolu and Mlilo, 1998) solves the problems inherent in standard cell formation models.

In the subsequent sections, we describe these two methods for cell formation: similarity order clustering, and similarity coefficient-distant measures.

Generalized machine cell formation

The systematic method used in constructing the generalized machine cell formation follows the four steps involved:

(a) A strategy for preliminary cell grouping using the similarity order-clustering (SOC) scheme.

(b) The preliminary cell grouping and base-data are used to determine machine utilization. The minimum number of each machine is inferred from the machine utilization consideration.
(c) Design for non-intergrouping material handling using single-pass heuristic.
(d) Implementation of final machine cells.

Similarity Order Clustering (SOC) method

The SOC-formulated model for preliminary cell grouping (Step (a) of the generalized machine cell formation) uses the $0-1$ machine-part incidence matrix as input and the Kusiak similarity coefficients matrices (Kusiak, 1987) for machines and parts as basis.

The steps involved are as follows:

(1) *Submit* $0-1$ *machine-part matrix to SOC scheme.* The routing sheet from a job shop production system is used for the input data. This is a simple two-dimensional matrix with machine identifications in rows and parts, which visit each machine for production entered in columns.
(2) *Calculate Kusiak similarity-coefficients matrices for machines and parts.* Pairwise rows (machines) are compared bit-wise for similarity, and the sum of all similar bits is summed up to give the similarity coefficient of the row (machine)-pair considered. Bits are similar if they are each "1" or "0", otherwise the contribution is zero. The diagonal values are all zero. Off-diagonal coefficient values are zero or integer values greater than zero. Similar matrix is generated for parts. These square matrices are easily used as input to computer packages for solving assignment, linear transportation and travelling salesman problems in linear programming.
(3) *Arrange the values of the Kusiak similarity matrix in descending order.* The square similarity coefficient matrix is scanned and the values are arranged in descending order. For the machine similarity coefficient matrix, the row and column indices that locate the values are stored in the topology matrix. A similarity topology matrix is obtained for the parts.
(4) *Machine and parts sequence logic scheme.* The machines and parts sequences logic scheme uses the topology generated in Step 3 to obtain the sequences of machines and parts, which when used to rearrange the initial machine-incidence matrix gives clusters of machines and parts that enables the user to identify cell groups.

Illustration

Let us see how the SOC procedure proposed by Onwubolu (1998) works. The operation routings information for a 9-machine 10-part problem in a job shop, which is to be redesigned to a manufacturing cell are shown in Table 8.2. The corresponding machine-incidence matrix is shown in Fig. 8.7.

Determining similarity

Similarity coefficients (SCs) define relationships between pairs of machines or parts. The closer the relationship is the higher the SC. The SCs are determined for both machines and parts from the machine-part incidence matrix. The SC used is the one by Kusiak (1987). Kusiak's SC for machines is as follows:

$$s_{ij} = \sum_{k=1}^{m} d_k; \quad \forall i \neq j \; d_k = \begin{cases} 1 & \text{if } a_{ik} = a_{ij} \\ 0 & \text{if otherwise} \end{cases}$$

$$s_{ij} = 0 \quad \forall j = i$$

where s_{ij} is the similarity coefficient between machines i and j, a_{ik} is the machine-part incidence matrix value which is non-zero if kth part visits ith machine and is zero if otherwise. The diagonal values of the square similarity coefficient matrix are all zero.

Based on Kusiak's coefficients, the SC between machines 3 and 4, s_{34} in Fig. 8.7 can be calculated as follows: machines 3 and 4 do not process parts

Table 7.2. Parts demand, operation routings and processing times.

Part n	Demand d_i	Operation routings	Processing time t_i
1	2500	M5 $\longrightarrow$ M6 $\longrightarrow$ M9	1, 5, 4
2	3000	M1 $\longrightarrow$ M3 $\longrightarrow$ M8	4, 1, 2
3	2000	M2 $\longrightarrow$ M4 $\longrightarrow$ M7	1, 5, 1
4	1500	M5 $\longrightarrow$ M6 $\longrightarrow$ M9	2, 3, 2
5	2300	M5 $\longrightarrow$ M6 $\longrightarrow$ M9	3, 2, 3
6	2700	M1 $\longrightarrow$ M3 $\longrightarrow$ M8	2, 2, 5
7	3000	M2 $\longrightarrow$ M4 $\longrightarrow$ M7	1, 2, 2
8	2900	M5 $\longrightarrow$ M6 $\longrightarrow$ M9	4, 4, 4
9	2100	M1 $\longrightarrow$ M2 $\longrightarrow$ M3 $\longrightarrow$ M8	1, 2, 3, 3
10	1300	M2 $\longrightarrow$ M4 $\longrightarrow$ M7	2, 1, 3

Machine	Part									
	1	2	3	4	5	6	7	8	9	10
1	0	1	0	0	0	1	0	0	1	0
2	0	0	1	0	0	0	1	0	1	1
3	0	1	0	0	0	1	0	0	1	0
4	0	0	1	0	0	0	1	0	0	1
5	1	0	0	1	1	0	0	1	0	0
6	1	0	0	1	1	0	0	1	0	0
7	0	0	1	0	0	0	1	0	0	1
8	0	1	0	0	0	1	0	0	0	0
9	1	0	0	1	1	0	0	1	1	0

Fig. 8.7. Machine-incidence matrix for Table 8.2.

Table 7.3. Similarity coefficients for SOC example ($a_{ij} = a_{ji}$).

	Machine								
Machine	1	2	3	4	5	6	7	8	9
1	0	5	10	4	3	3	4	9	4
2		0	5	9	2	2	9	4	3
3			0	4	3	3	4	9	4
4				0	3	3	10	5	2
5					0	10	3	4	9
6						0	3	4	9
7							0	5	2
8								0	3
9									0

1, 4, 5, and 6, and no two machines simultaneously process any part. The value of s_{34} is therefore 4 + 0 or 4. A s_{ij} is determined for all possible pairs of machines. The similarity coefficients for machines in Fig. 8.7 are given in Table 8.3.

Using the SOC to obtain machine and part sequences

SOC uses the order or rank of the similarity coefficient value (SCV) to determine the best sequence for a given problem. Three steps are involved:

(1) Order the SCVs in a decreasing manner.
(2) Obtain the topology matrix.
(3) Generate the sequence from the topology matrix.

Step 1 involves scanning the similarity coefficient matrix given in Table 8.3. It is useful to represent the similarity coefficient matrix as a half-triangle, having values contained in the upper part of the diagonal and the lower part of the diagonal being zero. The SCVs are 10, 9, 5, 4, 3, and 2.

Step 2 involves reading the row and column numbers of all ordered non-increasing SCVs into the topology matrix. For example, SCVs of 10 correspond to pair-wise labels of (1, 3), (4, 7) and (5, 6). The first value corresponds to the row number while the second value of the pair corresponds to the column number in the similarity coefficient matrix. The topology matrix $T(i, j)$ for the SOC example is given as follows:

$T_{(i,j)}$	1,3	4,7	5,6	1,8	2,4	2,7	3,8	·	7,8	·	·	6,8	·	·	8,9	·	·	4,9
SCV	10	10	10	9	9	9	9	·	5	·	·	4	·	·	3	·	·	2

In Step 3, the first pair in the topology matrix is taken into the machine sequence. Hence in the first iteration in this step, the partial sequence array Q holds two machine labels: $Q = \{1, 3\}$. A closed loop is obtained by searching through the topology matrix, so that $Q = \{1, 3, 8\}$. This corresponds to machines in one cell $C_1 = \{1, 3, 8\}$. In the second iteration, the labels 1, 3, 8 are avoided during the search but a new pair in the topology matrix is chosen so that the partial sequence array Q holds two machine labels: $Q = \{4, 7\}$. A closed loop is obtained by searching through the topology matrix, so that $Q = \{4, 7, 2\}$. This corresponds to machines in another cell $C_2 = \{4, 7, 2\}$. In the third iteration, the labels 1, 3, 8, 4, 7, 2 are avoided during the search but a new pair in the topology matrix are chosen so that the partial sequence array Q holds two machine labels: $Q = \{5, 6\}$. A closed loop is obtained by searching through the topology matrix so that $Q = \{5, 6, 9\}$. This corresponds to machines in a third cell $C_3 = \{5, 6, 9\}$. Since the total number of machines in all the cells equal to the given number of machines, the algorithm terminates generating sequences for the machines. We note that SOC determines the best sequence for machines within SCV range of 10 and 9. The next lower SCV values are 5, 4, 3 followed by the least value of 2. SOC ignores these because they are too low to give good machine clustering. Machines {1, 3 and 8} form the first cell, machines {4, 7, and 2} form the second cell and machines {5, 6, and 9} form the third cell.

The same procedure is followed for parts. Based on Kusiak's coefficients, the SC between parts 1 and 2, s_{12} in Fig. 8.7 can be calculated as follows: parts 1 and 2 are not processed by machines 2, 4, and 7, and no two parts are simultaneously processed by any machine. The value of s_{12} is therefore 3 + 0 or 3. A s_{ij} is determined for all possible pairs of machines. The similarity coefficients for parts in Fig. 8.7 are given as follows:

$$\begin{bmatrix} 0&3&3&9&9&3&3&9&4&3 \\ &0&3&3&3&9&3&3&6&3 \\ &&0&3&3&3&9&3&4&9 \\ &&&0&9&3&3&9&4&3 \\ &&&&0&3&3&9&4&3 \\ &&&&&0&3&3&6&3 \\ &&&&&&0&3&4&9 \\ &&&&&&&0&4&3 \\ &&&&&&&&0&4 \\ &&&&&&&&&0 \end{bmatrix}$$

The SCVs are 9, 6, 4, and 3. Following the same steps as in the case of machines, the corresponding parts that form the three part-families are {6, 2, 9}, {7, 10, 3} and {4, 1, 5, 8} respectively. There are three machine cells having machines {1, 3, 8}, {2, 4, 7} and {5, 6, 9}. This results in the final machine-part matrix in Fig. 8.8.

Machine utilization consideration

A standard model uses a binary incidence matrix and has solution of the form shown in Fig. 8.8. It ignores several practical manufacturing factors such as group sizes, part demand, machine utilization, and the cost of transporting a part between machine cells due to exceptional elements. Part 9 produced by machine 2 is an exceptional element which is now used to illustrate the drawback of the standard model. There are two possibilities in assigning Part 9. If Part 9 that is produced in machine 2 is extremely crucial to the product,

Machine	Part									
	6	2	9	7	10	3	4	1	5	8
1	1	1	1							
3	1	1	1							
8	1	1	1							
2				1	1	1				
4				1	1	1				
7				1	1	1				
5							1	1	1	1
6							1	1	1	1
9							1	1	1	1

Fig. 8.8. Preliminary cell grouping using SOC scheme.

then machine 2 has to be duplicated. Otherwise, Part 9 may be produced outside the three cells.

The cell formation method presented here considers the manufacturing factors discussed in the paper: group sizes, part demand, machine utilization and exceptional elements. To illustrate the machine utilization consideration by the approach presented, the part demand, operation routings and process times for the 9-machine, 10-part problem already considered are shown in Table 8.2.

The intention here is to replace the 1's in the machine-part matrix of the preliminary cell grouping obtained using the SOC scheme with actual machine utilization. The machine utilization, u is defined as:

$$u_i = (t_{im} + s_{im}/q_i) * d_i/a_m \tag{36}$$

where t_{im} is the processing time for Part i on machine type m, s_{im} is the setup time per batch obtained by adding Part i setup time to its apportioned share of family setup; q_i is the part batch sizes; d_i is the period demand for Part i; and a_m is the time available per machine per period.

The base data for the problem under consideration are $s_{im} = 0.2$, $q_i = 10$, $a_m = 8$ (working hours per day) and period $=$ 300 working days per annum.

The period for the demands is one year and the unit for the processing is in hours. From Table 8.3 and the base-data the utilization for machine 1 is obtained by summing the contributions made by Parts 6, 2 and 9 as shown in the first row of cell 1 in Fig. 8.8:

$$\text{Part 6 on machine 1 contribution} = (2 + 0.2/10) * 2700/(300 \times 8 \times 8)$$
$$= 0.284 = 0.3$$
$$\text{Part 2 on machine 1 contribution} = (4 + 0.2/10) * 3000/(300 \times 8 \times 8)$$
$$= 0.628 = 0.6$$
$$\text{Part 9 on machine 1 contribution} = (1 + 0.2/10) * 2100/(300 \times 8 \times 8)$$
$$= 0.11 = 0.1$$
$$\text{Total utilization for machine 1} = 0.3 + 0.6 + 0.1 = 1$$

The complete machine utilization matrix is given in Fig. 8.9. Summing across columns results in the total utilization values for each machine row. The lower bound on the number of machines of each type is obtained by summing up each machine total utilization values. The machine lower bound ml generated by the procedure described and the machine upper bound m_u decided by the organization are used by the decision rule for part assignment in the next section.

Designing for non-inter-grouping material handling

The classical cell grouping solution obtained in Fig. 8.8 has not taken machine utilization, cell sizes and exceptional elements into consideration. The single-pass heuristic (Askin and Stanbridge, 1993) uses the machine utilization matrix

Table 7.4. Assignments using single-pass heuristic logic (* means inadequate capacity).

Iteration	Part assigned	Group	Machine added	Resource update (machine, remaining time)
1	6	1	1, 3, 8	(1, 0.7), (3, 0.7), (8, 0.3)
2	2	1	—	(1, 0.1), (3, 0.5), (8.0.0)*
3	9	2	1, 3, 8, 2	(1, 0.9), (3, 0.7), (8, 0.7), (2, 0.8)
4	7	3	2, 4, 7	(2, 0.8), (4, 0.7), (7, 0.7)
5	10	3	—	(2, 0.7), (4, 0.6), (7, 0.5)
6	3	3	—	(2, 0.6), (4, 0.1), (7, 0.4)
7	4	4	5, 6, 9	(5, 0.8), (6, 0.8), (9, 0.8)
8	1	4	—	(5, 0.7), (6, 0.1)*, (9, 0.3)
9	5	5	5, 6, 9	(5, 0.6), (6, 0.8), (9, 0.6)
10	8	5	—	(5, 0.0), (6, 0.2), (9, 0.0)

and machine lower bounds shown in Fig. 8.9 to form the basis for resolving the problems. Parts are assigned in such a way that the number of machines required in a group does not exceed the upper limit of the number of machines decided by the management of the organization. A new group is formed when the machine upper limit is violated or when a machine does not have the capacity for additional part. For upper limit of machine, $m_u = 5$, the resource updates are shown in Table 8.4 and details of the steps in the allocation of parts for Fig. 8.9 to a group are shown as follows:

Machine	**Part**										**Total**	**Machine** Lower
	6	2	9	7	10	3	4	1	5	8	Utilization	Bound
1	0.3	0.6	0.1								1	1
3	0.3	0.2	0.3								0.8	1
8	0.7	0.3	0.3								1.3	2
2			0.2	0.2	0.1	0.1					0.6	1
4				0.3	0.1	0.5					0.9	1
7				0.3	0.2	0.1					0.6	1
5							0.2	0.1	0.4	0.6	1.3	2
6							0.2	0.7	0.2	0.6	1.7	2
9							0.2	0.5	0.4	0.6	1.7	2

Fig. 8.9. Machine utilization matrix and machine lower bounds.

$i = 1$. Part 6 is assigned to group 1, taking along machines 1, 3 and 8. This imposes a utilization of 0.3 on machines 1 and 3, and a utilization of 0.7 on machine 8. This leaves a slack of 0.7 for machines 1 and 3 and a slack of 0.3 on machine 8 as shown in Table 8.4.

$i = 2$. Part 2 fits into this group with no new machines required and, hence is added. The slack time is reduced to 0.1, 0.5 and 0 for machines 1, 3 and 8 respectively.

$i = 3$. Part 9 goes into a new group because machine 8 does not have the capacity to handle the added load. The slack time is now 0.9, 0.7, 0.7 and 0.8 for machines 1, 3, 8, and 2 respectively.

$i = 4$. Part 7 will require two new machines if it were added to group 2, and this will exceed the allowable group size; hence, a new group is started. The slack time is now 0.8, 0.7 and 0.7 for machines 2, 4 and 7 respectively.

$i = 5$. Part 10 fits into this group with no new machines required and, hence, is added. The slack time is reduced to 0.7, 0.6 and 0.5 respectively.

$i = 6$. Part 3 also fits into this group with no new machines required and, hence, is added. The slack time is reduced to 0.6, 0.1 and 0.4 respectively.

$i = 7$. Part 4 will require three new machines if it were added to group 3, hence, a new group is started. The slack is now 0.8 for machines 5, 6, and 9 respectively.

$i = 8$. Part 1 fits into this group with no new machine required. The slack time is reduced to 0.7, 0.1 and 0.3 for machines 5, 6 and 9.

$i = 9$. Part 5 goes into a new group because machines 6 and 9 do not have the capacity to handle the added load. The slack is now 0.6, 0.8 and 0.4 for machines 5, 6, and 9.

$i = 10$. Part 8 fits into this group and slack is reduced to 0, 0.2 and 0 for machines 5, 6, 9.

The new assignment shows that the number of machines required are 2 of machine 1; 2 of machine 2; 2 of machine 3; 1 of machine 4; 2 of machine 5; 2 of machine 6; 1 of machine 7; 2 of machine 8; and 2 of machine 9. Machines 1, 2 and 3 each exceed their lower bound by one (see Table 8.5). By applying the single-pass heuristic logic, it has been possible to deal with the intergroup-move that would have occurred when machine 2 produces Part 9. Moreover, we would have ascertained that the machine upper limit imposed by the management as well as machine capacities have been considered before the actual restructuring of the manufacturing system.

Now that the major manufacturing factors have been taken into consideration, the actual restructuring of the job-shop into cells can commence. One machine each of types 1, 2 and 3 may be purchased. There are other possibilities such as combining of groups. The systems engineer should explore various possibilities in consultation with management especially when suggesting the duplication of machines. Costs involved have to be considered at the point of

Table 7.5. Capacity check.

	1	2	3	4	5	6	7	8	9
Lower bound	1	1	1	1	2	2	1	2	2
Actual no. used	2	2	2	1	2	2	1	2	2
Excess required	1	1	1	0	0	0	0	0	0

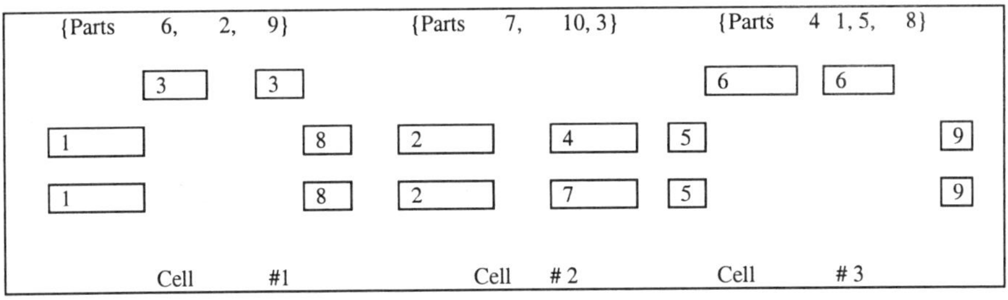

Fig. 8.10. Final cells that consider different manufacturing constraints.

implementation. The three cells of Fig. 8.9 with additional machines resulting from the single-pass heuristic logic are shown in Fig. 8.10.

A similarity coefficient-distance metric (SCDM) approach

The similarity coefficient-distance metric (SCDM) approach for cell grouping (Kerr and Balakrishnan, 1996; Onwubolu and Mlilo, 1998) is made up of three basic modules. The modules are: determine similarity coefficient matrix; for all possible sequences generated, deduce similarity coefficient-distance metric for each sequence, and retain sequence corresponding to the least similarity coefficient-distance metric, otherwise referred to as the *grouping sequence.* The procedure is carried out for both the machines and the parts. The *grouping sequences* obtained for the machines and for the parts are used to obtain the solution to the machine-part families grouping problem. The SCDM algorithm is presented and the different modules are now described in more details:

The SCDM algorithm

The machine and part similarity coefficient-distance matrices are computed. An initial machine part sequence is generated and the procedure performs a pair-wise interchange. The similarity coefficient-distance measures for all pair-wise interchanges are computed and the minimum associated sequence is retained. Similar steps are taken for the parts and the sequence that corresponds to the minimum SCDM value is also retained. These two sequences become the *grouping sequences.* The *grouping sequence* for the machines and the *grouping sequence* for the parts are used to form the final clustering matrix. There is provision for performing switches in the final matrix in order to enhance the quality of the clustering. The procedure then computes the bond energy and displays its value together with the final machine-part clustered matrix.

Let Q be a sequence, then the general outline of SCDM algorithm for solving machine-part families problem, can be represented in a pseudo code as follows:

```
procedure SCDM,
 begin
   generate an initial Q
   flag: = Q * d
   initialize counter
      for i: = 1 to count do
      begin
         generate Q
         β = Q * d
         if (β < flag) then flag = β and sequence: = Q;
      end
  until terminate.
```

Similarity coefficient matrix

The similarity coefficient between a pair of machines or parts determines how closely related they are. The higher the value the closer the similarity between the machines or parts. In machine-parts family problems, the similarity coefficient matrix for the machines and parts are separately computed and used for analysis. The input to a machine-parts grouping problem is usually a machine-part incidence matrix (see Fig. 8.11) in which cell entries for all values of i and j are $a_{ij} = 1$ or $a_{ij} = 0$. A cell entry for "1" means that part j visits machine i for processing and a cell entry of "0" (a blank) means that part j does not visit machine i for processing. The similarity coefficient method used here is the one defined by McAuley (1972) and is similar to the one defined by Carrie (1973) given as:

$$S_{ij} = \frac{n(m_i \cap m_j)}{n(m_i \cup m_j)} \tag{37}$$

where S_{ij} is the similarity coefficient between machine i and j and m_i and m_j are the sets of machines required to process parts i and j respectively. The

Machine	Part							
	3	8	6	4	1	7	2	5
5	1	1	1					
2	1	1	1					
1	1	1	1					
8				1	1	1		
6				1	1	1		
7							1	1
4							1	1
3							1	1

Fig. 8.11. Machine–part incidence matrix.

similarity coefficient between any pair of rows of machines in the machine-part incidence matrix is defined as the ratio of the number of common elements to the total number of elements found in either row. In matrix form,

$$[S] = \sum_{i=1}^{M} \sum_{j=1}^{M} S_{ij} \tag{38}$$

where there are M machines. A similar matrix is generated for the parts. The similarity coefficient matrix is a square matrix, which for M machines is an $M \times M$ matrix and for N parts is an $N \times N$ matrix. For example, from Fig. 8.11, $s_{12} = 3/3 = 1$; $s_{23} = 0/5 = 0$.

Permutative sequence generator and similarity coefficient-distance metric

The procedure presented here generates all the sequences for a given number of parts and machines, leading to $M!$ sequences for machines and $N!$ sequences for parts. The approach implemented avoids storing the sequences generated otherwise this would lead to computer memory allocation problem. For each sequence generated, the similarity coefficient-distance metric is computed. For each sequence generated there will be $M(M-1)/2$ computations of the similarity coefficient-distance metric, β defined as:

$$\beta_k = \sum_{i=1}^{M-1} \sum_{j=1}^{N} S_{i(j+1)} * d_{ij}\,; \quad k = 1, 2, \ldots, K \tag{39}$$

for M machines and N parts and $K = M!$, where $s_{i(j+1)}$ is the similarity coefficient between adjacent machines and the distance between them is $d_{ij} = |i - (j+1)|$.

Each time a similarity coefficient metric β_k is computed it is assigned to a flag variable. Subsequent value computed is compared with the previous and which ever is the minimum is assigned to the variable. Hence for all β_k computed the procedure tracks the minimum similarity coefficient-distance metric. Since a similarity coefficient-distance metric is computed for a given sequence generated, the sequence which gives the minimum β_k is tracked and becomes the *grouping sequence.*

To fix ideas, consider a 4-machine, 4-part incidence matrix which yield a similarity coefficient matrix:

$$[S] = \begin{bmatrix} s_{11} & s_{12} & s_{13} & s_{14} \\ s_{21} & s_{22} & s_{23} & s_{24} \\ s_{31} & s_{32} & s_{33} & s_{34} \\ s_{41} & s_{42} & s_{43} & s_{44} \end{bmatrix}.$$

These will be 4! (24) sequences given by {1, 2, 3, 4}, {3, 2, 4, 1}, etc. Applying the definition of the similarity coefficient-distance metric for the first sequence {1, 2, 3, 4},

$$\beta_1 = S_{12}{}^{*}1 + S_{13}{}^{*}2 + S_{14}{}^{*}3 + S_{22}{}^{*}1 + S_{24}{}^{*}2 + S_{34}{}^{*}1$$

$$flag = \beta_1\,.$$

Then for the sequence {3, 2, 4, 1},

$$\beta_2 = S_{32}{}^{*}1 + S_{34}{}^{*}2 + S_{31}{}^{*}3 + S_{24}{}^{*}1 + S_{21}{}^{*}2 + S_{41}{}^{*}1$$

$$\text{if } \beta_2 < flag \text{ then } flag = \beta_2\,.$$

The process is then continued for all the 4! sequences, thereby tracking the *group sequence* having the minimum SCDM. The same process is then carried out for the parts to obtain parts *grouping sequence.* In a way, the procedure reported here is a minimization technique because it guarantees that *group sequence* has minimum SCDM value. There is a relationship between a given sequence $Q = \{q_i\}$ and the similarity coefficient-distance metric as follows:

$$\beta_k = \sum_{i=1}^{M-1} \sum_{j=1}^{N} S_{i(j+1)} * |i - (j+1)| ; \quad k = 1, 2, \ldots, K . \tag{40}$$

This is the formulation used in procedure discussed here. For the problem shown in Fig. 8.11, the *group sequence* for the machines is {5, 2, 1, 8, 6, 7, 4, 3} and the *group sequence* for the parts is {3, 8, 6, 4, 1, 7, 2, 5}. Using these *group sequences*, the solution is given in Fig. 8.12.

Three manufacturing cells are identifiable. Cell 1 has three machines {5, 2, 1} that process three Parts {3, 8, 6}. Cell 2 has two machines {8, 6} that process three Parts {4, 1, 7}. Cell 3 has three machines {7, 4, 3} which process Parts {2, 5}.

Machine	Part							
	3	8	6	4	1	7	2	5
5	1	1	1					
2	1	1	1					
1	1	1	1					
8				1	1	1		
6				1	1	1		
7							1	1
4							1	1
3							1	1

Fig. 8.12. Final machine–part matrix grouping.

Performance measurements for manufacturing cells grouping

There is the need to measure the ability of an algorithm to cluster the "1"s of an initial machine-part incidence matrix together, and also to diagonalise it. Two measures that are useful are the grouping efficacy measure, η_g, derived by Kumar and Chandrasekharan (1990), and the bond energy measure, η_b, used by Miltenburg and Zhang (1991).

Grouping efficacy measure

Grouping efficacy is a measure used to evaluate the quality of a clustering procedure. A block-diagonalised machine-part matrix, which is the solution of the machine-part matrix-clustering problem, is given as:

$$\eta_g = \frac{n_1}{n_0 + \sum_{i=1}^{k} L_i B_i} \tag{41}$$

where n_1 is the number of "1"s in the diagonal blocks, no is the number of "1"s in the off-diagonal region, k is the total number of cells in the matrix, and L_i and B_i are the number of machines and parts in each cell i respectively. The higher the value of η_g the better the quality of the clustering. For Fig. 8.12, $\eta_g = 100\%$.

Bond energy measure

The closer the "1"s are to one another in each of the cells formed, the higher the bond energy. It measures the strength of clustering. The bond energy measure of a machine-part matrix is given as:

$$\eta_g = \frac{\sum_{i=1}^{M} \sum_{j=1}^{N-1} a_{ij} a_{i(j+1)} + \sum_{i=1}^{M} \sum_{j=1}^{N-1} a_{ij} a_{(i+1)j)}}{\sum_{i=1}^{M-1} \sum_{j=1}^{N} a_{ij}}\,. \tag{42}$$

The bond energy for Fig. 8.12 computed by the SCDM procedure is $\eta_g = 1.24$.

Several small-to-medium problems can be tested using the SCDM algorithm and it is found to have satisfactory solutions. The quality of solutions was judged in two ways:

(a) by calculating the grouping efficacy used by Kumar and Chandrasekharan (1990);
(b) by calculating the bond energy used by Miltenburg and Zhang (1991).

8.7.3. *Rationalization of part families and workloads*

Material handling cost consists of costs due to *intercell* and *intracell* movements of parts. Inter-cell cost is incurred when parts move from one cell to another to be fully processed. Intra-cell cost is the cost incurred due to movement of parts from one machine to the other but whilst within the same cell. In other words, it is the cost of the inter-machine movements of parts within a particular cell. If only one machine is assigned to each cell, then the intracell cost would be minimized, but the intercell cost due to intercell movements would be at its maximum. The optimal state therefore is to minimize the volume of intercellular movements as well as intracell movements.

The parts that move from cell to cell are called *exceptional parts* or *exceptional elements.* When the exceptional elements cost are optimized there is need to balance the machine workloads in each cell. The sum of the variation (known as cell load variation) of the workloads on each machine in a cell from the average workload in that cell should also be minimized.

When forming cells, the optimal cell performance condition should be aimed at, by attempting to minimize the volume of intercellular movements as well as intracell movements. This principle is helpful to the facility planner, in order to realize an effective and efficient cell.

8.7.4. *Selecting the type of cell layout*

The type of cell layout is decided, on the basis of the material handling used (Singh and Rajamani, 1996). In any manufacturing environment the processing sequence of parts may be partially different, or the same. Therefore, the arrangement of machines to process these parts would affect the material handling capacity, hence the cost. Several material handling systems commonly used (Askin and Standridge, 1993) are:

(1) automated guided vehicles (AGVs);
(2) material handling robots;
(3) gantry robot;
(4) conveyor system.

The five types of cellular layout commonly used in industry are shown in Fig. 8.13. Each of these is briefly described in this section.

Single cell robot

Machines are usually arranged in circular or U-shaped cell when material-handling robot is used to serve a work cell. In most cellular manufacturing layout facilities, a U-shaped layout for a cell (see Fig. 8.13(a)) is recommended for minimal intracell movements. U-cell having single robot for material handling has the following:

- *Employee flexibility.* Operators have similar skills and can work inter changeably;
- *Team work.* Cells form a sociological unit conducive to team work since the operators work as a group (Wemmerlov and Hyer, 1989);
- *Ease of material handling.* Ease of automation of material handling systems such as the use of automated guided vehicles and robots.

Straight line layout

Ideally, a separate straight line of machines (linear layout) should be set up for each and every part. Though this would reduce the material handling cost, machine idle time will be increased, it would not be cost-effective. A line set-up for a number of parts grouped together on the basis of similarity in their processing times and sequences would increase the utilization of machines and reduce the material handling cost. When automated guided vehicles are used for material handling, the machines are arranged in either (i) single row layout (see Fig. 8.13(b)) or (ii) double row layout (see Fig. 8.13(c)). Since AGVs operate best when the path is a straight line, using them will give the best performance to the facility.

Cell layout with gantry robot

Gantry robots are used to serve machines when there is space limitation layout (see Fig. 8.13(d)). In such a case the geometry of the layout is not an issue, since materials can be delivered to any part of the cell.

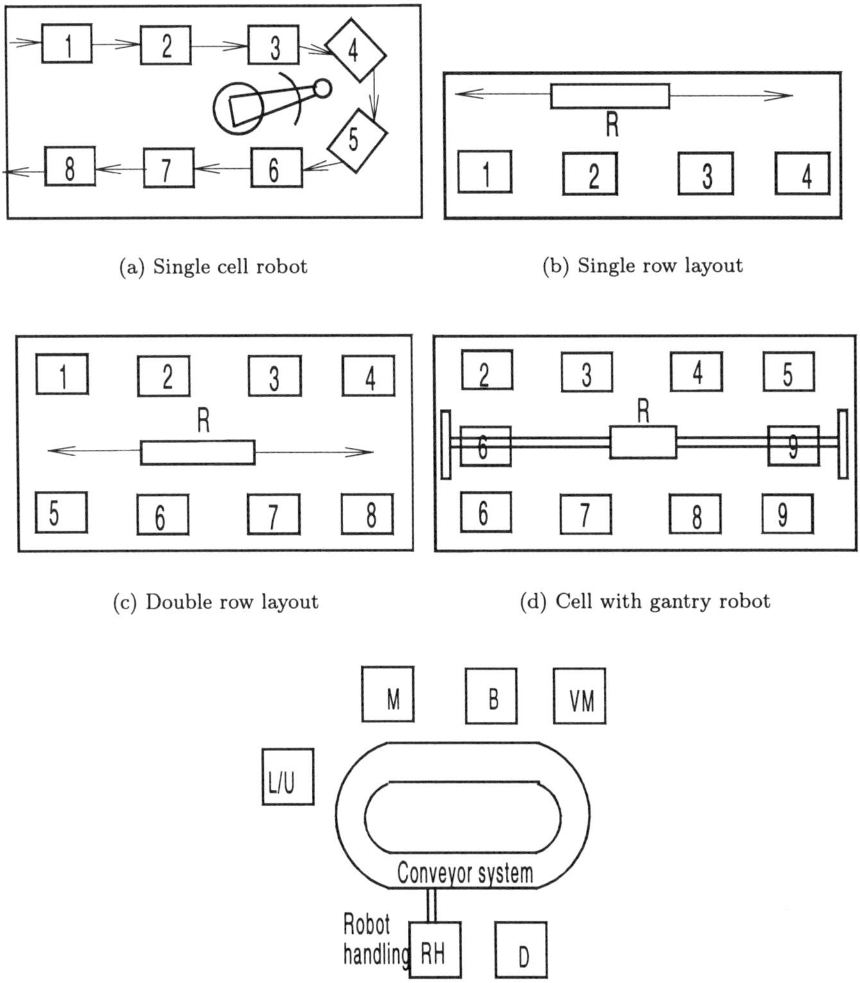

(a) Single cell robot

(b) Single row layout

(c) Double row layout

(d) Cell with gantry robot

(e) Flexible cell layout

Fig. 8.13. Layout of material handling devices.

Flexible cell layout

Conveyor systems are normally used for material handling in flexible cell layout (see Fig. 8.9(e)). Conveyor travel-path is uni-directional, and when used in a flexible cell layout, an auxiliary material-handling device is required at each workstation for loading and unloading of parts.

8.7.5. *Layout machines and auxiliary facilities in cell*

The next step in layout planning for cellular manufacturing, once the type of cell layout has been selected, is layout of machines and auxiliary facilities in cell. Layout planning in cellular manufacturing has several objectives (Hales, 1984):

(1) effective movement of material and personnel;
(2) effective utilization of facility space;
(3) adaptability to future unforeseen changes;
(4) ease of expansion;
(5) easy supervision and control;
(6) low cost;
(7) control of noise;
(8) security and safety;
(9) good appearance.

The most common layout planning constrains real-life include (Hales, 1984):

(1) material handling limitations;
(2) utility limitation;
(3) architectural limitations;
(4) organizational restrictions;
(5) identification of activities which must be separated;
(6) one or more fixed activities;
(7) budget;
(8) time.

The layout planner for cellular manufacturing facility has to be aware of the objectives that should be realized and the constraints that limited their realization, during the planning process. The layout process involves several stages (Singh and Rajamani, 1996):

(1) defining the problem of layout;
(2) selecting the solution methodology;
(3) generating alternative plans;
(4) modifying the selecting one or more plans;
(5) deciding the implementation procedure;
(6) readjustment and revision based on the implementation and initial problems encountered.

8.8. Application of Simulated Annealing to Cell Formation

Using all the concepts of the simulated annealing described in Sec. 8.3, we present the algorithm for solving the machine cell formation problem (Onwubolu and Mkandla, 1998). The simulated annealing algorithm accepts binary or fuzzy data for defining the initial machine-part incidence matrix. In the application of the simulated annealing procedure to solve machine-grouping problems, three significant items need to be addressed: (i) problem-representation, (ii) transition mechanism; and (iii) cooling schedule. Let us start by discussing the problem representation, transition mechanism, and cooling schedule, and then present a simulated annealing algorithm.

8.8.1. *Problem representation*

The problem representation includes two important aspects: a definition of the solution space or the neighbourhood, and the expression of the cost function. For the cell formation problem, the generation of a neighbourhood solution at each subsequent iteration is achieved by moving one machine from the current cell to another cell, thereby forming a new machine assignment. Several objective functions are available which must be chosen effectively with respect to the optimization objective.

8.8.2. *Transition mechanism*

A transition mechanism for transforming a current solution into a subsequent one, firstly generates a new configuration from a current one, secondly calculates the difference between the new cost and the current cost, and thirdly decides whether the new solution should be accepted or rejected. The generated solution (new part families and machine groups) is accepted, if the objective function value improves. If otherwise, the solution is accepted with a probability depending on the temperature, which is set to allow the acceptance of a large proportion of generated solutions at the beginning. The criteria for accepting a new solution is given as follows (reproduced from earlier discussion on simulated annealing for ease):

$$P_r = \begin{cases} 1 & \text{if } \Delta f \leq 0 \\ \exp\left(\dfrac{-\Delta f}{c}\right) & \text{if } \Delta f > 0 \end{cases}$$

where Δf is the difference in the cost between a new and the current solution and c is the control parameter, as already discussed.

8.8.3. *Cooling schedule*

A set of parameters must be specified to govern the convergence behaviour of the simulated annealing algorithm. The way of specifying the related parameters is called the cooling schedule which, includes:

(i) an initial value of the control parameter c_0
(ii) a decrement rule for changing the current value of the control parameter
(iii) a length of the homogenous Markov chain, and
(iv) a final value of the control parameter for terminating the execution.

Many kinds of cooling schedule have been proposed so far and details of some of them were discussed and summarized by Van Laarhoven and Aarts (1987).

Initial control parameter c_0

An initial acceptance ratio is defined as the number of accepted transitions divided by the number of proposed transitions. The value of the initial control parameter c_0 is set in such a way that the initial acceptance ratio is close to unity, and is of the order of the expected objective function value f. The value of the initial control parameter, c_0, is adjusted to bring the acceptance ratio obtained within the first ten iterations to be in the range of 0.95 to 1.0.

Decrement rule for changing control parameter

The control parameter decrements should be gradual to ensure slow cooling. A frequently used decrement rule is $c_i = \alpha c_i - 1$, where $0 < \alpha < 1$. We used another version that depends on the standard error of cost generated at c_i and small, real positive number.

Length of the homogenous Markov chain

The length of the Markov chains are controlled in such a way that for each value of control parameter c_i a minimum number of transitions should be acceptable. The length of Markov chain affects the cooling rate. By choosing longer Markov chains and smaller decrement rule constant α, faster cooling is achieved.

Stopping criteria

The annealing process is allowed to continue until the system is frozen. This is normally achieved in two ways: the acceptance ratio being less than a lower bound for a consecutive number of control parameters or the control parameter being less than a final value. The later case is used in the implementation described here ($c < c_f$). Below is a pseudo code for a generic simulated annealing algorithm:

```
Step 1    Generate an initial solution
Step 2    Generate an initial control parameter c > 0
Step 3    While not yet frozen do
                While not yet equilibrium at control parameter c do
                      Generate a random neighbour j of i
                      Calculate difference in costs Δf_ij = f_j − f_i
                      If Δf_ij ≤ 0
                            Set i = j
                      Else
                            Set i = j with probability exp(−Δf_ij/c)
                Set c = f(c, f)
          Return solution i.
```

8.8.4. *The simulated annealing algorithm*

The cell formation problem is formulated as follows:

$$\text{minimize } (P) = \sum_{i=1}^{N} \sum_{l=1}^{C} y_{il} \|X_i - Z_i\|^2 \tag{43}$$

subject to

$$\sum_{l=1}^{C} y_{il} = 1\,, \qquad \forall_i = 1,2,\ldots,N$$
$$y_{il} = 0 \text{ or } 1\,, \quad \forall_i = 1,2,\ldots,N;\ l = 1,2,\ldots,C$$

where

$$\|X_i - Z_i\|^2 = \left[\sum_{k=1}^{N} |x_{ik} - z_{lk}|\right]^{\frac{1}{2}}$$

In this formulation, X_i represents the ith machine (or part), and Z_l which, represents the lth machine group (or part family) and known as the centre of cluster l is defined as:

$$Z_l^k = \begin{cases} \sum_{i \in l_l^{k-1}} X_i/M_l^{k-1} & \text{if } M_l^{k-1} > 0 \\ 0 & \text{if otherwise} \end{cases} \tag{44}$$

where M_L^k is the number of machines (or parts) assigned to cluster l.

The steps involved in the cell formation problem are:

(1) Configuration initialisation: the initial configuration of machines to cells is done by random assignment, while avoiding empty cell condition.
(2) Computation of the energy: the total cell load variation of the currently assigned machines to cells, f_i, is computed using the objective function of Eq. (43).
(3) Generation of a neighbourhood: two machines are randomly generated and their cell membership are exchanged, so that a new state, j, is obtained. This results in a new cost f_j, and a cost difference $\Delta f_{ij} = f_j - f_i$.
If $\Delta f_{ij} > 0$ then accept move with probability $p(\Delta f) = \exp(-\Delta f_{ij}/c)$.
If $\Delta f_{ij} \leq 0$, accept the new configuration and continue.
(4) Test for equilibrium while state is not frozen.

Illustration

Let us now see how to apply simulated annealing to the cell formation problem. In practical terms, the incidence matrix may not necessarily contain 1's and 0's as in most published literature. This is because the processing times for a job on different machines normally differ. Consequently, we use the actual

	1	2	3	4	5	6	7	8
1	0.3	0.0	0.0	0.4	0.0	0.0	0.5	0.0
2	0.0	0.3	0.1	0.0	0.5	0.2	0.0	0.6
3	0.0	0.3	0.3	0.0	0.1	0.2	0.0	0.2
4	0.6	0.0	0.0	0.7	0.0	0.0	0.8	0.0
5	1.1	0.0	0.0	0.2	0.0	0.0	0.3	0.0

Fig. 8.14. Incidence matrix for a 5 × 8 problem.

times of processing of jobs on machines rather than 1's and 0's. Consider an incidence matrix given in Fig. 8.14.

Step 1 involves configuration initialization, in which the initial configuration of machines to cells is done by random assignment while avoiding empty cell condition. We assume that we have two cells. There are several techniques for determining the initial number of cells. One simple method is to divide the size of the incidence matrix by the number of the non-zero entries. For example, in the incidence matrix under consideration in Fig. 8.14, the approximate number of cells = (5*8)/19 = 2.1 ∼ 2. Let us then take this value as the number of cells. Then, the machines and parts are randomly assigned to these two cells. One such assignment is {4, 3} and {2, 1, 5} for machines; {2, 3, 5, 8, 6} and {1, 4, 7} for parts. In this assignment, empty cell condition is successfully avoided. The two cells are not empty. The configuration of this solution is shown in Fig. (8.14(a)):

	2	3	5	8	6	1	4	7
4	0.0	0.0	0.0	0.0	0.0	0.6	0.7	0.8
3	0.3	0.3	0.1	0.2	0.2	0.0	0.0	0.0
2	0.3	0.1	0.5	0.6	0.2	0.0	0.0	0.0
1	0.0	0.0	0.0	0.0	0.0	0.3	0.4	0.5
5	0.0	0.0	0.0	0.0	0.0	0.1	0.2	0.3

Fig. 8.14(a). One solution-configuration for a 5 × 8 problem.

Step 2 involves computation of the energy: the total cell load variation of the currently assigned machines to cells, f_i, is computed using the objective function of Eq. (43). Note that in this configuration, there are five exceptional elements from machine {2}, parts {2, 3, 5, 8, and 6}; and three exceptional elements from machine {4}, parts {1, 4, and 7}. These have to be excluded from the computation when considering only, the first part of the objective function, i.e. the total cell load variation. You will see the reason for including another component in the objective function, i.e. inter-cellular movement. Using Eqs. (43) and (44):

$$
\begin{aligned}
Z_1^1 &= (0+0.3)/2 = 0.15\\
Z_2^1 &= (0+0.3)/2 = 0.15\\
Z_3^1 &= (0+0.1)/2 = 0.05\\
Z_4^1 &= (0+0.2)/2 = 0.10\\
Z_5^1 &= (0+0.2)/2 = 0.10\\
Z_1^2 &= (0+0.3+0.1)/3 = 0.1333\\
Z_2^2 &= (0+0.4+0.2)/3 = 0.2000\\
Z_3^2 &= (0+0.5+0.3)/3 = 0.2666\\
Z^1 &= (0.0-0.15)^2 + (0.3-0.15)^2 + (0.0-0.15)^2\\
&\quad + (0.3-0.15)^2 + (0.0-0.05)^2 + (0.1-0.05)^2\\
&\quad + (0.0-0.1)^2 + (0.2-0.1)^2 + (0.0-0.1)^2\\
&\quad + (0.2-0.1)^2 + (0.0-0.1333)^2 + (0.3-0.1333)^2\\
&\quad + (0.1-0.1333)^2 + (0.0-0.2)^2 + (0.4-0.2)^2\\
&\quad + (0.2-0.2)^2 + (0.0-0.2666)^2\\
&\quad + (0.5-0.2666)^2 + (0.3-0.2666)^2\\
&= 0.0225 + 0.0225 + 0.0225 + 0.0225 + 0.0225\\
&\quad + 0.0225 + 0.01 + 0.01 + 0.01 + 0.01 + 0.01\\
&\quad + 0.01 + 0.017769 + 0.027789 + 0.0011089 + 0.04\\
&\quad + 0.04 + 0.0 + 0.071076 + 0.054476 + 0.0011156\\
&= 0.388344\\
Z_2 &= 0.3 + 0.1 + 0.5 + 0.6 + 0.2 + 0.6 + 0.7 + 0.8 = 3.8\\
Z_1 &= Z_1 + Z_2 = 0.388344 + 3.8 = 4.188334\,.
\end{aligned}
$$

Step 3 involves generation of a neighbourhood: two machines are randomly generated and their cell membership are exchanged, so that a new state, j, is obtained. This results in a new cost f_j, and a cost difference $\Delta f_{ij} = f_i - f_j$. The algorithm generates a neighbourhood. One such neighbourhood is achieved by swapping machines 2 and 4 between cells 1 and 2 respectively. This results in a new configuration, which is {2, 3}, {4, 1, 5} for machines; {2, 3, 5, 8, 6} and {1, 4, 7} for parts respectively. In this assignment, empty cell condition is successfully avoided. The two cells are not empty. The configuration of this solution is shown in Fig. (8.14(b)):

	2	3	5	8	6	1	4	7
2	0.3	0.1	0.5	0.6	0.2	0.0	0.0	0.0
3	0.3	0.3	0.1	0.2	0.2	0.0	0.0	0.0
4	0.0	0.0	0.0	0.0	0.0	0.6	0.7	0.8
1	0.0	0.0	0.0	0.0	0.0	0.3	0.4	0.5
5	0.0	0.0	0.0	0.0	0.0	0.1	0.2	0.3

Fig. 8.14(b). A neighbourhood for the 5 × 8 problem.

For this neighbourhood, using Eqs. (43) and (44):

$$Z_1^1 = (0.3 + 0.3)/2 = 0.3$$

$$Z_2^1 = (0.1 + 0.3)/2 = 0.2$$

$$Z_3^1 = (0.5 + 0.1)/2 = 0.3$$

$$Z_4^1 = (0.6 + 0.2)/2 = 0.4$$

$$Z_5^1 = (0.2 + 0.2)/2 = 0.2$$

$$Z_1^2 = (0.6 + 0.3 + 0.1)/3 = 0.3333$$

$$Z_2^2 = (0.7 + 0.4 + 0.2)/3 = 0.4333$$

$$Z_3^2 = (0.8 + 0.5 + 0.3)/3 = 0.5333$$

$$\begin{aligned} Z^1 &= (0.3 - 0.3)^2 + (0.3 - 0.3)^2 + (0.1 - 0.2)^2 + (0.3 - 0.2)^2 \\ &\quad + (0.5 - 0.3)^2 + (0.1 - 0.3)^2 + (0.6 - 0.4)^2 \\ &\quad + (0.2 - 0.4)^2 + (0.2 - 0.2)^2 + (0.2 - 0.2)^2 \\ &\quad + (0.6 - 0.3333)^2 + (0.3 - 0.3333)^2 + (0.1 - 0.3333)^2 \end{aligned}$$

$$
\begin{aligned}
&+ (0.7 - 0.4333)^2 + (0.4 - 0.4333)^2 + (0.2 - 0.4333)^2 \\
&+ (0.8 - 0.5333)^2 + (0.5 - 0.5333)^2 + (0.3 - 0.5333)^2 \\
&= 0.0 + 0.0 + 0.01 + 0.01 + 0.04 + 0.04 \\
&+ 0.04 + 0.04 + 0.0 + 0.0 \\
&+ 0.071129 + 0.001109 + 0.0054429 \\
&+ 0.071129 + 0.001109 + 0.0054429 \\
&+ 0.071129 + 0.001109 + 0.0054429 \\
&= 0.559952 \\
Z_2 &= 0.0 \\
Z_1 &= Z_1 + Z_2 = 0.559952 + 0.559952\,.
\end{aligned}
$$

Table 8.6(a). The sequence of operations of parts.

Part i	Operation routings for machines
1	M6 $\longrightarrow$ M7
2	M1 $\longrightarrow$ M3
3	M1 $\longrightarrow$ M2 $\longrightarrow$ M4 $\longrightarrow$ M7 $\longrightarrow$ M8
4	M2 $\longrightarrow$ M4 $\longrightarrow$ M7 $\longrightarrow$ M8
5	M5 $\longrightarrow$ M6
6	M2 $\longrightarrow$ M4 $\longrightarrow$ M5 $\longrightarrow$ M7 $\longrightarrow$ M8
7	M2 $\longrightarrow$ M4 $\longrightarrow$ M7 $\longrightarrow$ M8
8	M1 $\longrightarrow$ M3
9	M1 $\longrightarrow$ M3 $\longrightarrow$ M6
10	M4 $\longrightarrow$ M5 $\longrightarrow$ M6
11	M1 $\longrightarrow$ M3 $\longrightarrow$ M7
12	M5 $\longrightarrow$ M6 $\longrightarrow$ M7
13	M1 $\longrightarrow$ M3
14	M1 $\longrightarrow$ M2 $\longrightarrow$ M3
15	M5 $\longrightarrow$ M6
16	M1 $\longrightarrow$ M3
17	M1 $\longrightarrow$ M3 $\longrightarrow$ M5
18	M2 $\longrightarrow$ M4 $\longrightarrow$ M7 $\longrightarrow$ M8
19	M1 $\longrightarrow$ M3
20	M2 $\longrightarrow$ M4 $\longrightarrow$ M6 $\longrightarrow$ M7 $\longrightarrow$ M8

Table 8.6(b). The incidence matrix for Table 8.6(a).

Machine	Part 1	2	3	4	5	6	7	8	9	10	11	12	13	14	15	16	17	18	19	20
1	0	1	1	0	0	0	0	1	1	0	1	0	1	1	0	1	1	0	1	0
2	0	0	1	1	0	1	1	0	0	0	0	0	0	1	0	0	0	1	0	1
3	0	1	0	0	0	0	0	1	1	0	1	0	1	1	0	1	1	0	1	0
4	0	0	1	1	0	1	1	0	0	1	0	0	0	0	0	0	0	1	0	1
5	1	0	0	0	1	1	0	0	0	1	0	1	0	0	1	0	1	0	0	0
6	1	0	0	0	1	0	0	0	1	1	0	1	0	0	1	0	0	0	0	1
7	0	0	1	1	0	1	1	0	0	0	1	1	0	0	0	0	0	1	0	1
8	0	0	1	1	0	1	1	0	0	0	0	0	0	0	0	0	0	1	0	1

In this present move, $\Delta f_{ij} = (0.559952 - 4.188334) = -3.628382 < 0$. Applying the condition:

If $\Delta f_{ij} > 0$ then accept move with probability $p(\Delta f) = \exp(-\Delta f_{ij}/c)$;

If $\Delta f_{ij} \leq 0$, accept the new configuration and continue;

means that the new configuration is accepted and the algorithm continues. Incidentally, this is the optimum solution. In actual simulation, it takes several hundreds of iterations to reach an optimum solution.

Let us apply the same steps to another problem. Table 8.6(a) shows the sequence of operations of parts for the 8-machine, 20-part problem. Table 8.6(b) shows the incidence matrix for the same problem. Following the steps previously described for the simulated annealing-algorithm, leads to the solution in Fig. 8.15.

	3	4	6	7	18	20	02	08	09	11	13	14	16	17	19	01	05	10	12	15
2	1	1	1	1	1	1						1								
4	1	1	1	1	1	1												1		
7	1	1	1	1	1	1				1									1	
8	1	1	1	1	1	1														
1	1						1	1	1	1	1	1	1	1	1					
3							1	1	1	1	1	1	1	1	1					
5			1											1		1	1	1	1	1
6						1				1						1	1	1	1	1

Fig. 8.15. Solution to Tables 8.17(a) and 8.17(b).

For the well-formed problems, the simulated annealing procedure discussed in this chapter compare well with the result of the similarity order clustering (SOC) (Onwubolu, 1998) and the result of ideal seed method of Chandrasekharan and Rajagopalan (1986) that is so far the only known non-hierarchical method reported in the literature. The simulated annealing algorithm formed three machine grouping and part families as shown in Table 8.7. The similarity order clustering and ideal seed (Chandrasekharan and Rajagopalan, 1986b) algorithms also formed three machine groupings and part families as shown in Tables 8.9 and 8.10 respectively. The simulated annealing has the same clustering efficiency $\eta = 85.24\%$ which, agrees with the value obtained using both ideal seed and similarity order clustering methods. However, as the number of machines and parts grow into very large numbers (say machines = 100; parts = 1000), typical of practical manufacturing environments, simulated annealing becomes distinctively advantageous than most cell grouping techniques.

Table 8.7. Machine groups/part families using simulated annealing.

Group	Machines	Parts
1	2, 4, 7, 8	4, 7, 3, 18, 20, 6
2	3, 1	11, 2, 8, 14, 17, 9, 16, 19, 13
3	6, 5	5, 10, 1, 12, 15

Table 8.8. Machine groups/part families using similarity order clustering.

Group	Machines	Parts
1	2, 8, 4, 7	7, 4, 3, 6, 18, 20
2	1, 3	8, 2, 13, 16, 19, 9, 11, 14, 17
3	5, 6	5, 1, 15, 10, 12

Table 8.9. Machine group/part families using ideal seed method.

Group	Machines	Parts
1	2, 4, 7, 8	3, 4, 6, 7, 18, 20
2	1, 3	2, 8, 9, 13, 14, 16, 17, 19, 11
3	6, 5	1, 5, 10, 12, 15

Table 8.10. Routing information for Question 17.

Machine	Part														
	01	02	03	04	05	06	07	08	09	10	11	12	13	14	15
1	0	1	0	0	0	0	0	1	0	1	1	1	0	0	0
2	0	0	1	0	1	0	0	1	0	0	0	0	1	0	1
3	1	0	0	0	0	1	0	0	1	0	0	0	0	1	0
4	1	0	0	1	0	0	0	0	1	0	0	0	0	1	0
5	0	0	1	0	1	0	0	1	0	0	0	0	1	0	1
6	1	0	0	1	0	1	0	1	1	0	0	0	0	1	0
7	0	1	0	0	0	0	1	0	0	1	1	1	0	0	0
8	0	0	1	0	1	0	0	1	0	0	0	0	1	0	1
9	0	0	0	1	0	1	0	0	1	0	0	0	0	1	0
10	0	1	0	0	0	0	1	0	0	1	1	1	0	0	0

8.9. Summary

The fundamental aspects of simulated annealing are discussed in this chapter. Production layout planning is an aspect that has not been emphasized in production planning and control. This chapter, bridges the gap by first reviewing articles in the area, and then discussing in details the hierarchical process involved in production layout planning process. Two cell formation techniques (SOC and SCDM) are discussed in detail. This is then followed by the application of simulated annealing (SA) to the cellular manufacturing cell problem. This chapter presents the modern approach in converting a job shop into manufacturing cells using simulated annealing technique. Simulated annealing, which is one of the meta-stategic approaches for solving combinatorial optimization problems in industrial engineering, is a powerful and general random neighbourhood-search method. Simulated annealing is suitable for solving large scale-problems which are characteristic of manufacturing firms where routing sheets contain hundreds of parts which are produced by different machines. For the small-sized problems used for comparison, the simulated annealing has the same clustering efficiency (with the value obtained using both ideal seed and similarity order clustering methods. However as the number of machines and parts grow into very large numbers (say machines = 100; parts = 1000), typical of practical manufacturing environments, simulated annealing becomes distinctively advantageous than most cell grouping techniques. Simulated annealing is an efficient approach to solving combinatorial problems and it is very flexible in that an appropriate objective function relevant to the problem being solved is optimized. Simulated annealing has the capability to escape from local minima and hence it can yield solutions that are very near optimal. It is problem-independent, hence it has found wide applications in industrial and manufacturing application. The chapter highlights the advantages of converting currently job shops into manufacturing cells and the advantages of such strategy in productivity improvement in the manufacturing sector.

Exercises

1. Explain the relationship between real and simulated annealing.
2. What is the role of Monte Carlo model for simulating the transition from a given state to the thermal equilibrium-state of a solid for a fixed value of temperature?
3. How does the probabilistic acceptance criterion of simulated annealing affect the final solution for a given problem?
4. What do you understand by the stationary distribution of the Markov chain in relation to simulated annealing?
5. How is simulated quenching achieved and what are its disadvantages?
6. What are the merits and demerits for the step-wise, continuous, and non-monotonic schemes presented in this chapter?
7. Is it preferred to use problem-dependent Markov chain or one that is dependent of neighbourhood size definition?
8. What modifications would you propose for making the basic simulated annealing more effective and efficient?
9. Would you obtain a better solution by submitting a seed-solution to simulated annealing? Justify your answer.
10. Itemize the key issues involved in production layout strategies.
11. How would you position cellular manufacturing in the product variety and production volume relations?
12. Clearly distinguish the characteristic differences for the following: group technology (GT) flow-line; GT-cell; and GT-centre.
13. What are the benefits of group technology in production layout planning?
14. Clearly state the steps involved in cell formation.
15. A manufacturing company is involved in job shop operations, but would want to switch over to cellular manufacturing. What steps should be taken to realize the new change?
16. How does the inclusion of inter-cellular move to the total cell load variation-model improve the cell formation solution?

17. Tables 8.7–8.10 show the routing information of a manufacturing process. Use the total cell load variation model and simulated annealing technique to solve the cell-formation problem.
18. Use the inter-cellular movement-model and simulated annealing technique to solve the cell formation problem of problem 17.
19. Apply the combined total cell-load and inter-cellular movement-model to solve the cell formation problem, using the simulated annealing technique for problem 17.
20. Develop a model for alternate routing and apply simulated annealing.

References

Aarts, E. H. L. and Van Laarhoven, P. J. M., 1985, Statistical cooling: A general approach to combinatorial optimization problems, *Phillips Journal of Research*, **40** 193–226.

Askin, R. G. and Standridge, C. R., 1993, *Modelling and analysis of manufacturing systems* (John Wiley and Sons, Inc).

Askin, R. G. and Subramanian, S. P., 1987, A cost based heuristic for group technology configuration, *International Journal of Production Research*, **25**(1) 101–113.

Ballakur, A. and Steudel, H. J., 1987, A within-cell utilization based heuristic for designing cellular manufacturing systems, *International Journal of Production Research*, **25**(5) 639–665.

Bitran, G. R. and Hax, A. C., 1981, On the design of hierarchical production planning systems, *Decision Sciences*, **8** 717–743.

Burbidge, J. L., 1989, *Production Flow Analysis for Planning Group Technology* (Oxford Science Publications, Clarendon Press, Oxford).

Burbidge, J. L., 1991, Comments on clustering methods for finding GT groups and families, *Journal of Manufacturing Systems*, **12**(5) 428–429.

Chandrasekharan, M. P. and Rajagopalan, R., 1986a, MODROC: An extension of rank order clustering for group technology, *International Journal of Production Research*, **24**(5) 1221–1233.

Chandrasekharan, M. P. and Rajagopalan, R., 1986b, An ideal seed non-hierarchical clustering algorithm for cellular manufacturing, *International Journal of Production Research*, **24**(2) 451–464.

Chandrasekharan, M. P. and Rajagopalan, R., 1987, ZODIAC — an algorithm for concurrent formulation of part families and machine cells, *International Journal of Production Research*, **26**(6) 835–850.

Hsu, C.-M. and Su, C.-T., 1998, Multi-objective machine-component groupitn in cellular manufacturing: A genetic algorithm, *Production Planning and Control*, **9**(2) 155–166.

De-Witte, J., 1980, The use of similarity coefficients in production flow analysis, *International Journal of Production Research*, **18**(4) 503–514.

Escoto, P. P., McDonnell, L. R., Salort, E. V. and Esteban, F. C. L., 1998, Development of an algorithm for the group technology *n* the design of manufacturing systems, *Production Planning and Control*, **9**(3) 276–274.

Hales, H. L., 1984, *Computer-Aided Facilities Planning* (Marcel Dekker).

Kerr, D. C. and Balakrishnan, J., 1996, Manufacturing cell formation using spreadsheets, *International Journal of Operations and Production Management*, **16**(9) 60–73.

King, J. R., 1980, Machine component grouping in production flow analysis: An approach using rank order clustering, *International Journal of Production Research*, **18** 213–232.

King, J. R. and Nakornchai, V., 1982, Machine component group formation in group technology — review and extension, *International Journal of Production Research*, **20** 117–133.

Kirkpatrick, S., Gelatt, Jr. C. D. and Vecchi, M. P., 1983, Optimization by simulated annealing, *Science*, **220** 671–680.

Kumar, C. S. and Chandrasekharan, M. P., 1990, Grouping efficacy: A quantitative criterion for goodness of block diagonal forms of binary matrices in group technology, *International Journal of Production Research*, **28**(2) 233–243.

Kumar, K. R., Kusiak, A. and Vannelli, A., 1986, Grouping of parts and components in flexible manufacturing systems, *European Journal of Operational Research*, **24**(2) 387–397.

Kumar, K. R. and Vannelli, A., 1987, Strategic subcontracting for efficient disaggregated manufacturing, *International Journal of Production Research*, **25**(12) 1715–1728.

Kusiak, A., 1987, A generalised group technology concept, *International Journal of Production Research*, **25** 561–569.

Lundy, M. and Mees, A., 1986, Convergence of an annealing algorithms, *Mathematical Programming*, **34** 111–124.

MacQueen, J. B., 1967, Some methods for classification and analysis of multivariate observations, *Proceedings of the Fifth Symposium on Mathematical Statistics and Probability*, University of California, Berkeley, **1** 281.

McAuley, J., 1972, Machine grouping for efficient production, *Production Engineer*, **51** 53.

McCormick, W. T., Schweitzer, P. J. and White, T. W., 1972, Problem decomposition and data reorganisation by a clustering technique, *Operation Research*, **20** 993–1009.

Metropolis, N., Rosenbluth, N., Rosenbluth, M., Teller, M. and Teller, E., 1956, Equation of steady-state calculation by fast computing machines, *Journal of Chemical Physics*, **21** 1087–1092.

Miltenburg, J. and Zhang, W., 1991, A comparative evaluation of nine well-known algorithms for solving the cell formation problem in group technology, *Journal of Operations Management*, **10**(1) 44–69.

Mulvey, J. M. and Crowder, H. P., 1979, Clustering analysis: An application of Lagrangian relaxation, *Management Science*, **25**(4) 329–340.

Onwubolu, G. C. and Mkandla, F., 1998, Manufacturing systems construction using meta-strategic approach: Simulated annealing, *Towards Higher Technological Achievement*, eds. Kanhukamwe, Q. C. and Onwubolu, G. C. (Belmont Press).

Onwubolu, G. C. and Mlilo, P., 1998, Manufacturing cell grouping using similarity coefficient-distance measure, *Journal of Production Planning and Control*, **9**(5) 489–493.

Onwubolu, G. C., 1998, Redesigning job-shop to cellular manufacturing system, *International Journal of Manufacturing Systems*, **9**(6) 377–382.

Onwubolu, G. C. and Songore, V., 2000, A tabu search-approach to cellular manufacturing systems, *Production Planning and Control*, **11**(2) 153–164.

Osman, I. H. and Christofides, N., 1994, Capacitated clustering problems by hybrid simulated annealing and tabu search, *International Transaction in Operations Research*, **1**(3) 317–336.

Ravikumar, K., Kusiak, A. and Vannelli, A., 1986, Grouping of parts and components in flexible manufacturing systems, *European Journal of Operational Research*, **24** 387–397.

Romeo, F. and Sangiovanni-Vicentelli, 1985, Probabilistic hill climbing algorithms: Properties and application, *Proc. 1985 Chapel Hill Conference on VLSI*, May 1985, 393–417.

Sarker, B. R. and Xu, Y., 1998, Operations sequences-based cell formation methods: A critical survey, *Production Planning and Control*, **9**(8) 771–783.

Seifoddini, H. and Wolfe, P. M., 1986, Application of the similarity coefficient method in group technology, *IIE Transactions*, **18**(3) 271–277.

Singh, N. and Rajamani, D., 1996, *Cellular Manufacturing Systems* (Chapman and Hall, Inc. London).

Srinivasan, G., Narendran, T. T. and Mahadevan, B., 1990, An assignment model for the part-families problem in group technology, *International Journal of Production Research*, **28**(1) 145–152.

Steudel, H. J. and Desruelle, P. L., 1992, *Manufacturing in the Nineties: How to Become a Mean, Lean, World-class Competitor* (Van Nostrand Reinhold, New York).

Taboun, S. M., Merchawi, N. S. and Ulger, T., 1998, Part family and machine cell formation inmulti-period planning horizons of cellular manufacturing systems, *Production Planning and Control*, **9**(6) 561–571.

Tabucanon, M. T. and Ojha, R., 1987, ICRMA — a heuristic approach for intercell flow reduction in cellular manufacturing systems, *Material Flow*, **4** 189–197.

Tam, K. Y., 1990, An operation sequence based similarity coefficient for part families formations, *Journal of Manufacturing Systems*, **9**(1) 55–68.

Van Laarhoven, R. J. M. and Aarts, E. H. L., 1987, *Simulated Annealing: Theory and applications* (Kluwer Academic Publishers).

Waghodekar, P. H. and Sahu, S., 1984, Machine-component cell formation in group technology: MACE, *International Journal of Production Research*, **22**(6) 937–948.

Wemmerlov, U. and Hyer, N. L., 1989, Cellular manufacturing in the US industry: A survey of user, *International Journal of Production Research*, **27** 1511–1530.

Wemmerlov, U. and Johnson, D. J., 1997, Cellular manufacturing at 46 user plants: Implementation experiences and performance improvements, *International Journal of Production Research*, **35**(1) 29–49.

Chapter 9
Programming Techniques

In one of my articles on tabu search approach to the theory of constraints product mix-problem, one of the reviewers asked the question, "how long does it take a researcher involved in program-development to develop the tabu search problem?" My reaction to the question raised was that, many users of tabu search are of the opinion that it takes much expertise to write a computer code. This chapter attempts to correct this notion by presenting not necessarily a "smart" tabu search code, rather, we present a template for "a simple tabu search" which users can adapt to their specific needs. Some readers who are good at programming may find the structure presented, boring. There are many smart computer codes available for specific problems. My experience with some of these codes is that one runs into a lot of problem attempting to modify them to solve other similar problems. To address this problem, we attempt to present in this chapter a template that can be used as an initial framework for designing more robust tabu search package. Therefore, we call it "a simple tabu search". We first construct the data structure and algorithms necessary to implement the simple tabu search algorithm earlier discussed. We write a Pascal code called tabu search for the theory of constraint problem (*ts_toc*). This code is robust because we only need to change the objective function in order to solve another problem of a different nature. This code is versatile because we can choose from the menu the particular neighbourhood-type of interest in order to solve the problem at hand. The Pascal programming language is chosen because it is simple enough for both researchers and practitioners, to understand. For new comers to Pascal Appendix-A presents a programming course layout containing the basic structure and functions of the language. A full listing of the tabu search program is given in Appendix-B. Let us now commence with the data structure and then discuss the different components of our simple tabu search algorithm.

9.1. Data Structure

The main objective of the theory of constraints, product mix-problem is to determine feasible number of product types without violating resource capacity. This means that we need to initially, generate 'reasonable' numbers of product types or market demand. These are only seed solutions. Neighbourhood moves are required to advance the solutions to an optimum value. Other useful information, include selling price and material price. It is only logical therefore, to have a data structure such as the one shown in Fig. 9.1.

The processing times are stored in an array, and the available-capacities for each part-type are stored in a vector. Using the processing times and

Sequence	0	1	2	3	....	n	sigmaT
Part-type	100	50	10	20	....		
Selling price	90	100	35	58	....		
Material price	45	40	10	10	....		

Fig. 9.1. Data structure of the simple tabu search.

the part-types, the load for each resource can be calculated and compared with the available capacity in order to determine whether or not there is any violation. Figure 9.2 shows the data type declaration in Pascal for the simple tabu search. Some constants are declared: *mnom* is the maximum number of resources, *mnop* is the maximum number of part-type. The maximum length of tabu list is *mtl*. What these declarations mean, is that we can deal with problem-sizes up to the magnitude of 30 resources and 90 product-types. These declarations specify the size of problems that can be solved using our simple tabu search algorithm. It can actually handle quite a large size given that most problems in the literature are less than 10×10. The number of resources is not

```
uses crt,dos;
const
  mnom = 30;{max number of machines ,m}
  mnop  = 90; {max number of parts ,n}
  mtl     = 76;  {max length of tabu list}
type
  tabulistData = record
                rno1 :  integer;
                rno2 :  integer;
               end;{record}
type
 data2 = array[1..mnom,1..mnop] of real;
 datap = array[1..mnop] of integer;
 datar = array[1..mnop] of real;

      partData = record
                partNo       : longint;
                partype      : integer;
                throughput  : real;
                optime       : real;
                turdy          : real;
end; { record of solution data }

    type
      solutionData = record
                sequence : array[0..mnop] of partData;
                zigmaT   :  real;
    end; {record}
```

Fig. 9.2. Data type declaration in Pascal for the simple tabu search.

relevant in the data structure of Fig. 9.1; what is important is the number of product types that can be dealt with. The number of resources only limits the size of the array *data*2 that stores the processing times. This array is an array of real numbers. As can be seen, type *solutionData* is a record of *sequence*, *zigmaT*, and *throughput*. Further, *sequence* is an array of *partData*. We notice that *partData* is a record of *partNo*, which represents the market demand and *partype*, which represents the sequence for generating neighbourhoods. Each of these, is an integer. The solution to our product-mix problem is stored in *zigmaT*, which is real.

There are some important global program-variables that affect the operation of the entire simple tabu search-code. These are found in Fig. 9.3. In the simple tabu search algorithm, we require *current-best* solution of product-mix, and *best-so-far*. The *current-best* solution corresponds to the best amongst the solutions contained in the tabu list at any point in time. The *best-so-far* as the name implies is the best solution found so far at any point in time. These information are captured using the following declarations: *currentsequence*, *currentbest*, and *bestsequence*, which are of type *solutionData* as shown together with the global variables in Fig. 9.3. When a move is declared tabu, it is stored in the *tabu* array, which is of type *tabulistData*. There are some arrays

```
var j, n, g, n1, i, i2,  m, nz, l, l2, k, jk, decoloption, count : integer;
   tabu : array[1..mtl] of tabulistData;
   tabuentry  fullsearch, tabutenure, diverse, rand_num: integer;
   isbestnoChange :boolean;
   newNeighbourhood :boolean;
   noChangeEntry, another : integer;
   currentsequence, currentbest : solutionData;
   bestsequence     :solutionData;
   processt: data2;
   price, mat_cost, c_limit: datar;
   mktdem : datap;
   ch :char;
   rn1, rn2, response, operate_cost : integer;
   start_time, end_time : :real;
   filenam1   : string[20];
   filenam2   : string[20];
   f1, f2        : text;
```

Fig. 9.3. Global variable in Pascal for the simple tabu search.

of type *datar*, which are of type real: *price*, *mat_cost*, *c_limit*, which store the market price, material cost, and resource capacity respectively. The market demand is stored in an integer array, *mktdem* of type *datap*, which in turn is of type integer. The processing times of part-types on resources are stored in the array *data*2 of type real. A number of important integer variables are *tabuentry*, *tabutenure*, *fullsearch*, *diverse*, and *rand_num*. The first two relate to the tabu status in terms of whether a tabu flag is on, and tabu length respectively. The other three variables relate to whether partial or full neighbourhood search is considered, diversification counter, and random number generation respectively. The input data are stored in an input file, while the solution to the product-mix problem is stored in the output file, so we have the following declarations related to file: *filenam*1: *string*[20]; *filenam*2: *string*[20]; *f*1, *f*2: text. We notice that these file declarations are of string-types since they are referenced by text names.

9.2. Modular Design

The manner in which a program is designed determines whether it will be easy to maintain or not. Maintainability is the ease with which a program can be expanded or adapted in the future without having to overhaul the structure of the program. Modular-design will enhance maintainability. To achieve modularity, we have designed our tabu search algorithm in such a way that specific modules are stored in specific files, which are then included during compilation. The declaration {$I*filename*} will include a segment of a program stored in a file names filename in the appropriate order. Such file will then contain several procedures and functions. The order of inclusion using the declaration {$I*filename*} is important in Pascal because a procedure can only *call* another procedure that is contained in an earlier included file. To explain this, suppose a procedure, *proc*1 is in a file, which is included in a program using {$I*file*1} and another procedure, *proc*2 is in a file, which is included in the same program using {$I*file*2}. Then *proc*2 can only call or access *proc*1 if {$I*file*1} is listed before {$I*file*2}. A programmer need to understand this point otherwise the system will return an error message, such as "*proc*1 not found". These declarations for inclusion for the simple tabu search are shown in Fig. 9.4. We notice that the file names do not have the extension, *.pas* when {$I} is used. What this means is that all procedures listed in Fig. 9.4 can access *cosmetix.pas*, all except *cosmetix.pas* can access *data_in.pas*, etc.

```
{$I cosmetix} {deco1, deco2, deco3, lines}
{$I data_in}
{$I dataexit}
{$I obj_vals}
{$I istabu  }
{$I initsol }
{$I driver }
{$I schedule }
```

Fig. 9.4. Include {$*I*} declaration in Pascal for the simple tabu search.

For example, the objective function for the product mix problem is held in file *obj_vals.pas.* Then, in future if we are going to adapt the simple tabu search program to solve another type of problem, such as scheduling, all we need is to concentrate our effort re-coding the objection functions in file *obj_vals.pas.* This kind of programming management technique is useful to enable maintainability in the near and far future. For example, the simple tabu search program being discussed in this chapter was adapted from the tabu search previously used to solve other problems. There was no need to start re-designing *ab initio.*

Now that the data structure and programming management have been dealt with, for the simple tabu search, we need to consider the specific features of tabu search algorithm and their coding. The foundation has been laid, the building must now continue. It is like construction-work in a building site. The foundation takes time. When it is completed, bricks are laid one upon another; beams, columns, and slabs are put in place and the building takes shape within a short time. Programmers need to take time to design good data structure for an efficient and effective package. What we need is not a very "smart short-cut" that does one job, but a well planned program that has the structure, which gives it versatility and robustness. Our simple tabu search is not very smart, but it is versatile and robust. This is the main objective of this chapter. The main program of the simple tabu search is described in Fig. 9.5.

Every Pascal main program commences with a *begin* construct and terminates with an *end. construct.* Notice that there is a dot (.) after the *end* to show that this is the main program. We assign an integer variable, *another* equal to 1. The *while* construct controls the program so that we can terminate if we decide not to make another run. We simply respond with a value of zero (No) after the prompt. Otherwise the program will be executed

```
Begin
 another := 1;
 while (another = 1) do begin
   Duedate_optime_input;
   scheduler;
   gotoxy(10,21);write('Making another run ?[Yes = 1][No = 0] ');
   readln(another);
 end;
End.
```

Fig. 9.5. Main program for the simple tabu search.

once again. There are only two main procedures to be accessed. These are *Duedate_optime_input* and *scheduler* respectively. Figure 9.6 shows the cosmetic routines that are accessed by *Duedate_optime_input.* We do not want to spend time on these cosmetic routines as they are listed in full in Appendix B.

Let us consider what happen in *Duedate_optime_input* routine, which is shown in Fig. 9.7. The routine first accesses *drawframe* and *deco*1, which are the major cosmetic functions. The routine *Namefile* is accessed next. This routine prompts the user to enter the name of the input file, which is then assigned to $filenam1$, a global variable character type. Using the assign construct, $filenam1$ is assigned to $f1$ and the file pointer points to the top of the file. The user is then prompted to enter the name of the output file, which is then assigned to $filenam2$, a global variable character type. Using the *assign* construct, $filenam2$ is assigned to $f2$. If the file already exists, the current information is then appended to the information in the existing file. The other routine to be accessed is *Readtoc.* This routine reads the number of resources, m, and the number of product-types, n from the input file, $f1$. Then using the *for*-construct, for the mxn problem, the processing times are read from the input file into the array, *processt* of type *real.* For each resource, the resource limit is read from the input file into the array *c_limit* of type real. Using the *for*-construct for the product types of size n, the market demand, market price and material cost are read from the input file into arrays *mktdem*, *price*, and *mat_cost* of type *real*, respectively. Finally, the operating cost is read from the file and assigned to a real variable, *operate_cost*. The input file is then closed.

The other routing that is accessed in the main program is the *scheduler*. This is the engine of our simple tabu search program. The main components

```
{file Cosmetix.pas : contains drawframe,lowerlimit, rnd, parameters}
     procedure drawframe;
     const   scrhoriz = 79;
             scrvet   = 23;
      var
      fr1, fr2 :string[scrhoriz];
      x, y : integer;
     begin

      clrscr;
      gotoxy(1,1);
      for x:=1 to scrhoriz do
       write(chr(205)); {top  horizontal}
      for y:=2 to scrvet do begin
       gotoxy(1, y);
       write(chr(186)); {left vertical}
       gotoxy(scrhoriz, y);
       write(chr(186)); {right vertical}
       gotoxy(1,1);write(chr(201)); {top left}
       gotoxy(scrhoriz,1);write(chr(187)); {top right}
      end;
      gotoxy(1,scrvet);
      for x:=1 to scrhoriz do
        write(chr(205)); {bottom horizontal}
        gotoxy(1, scrvet); write(chr(200)); {bottom left}
        gotoxy(scrhoriz,scrvet); write(chr(188)); {bottom right}
     end;

     procedure deco1;
      var
      title1,title2,title3,str1,str2 :string[80];

      begin
      title1:='T H E O R Y   O F  C O N S T R A I N T S';
      title2:='USING A METAHEURISTIC APPROACH:';
      title3:='TABU SEARCH !';
      str1:='Diversification count==>';
      str2:='Iteration count=======>';
```

Fig. 9.6. Cosmetic functions.

```
gotoxy(17,3); write(title1);
gotoxy(22,5); write(title2);
gotoxy(31,7); write(title3);
if deco1option=1 then begin
 gotoxy(20,11); write(str1);
 gotoxy(20,15); write(str2);
 end;{if}
end;

procedure lines( var length ,factor :integer);
 var s :integer;
 begin
  writeln(f2,'');
  for s:=1 to length*factor do
   write(f2,chr(205));
   writeln(f2,'');
 end; {lines}

function initdeco :integer;
    begin
     drawframe;
     deco1option:=0;
     deco1;
     deco3;
    end;

function Time:real;
var hr,min,sec,hunsec :word;
begin
 gettime(hr,min,sec,hunsec);
 time := (hr*60.0 + min)*60.0 + sec*1.0 + hunsec*0.01
end; {end of time}

procedure parameters ;
begin
  fullsearch:=trunc(m*n*n/4);
  tabutenure:= trunc(m+(m+1)/2);
  if tabutenure < 7 then tabutenure := 5 ;
  {diverse:=trunc(1.5*m)}diverse := 100; {if diverse > 10 then diverse := 10 ;}
 end;
```

Fig. 9.6 (*Continued*)

```
function flip(probability:real):boolean;
{ Flip a biased coin - true if heads }
begin
 if probability = 1.0 then flip := true
 else flip := (random <= probability);
end;

function rnd(low,high:integer):integer;
 { pick a random integer between low and high }
 var rn : integer;
 begin
  if low >= high then rn := low
   else begin
    rn := trunc( random * (high-low+1) + low);
   if rn > high then rn := high;
  end;
   rnd := rn;
 end;

function lowerlimit mktdemand:integer):integer;
var i   : boolean;
   low : integer;
BEGIN
 i := false;
 while(i = false) do begin
  low := rnd(1,mktdemand); i := flip(low/mktdemand);
  if (i) then lowerlimit := low ;
 end;
END;
```

Fig. 9.6 (*Continued*)

```
{file : Data_in.pas: contains NameFiles, Fileexists, Readtoc, & Duedate_optime_input}

 procedure Readtoc;
 { Reads the processing times,market demand, price, material cost }
  var  i,j        : integer;
 BEGIN
  readln(f1, m, n);
   for i : = 1 to m do begin
    for j : = 1 to n do begin
     read(f1,processt[i][j]);
    end;                 { End j }
    readln(f1, c_limit[i]);
   end;                  { End i }
    for j := 1 to n do read(f1, mktdem[j]);
    for j := 1 to n do read(f1, price[j]);
    for j := 1 to n do read(f1, mat_cost[j]);
     read(f1, operate_cost);
    close(f1);            { Close file }
END;                      { End Readtoc }


  function FileExists FileName:string):Boolean;
    var F:file;
   begin
    {$I-}
    Assign(F,FileName);
    FileMode := 0;              { Set file access to read only }
    Reset(F);
    Close(F);
    {$I+}
    FileExists := (IOResult = 0) and (FileName<> ' ');
   end;                      { End FileExists }
```

Fig. 9.7. Procedure *data_in* for input data from file.

```
procedure NameFiles; { Assigns input and output files }
  BEGIN
    gotoxy(5,11); write('ENTER THE NAME OF INPUT FILE -----> ');
    readln(filenam1);
    assign(f1,filenam1);
    reset(f1);
    gotoxy(5,13); write('ENTER NAME OF THE OUTPUT FILE ----> ');
    readln(filenam2);
    assign(f2,filenam2);
    if fileexists(filenam2) then append(f2)
    else rewrite(f2);
  END;              { end Namefiles }


Procedure Duedate_optime_input;
  BEGIN
    drawframe;
    deco1;
    NameFiles;
    Readtoc;
  END;
```

Fig. 9.7 (*Continued*)

of the *scheduler* are:

- *driver* routine;
- *dataexit* routine;
- *toc_value* routine.

The *dataexit* routine merely outputs the solution from our simple tabu search programs as shown in Fig. 9.8, while *toc_value* routine computes the objective function value for a particular solution by accessing *obj_toc* as shown in Fig. 9.9. Remember that our main objective is to design a modular program rather than designing a "get solution quick" program.

The routine, *driver* in the *scheduler* routine is like the driver of a car who selects the appropriate gears, triggers the appropriate signals to other road users, and steers the car in the right direction. Therefore, let us explore the main components of the *driver* routine. These main components are:

- *istabu* routine;
- *updatetabulist* routine;

```
{ file Dataexit.pas: contains dataexit for final solution}
    procedure dataexit(var solution :solutionData) ;
    var throughput,x1,x2 : real;
    begin
    throughput:=0;
    for i := 1 to n do begin
    x1 := price[solution.sequence[i].partype];
    x2 := mat_cost[solution.sequence[i].partype];
    throughput := throughput + (x1-x2)*solution.sequence[i].partNo;
    end;
    throughput := throughput - operate_cost;
    writeln(f2,' ');
    if (response=1) or (response = 2) then writeln(f2,'      T H E O R Y  OF  C O N S T R A I N TS');
    writeln(f2,'');
    write(f2,'Solution time        =  ');
    writeln(f2,(end_time-start_time):10:4,'sec');
     write(f2,'Product type          ');
    for i:=1 to n do
     write(f2, solution.sequence[i].partype :8);
     writeln(f2,' ');
     write(f2,'Mix contribution      ');
    for i:=1 to n do
    {  write(f2, solution.sequence[i].partNo :8);}
     write(f2,  price[solution.sequence[i].partype] -
      mat_cost[solution.sequence[i].partype] :8:1);
     writeln(f2,' ');
     write(f2,'Product Mix           ');
    for i:=1 to n do
     write(f2, solution.sequence[i].partNo :8);
     writeln(f2,' ');
     if (response = 1) or (response = 2) then  {Total throughput }
       begin
       write(f2,'Total throughput      = ');
      { writeln(f2, solution.zigmaT*100 :8:2);} {flow_time};
       writeln(f2, throughput :8:2); {flow_time};
       end;
   end; {Dataexit}
```

Fig. 9.8. Procedure *dataexit* for printing solution into file.

- *updatesol* routine;
- *insertsol* routine;
- *constraints* routine;
- *bestsofar* routine;
- *initsol* routine.

Each of these is now discussed in some details. The *istabu* routine shown in Fig. 9.10 declares a *move* tabu or taboo if the *Boolean*-function *istabu* is false.

```
{TOC_OBJ.PAS : Contains Procedure For TOC Objective Function Calculations }

 function obj_toc(var sol:solutionData):real;
 var th : real;
 BEGIN
  th := 0.0;
   for j := 1 to n do begin
    th := th + (price[j] - mat_cost[j])   sol.sequence[j].partNo ;
   { th := th + sol.sequence[j].throughput* sol.sequence[j].partNo ;}
   end; { End j }
   sol.sequence[j].throughput := th/100 ;
   obj_toc := th/100 ;
   sol.zigmaT := th/100;
END;

procedure toc_value( var solution :solutionData);
    begin
      solution.zigmaT := obj_toc(solution);
    end; {toc}
```

Fig. 9.9. Procedures *obj_toc* and *toc_val* compute objective function values.

In this case, the move is penalized for some discrete time equal to the *tabu tenure* (length) by updating the *tabu list* to avoid possible cycling. Figure 9.11 shows the updating of the tabu list when the *updatetabulist* routine is accessed. A *stack* construct is utilized based on come-come-first-served principle.

The routines *updatesol* routines *insertsol*, *constraints*, *bestsofar*, and *initsol* work together hence, they are grouped together into Fig. 9.12. In fact, when the simple tabu search program is executed, after initialization, the first main routine to be accessed is *initsol*. The main function of this procedure is to generate randomly, initial market demand of part-types and then test using the *constraints* routine to ensure that the resource capacity limit is not violated. An 'initial sequence' is then generated. This is because we need this 'initial-sequence' to generate subsequent neighbourhoods. If no diversifaction has been done, the initial solution is taken as the best solution so far, otherwise the routine *bestsofar* is accessed to determine the best solution so far. The initial solution generated randomly from *initsol* is the current solution, held in the global variable *currentsequence*, of type *solutionData*. When the routine *bestsofar* is accessed from *initsol*, the objective function value is computed

```
{file: ISTABU.PAS: contains istabu, updatetabulist routines}

procedure constraints(var strng:solutionData);
    var i,j: integer;   tot_load : real;
  { ******* testing for unique parts indices ******* }
  BEGIN
    tot_load := 0.0; count := 0;
    for i := 1 to m do begin
     for j := 1 to n do begin
      tot_load := tot_load + processt[i][j]  strng.sequence[j].partNo;
      end;  {j}
      if (tot_load > c_limit[i]) then exit; { Increment string bit counter if constraints are satisfied }
      tot_load := 0.0;
    end; {i}
    count := count + 1;
  END;

 function istabu(var solution :solutionData) :boolean;
  var
  i,upper,lower : integer;
  temp :solutionData;
  begin
     istabu:=false;

{1}    if (rn1<>0) or (rn2<>0) then begin
       { if ( solution.sequence[rn1].partNo = solution.sequence[rn2].partNo ) then begin
          istabu := true;
            exit;
         end;}

{2}       if (rn1=rn2) then begin    {the same job indices}
        istabu :=true;
        exit;
       end;

{3}    if (count = 0) then begin
     istabu := true;
     exit;
     end;
```

Fig. 9.10. Procedure *istabu* for tabu classification.

```
{4}      if tabuentry<>0 then begin
      for i:=1 to tabuentry do begin
       if( rn1=tabu[i].rno1) or  ( rn1=tabu[i].rno2) then begin {in tabulist}
        if( rn2=tabu[i].rno1) or  ( rn2=tabu[i].rno2) then begin    {in tabulist}
          istabu := true;
          exit;
        end;
       end; {if rn1}
      end;{for i}
     end; {if tabuentry
   end; { if rn1<>0 or rn2<>0 }
  END; {istabu}
```

Fig. 9.10 (*Continued*)

```
function updatetabulist :integer;
     begin
      if tabuentry    tabutenure then begin
       tabuentr  :=tabuentry + 1;
       tabu[tabuentry].rno1 := rn1;
       tabu[tabuentry].rno2 := rn2;
       end
       else
       begin
       for I :=1 to (tabutenure-1) do
       tabu[i] := tabu[i+1];
       tabu[i+1].rno1:= rn1;
       tabu[i+1].rno2:= rn2;
     end; {if}
      write(f2,tabu[tabuentry].rno1 :5,tabu[tabuentry].rno2 :5);
      writeln(f2,currentbest.zigmaT :10:3);
    END; {updatetabulist}
```

Fig. 9.11. Procedure *updatetabulist* to prevent cycling.

```
{ file: INITSOL.PAS : containing updatsol, initsol, bestsofar}
     procedure updatesol(var solution : solutionData);
      var
       temp :solutionData;
       begin
        temp.sequence[0] := solution.sequence[rn1];
        solution.sequence[rn1]:= solution.sequence[rn2];
        solution.sequence[rn2]:= temp.sequence[0];
        constraints(solution);
       end;

  procedure inversesol var solution : solutionData);
      var
     temp :solutionData;
       begin
        temp.sequence[0] := solution.sequence[rn1];
        solution.sequence[rn1]:= solution.sequence[rn2];
        solution.sequence[rn2]:= temp.sequence[0];
     end;

     procedure insertsol(var solution : solutionData);
      var   i,lower,upper   : integer;
            temp : solutionData;
       begin
        if (rn1<>rn2) then begin
        if (rn1 > rn2) then  begin upper := rn1; lower := rn2;
        end
        else  begin upper := rn2; lower := rn1;
        end; {if}
        temp.sequence[0] := solution.sequence[upper];
        for i := (upper-1) downto  lower do begin
        solution.sequence[i+1]:= solution.sequence[i];
        end;
        solution.sequence[lower]:= temp.sequence[0];
        end;
        constraints(solution);
     end;
```

Fig. 9.12. Procedures *initsol*, *bestsofar*, *insertsol*, *updatesol* and *reversesol*.

```
procedure reversesol(var solution : solutionData);
  var   i,lower,upper   : integer;
                  temp : solutionData;
   begin
    if (rn1<>rn2) then begin
    if rn1 > rn2 then  begin upper := rn1; lower := rn2;
    end
    else  begin upper := rn2; lower := rn1;
    end; {if}
    temp.sequence[0] := solution.sequence[lower];
    for i := (lower+1) to  upper do begin
    solution.sequence[i-1]:= solution.sequence[i];
    end;
    solution.sequence[upper]:= temp.sequence[0];
    end;
  end;

procedure bestsofar(var solution,best :solutionData);
 begin
   obj_toc(solution);
   if(best.zigmaT > solution.zigmaT) then begin
     best:=solution;
     updatetabulist;
     if bestsequence.zigmaT > best.zigmaT then bestsequence:=best;
      noChangeEntry:=0;
   end
     else begin
      if (response = 1) then inversesol(solution);  {rejecting the move:Swapping}
      if (response = 2) then reversesol(solution); { rejecting the move:Insertion }
      noChangeEntry := noChangeEntry +1;
     end;
 end;{bestsofar}
```

Fig. 9.12 (*Continued*)

```
procedure initsol(var solution,best :solutionData);
  var j1 : integer;
   begin
    count := 0;
    repeat
     for j1:= 1 to n do begin  {random number}
     solution.sequence[j1].partNo :=
     rnd(lowerlimit mktdem[j1]),mktdem[j1]) ;
     end;
     constraints(solution); {test for a valid random number}
    until (count > 0);    {repeat}
     for j1:= 1 to n do begin
      solution.sequence[j1].partype := j1 ;
      end;

      best:=solution;
      obj_toc(best);

       if g=1 then  begin
        bestsequence:=best;
        exit;
       end; { if g=1 }
      bestsofar(solution,best);

  END;
```

Fig. 9.12 (*Continued*)

by accessing *obj_toc*. This objective function value is compared with the best solution for so far and the best amongst them becomes the best so far. The 'best sequence' is also updated. If worse solution is obtained the *move* is rejected by either accessing *inversesol* or *reversesol* depending on whether the swapping or insertion neighobourhood is active.

These key procedures are enabling-routines that support the *driver* routine. They are like the steering, steering rod, gear-box, gear selector, differential axle, tie-rods, and wheels of a vehicle. When the engine is switched on, the power transmission system drives the vehicle. The overall schema of the driver routine is shown in Fig 9.13.

Now that we have seen the main components of our simple tabu search program, we finish with the *scheduler* routine. It is simply a menu that prompts

```
{file: DRIVER.PAS : contains driver routine}
   procedure driver;
    var search, search2,s3 : integer;
   BEGIN
     noChangeEntry:=0;
     isbestnoChange:=false;
    { drawframe;
      deco1option:=1;
      deco1; }
      parameters;
      writeln(f2,'T A B U  L I S T');
      start_time:=time;
      for g:=1 to diverse do   { diversification loop }
       begin
        tabuentry:=0;
        writeln(f2,'');
       { rn1:= 0; rn2 := 0;}
        for j:= 1 to n do currentsequence.sequence[j].partNo := 0;
         newNeighbourhood := true;
         initsol(currentsequence, currentbest);
         gotoxy(17,18); write('diversification ===> ');write(g:6);
         if newNeighbourhood then
          begin  { search start here }
          s3:=0;
          for rn1 :=1 to n  do  {intensification search loop}
           begin
          for rn2 :=1 to n  do
          begin
          s3:=s3+1;
           gotoxy(17, 19);write('intensification ===> ');write(rn1  :6);
           gotoxy(17, 20);write('search count    ===> ');write(s3{search} :6);
           if not istabu(currentsequence) then
```

Fig. 9.13. Procedure Driver that drives several modules.

```
begin
if response = 1 then updatesol(currentsequence);  { swapping jobs in string }
if response = 2 then insertsol(currentsequence);  { inserting jobs in string }
 constraints(currentsequence);
 bestsofar(currentsequence,currentbest);
end;{if not tabu}
end;{seach2}
end; {search}
end; {if ne   neighbourhood}
end; {diversification loop}
end_time:=time;
END;
```

Fig. 9.13 (*Continued*)

the user to decide whether, swap-neighbourhood or insertion-neighbourhood structure should be used. This way, it becomes easy to compare results using either of the two neighbourhood structures. The driver routine is accessed, next the routine *toc_value* is accessed, and finally *dataexit* is accessed. This flow of operations is shown in Fig. 9.14.

But the way, Pascal works is that the main program should come last and all procedures and functions that are accessed must be placed before the procedure that is accessing them. It is only therefore, reasonable to list the main at this point as shown in Fig. 9.15. All that we have discussed so far are now tied together in a modular manner. We reiterate that the whole idea of this chapter is to demonstrate a good programming approach that is modular, versatile and robust. This enhances maintainability and reliability.

9.3. Simple Tabu Search Run

The code for the simple tabu search program is given, but how does it fare in the product mix problem? We can only know by running the program. The problem we are using for testing our simple tabu search program is shown in Fig. 9.16. It is a 4-resource, 3-product type-problem, with resource-limit of 2400 for each resource. The market demand, selling price and material cost are given in the last three rows for each product type. For example, product type-1 has market demand of 100, selling price of \$100, and material cost of \$45.

```
{file: SCHEDULE.PAS: contains scheduler routine }
    procedure scheduler;
     BEGIN
      clrscr;
      drawframe;
      randomize;
      gotoxy(12,3);write('=========================================');
      gotoxy(12,4);write('| THEORY OF CONSTRAINTS                  |');
      gotoxy(12,5);write('| USING <TABU SEARCH> META-HEURISTIC     |');
      gotoxy(12,6);write('| (c) Professor G C ONWUBOLU             |');
      gotoxy(12,7);write('| All rights reserved. v1.0, 1999        |');
      gotoxy(12,8);write('========================================= ');
      gotoxy(15,9);write('Select Method of Execution');
      gotoxy(15,10);write(' 1...TOC using Swap technique:');
      gotoxy(15,11);write(' 2...TOC using Insertion technique:');
      gotoxy(15,12);write(' 3...Exit:');
      gotoxy(15,13);write('Your Response (as 1,...,2)  ===>  ');
      read(response);
      start_time:=time;
      case response of
      1: begin
        driver;
        toc_value(bestsequence);
        dataexit(bestsequence);
        close(f2);
        end;

      2: begin
        driver;
        toc_value(bestsequence);
        dataexit(bestsequence);
        close(f2);
        end;

     3: begin
        exit;
        end;
     end; {case}
   END; {scheduler}
```

Fig. 9.14. Procedure *scheduler*, which is a menu.

```
Program ts_toc; {Developed by Professor Godfrey C Onwubolu}
      uses crt,dos;  {Version 0.1 :  August 1999 }
      const
       mnom = 30;{max number of machines ,m}
       mnop  = 90; {max number of parts ,n}
       mtl     = 76;  {max length of tabu list}
      type
       tabulistData = record
                  rno1 :  integer;
                  rno2 :  integer;
     end;{record}
      type
       data2 = array[1..mnom,1..mnop] of real;
       datap = array[1..mnop] of integer;
       datar = array[1..mnop] of real;
          partData = record
                  partNo        : longint;
                  partype       : integer;
                  throughput   : real;
                  optime        : real;
                  turdy          : real;
      end; { record of solution data }

        type
          solutionData = record
                  sequence : array[0..mnop] of partData;
                  zigmaT   :  real;
        end; {record}

    var j, n, g, n1, i, i2,  m, nz, l, l2, k, jk, deco1option, count : integer;
      tabu : array[1..mtl] of tabulistData;
      tabuentry  fullsearch, tabutenure, diverse, rand_num: integer;
      isbestnoChange :boolean;
      newNeighbourhood :boolean;
      noChangeEntry, another : integer;
      currentsequence, currentbest : solutionData;
      bestsequence    :solutionData;
      processt: data2;
```

Fig. 9.15. Main program for the simple tabu search.

```
price, mat_cost, c_limit: datar;
mktdem : datap;
ch :char;
rn1, rn2, response, operate_cost : integer;
start_time, end_time : :real;
filenam1   : string[20];
filenam2   : string[20];
f1, f2     : text;
{$I cosmetix} {deco1, deco2, deco3, lines}
{$I data_in}
{$I dataexit}
{$I obj_vals}
{$I istabu  }
{$I initsol }
{$I driver }
{$I schedule }
 Begin
   another := 1;
   while (another = 1) do begin
     Duedate_optime_input;
     scheduler;
     gotoxy(10,21);write('Making another run ?[Yes = 1][No = 0] ');
     readln(another);
   end;
 End.
```

Fig. 9.15 (*Continued*)

```
4
3
15  10  15  2400
15  30  20  2400
15   5  10  2400
10   5   0  2400
100 50  20
90  100 58
45  40  10
```

Fig. 9.16. Input data for the simple tabu search program.

TABU LIST

1	2	21.420
3	5	25.320
1	2	18.300
3	4	18.400
3	5	19.500
1	2	19.930
4	5	20.530
1	2	20.430
1	5	22.690
2	3	24.730
3	5	25.430
1	2	24.250
1	5	26.380
2	3	27.400
3	4	28.800
4	5	29.200
1	2	15.050
2	3	15.220
3	4	16.220
3	5	17.820
1	2	17.760
3	5	19.160
4	3	19.260

Fig. 9.17. Sample output for the simple tabu search program.

1	2	19.870
1	5	22.140
2	3	23.670
1	2	21.600
1	5	23.340
2	3	24.700
3	4	25.100
1	2	19.700
1	5	20.630
2	3	20.800
3	4	22.500
1	2	23.300
1	5	25.090
2	3	26.450
3	4	27.250
4	5	27.350
1	2	22.110
1	5	23.490
2	3	25.360
3	4	26.460
1	2	19.860
1	5	22.580
2	3	24.960
3	4	25.260

Fig. 9.17 (*Continued*)

1	2	21.890
1	5	23.030
2	3	25.240
3	4	26.240
1	2	23.480
3	5	27.580
1	2	16.000
1	5	19.370
2	3	20.050
2	5	20.270
3	4	20.670
4	2	20.950
1	2	20.160
1	5	23.560
2	3	24.410
3	4	25.110
3	5	25.610
1	2	22.620
1	5	23.870
2	3	24.720
3	4	26.220
4	5	26.520
1	2	19.800
1	5	23.360
2	3	23.700
3	4	24.100
3	5	25.300

Fig. 9.17 (*Continued*)

1	2	20.490
1	5	22.440
2	3	24.140
3	4	25.240
1	2	18.430
3	4	18.930
3	5	23.230
1	2	20.040
1	5	21.450
2	3	25.190
3	4	25.390
1	2	15.070
1	4	15.660
1	5	17.090
2	3	18.110
1	2	23.400
1	5	24.210
2	3	25.400
3	4	26.600
4	5	26.700
1	2	22.390
1	5	22.420
2	3	25.990
3	4	26.590

Fig. 9.17 (*Continued*)

1	2	16.210
1	5	19.100
2	3	22.160
4	5	24.360
1	2	23.260
1	5	25.550
2	3	27.760
3	4	28.660
1	2	21.060
1	5	23.270
2	3	25.310
4	5	25.510
1	2	17.480
1	5	21.150
2	3	23.530
1	2	21.550
1	5	24.680
2	3	24.850
3	4	26.750
1	2	20.180
1	5	21.850
2	3	25.080
1	2	22.850
1	5	24.900
2	3	25.750
3	4	26.850
4	5	27.350

Fig. 9.17 (*Continued*)

1	2	22.700
1	5	25.400
2	3	27.100
3	4	27.600
1	2	22.540
3	5	26.440
1	2	20.250
1	5	20.420
2	3	23.650
3	4	24.750
1	5	21.300
2	3	22.320
3	4	24.120
3	5	25.320
1	2	24.770
1	5	26.500
2	3	28.370
3	4	29.270
1	2	24.640
1	5	25.750
2	3	27.790
3	4	28.790
1	2	20.910
1	5	21.100
2	3	25.010
3	4	25.410

Fig. 9.17 (*Continued*)

1	2	23.650
1	5	25.050
2	3	26.750
3	4	27.350
1	2	20.530
1	5	22.320
2	3	25.380
3	4	25.480
1	5	21.350
2	3	23.560
3	4	25.460
4	5	26.160
1	2	21.490
1	5	24.080
2	3	26.290
3	4	27.690
1	2	21.650
1	5	23.940
2	3	24.450
3	4	26.050
1	4	17.170
1	5	21.100
2	3	22.970
1	5	23.220
2	3	23.560
3	4	24.560
3	5	25.760

Fig. 9.17 (*Continued*)

1	2	20.250
3	4	20.550
3	5	22.050
1	5	21.510
3	4	22.610
3	5	23.310
1	2	21.070
1	5	23.110
2	3	23.620
3	4	24.720
3	5	25.020
1	5	22.420
2	3	23.440
3	4	24.640
4	5	24.740

Solution time	=	0.2200sec				
Product type		5	1	3	2	4
Mix contribution		5.0	20.0	25.0	8.0	15.0
Product Mix		60	19	39	30	30
Total throughput	=	2345.00				

Fig. 9.17 (*Continued*)

The report of the simple tabu search program when run is shown in Fig. 9.17. The *moves* that are tabooed and held in the *tabu list* are first listed in the file. The actual values are multiplied by a factor of 1000. Before the content of each tabu list is emptied, it is observed that the best objective function value is at the bottom of each tabu list.

9.4. Summary

This chapter takes the reader through a modular design of our simple tabu search program, which has been written in Pascal programming language. We do not claim that it is 'smart' but it is apparent that it is modular, versatile and robust. This code is robust because we only need to change the objective function in order to solve another problem of different a nature. This code is versatile because we can choose from the menu the particular neighbourhood-type of interest in order to solve the problem at hand. The Pascal programming language is chosen because it is simple enough for both researchers and practitioners, to understand. For new comers to Pascal Appendix-A presents a programming course layout containing the basic structure and functions of the language. A full listing of the tabu search program is given in Appendix-B.

Exercises

1. Rewrite the simple tabu search code using C-programming language.
2. Rewrite the simple tabu search code using C++-programming language.
3. Discuss the merits and demerits of the three programming languages (Pascal C, and C++) for this application.
4. Describe the weaknesses of our simple tabu search program.
5. Rewrite the simple tabu search code to constitute all the elements of short term memory (STM) as discussed in Chapter 6 of this book.
6. Rewrite the simple tabu search code to constitute all the elements of long term memory (LTM) as discussed in Chapter 6 of this book.
7. Explore other neighbourhood structures that are preferred to the two used in our simple tabu search program.
8. To test the robustness of the framework used in the design of our simple tabu search program, use it as a template for re-designing another application, and measure how much re-organizing is required.

References

Borland International, Inc. 1992, Turbo Pascal Version 7.0 Reference Manual.

Appendix

A. Pascal Fundamentals

This appendix gives an overview of Pascal fundamentals, which are made available to the public by Borland (1992). Beginners in Pascal will find this appendix useful. We also include a simple code for generating randomly data for the product mix problem, so that users can see how the various Pascal fundamentals are further applied.

array

The *array* construct has the syntax:

array [index-type] of element-type

The index type must be an ordinal type but the element type can be any type.

Example

```
type
  integerlist = array [1..100] of integer;
  matrix = array [0..19, 0..89] of real;
  charData = array ['a'..'z'] of byte;
```

begin...end

The *begin* and *end* construct is a compound statement, which serves as statement brackets. The syntax is:

```
begin
 statement;
 statement;
 ...
```

```
 statement
end;
```

When bracketed in this way, any number of consecutive statements are treated as a single statement as shown in the following example.

```
 { Compound statement used within an 'if' statement }
if first < last then
begin
 temp := first;
 first := last;
 last := temp;
end;
```

case

The *case* construct consists of an expression and a list of statements, each prefixed with a case. The syntax is:

```
case expression of
 case: statement;
 ...
 case: statement;
end
```

The *case* construct can take another form, in which it consists of one or more constants or ranges, separated by commas, while the else part is optional:

```
case expression of
 case: statement;
...
 case: statement;
else
 statement
end
```

Example

```
case Ch of
```

```
 'A'..'Z', 'a'..'z': Writeln ('letter');
 '0'..'9':    Writeln ('digit');
 '+', '−', '*', '/': Writeln ('operator');
else
 Writeln ('Special character');
end;
```

const

The *constant* construct normally defines an identifier that denotes a constant value within the block containing the declaration. The syntax is:

```
const
 identifier = expression;
 ...
 identifier = expression;
```

Examples

```
{ Constant Declarations }
const
 maxData = 100;
 numChars = Ord ('Z') - Ord ('A') + 1;
 message = 'Getting to understand Pascal...';
const
 identifier: type = value;
 ...
 identifier: type = value;
```

do

The *do* construct is used in *while*, *for*, and *with* statements.

Example

```
while Ch = ' ' do Ch := GetChar;
for Ch := 1 to 100 do Ch := GetChar;
with pop[i] do extension := stg[j];
```

else

The *else* construct is used in *if* and *case* statements.

Example

```
{ using 'if' statement }

if (i < min) or (i > max) then i := 0;

if counter <> 2 then
begin
  writeln ('Bad command line');
  Exit (1);
end
else
begin
  readfile (ParamStr (1));
  writefile (ParamStr (2));
end;

{ using 'case' statement }
case Ch of
  'A'..'Z', 'a'..'z': Writeln ('letter');
  '0'..'9':     Writeln ('digit');
  '+', '-', '*', '/': Writeln ('operator');
else
  Writeln ('Special character');
end;
```

end

The *end* construct is used with *begin* to form compound statements, *case* to form case statements, *record* to declare record types, and *object* to declare object types.

Examples

```
{ with 'begin' to form compound statement }

if first < last then
begin
  temp := first;
  first := last;
  last := temp;
end;
```

```
{ with 'case' statement }
case Ch of
  'A'..'Z', 'a'..'z': writeln ('letter');
  '0'..'9':    writeln ('digit');
  '+', '-', '*', '/': writeln ('operator');
else
  writeln ('Special character');
end;
```

```
{ with record type definitions }

type
  Class = (num, dat, str);
  Date = record
  D, M, Y: integer;
  end;
  information = record
    name: string [10];
    case kind: Class of
      num: (N: real);
      dat: (D: Date);
      str: (S: string);
  end;
```

```
{ with object type definitions }

type
  locationPtr = ^location;
  location = object
    x, y: integer;
    procedure init (px, py: integer);
    function getX: integer;
    function getY: integer;
  end;
```

file

The *file* type consists of a linear sequence of components of the component type, which can be any type except a file type. The syntax is: *file of type*, or *file*.

Example

```
{ File type declarations }
type
  employee = record
  firstName: string [20];
  lastName: string [30];
  address: string [50];
end;
employeeFile = file of employee;
numberFile = file of integer;
```

for...to, for...downto

The *for* statement will cause the statement after *do* to be executed once for each value in the range first to last. There are two ways of presenting the syntax:

for var := first *to* last do
 statement

Alternatively, the syntax is:

for var := first *downto* last do
 statement

The control variable and the initial and final values must be of an ordinal type. With *to*, the value of the control variable is incremented by 1 for each loop. With *downto*, the value of the control variable is decremented by 1 for each loop.

Example

```
{ for...to, for...downto }
for i := 1 to counter do
  writeln (ParamStr (i);
for i := 1 to 25 do
  for j := 1 to 25 do
  begin
    x := 0;
```

```
  for k := 1 to 25 do
    x := x + array1 [i, k]* array2 [k, j];
  array [i, j] := x;
end;
```

function

The *function* construct is a program part that computes and returns a value, with the syntax:

function ident: type;

Alternatively, it has the syntax:

function ident (parameters): type;

A *function* is activated by the evaluation of a function call in an expression.

The function heading is followed by: a declaration part that declares local objects, and the statement part, which specifies the statements to be executed when the function is called. The statement part should contain at least one statement that assigns a value to the function identifier; the result of the function is the last value assigned.

Example

```
{ Function declaration }
function upCaseString (Stg: string): string;
var
  i: integer;
begin
  for i := 1 to length (S) do
  if (Stg [i] >= 'a') and (Stg [i] <= 'z') then
  Dec (Stg [i], 32);
upCaseString := Stg;
end;
```

goto

The *goto* construct transfers program execution to the statement prefixed by the label referenced in the statement, and has the syntax:

goto label

Unless it cannot be avoided, the *goto* construct should be minimally used. The label must be in the same block as the *goto* statement. It is not possible to jump out of a procedure or a function.

Example

```
label 1, 2;
goto 1
  .
  .
  .
1: writeln ('Abnormal program termination');
2: writeln ('Normal program termination');
```

if...then...else

The if...then...else construct uses the *if*, *then*, and *else* to specify the conditions under which a statement will be executed. The syntax is:

if expression *then* statement

Alternatively, the syntax is:

if expression *then* statement *else* statement

If the Boolean expression after *if* is True, the statement after *then* is executed. Otherwise, if the *else* part is present, the statement after *else* is executed.

Example

```
{ 'if' statements }

if (i < min) or (i > max) then i := 0;

if counter <> 2 then
begin
  writeln ('Unaccepted command line');
  Exit (1);
end
else
```

```
begin
 readFile (ParamStr (1));
 writeFile (ParamStr (2));
end;
```

Pointer

The *pointer* construct contains the memory address of a dynamic variable of a specified base type. A value can be assigned to a *pointer* variable with: *New* or *GetMem* procedures, @ operator, and *Ptr* function.

The reserved word *nil* means a pointer constant that does not point to anything. The predefined type *PChar* denotes a pointer to a null-terminated string.

PChar is declared as:

```
 type PChar = ^Char;
```

Example

```
{ Pointer Type Declaration }

type
  bytePtr = ^byte;
  wordPtr = ^word;
  identPtr = ^identRec;
  identRec = record
    ident: string [15];
    refCount: word;
    next: identPtr;
end;
```

procedure

The *procedure* construct is a program part that performs a specific action, often based on a set of parameters. The syntax is:

procedure identifier;

Alternatively we can use the syntax:

procedure identifier (parameters);

A *procedure* is activated by a *procedure*-statement. The *procedure* heading specifies the identifier for the procedure and the formal parameters, if there is any. The procedure heading is followed by: a declaration part that declares local objects, and the statements between begin and end, which specify what is to be executed when the procedure is called.

Example

```
{ Procedure Declaration }

procedure updatesol (var solution : solutionData);
   var
    temp: solutionData;
   begin
    temp.sequence [0] := solution.sequence [rn1];
    solution.sequence [rn1]:= solution.sequence [rn2];
    solution.sequence [rn2]:= temp.sequence [0];
    constraints (solution);
   end;
```

program

Th *program* construct is the general layout of a Turbo Pascal program:

```
program      ...; { Program heading }
uses         ...; { Uses clause }
label        ...; { Labels }
const        ...; { Constants }
type         ...; { Types }
var          ...; { Variables }
procedure    ...; { Procedures }
function     ...; { Functions }
begin
  statement;      { Statements }
  ...
end.
```

The program heading specifies the program's name and parameters. It is merely informative and has no meaning to the program itself.

The *uses* clause lists the units used by the program, e.g. crt, dos.

The *label*, *constant*, *variable*, *procedure*, and *function* declaration parts can be listed in any order, and repeated any number of times.

The *statement* part specifies the statements to be executed when the program runs.

Randomize

The *randomize* construct initializes the built-in random number generator with a random value (obtained from the system clock). The syntax is:

procedure *randomize*;

If a range is not specified, the result is a real-type random number within the range defined as ($0 <= x < 1$). If a range is specified, it must be an expression of type *word*, and the result is a word-type random number within the range ($0 <= x <$ range). If range equals 0, a value of 0 is returned. The random number generator should be initialized by making a call to *randomize*, or by assigning a value to *randseed*.

Example

```
Program random.pas; {Sample code for the random function: Borland Inc.}
uses Crt; { For Windows: uses WinCrt; }

begin
 randomize;
 repeat
   { Write random numbers }
   writeln (Random (1000));
 until keyPressed;
end.
```

record

The *record* construct contains a number of components, or fields, that can be of different types. The syntax is:

record
  fields;

```
  fields;
  ...
  fields
end;
```

Alternatively another syntax used is:

```
record
  fields;
  ...
  case tag: type of
  case: ( fields );
  ...
  case: ( fields )
end;
```

Example

```
{ Record Type Definitions }
type
  Class = (num, dat, str);
  Date = record
  D, M, Y: Integer;
  end;
  information = record
    name: string [10];
    case kind: Class of
     num: (N: real);
     dat: (D: Date);
     str: (S: string);
  end;
```

repeat...until

The *repeat...until* construct executes the statements between *repeat* and *until* in sequence, until at the end of a sequence, the Boolean expression is True. The syntax is:

```
  repeat
  statement;
```

```
  statement;
  ...
  statement
until expression
```

Example

```
  { Repeat Statements }
    repeat ch := getChar until ch <> '';

    repeat
      write ('Enter any value of your choice: ');
      readln (i);
    until (i >= 0) and (i <= 9);
```

type

The *type* construct specifies an identifier that denotes a type, with the syntax:

```
type
  identifier = anyType;
  ...
  identifier = anyType;
```

AnyType is a type identifier of any of the following types: array, file, object, ordinal, pointer, real, record, set, and string.

Example

```
type
  tabulistData = record
      rno1: integer;
      rno2: integer;
end; {record}

type
  data2 = array [1..mnom,1..mnop] of real;
  datap = array [1..mnop] of integer;
  datar = array [1..mnop] of real;

  partData = record
        partNo      : longint;
        partype     : integer;
```

```
        throughput : real;
        optime     : real;
end; { record of solution data }

type
    solutionData = record
    sequence: array [0..mnop] of partData;
    zigmaT: real;
end; {record}
```

uses

The *uses* construct identifies the name of a unit used by the current program or unit. The syntax is:

uses identifier, ... identifier;

Example

```
Program personalProgram;
uses WinDos, WinCrt; { program can now use routines in WinDos and
WinCrt }

Program personalProgram;
uses Dos, Crt; { program can now use routines in Dos and Crt }
```

var

The variable construct associates an identifier and a type with a location in memory where values of that type can be stored. The syntax is:

```
var
  identifier, ... identifier: type;
  ...
  identifier, ... identifier: type;
```

Examples

```
{ Variable Declarations }
  var
    w, x, y, z: real;
    i, j, k: integer;
```

```
  istabu: Boolean;
  vector: array [1..20] of real;
  matrix: array [1..20, 1..20] of real;
  name: string [30
  inputFile, outputFile: text;
  letters: set of 'A'..'Z';
```

while

The *while* construct contains an expression that controls the repeated execution of a singular or compound statement. The syntax is:

while expression *do* statement

The statement after do is executed repeatedly as long as the Boolean expression is True. The expression is evaluated before the statement is executed, so if the expression is False at the beginning, the statement is not executed.

Example

```
{ 'while' statements }
while not Eof (inputFile) do
begin
 readLn (inputFile, Line);
 writeln (outputFile, Line);
 inc (lineCounter);
end;
```

with

The *with* construct is a shorthand method for referencing the fields of a record, with the syntax:

with var, var, ... var do statement

Example

```
{ 'with' statement }
 with date [i] do begin
   day: = 1;
```

```
  month: = 1;
  year: = year + 1;
end;
```

This *with* construct is equivalent to:

```
date [i].day: = 1;
date [i].month: = 1;
date [i].year: = date [i].year + 1;
```

A.1. *Putting Pascal fundamentals to use*

Now that all the ingredients for the understanding of Pascal have been given, we attempt to apply these fundamentals to a simple code. This code, known as *tocdata* is used to generate randomly, data for the tabu search presented in this book. All the randomly generated data used for running our tabu search program were obtained using the source code, *tocdata* shown in Fig. A.1.

The program, *tocdata* commences with the program header, followed by the type of unit used by the program, which in this case, are *crt* and *dos*. Some constants are defined: maxmc and maxpart, each taking value of 50 and 150 respectively. This means we can deal with up to 50 resources and 150 part-types. Some global variable are then defined: *mc*, *pc*, *i*, *j*, *low_process*, *up_process*, *low_selling*, *up_selling*, *low_demand*, *up_demand*, *low_material*, *up_material*, *max_capacity* of type integer; *capacity* of type real; wij of integer array of matrix-type; p, demand, selling, material of integer array of vector-type; f1, f2 of file type.

We then describe two function, *rnd* and *fileExists* for generating random numbers and checking if files exist respectively. One procedure, *Namefiles* opens existing files for reading data or creates a new file for output solutions. The main program then described with a begin..end construct. Within the main program, there is a *clrscr* command that clears the screen. Then, *Namefiles* is accessed, allowing the user to specify the file name for the random data to be generated. Then, the user is prompted to input the lower and upper bounds of the processing times, market demand, selling price, and material cost, as well as the number of resources and product mix types. The user is advised on a reasonable resource-limit, but prompted to make a decision on

```
program tocdata; { Developed by Professor Godfrey C Onwubolu : December 1998}
uses crt,dos;  { Program randomly generates 1) resource processing times}
const                                    {     2) demand        }
                                         {     3) selling price }
   maxmc  = 50;                          {     4) market price  }
   maxpart = 150;
   var mc, pc, i, j, ch : integer;
   low_process, up_process, low_selling, up_selling : integer;
   low_demand, up_demand, low_material, up_material, max_capacity  : integer;
   capacity : real;
   wij : array [1. maxmc,1..maxpart] of integer;
 p, demand, selling, material : array [1..maxpart] of integer;
 filenam1  : string[20];
 filenam2  : string[20];
 f1, f2     : text;

 function rnd(low  high:integer):integer;
 var i :integer;
 begin
 if low >= high then i:= low
  else begin
   i := trunc(random*(high-low+1) + low);
    if i > high then i := high;
  end; {else}
   rnd := i;
 end; {if}

 function FileExists(FileName:string) :Boolean;
 var F:file;
 begin
 {$I-}
 Assign(F,FileName);
 FileMode := 0;      {Set file access to read only}
 Reset(F);
 Close(F);
 {$I+}
 FileExists := (IOResult = 0) and (FileName<> ' ');
 end;       {End FileExists}

 procedure NameFiles;
 begin
    gotoxy(10,4); write ('Enter Randomized Output File Name ....');
    readln(filenam2);
    assign(f2,filenam2);
    if FileExists(filenam2) then append(f2)
    else rewrite(f2);
 end;  {end NameFiles}
```

Fig. A.1. Pascal program *tocdata*.

```
BEGIN
clrscr;
NameFiles;
gotoxy(10, 6);write('Process times: low / upper bounds ? ');read(low_process);read(up_process
gotoxy(10, 8);write('Demand: low / upper bounds ? ');read(low_demand);read(up_demand);
gotoxy(10,10);write('Selling price: low / upper bounds ? ');read(low_selling);read(up_selling);
gotoxy(10,12);write('Material price: low / upper bounds ? ');read(low_material);read(up_materi
gotoxy(10,14);write('How many resources ? ');read(mc);
gotoxy(10,16);write('How many product mixes ? ');read(pc);
capacity := up_demand*up_process*pc*0.75;
gotoxy(10,18);write('Resources capacity ?',' [approx.] ', capacity :10:2,' ');read(max_capacity);

{generate incidence matrix}
clrscr;
randomize;
for i:= 1 to mc do begin
 for j:= 1 to pc do begin
 wij[i][j] := 0
  end; {j}
end; {i}

for i:= 1 to mc do begin
 for j:= 1 to pc do begin
  wij[i][j] := rnd(low_process,up_process);
end; {j}
end; {i}

{generate demand, selling price, and material cost}
for j:= 1 to pc do begin
 demand[j] := rnd(low_demand, up_demand);
 selling[j] := rnd(low_selling, up_selling);
 material[j] := rnd(low_material, up_material);
end; {j}

{write Incidence Matrix into File}
writeln(''); writeln('');
writeln(f2, mc);
writeln(f2, pc);
```

Fig. A.1 (*Continued*)

```
{write processing times into file}
 for i := 1 to mc do begin
 for j := 1 to pc do begin
   write(f2, wij[i][j]:5 );
  end; {j}
write(f2, max_capacity:8 );

writeln(f2);
end; {i}

{write demands into file}
 for j := 1 to pc do begin
    write(f2, demand[j]:5);
 end; {j}
writeln(f2);

{write selling prices into file}
for i := 1 to pc do begin
write(f2, selling[i]:5);
end;
writeln(f2);

{write raw material prices into file}
for i := 1 to pc do begin
write(f2, material[i]:5);
end;

close(f2);
writeln( ' Program execution complete.');
Writeln('');
Write( 'Enter any number to terminate!  ');
read(ch);
END.
```

Fig. A.1 (*Continued*)

the eventual resource-limit. The monitor screen is once more cleared using the *clrscr* command. The *randomize*-construct then initializes the built-in random number generator with a random value (obtained from the system clock). We initialize the array *wij*. The program then automatically generates randomly the processing times, market demand, selling price and material cost by accessing the function *rnd*, which generates values between the specified lower and upper bounds specified.

The program then writes the following data into the file specified by the user: number of resources, number of part types, processing times, resource capacity, market demand, selling price and material cost. The file is then closed, and then program terminates. Readers who are interested in in-depth understanding of Pascal programming language need to dive deeper into more comprehensive coverage of the subject matter, using comprehensive textbooks, and consulting Borland Inc. library.

The program randomly generates data shown in Fig. A.3 and dumps it onto the file $f15 \times 10.dat$. As mentioned, this becomes the input data for our simple tabu search program. So, when we respond to the input prompts of our program, *tocdata*, with the values shown in Fig. A.2, what we see in our file, $f15 \times 10.dat$. is Fig. A.3. We then use f15x10.dat. to run our simple tabu search program. Running the Pascal program tocdata, several times for a wide range of problem configurations, extensive experimentation can be done by the user.

A.2. *Getting something from Pascal fundamentals*

Now let us run our simple Pascal program *tocdata*. When we run it, the program prompts the user for a few input data shown in Fig. A.2. The name of the file specified by the user becomes the input data file name to be used later for the simple tabu search program already discussed in Chapter. 9. In this case, the user specifies that the name of the file is $f15 \times 10.dat$. The name is chosen to reflect the size of the problem. This is useful during simulation when several problems of varying sizes need to generate randomly. The lower limit for the processing times is 0, while the upper limit is 10. This means that the function *rnd* generates random values within these limits. The lower limit for the market demand is 30, while the upper limit is 100. This means that the function *rnd* generates random values within these limits. The lower limit for the selling price is 20, while the upper limit is 60. This means that the function *rnd* generates random values within these limits. The lower limit for the material price is 50, while the upper limit is 25. This means that the function *rnd* generates random values within these limits. There are 15 resources, and 10 product types. The resources capacity or limit is 8000.

Enter Randomized Output File Name ....> *f15x10.dat*
Process times: low / upper bounds ? > *0, 10*
Demand: low / upper bounds ? > *30, 100*
Selling price: low / upper bounds ? > *20, 60*
Material price: low / upper bounds ? > *5, 25*
How man   resources > *15*
How many product mixes > *10*
Resource capacity ? > *8000*

Fig. A.2. Input of Pascal program *tocdata*.

```
15
10
  3   6   2   5  10   0   5   0   5   2  8000
  3   1   3   2   4   4   2   9   5   9  8000
  6   1   3   6   8   1   7   3   6  10  8000
  5   9   4   9  10   4   3  10   3   4  8000
  5   2   2   9   8   0   3  10   1   7  8000
  9   3   4   4   7   0   3   8  10  10  8000
  4   4   0   2   1   3   6   3   7   9  8000
 10   9   7   5   9   4   9  10   8  10  8000
  7   3  10   0   6   3   3   3   2   7  8000
  0   8   8   7   7  10   1   3   2   6  8000
  9   8  10   5   4   8   2   1   5   3  8000
 10   1   3   9   9   0   8   7   6   6  8000
  9   9   7   5   9   8   9   3   6   2  8000
  2   4   2  10   0   6  10   4  10   0  8000
  5   3  10   4   6   0   7   9   5   1  8000
 62  36  27  37  41  76 100  70  56  51
 47  55  42  24  28  21  45  23  58  58
 20  20  13  16  15   9  20  19  11  15
```

Fig. A.3. Output of Pascal program *tocdata*.

A.3. *Summary*

This appendix illustrates the essential ingredients of Pascal programming language by considering a number of constructs, which programmers need to be familiar with. Each construct is explained, and one or more examples are given so that readers can understand how the constructs work. Then we worked through a simple program that generates randomly, input data for our simple tabu search program. These two aspects of this appendix are useful for the understanding of the tabu search code of this book.

Exercises

1. Rewrite the program, *tocdata* so that the functions *rnd* and *fileExists* are contained in a file called cosmetic.pas, and then use {$I} to include this file.
2. In addition to problem 1, rewrite the program, *tocdata* so that the part that generates randomly, processing times, resource capacity, market demand, selling price and material cost, are contained in a file called rnd_data.pas, and then use {$I} to include this file.
3. In addition to problem 2, rewrite the program, *tocdata* so that the part that writes number of resources, number of product types, processing times, resource capacity, market demand, selling price and material cost into the file, are contained in a file called rnd_data.pas. Use {$I} to include this file.
4. Explore further modification that makes our program, *tocdata* more modular.
5. Integrate the modified program, *tocdata* into our simple tabu search program. What is the advantage of doing this? What is the disadvantage of doing this?

References

bert Borland International, Inc. 1992, Turbo Pascal Version 7.0 Reference Manual.

B. A Simple Tabu Search in Pascal

The complete program code for the simple tabu search algorithm for the product mix problem is given in Figs. B.1–9. The code consists of eight files:

Cosmetix.pas

DDate_in.pas

Dataexit.pas

Obj_vals.pas

Istabu.pas

Initsol.pas

Driver.pas

Schedule.pas

Ts_toc.pas

```
{file Cosmetix.pas : contains drawframe,lowerlimit, rnd, parameters}
      procedure drawframe;
      const   scrhoriz=79;
                 scrvet  =23;
       var
        fr1,fr2 :string scrhoriz];
        x,y : integer;
        begin

        clrscr;
        gotoxy(1,1);
        for x:=1 to scrhoriz do
         write(chr(205)); {top  horizontal}
        for y:=2 to scrvet do begin
         gotoxy(1,y);
         write(chr(186)); {left vertical}
         gotoxy(scrhoriz,y);
         write(chr(186)); {right vertical}
         gotoxy(1,1);write(chr(201)); {top left}
         gotoxy(scrhoriz,1);write(chr(187)); {top right}
       end;
       gotoxy(1,scrvet);
       for x:=1 to scrhoriz do
         write(chr(205)); {bottom horizontal}
         gotoxy(1,scrvet);write(chr(200)); {bottom left}
         gotoxy(scrhoriz,scrvet);write(chr(188)); {bottom right}
      end;

      procedure deco1;
       var
       title1,title2,title3,str1,str2 :string[80];
```

Fig. B.1. Cosmetic functions.

```
begin
title1:='T H E O R Y   O F  C O N S T R A I N T S';
title2:='USING A METAHEURISTIC APPROACH:';
title3:='TABU SEARCH !';
str1:='Diversification count==>';
str2:='Iteration count========>';

gotoxy(17,3); write(title1);
gotoxy(22,5); write(title2);
gotoxy(31,7); write(title3);
if deco1option=1 then begin
 gotoxy(20,11); write(str1);
 gotoxy(20,15); write(str2);
 end;{if}
end;


procedure lines( var length ,factor :integer);
 var s :integer;
 begin
  writeln(f2,'');
  for s:=1 to length*factor do
   write(f2,chr(205));
  writeln(f2,'');
 end; {lines}


   function initdeco :integer;
   begin
    drawframe;
    deco1option:=0;
    deco1;
    deco3;
   end;

function Time:real;
var hr,min,sec,hunsec :word;
begin
 gettime(hr,min,sec,hunsec);
 time:=(hr*60.0 + min)*60.0 + sec*1.0 + hunsec*0.01
end; {end of time}
```

Fig. B.1 (*Continued*)

```
procedure parameters ;
begin
  fullsearch:=trunc(m*n*n/4);
  tabutenure:= trunc(m+(m+1)/2);
  if tabutenure < 7 then tabutenure := 5 ;
  {diverse:=trunc(1.5*m)}diverse := 100; {if diverse > 10 then diverse := 10 ;}
 end;

function flip(probability:real):boolean;
{ Flip a biased coin - true if heads }
begin
 if probability = 1.0 then flip := true
 else flip := (random <= probability);
end;


function rnd(low,high:integer):integer;
{ pick a random integer between low and high }
var rn : integer;
begin
 if low >= high then rn := low
  else begin
   rn := trunc( random * (high-low+1) + low);
  if rn > high then rn := high;
 end;
  rnd := rn;
end;

function lowerlimit mktdemand:integer):integer;
var i   : boolean;
   low : integer;
BEGIN
 i := false;
 while(i = false) do begin
  low := rnd(1,mktdemand); i := flip(low/mktdemand);
  if (i) then lowerlimit := low ;
 end;
END;
```

Fig. B.1 (*Continued*)

```
{file : DDate_in.pas: contains NameFiles, Fileexists, Readtoc, &
  Duedate_optime_input}

procedure Readtoc;
{ Reads the processing times,market demand, price, material cost }
 var  i,j       : integer;

BEGIN
 readln(f1,m,n);
  for i := 1 to m do begin
   for j := 1 to n do begin
     read(f1,processt[i][j]);
   end;                 { End j }
   readln(f1,c_limit[i]);
 end;                   { End i }
   for j := 1 to n do read(f1,mktdem[j]);
   for j := 1 to n do read(f1,price[j]);
   for j := 1 to n do read(f1,mat_cost[j]);
    read(f1,operate_cost);
   close(f1);           { Close file }
END;                    { End Readtoc }


    function FileExists(FileName:string):Boolean;
    var F:file;
    begin
    {$I-}
    Assign(F,FileName);
    FileMode := 0;             { Set file access to read only }
    Reset(F);
    Close(F);
```

Fig. B.2. Procedure Data_in for input data from file.

```
{$I+}
 FileExists := (IOResult = 0) and (FileName<> ' ');
 end;                           { End FileExists }


procedure NameFiles;
 { Assigns input and output files }
BEGIN

  gotoxy(5,11); write('ENTER THE NAME OF INPUT FILE -----> ');
  readln(filenam1);
  assign(f1,filenam1);
  reset(f1);
  gotoxy(5,13); write('ENTER NAME OF THE OUTPUT FILE ----> ');
  readln(filenam2);
  assign(f2,filenam2);
  if fileexists(filenam2) then append(f2)
  else rewrite(f2);
END;                   { end Namefiles }


  Procedure Duedate_optime_input;
  BEGIN
   drawframe;
   deco1;
   NameFiles;
   Readtoc;
  END;
```

Fig. B.2 (*Continued*)

```
{ file Dataexit.pas: contains dataexit for final solution}
    procedure dataexit(var solution :solutionData) ;
    var throughput,x1,x2 : real;
    begin
    throughput:=0;
    for i := 1 to n do begin
    x1 := price[solution.sequence[i].partype];
    x2 := mat_cost[solution.sequence[i].partype];
    throughput := throughput + (x1-x2)*solution.sequence[i].partNo;
    end;
    throughput := throughput - operate_cost;
    writeln(f2,' ');
    if (response=1) or (response = 2) then writeln(f2,'        T H E O R Y  OF  C O N S T R A I N TS');
    writeln(f2,'');
    write(f2,'Solution time        =  ');
    writeln(f2,(end_time-start_time):10:4,'sec');
     write(f2,'Product type          ');
    for i:=1 to n do
     write(f2, solution.sequence[i].partype :8);
     writeln(f2,' ');
     write(f2,'Mix contribution      ');
    for i:=1 to n do
    {  write(f2, solution.sequence[i].partNo :8);}
     write(f2,  price[solution.sequence[i].partype] -
      mat_cost[solution.sequence[i].partype] :8:1);
     writeln(f2,' ');
     write(f2,'Product Mix           ');
    for i:=1 to n do
     write(f2, solution.sequence[i].partNo :8);
     writeln(f2,' ');
    if (response = 1) or (response = 2) then  {Total throughput }
      begin
      write(f2,'Total throughput     =');
     { writeln(f2, solution.zigmaT*100 :8:2);} {flow_time};
      writeln(f2, throughput :8:2); {flow_time};
      end;
   end; {Dataexit}
```

Fig. B.3. Procedure Dataexit for printing solution into file.

```
{TOC_OBJ.PAS : Contains Procedure For TOC Objective Function Calculations }
function obj_toc(var sol:solutionData):real;
var th : real;
BEGIN
 th := 0.0;
  for j := 1 to n do begin
   th := th + (price[j] - mat_cost[j])   sol.sequence[j].partNo ;
  { th := th + sol.sequence[j].throughput* sol.sequence[j].partNo ;}
  end; { End j }
  sol.sequence[j].throughput := th/100 ;
  obj_toc := th/100 ;
  sol.zigmaT := th/100;
END;

    procedure toc_value( var solution :solutionData);
      begin
       solution.zigmaT := obj_toc(solution);
    end; {toc}
```

Fig. B.4. Procedure Obj_vals computed objective function values.

```
{file: ISTABU.PAS: contains istabu, updatetabulist routines}
 procedure constraints(var strng:solutionData);
   var i,j: integer;  tot_load : real;
  { ******* testing for unique parts indices ******* }
  BEGIN
    tot_load := 0.0; count := 0;
    for i := 1 to m do begin
     for j := 1 to n do begin
      tot_load := tot_load + processt[i][j]  strng.sequence[j].partNo;
      end;  {j}
      if (tot_load > c_limit[i]) then exit; { Increment string bit counter if constraints are satisfied }
      tot_load := 0.0;
    end; {i}
    count := count + 1;
  END;

function istabu(var solution :solutionData) :boolean;
 var
 i,upper,lower : integer;
 temp :solutionData;
 begin
    istabu:=false;
```

Fig. B.5. Procedure Istabu for tabu classification.

```
{1}   if (rn1<>0) or (rn2<>0) then begin
      { if ( solution.sequence[rn1].partNo = solution.sequence[rn2].partNo ) then begin
        istabu := true;
          exit;
       end;}

{2}      if (rn1=rn2) then begin    {the same job indices}
       istabu :=true;
       exit;
      end;

{3}      if tabuentry<>0 then begin
      for i:=1 to tabuentry do begin
      if( rn1=tabu[i].rno1) or  ( rn1=tabu[i].rno2) then begin {in tabulist}
       if( rn2=tabu[i].rno1) or  ( rn2=tabu[i].rno2) then begin    {in tabulist}
         istabu := true;
         exit;
       end;
      end; {if rn1}
     end;{for i}
    end; {if tabuentry
  end; { if rn1<>0 or rn2<>0 }
 END; {istabu}

 function updatetabulist :integer;
    begin
     if tabuentry    tabutenure then begin
      tabuentr  :=tabuentry + 1;
      tabu[tabuentry].rno1 := rn1;
      tabu[tabuentry].rno2 := rn2;
      end
      else
      begin
      for i:=1 to (tabutenure-1) do
      tabu[i]:=tabu[i+1];
      tabu[i+1].rno1:=rn1;
      tabu[i+1].rno2:=rn2;
    end; {if}
     write(f2,tabu[tabuentry].rno1 :5,tabu[tabuentry].rno2 :5);
     writeln(f2,currentbest.zigmaT :10:3);
  END; {updatetabulist}
```

Fig. B.5 (*Continued*)

```
{ file: INITSOL.PAS : containing updatsol, initsol, bestsofar}
        procedure updatesol(var solution : solutionData);
         var
          temp :solutionData;
          begin
           temp.sequence[0] := solution.sequence[rn1];
           solution.sequence[rn1]:= solution.sequence[rn2];
           solution.sequence[rn2]:= temp.sequence[0];
           constraints(solution);
          end;

   procedure inversesol var solution : solutionData);
         var
        temp :solutionData;
          begin
           temp.sequence[0] := solution.sequence[rn1];
           solution.sequence[rn1]:= solution.sequence[rn2];
           solution.sequence[rn2]:= temp.sequence[0];
        end;

        procedure insertsol(var solution : solutionData);
         var    i,lower,upper    : integer;
              temp : solutionData;
         begin
          if (rn1<>rn2) then begin
          if (rn1 > rn2) then  begin upper := rn1; lower := rn2;
          end
          else  begin upper := rn2; lower := rn1;
          end; {if}
          temp.sequence[0] := solution.sequence[upper];
          for i := (upper-1) downto  lower do begin
          solution.sequence[i+1]:= solution.sequence[i];
          end;
          solution.sequence[lower]:= temp.sequence[0];
          end;
          constraints(solution);
        end;
```

Fig. B.6. Procedures Initsol that generates initial solution.

```
procedure reversesol(var solution : solutionData);
 var   i,lower,upper   : integer;
                 temp : solutionData;
  begin
   if (rn1<>rn2) then begin
   if rn1 > rn2 then  begin upper := rn1; lower := rn2;
   end
   else  begin upper := rn2; lower := rn1;
   end; {if}
   temp.sequence[0] := solution.sequence[lower];
   for i := (lower+1) to  upper do begin
   solution.sequence[i-1]:= solution.sequence[i];
   end;
   solution.sequence[upper]:= temp.sequence[0];
   end;
 end;

procedure bestsofar(var solution,best :solutionData);
 begin
   obj_toc(solution);
   if(best.zigmaT > solution.zigmaT) then begin
    best:=solution;
    updatetabulist;
    if bestsequence.zigmaT > best.zigmaT then bestsequence:=best;
     noChangeEntry:=0;
   end
    else begin
     if (response = 1) then inversesol(solution);  {rejecting the move:Swapping}
     if (response = 2) then reversesol(solution); { rejecting the move:Insertion }
     noChangeEntry:=noChangeEntry +1;
    end;
 end;{bestsofar}
```

Fig. B.6 (*Continued*)

```
procedure initsol(var solution,best :solutionData);
   var j1 : integer;
    begin
     repeat
      for j1:= 1 to n do begin  {random number}
      solution.sequence[j1].partNo :=
      rnd(lowerlimit mktdem[j1]),mktdem[j1]) ;
      end;
      constraints(solution); {test for a valid random number}
     until (count > 0);     {repeat}
      for j1:= 1 to n do begin
       solution.sequence[j1].partype := j1 ;
       end;

       best:=solution;
       obj_toc(best);

        if g=1 then  begin
         bestsequence:=best;
         exit;
        end; { if g=1 }
       bestsofar(solution,best);
 END;
```

Fig. B.6 (*Continued*)

```
{file: DRIVER.PAS : contains driver routine}
    procedure driver;
     var search, search2,s3 : integer;
    BEGIN
      noChangeEntry:=0;
      isbestnoChange:=false;
     { drawframe;
```

Fig. B.7. Procedure Driver that drives several modules.

```
  deco1option:=1;
  deco1; }
  parameters;
  writeln(f2,'T A B U  L I S T');
  start_time:=time;
  for g:=1 to diverse do  { diversification loop }
   begin
    tabuentry:=0;
    writeln(f2,'');
   { rn1:=0; rn2:=0;}
    for j:= 1 to n do currentsequence.sequence[j].partNo := 0;
     newNeighbourhood:=true;
     initsol(currentsequence,currentbest);
     gotoxy(17,18); write('diversification ===> ');write(g:6);
     if newNeighbourhood then
      begin  { search start here }
      s3:=0;
      for rn1 :=1 to n  do  {intensification search loop}
       begin
      for rn2 :=1 to n  do
      begin
      s3:=s3+1;
       gotoxy(17,19);write('intensification ===> ');write(rn1  :6);
       gotoxy(17,20);write('search count    ===> ');write(s3{search} :6);
       if not istabu(currentsequence) then
       begin
       if response = 1 then updatesol(currentsequence);  { swapping jobs in string }
       if response = 2 then insertsol(currentsequence);  { inserting jobs in string }
        constraints(currentsequence);
        bestsofar(currentsequence,currentbest);
       end;{if not tabu}
       end;{seach2}
       end; {search}
     end; {if ne   neighbourhood}
    end; {diversification loop}
    end_time:=time;
END;
```

Fig. B.7 (*Continued*)

```
{file: SCHEDULE.PAS: contains scheduler routine }
  procedure scheduler;
   BEGIN
    clrscr;
    drawframe;
    randomize;
    gotoxy(12,3);write('============================================a');
    gotoxy(12,4);write('| THEORY OF CONSTRAINTS          |');
    gotoxy(12,5);write('| USING <TABU SEARCH> META-HEURISTIC|');
    gotoxy(12,6);write('| (c) Professor G C ONWUBOLU|              ');
    gotoxy(12,7);write('| All rights reserved. v1.0, 1999   |');
    gotoxy(12,8);write('============================================ ');
    gotoxy(15,9);write('Select Method of Execution');
    gotoxy(15,10);write(' 1...TOC using Swap technique:');
    gotoxy(15,11);write(' 2...TOC using Insertion technique:');
    gotoxy(15,12);write(' 3...Exit:');
    gotoxy(15,13);write('Your Response (as 1,...,2) ===> ');
    read(response);
    start_time:=time;
    case response of
    1: begin
      driver;
      toc_value(bestsequence);
      dataexit(bestsequence);
      close(f2);
      end;

    2: begin
      driver;
      toc_value(bestsequence);
      dataexit(bestsequence);
      close(f2);
      end;

   3: begin
      exit;
      end;
    end; {case}
 END; {scheduler}
```

Fig. B.8. Procedure Schedule, which is a menu.

```
Program ts_toc; { Developed by Professor G C Onwubolu }
    uses crt,dos;    { Version 0.1 : August 1999 }
    const
     mnom = 30;{max number of machines ,m}
     mnop = 90; {max number of parts ,n}
     mtl  = 76; {max length of tabu list}
    type
     tabulistData = record
            rno1 : integer;
            rno2 : integer;
           end;{record}

    type
     data2 = array[1..mnom,1..mnop] of real;
     datap = array[1..mnop] of integer;
     datar = array[1..mnop] of real;

       partData = record
            partNo       : longint;
            partype      : integer;
            throughput    : real;
            optime       : real;
            turdy        : real;
           end; { record of solution data }

      type
      solutionData = record
            sequence : array[0..mnop] of partData;
            zigmaT   : real;
            end; {record}

   var j,n,g,n1,i,i2,m,nz,l,l2,k,jk,deco1option,count :integer;
    tabu : array[1..mtl] of tabulistData;
    tabuentr  ,fullsearch,tabutenure,diverse,rand_num: integer;
    isbestnoChange :boolean;
    newNeighbourhood :boolean;
    noChangeEntry,another : integer;
    currentsequence,currentbest :solutionData;
```

Fig. B.9. Main program for a simple tabu search: product mix problem.

```
    bestsequence   :solutionData;
    Citime,Wait : data2;
    processt: data2;
    price,mat_cost,c_limit: datar;
    mktdem : datap;
    ch :char;
    rn1,rn2,response,operate_cost :integer;
    start_time,end_time :real;
    filenam1  : string[20];
    filenam2  : string[20];
    f1,f2     : text;

 {$I cosmetix} {deco1 ,deco2, deco3,lines}
 {$I data_in}
 {$I dataexit}
 {$I obj_vals}
 {$I istabu  }
 {$I initsol }
 {$I driver }
 {$I schedule }

Begin
 another := 1;
 while (another = 1) do begin
 Duedate_optime_input;
 scheduler;
 gotoxy(10,21);write('Making another run ?[Yes = 1][No = 0] ');
 readln(another);
 end;
End.
```

Fig. B.9 (*Continued*)

Author Index

Subject Index